Student Solutions Manual

Intermediate Algebra
An Applied Approach

EIGHTH EDITION

Richard N. Aufmann

Joanne S. Lockwood

Prepared by

Ellena Reda
Dutchess Community College

BROOKS/COLE
CENGAGE Learning

Australia • Brazil • Japan • Korea • Mexico • Singapore • Spain • United Kingdom • United States

For product information and technology assistance, contact us at **Cengage Learning Customer & Sales Support, 1-800-354-9706**

For permission to use material from this text or product, submit all requests online at **www.cengage.com/permissions** Further permissions questions can be emailed to **permissionrequest@cengage.com**

ISBN-13: 978-0-538-49392-5
ISBN-10: 0-538-49392-5

Brooks/Cole
20 Davis Drive
Belmont, CA 94002-3098
USA

Cengage Learning is a leading provider of customized learning solutions with office locations around the globe, including Singapore, the United Kingdom, Australia, Mexico, Brazil, and Japan. Locate your local office at: **www.cengage.com/global**

Cengage Learning products are represented in Canada by Nelson Education, Ltd.

To learn more about Brooks/Cole, visit **www.cengage.com/brookscole**

Purchase any of our products at your local college store or at our preferred online store **www.CengageBrain.com**

Printed in the United States of America
1 2 3 4 5 6 7 14 13 12 11 10

Contents

Chapter 1: Review of Real Numbers

Prep Test

1. $\dfrac{5}{12} + \dfrac{7}{30} = \dfrac{25}{60} + \dfrac{14}{60} = \dfrac{39}{60} = \dfrac{3 \cdot 13}{3 \cdot 30} = \dfrac{13}{20}$

2. $\dfrac{8}{15} - \dfrac{7}{20} = \dfrac{32}{60} - \dfrac{21}{60} = \dfrac{11}{60}$

3. $\dfrac{5}{6} \cdot \dfrac{4}{15} = \dfrac{5 \cdot 2 \cdot 2}{3 \cdot 2 \cdot 5 \cdot 3} = \dfrac{2}{9}$

4. $\dfrac{4}{15} \div \dfrac{2}{5} = \dfrac{4}{15} \cdot \dfrac{5}{2} = \dfrac{2 \cdot 2 \cdot 5}{5 \cdot 3 \cdot 2} = \dfrac{2}{3}$

5. $8.000 + 29.340 + 7.065 = 44.405$

6. $92.00 - 18.37 = 73.63$

7. $2.19(3.4) = 7.446$

8. $32.436 \div 0.6 = 324.36 \div 6 = 54.06$

9.
   a) $-6 > -8$ Yes
   b) $-10 < -8$ No
   c) $0 > -8$ Yes
   d) $8 > -8$ Yes

10.
    $\dfrac{1}{2} = 0.5$ C

    $\dfrac{7}{10} = 0.7$ D

    $\dfrac{3}{4} = 0.75$ A

    $\dfrac{89}{100} = 0.89$ B

Section 1.1

Objective A Exercises

1. a) integers: $0, -3$

   b) rational numbers: $-\dfrac{15}{2}, 0, -3, 2.\overline{33}$

c) irrational numbers:

     $\pi, 4.232232223\ldots, \dfrac{\sqrt{5}}{4}, \sqrt{7}$

d) real numbers: all

3. -27

5. $-\dfrac{3}{4}$

7. 0

9. $\sqrt{33}$

11. 91

13. Replace y with each element in the set to determine if the inequality is true or false.
    $-6 > -4$ False
    $-4 > -4$ False
    $7 > -4$ True

15. Replace w with each element in the set to determine if the inequality is true or false.
    $-2 \le -1$ True
    $-1 \le -1$ True
    $0 \le -1$ False
    $1 \le -1$ False

17. Replace b with each element in the set then evaluate the expression.
    $-(-9) = 9$
    $-(0) = 0$
    $-(9) = -9$

19. Replace c with each element in the set then evaluate the expression.
    $|-4| = 4$
    $|0| = 0$
    $|4| = 4$

21. $-x > 0$ whenever the value of x is a negative real number.

Objective B Exercises

23. $\{-2, -1, 0, 1, 2, 3, 4\}$

25. $\{2, 4, 6, 8, 10, 12\}$

27. $\{3, 6, 9, 12, 15, 18, 21, 24, 27, 30\}$

29. $\{x \mid x > 4, x \in \text{integers}\}$

31. $\{x \mid x \geq -2\}$

33. $\{x \mid 0 < x < 1\}$

35. $\{x \mid 1 \leq x \leq 4\}$

37. Yes

39. No

41. No

43. $\{x \mid x < 2\}$

45. $\{x \mid x \geq 1\}$

47. $\{x \mid -1 < x < 5\}$

49. $\{x \mid 0 \leq x \leq 3\}$

51. $(-2, 4)$

53. $[-1, 5]$

55. $(-\infty, 1)$

57. $[-2, \infty)$

59. $\{x \mid 0 < x < 8\}$

61. $\{x \mid -5 \leq x \leq 7\}$

63. $\{x \mid -3 \leq x < 6\}$

65. $\{x \mid x \leq 4\}$

67. $\{x \mid x > 5\}$

69. $(-2, 5)$

71. $[-1, 2]$

73. $(-\infty, 3]$

75. $[3, \infty)$

Objective C Exercises

77. $A \cup B = \{1, 2, 4, 6, 9\}$

79. $A \cup B = \{2, 3, 5, 8, 9, 10\}$

81. $A \cup B = \{-4, -2, 0, 2, 4, 8\}$

83. $A \cup B = \{1, 2, 3, 4, 5\}$

85. $A \cap B = \{6\}$

87. $A \cap B = \{5, 10, 20\}$

89. $A \cap B = \emptyset$

91. $A \cap B = \{4, 6\}$

93. $[5, \infty) \cap (0, 5)$

95. $\{x \mid x > 1\} \cup \{x \mid x < -1\}$

97. $\{x \mid x \leq 2\} \cap \{x \mid x \geq 0\}$

99. $\{x|\ x>1\} \cap \{x|x\ge -2\}$

-5 -4 -3 -2 -1 0 1 2 3 4 5

101. $\{x|\ x>-3\}\cup \{x|\ x<1\}$

-5 -4 -3 -2 -1 0 1 2 3 4 5

103. $[-5, 0)\cup (1,4]$

-5 -4 -3 -2 -1 0 1 2 3 4 5

105. $[-3, 3]\cap [0,5]$

-5 -4 -3 -2 -1 0 1 2 3 4 5

Applying the Concepts

107. $A\cup B = A$

109. $B\cap B = B$

111. $A\cap R = A$

113. $B\cup R = R$

115. $R\cup R = R$

117. $B\cap C = \{0\}$

Section 1.2

Objective A Exercises

1. a) To add two numbers with the same sign add the absolute values of the numbers. Then attach the sign of the addends.

b) To add two numbers with different signs find the absolute value of each number. Subtract the smaller of these two numbers from the larger. Then attach the sign of the larger absolute value.

3. $-18 + (-12) = -30$

5. $5 - 22 = 5 + (-22) = -17$

7. $3\cdot 4\cdot (-8) = 12\cdot (-8) = -96$

9. $18 \div (-3) = -6$

11. $-60 \div (-12) = 5$

13. $-20(35)(-16) = -700(-16) = 11,200$

15. $8 - (-12) = 8 + 12 = 20$

17. $|12(-8)| = |-96| = 96$

19. $|15 - (-8)| = |15 + 8| = |23| = 23$

21. $|-56 \div 8| = |-7| = 7$

23. $|-153 \div (-9)| = |17| = 17$

25. $-|-8| + |-4| = -8 + 4 = -4$

27. $-30 + (-16) - 14 - 2$
$= -30 + (-16) + (-14) + (-2)$
$= -46 + (-14) + (-2)$
$= -60 + (-2)$
$= -62$

29. $-2 + (-19) - 16 + 12$
$= -2 + (-19) + (-16) + 12$
$= -21 + (-16) + 12$
$= -37 + 12$
$= -25$

31. $13 - |6 - 12| = 13 - |6 + (-12)|$
$= 13 - |-6|$
$= 13 - 6$
$= 13 + (-6)$
$= 7$

33. $738 - 46 + (-105) - 219$
$= 738 + (-46) + (-105) + (-209)$
$= 692 + (-105) + (-209)$
$= 587 + (-209)$
$= 368$

35. $-442 \div (-17) = 26$

37. $-4897 \div 59 = -83$

39. The sign of the product of an odd number of negative factors will always be negative.

Objective B Exercises

41. a) The least common multiple of two numbers is the smallest number that is a multiple of each of the two numbers.

b) The greatest common factor of two numbers is the largest integer that divides evenly into both numbers.

43. $\dfrac{7}{12} - \left(-\dfrac{5}{16}\right) = \dfrac{7}{12} + \dfrac{5}{16} = \dfrac{28}{48} + \dfrac{15}{48}$
$= \dfrac{28 + 15}{48} = \dfrac{43}{48}$

45. $-\dfrac{5}{9} - \dfrac{14}{15} = -\dfrac{25}{45} - \dfrac{42}{45} = -\dfrac{67}{45}$

47. $-\dfrac{1}{3} + \dfrac{5}{9} - \dfrac{7}{12} = -\dfrac{12}{36} + \dfrac{20}{36} - \dfrac{21}{36}$
$= \dfrac{-12 + 20 - 21}{36} = -\dfrac{13}{36}$

49. $\dfrac{2}{3} - \dfrac{5}{12} + \dfrac{5}{24} = \dfrac{16}{24} - \dfrac{10}{24} + \dfrac{5}{24}$
$= \dfrac{16 - 10 + 5}{24} = \dfrac{11}{24}$

51. $\dfrac{5}{8} - \dfrac{7}{12} + \dfrac{1}{2} = \dfrac{15}{25} - \dfrac{14}{24} + \dfrac{12}{24}$
$= \dfrac{15 - 14 + 12}{24} = \dfrac{13}{24}$

53. $\left(\dfrac{6}{35}\right)\left(-\dfrac{5}{16}\right) = -\dfrac{6 \cdot 5}{35 \cdot 16}$
$= -\dfrac{2 \cdot 3 \cdot 5}{5 \cdot 7 \cdot 2 \cdot 2 \cdot 2 \cdot 2} = -\dfrac{3}{56}$

55. $-\dfrac{8}{15} \div \dfrac{4}{5} = -\dfrac{8}{15} \cdot \dfrac{5}{4} = -\dfrac{8 \cdot 5}{15 \cdot 4}$
$= -\dfrac{2 \cdot 2 \cdot 2 \cdot 5}{3 \cdot 5 \cdot 2 \cdot 2} = -\dfrac{2}{3}$

57. $-\dfrac{11}{24} \div \dfrac{7}{12} = -\dfrac{11}{24} \cdot \dfrac{12}{7} = -\dfrac{11 \cdot 12}{24 \cdot 7}$
$= -\dfrac{11 \cdot 2 \cdot 2 \cdot 3}{2 \cdot 2 \cdot 2 \cdot 3 \cdot 7} = -\dfrac{11}{14}$

59. $\left(-\dfrac{5}{12}\right)\left(\dfrac{4}{35}\right)\left(\dfrac{7}{8}\right) = -\dfrac{5 \cdot 4 \cdot 7}{12 \cdot 35 \cdot 8}$
$= -\dfrac{5 \cdot 2 \cdot 2 \cdot 7}{2 \cdot 2 \cdot 3 \cdot 5 \cdot 7 \cdot 2 \cdot 2 \cdot 2}$
$= -\dfrac{1}{24}$

61. $-14.270 + 1.296 = -12.974$

63. $1.832 - 7.840 = 1.832 + (-7.840) = -6.008$

65. $(0.03)(10.5)(6.1) = (0.315)(6.1) = 1.9215$

67. $5.418 \div (-0.9) = -6.02$

69. $-0.4355 \div 0.065 = -6.7$

71. If the product of six numbers is negative that means that 1 or 3 or 5 of those numbers could be negative.

73. $38.241 + [-(-6.027)] - 7.453$
$= 38.241 + 6.027 + (-7.453)$
≈ -1.11

75. $-287.3069 \div (0.1415) \approx -2030.44$

Objective C Exercise

77. $5^3 = 5 \cdot 5 \cdot 5 = 125$

79. $-2^3 = -(2 \cdot 2 \cdot 2) = -8$

81. $(-5)^3 = (-5) \cdot (-5) \cdot (-5) = -125$

83. $2^2 \cdot 3^4 = (2 \cdot 2) \cdot (3 \cdot 3 \cdot 3 \cdot 3) = 324$

85. $-2^2 \cdot 3^2 = -(2 \cdot 2) \cdot (3 \cdot 3) = -36$

87. $(-2)^3 \cdot (-3)^2 = (-2) \cdot (-2) \cdot (-2) \cdot (-3) \cdot (-3) = -72$

89. $2^3 \cdot 3^3 = (2 \cdot 2 \cdot 2) \cdot (3 \cdot 3 \cdot 3) = 216$

91. $-2^2 \cdot (-2)^2 = -(2 \cdot 2) \cdot (-2) \cdot (-2) = -16$

93. $\left(-\dfrac{2}{3}\right)^2 \cdot 3^3 = \left(-\dfrac{2}{3}\right)\left(-\dfrac{2}{3}\right) \cdot 3 \cdot 3 \cdot 3 = 12$

95. $2^5(-3)^4 \cdot 4^5 = 32 \cdot 81 \cdot 1024 = 2{,}654{,}208$

97. Negative

99. Positive

Objective D Exercises

101. We need to use Order of Operation Agreement in order to ensure that there is not more than one answer to the same problem.

103. $5 - 3(8 \div 4)^2 = 5 - 3(2)^2 = 5 - 3(4)$
$= 5 - 12 = -7$

105. $16 - \dfrac{2^2 - 5}{3^2 + 2} = 16 - \dfrac{4 - 5}{9 + 2} = 16 - \dfrac{-1}{11}$
$= 16 + \dfrac{1}{11} = \dfrac{177}{11}$

107. $\dfrac{3 + \dfrac{2}{3}}{\dfrac{11}{16}} = \dfrac{\dfrac{11}{3}}{\dfrac{11}{16}} = \dfrac{11}{3} \cdot \dfrac{16}{11} = \dfrac{16}{3}$

109. $5[(2 - 4) \cdot 3 - 2] = 5[(-2) \cdot 3 - 2]$
$= 5[(-6) - 2]$
$= 5(-8) = -40$

111. $16 - 4\left(\dfrac{8 - 2}{3 - 6}\right) \div \dfrac{1}{2} = 16 - 4\left(\dfrac{6}{-3}\right) \div \dfrac{1}{2}$
$= 16 - 4(-2) \div \dfrac{1}{2} = 16 + 8 \div \dfrac{1}{2}$
$= 16 + 8 \cdot 2 = 16 + 16 = 32$

113. $6[3 - (-4 + 2) \div 2] = 6[3 - (-2) \div 2]$
$= 6[3 - (-1)] = 6[3 + 1]$
$= 6(4) = 24$

115. $\dfrac{1}{2} - \left(\dfrac{2}{3} \div \dfrac{5}{9}\right) + \dfrac{5}{6} = \dfrac{1}{2} - \left(\dfrac{2}{3} \cdot \dfrac{9}{5}\right) + \dfrac{5}{6}$
$= \dfrac{1}{2} - \dfrac{6}{5} + \dfrac{5}{6} = \dfrac{15}{30} - \dfrac{36}{30} + \dfrac{25}{30}$
$= \dfrac{15 - 36 + 25}{30} = \dfrac{4}{30} = \dfrac{2}{15}$

117. $\dfrac{1}{2} - \dfrac{\dfrac{17}{25}}{4 - \dfrac{3}{5}} \div \dfrac{1}{5} = \dfrac{1}{2} - \dfrac{\dfrac{17}{25}}{\dfrac{17}{5}} \div \dfrac{1}{5} = \dfrac{1}{2} - \left(\dfrac{17}{25} \cdot \dfrac{5}{17}\right) \div \dfrac{1}{5}$
$= \dfrac{1}{2} - \dfrac{1}{5} \div \dfrac{1}{5} = \dfrac{1}{2} - \dfrac{1}{5} \cdot \dfrac{5}{1}$
$= \dfrac{1}{2} - 1 = -\dfrac{1}{2}$

119. $\dfrac{2}{3} - \left(\dfrac{3}{8} + \dfrac{5}{6}\right) \div \dfrac{3}{5} = \dfrac{2}{3} - \left(\dfrac{9}{24} + \dfrac{20}{24}\right) \div \dfrac{3}{5}$
$= \dfrac{2}{3} - \left(\dfrac{29}{24}\right) \div \dfrac{3}{5} = \dfrac{2}{3} - \dfrac{29}{24} \cdot \dfrac{5}{3}$
$= \dfrac{2}{3} - \dfrac{145}{72} = \dfrac{48}{72} - \dfrac{145}{72} = -\dfrac{97}{72}$

121. $0.4(1.2 - 2.3)^2 + 5.8 = 0.4(-1.1)^2 + 5.8$
$= 0.4(1.21) + 5.8 = 0.484 + 5.8 = 6.284$

123. $1.75 \div 0.25 - (1.25)^2$
$= 1.72 \div 0.25 - 1.5625$
$= 7 - 1.5625 = 5.4375$

125. $27.2322 \div (6.96 - 3.27)^2 = 27.2322 \div (3.69)^2$
$= 27.2322 \div 13.6161 = 2$

127. $8^2 - 2^2(5 - 3)^3 = 32$
(i) $6^2(2)^3 = 288$ Not equivalent
(ii) $64 - 4(8) = 32$ Equivalent
(iii) $60(2)^3 = 480$ Not equivalent
(iv) $64 - 8^3 = -448$ Not equivalent

Applying the Concepts

129.   0

131. No, 0 does not have a multiplicative inverse.

133. $7^1 = 7$; $7^2 = 49$; $7^3 = 343$; $7^4 = 2401$;
$7^5 = 16,807$; $7^6 = 117,649$; $7^7 = 823,543$;
$7^8 = 5,764,801$
There is a repeating pattern of the digits in the ones place: 7, 9, 3, 1, 7, 9, 3, 1,…
There are 4 numbers in the pattern. 18 has a remainder of 2 when divided by 4. The ones digit in 7^{18} is the second number in the pattern, which is 9.

135. 5^{234} is a very large number. The last three Digits are 625. $5^3 = 125$; $5^4 = 625$; $5^5 = 3125$;
$5^6 = 15,625$; $5^7 = 78,125$; $5^8 = 390,626$;
$5^9 = 1,953,125$; $5^{10} = 9,765,625$
When 5 is raised to an odd power, the last three digits are 125. When 5 is raised to an even power the last three digits are 625. 5^{234} is an even power. The last three digits in 5^{234} are 625.

137. First find b^c and then find $a^{(b^c)}$

Section 1.3

Objective A Exercises

1. $3 \cdot 4 = 4 \cdot 3$

3. $(3 + 4) + 5 = 3 + (4 + 5)$

5. $\dfrac{5}{0}$ is undefined.

7. $3(x + 2) = 3x + 6$

9. $\dfrac{0}{-6} = 0$

11. $\dfrac{1}{mn} \cdot mn = 1$

13. $2(3x) = (2 \cdot 3)x$

15. The Division Property of Zero

17. The Inverse Property of Multiplication

19. The Addition Property of Zero

21. The Division Property of Zero

23. The Distributive Property

25. The Associative Property of Multiplication

27. When the sum of a positive number n and its additive inverse are multiplied by the reciprocal of the number n the result is 0.

Objective B Exercises

29. $ab + dc$
$2(3) + (-4)(-1) = 6 + 4 = 10$

31. $4cd \div a^2$
$4(-1)(-4) \div (2)^2 = 4(-1)(-4) \div 4$
$= (-4)(-4) \div 4 = 16 \div 4 = 4$

33. $(b - 2a)^2 + c$

$(3 - 2(2))^2 + (-1)$

$= (3 - 4)^2 + (-1)$

$= (-1)^2 + (-1) = 1 + (-1) = 0$

35. $(bc + a)^2 \div (d - b)$

$(3(-1) + 2)^2 \div (-4 - 3) = (-3 + 2)^2 \div (-4 - 3)$

$= (-1)^2 \div (-7) = 1 \div (-7) = -\dfrac{1}{7}$

37. $\dfrac{1}{4}a^4 - \dfrac{1}{6}bc$

$\dfrac{1}{4}(2)^4 - \dfrac{1}{6}(3)(-1) = \dfrac{1}{4}(16) - \dfrac{1}{6}(-3)$

$= 4 + \dfrac{1}{2} = \dfrac{9}{2}$

39. $\dfrac{3ac}{-4} - c^2$

$\dfrac{3(2)(-1)}{-4} - (-1)^2 = \dfrac{3(2)(-1)}{-4} - 1$

$= \dfrac{6(-1)}{-4} - 1 = \dfrac{-6}{-4} - 1$

$= \dfrac{6}{4} - \dfrac{4}{4} = \dfrac{2}{4} = \dfrac{1}{2}$

41. $\dfrac{3b - 5c}{3a - c}$

$\dfrac{3(3) - 5(-1)}{3(2) - (-1)} = \dfrac{9 + 5}{6 - (-1)}$

$= \dfrac{14}{6 + 1} = \dfrac{14}{7} = 2$

43. $\dfrac{a - d}{b + c}$

$\dfrac{2 - (-4)}{3 + (-1)} = \dfrac{2 + 4}{2} = \dfrac{6}{2} = 3$

45. $-a|a + 2d|$

$-2|2 + 2(-4)| = -2|2 + (-8)|$

$= -2|-6| = -2(6) = -12$

47. $\dfrac{2a - 4d}{3b - c}$

$\dfrac{2(2) - 4(-4)}{3(3) - (-1)} = \dfrac{4 - (-16)}{9 + 1}$

$= \dfrac{4 + 16}{10} = \dfrac{20}{10} = 2$

49. $-3d \div \left| \dfrac{ab - 4c}{2b + c} \right|$

$-3(-4) \div \left| \dfrac{2(3) - 4(-1)}{2(3) + (-1)} \right| = 12 \div \left| \dfrac{6 + 4}{6 - 1} \right|$

$= 12 \div \left| \dfrac{10}{5} \right| = 12 \div |2| = 12 \div 2 = 6$

51. $2(d - b) \div (3a - c)$

$2(-4 - 3) \div (3(2) - (-1))$

$= 2(-7) \div (6 - (-1))$

$= 2(-7) \div (6 + 1) = 2(-7) \div 7$

$= -14 \div 7 = -2$

53. $-d^2 - c^3 a$

$-(-4)^2 - (-1)^3(2) = -16 - (-1)(2)$

$= -16 - (-2) = -16 + 2 = -14$

55. $-d^3 + 4ac$

$-(-4)^3 + 4(2)(-1) = -(-64) + 4(2)(-1)$

$= -(-64) + (8)(-1) = -(-64) + (-8)$

$= 64 + (-8) = 56$

57. $4^{(a^2)}$

$4^{(2^2)} = 4^4 = 256$

59. If $a = -38$, $b = -52$ and $c > 0$, the numerator is positive and the denominator is negative therefore, the expression is negative.

Objective C Exercises

61. $5x + 7x = 12x$

63. $3x - 5x + 9x = 7x$

65. $5b - 8a - 12b = -8a - 7b$

67. $\dfrac{1}{3}(3y) = y$

69. $-\dfrac{2}{5}\left(-\dfrac{5}{2}z\right) = z$

71. $3(a - 5) = 3a - 15$

73. $-5(x - 9) = -5x + 45$

75. $-(x + y) = -x - y$

77. $4x - 3(2y - 5) = 4x - 6y + 15$

79. $25x + 10(9 - x) = 25x + 90 - 10x$
$= 15x + 90$

81. $3[x - 2(x + 2y)] = 3[x - 2x - 4y]$
$= 3[-x - 4y] = -3x - 12y$

83. $3[a - 5(5 - 3a)] = 3[a - 25 + 15a]$
$= 3[16a - 25] = 48a - 75$

85. $-2(x - 3y) + 2(3y - 5x) = -2x + 6y + 6y - 10x$
$= -12x + 12y$

87. $5(3a - 2b) - 3(-6a + 5b)$
$= 15a - 10b + 18a - 15b$
$= 33a - 25b$

89. $3x - 2[y - 2(x + 3[2x + 3y])]$
$= 3x - 2[y - 2(x + 6x + 9y)]$
$= 3x - 2[y - 2(7x + 9y)]$
$= 3x - 2[y - 14x - 18y]$
$= 3x - 2[-14x - 17y] = 3x + 28x + 34y$
$= 31x + 34y$

91. $4 - 2(7x - 2y) - 3(-2x + 3y)$
$= 4 - 14x + 4y + 6x - 9y$
$= 4 - 8x - 5y$

93. $\dfrac{1}{3}[8x - 2(x - 12) + 3] = \dfrac{1}{3}[8x - 2x + 24 + 3]$

$= \dfrac{1}{3}[6x + 27]$

$= 2x + 9$

95. Simplify $31a - 102b + 73 - 88a + 256b - 73$
to $-57a + 154b$
a) The coefficient of a is negative
b) The coefficient of b is positive
c) The constant term is 0

Applying the Concepts

97. $0.052x + 0.072(x + 1000)$
$= 0.052x + 0.072x + 72 = 0.124x + 72$

99. $\dfrac{t}{20} + \dfrac{t}{30} = \dfrac{3t}{60} + \dfrac{2t}{60} = \dfrac{5t}{60} = \dfrac{t}{12}$

Section 1.4

Objective A Exercises

1. Let n represent the unknown number.
Eight <u>less than</u> a number: $n - 8$

3. Let n represent the unknown number.
Four-fifths <u>of</u> a number : $\dfrac{4}{5}n$

5. Let n represent the unknown number.
The <u>quotient</u> of a number and fourteen: $\dfrac{n}{14}$

7. Let n represent the unknown number.
Five <u>subtracted</u> from the <u>product</u> of the cube of eight and a number: $8^3 n - 5$
No. This is not the same as the given expression $8n^3 - 5$.

9. Let n represent the unknown number.
A number <u>minus</u> the <u>sum</u> of the number and two: $n - (n + 2)$
$n - (n + 2) = n - n - 2 = -2$

11. Let n represent the unknown number.
Five <u>times</u> the <u>product</u> of eight and a number:
$5(8n) = 40n$

13. Let n represent the unknown number.
The <u>difference</u> <u>between</u> seventeen <u>times</u> a number and <u>twice</u> the number:
$17n - 2n = 15n$

15. Let n represent the unknown number.
The <u>difference</u> between the <u>square</u> of a number and the <u>total</u> of twelve and the <u>square</u> of the number:
The square of the number: n^2
The total of twelve and the square of the number: $(12 + n^2)$
$n^2 - (12 + n^2) = n^2 - 12 - n^2 = -12$

17. Let n represent the unknown number.
The <u>sum</u> of five <u>times</u> a number and twelve is <u>added</u> to the <u>product</u> of fifteen and the number.
Five times the number and twelve: $5n + 12$
The product of fifteen and the number: $15n$
$(5n + 12) + 15n = 20n + 12$

19. The <u>sum</u> of two numbers is fifteen. Let x represent the smaller of the two numbers.
The larger number is $15 - x$.
The <u>sum</u> of two <u>more than</u> the larger number and <u>twice</u> the smaller number:
$(15 - x + 2) + 2x = 17 + x$

21. The <u>sum</u> of two numbers is thirty-four. Let x represent the larger of the two numbers.
The smaller number is $34 - x$.
The <u>quotient</u> of five <u>times</u> the smaller number and the <u>difference</u> <u>between</u> the larger number and three:
$$\frac{5(34 - x)}{x - 3}$$

Objective B Exercises

23. a) Let t represent the historical average temperature of the Arctic Ocean.
2007 historical average temperature: $t + 3.5$
b) Let t represent the historical maximum temperature of the Arctic Ocean.

2007 historical maximum temperature: $t + 1.5$

25. Let d represent the distance from Earth to the moon.
Distance from the Earth to the sun is 390 times the distance from the Earth to the moon: $390d$

27. Let x represent the amount in the first account.
The total amount in both accounts is $10,000.
The amount in the second account: $10,000 - x$

29. Let x represent the measure of angle B.
The measure of angle A is twice the measure of angle B: $2x$
The measure of angle C is twice the measure of angle A: $2(2x) = 4x$

31. In 2010, a house sold for $30,000 more than the same house in sold for in 2005.
s represents the selling price of the house in 2005.
$s + 30,000$ represents the selling price of the house in 2020.

Applying the Concepts

33. The sum of twice x and 3

35. Twice the sum of x and 3

37. a) The product of mass m and acceleration a:
ma
b) The product of area A and the square of the velocity v: Av^2

Chapter 1 Review Exercises

1. $\{-2, -1, 0, 1, 2, 3\}$

2. $A \cap B = \{2, 3\}$

3. $(-2, 4]$

4. The Associative Property of Multiplication

5. $-4.07 + 2.3 - 1.07 = -1.77 - 1.07 = -2.84$

6. $(a - 2b^2) \div (ab)$

$(4 - 2(-3)^2) \div (4(-3))$

$= (4 - 2(9)) \div (4(-3))$

$= (4 - 18) \div (-12) = -14 \div (-12)$

$= \dfrac{-14}{-12} = \dfrac{7}{6}$

7. $-2 \cdot (4^2) \cdot (-3)^2 = -2 \cdot 16 \cdot 9 = -288$

8. $4y - 3[x - 2(3 - 2x) - 4y]$

$= 4y - 3[x - 6 + 4x - 4y]$

$= 4y - 3[5x - 4y - 6]$

$= 4y - 15x + 12y + 18 = 16y - 15x + 18$

9. The additive inverse of $-\dfrac{3}{4}$ is $\dfrac{3}{4}$.

10. $\{x \,|\, x < -3\}$

11. $\{x \,|\, x < 1\}$

12. $-10 - (-3) - 8 = -10 + 3 - 8 = -15$

13. $-\dfrac{2}{3} + \dfrac{3}{5} - \dfrac{1}{6} = -\dfrac{20}{30} + \dfrac{18}{30} - \dfrac{5}{30}$

$= \dfrac{-20 + 18 - 5}{30} = \dfrac{-7}{30} = -\dfrac{7}{30}$

14. $3 + (4 + y) = (3 + 4) + y$

15. $-\dfrac{3}{8} \div \dfrac{3}{5} = -\dfrac{3}{8} \cdot \dfrac{5}{3} = -\dfrac{3 \cdot 5}{8 \cdot 3} = -\dfrac{5}{8}$

16. Replace x with each element in the set to determine if the inequality is true or false.

a) $-4 > -1$ False

b) $-2 > -1$ False

c) $0 > -1$ True

d) $2 > -1$ True

17. $2a^2 - \dfrac{3b}{a}$

$2(-3)^2 - \dfrac{3(2)}{-3} = 2(9) - \dfrac{6}{-3}$

$= 18 - (-2) = 18 + 2 = 20$

18. $18 - |-12 + 8| = 18 - |-4| = 18 - 4 = 14$

19. $20 \div \dfrac{3^2 - 2^2}{3^2 + 2^2} = 20 \div \dfrac{9 - 4}{9 + 4} = 20 \div \dfrac{5}{13}$

$= 20 \cdot \dfrac{13}{5} = 52$

20. $[-3, \infty)$

21. $A \cup B = \{1, 2, 3, 4, 5, 6, 7, 8\}$

22. $-204 \div (-17) = 12$

23. $[x \,|\, -2 \le x \le 3\}$

24. $\dfrac{3}{4}\left(-\dfrac{8}{21}\right)\left(-\dfrac{7}{15}\right) = \dfrac{3 \cdot 10 \cdot 7}{5 \cdot 21 \cdot 15}$

$= \dfrac{3 \cdot 2 \cdot 2 \cdot 2 \cdot 7}{2 \cdot 2 \cdot 3 \cdot 7 \cdot 3 \cdot 5} = \dfrac{2}{15}$

25. $6x - 21y = 3(2x - 7y)$

26. $\{x| \ x \le -3\} \cup \{x| \ x > 0\}$

27. $-2(x - 3) + 4(2 - x) = -2x + 6 + 8 - 4x$

$= -6x + 14$

28. Replace p with each element in the set to determine the value of $|-p|$

a) $-|-4| = -4$

b) $-|0| = 0$

c) $-|7| = -7$

29. The Inverse Property of Addition

30. $-3.286 \div (-1.06) = 3.1$

31. The additive inverse of -87 is 87.

32. Replace y with each element in the set to determine if the inequality is true or false.

a) $-4 > -2$ False

b) $-1 > -2$ True

c) $4 > -2$ True

33. $\{-3, -2, -1, 0, 1\}$

34. $\{x|\, x < 7\}$

35. $A \cup B = \{-4, -2, 0, 2, 4, 5, 10\}$

36. $A \cap B = \emptyset$

37. $\{x|\, x \le 3\} \cap \{x|\, x > -2\}$

38. $(-3, 4) \cup [-1, 5]$

39. $9 - (-3) - 7 = 9 + 3 - 7 = 12 - 7 = 5$

40. $-\dfrac{2}{3} + \left(-\dfrac{1}{4}\right) - \left(-\dfrac{5}{12}\right) = -\dfrac{8}{12} + \left(-\dfrac{3}{12}\right) - \left(-\dfrac{5}{12}\right)$

$= \dfrac{-8 + (-3) + 5}{12} = \dfrac{-6}{12} = -\dfrac{1}{2}$

41. $\left(\dfrac{2}{3}\right)^3 (-3)^4 = \dfrac{8}{27} \cdot 81 = 24$

42. $(-3)^3 - (2 - 6)^2 \cdot 5 = (-3)^3 - (-4)^2 \cdot 5$

$= -27 - 16 \cdot 5 = -27 - 80 = -107$

43. $-8ac \div b^2$

$-8(-1)(-3) \div 2^2 = -8(-1)(-3) \div 4$

$= -24 \div 4 = -6$

44. $-(3a + b) - 2(-4a - 5b) = -3a - b + 8a + 10b$

$= 5a + 9b$

45. Let x represent the unknown number.
Four <u>times</u> the <u>sum</u> of a number and four:
$4(x + 4) = 4x + 16$

46. Let t represent the flying time between San Diego to New York.
Total flying time is 13 hrs.
The flying time between New York and San Diego: $13 - t$

47. Let C represent the number of calories burned by walking at 4 mph for one hour.

The number of calories burned cross country skiing for one hour is 396 more than the number of calories burned walking: $C + 396$

48. Let x represent the unknown number.
Eight <u>more than</u> twice the <u>difference</u> between a number and two: $2(x - 2) + 8$
$2(x - 2) + 8 = 2x - 4 + 8 = 2x + 4$

49. Let x represent the first integer.
The second integer is 5 more than four times the first integer.
The second integer: $4x + 5$

50. Let x represent the unknown number.
Twelve <u>minus</u> the <u>quotient</u> of three <u>more than</u> a number and four. $12 - \dfrac{x + 3}{4}$

$12 - \dfrac{x + 3}{4} = \dfrac{48}{4} - \dfrac{x + 3}{4}$

$= \dfrac{48 - x + 3}{4} = \dfrac{45 - x}{4}$

51. The sum of two numbers is forty.
Let x represent the smaller number.
Let $40 - x$ represent the larger number.
The <u>sum</u> of <u>twice</u> the smaller number and five <u>more than</u> the larger number: $2x + (40 - x) + 5$
$2x + (40 - x) + 5 = 2x + 40 - x + 5 = x + 45$

52. Let w represent the width of the rectangle.
The length is three feet <u>less than</u> three <u>times</u> the width: $3w - 3$

Chapter 1 Test

1. $(-2)(-3)(-5) = 6(-5) = -30$

2. $A \cap B = \{5, 7\}$

3. $(-2)^3 (-3)^2 = (-8)(9) = -72$

4. $(-\infty, 1]$

5. $A \cap B = \{-1, 0, 1\}$

6. $(a-b)^2 \div (2b+1)$

$(2-(-3))^2 \div (2(-3)+1)$

$= (2+3)^2 \div (-6+1)$

$= (5)^2 \div (-5)$

$= 25 \div (-5) = -5$

7. $|-3-(-5)| = |-3+5| = |2| = 2$

8. $2x - 4[2 - 3(x+4y) - 2]$

$= 2x - 4[2 - 3x - 12y - 2]$

$= 2x - 4[-3x - 12y] = 2x + 12x + 48y$

$= 14x + 48y$

9. The additive inverse of -12 is 12.

10. $-5^2 \cdot 4 = -25 \cdot 4 = -100$

11. $\{x|\, x < 3\} \cap \{x|\, x > -2\}$

12. $2 - (-12) + 3 - 5 = 2 + 12 + 3 - 5 = 12$

13. $\dfrac{2}{3} - \dfrac{5}{12} + \dfrac{4}{9} = \dfrac{24}{36} - \dfrac{15}{36} + \dfrac{16}{36}$

$= \dfrac{24 - 15 + 16}{36} = \dfrac{25}{36}$

14. $(3+4) + 2 = (4+3) + 2$

15. $\left(-\dfrac{2}{3}\right)\left(\dfrac{9}{15}\right)\left(\dfrac{10}{27}\right) = -\dfrac{2 \cdot 3 \cdot 3 \cdot 2 \cdot 5}{3 \cdot 3 \cdot 5 \cdot 3 \cdot 3 \cdot 3} = -\dfrac{4}{27}$

16. Replace x with each element in the set to determine if the inequality is true or false.

$-5 < -1$ True

  $3 < -1$ False

  $7 < -1$ False

17. $\dfrac{b^2 - c^2}{a - 2c}$

$\dfrac{(3)^2 - (-1)^2}{2 - 2(-1)} = \dfrac{9-1}{2+2} = \dfrac{8}{4} = 2$

18. $-180 \div 12 = -15$

19. $12 - 4\left(\dfrac{5^2 - 1}{3}\right) \div 16 = 12 - 4\left(\dfrac{25-1}{3}\right) \div 16$

$= 12 - 4\left(\dfrac{24}{3}\right) \div 16 = 12 - 4(8) \div 16$

$= 12 - 32 \div 16 = 12 - 2 = 10$

20. $(3, \infty)$

21. $A \cup B = \{1, 2, 3, 4, 5, 7\}$

22. $3x - 2(x-y) - 3(y-4x)$

$= 3x - 2x + 2y - 3y + 12x$

$= 13x - y$

23. $8 - 4(2-3)^2 \div 2 = 8 - 4(-1)^2 \div 2$

$= 8 - 4(1) \div 2 = 8 - 4 \div 2$

$= 8 - 2 = 6$

24. $\dfrac{3}{5}\left(-\dfrac{10}{21}\right)\left(-\dfrac{7}{15}\right) = \dfrac{3 \cdot 2 \cdot 5 \cdot 7}{5 \cdot 3 \cdot 7 \cdot 3 \cdot 5} = \dfrac{2}{15}$

25. The Distributive Property

26. $\{x|\, x \leq 3\} \cup \{x|\, x < -2\}$

27. $4.27 - 6.98 + 1.3 = -1.41$

28. $A \cup B = \{-2, -1, 0, 1, 2, 3\}$

29. The sum of two numbers is nine.

Let x represent the larger of the two numbers.

Let $9 - x$ represent the smaller number.

The <u>difference</u> <u>between</u> one <u>more than</u> the larger number and <u>twice</u> the smaller number:

$(x+1) - 2(9-x)$

$(x+1) - 2(9-x) = x + 1 - 18 + 2x = 3x - 17$

30. Let x represent the amount of cocoa produced in Ghana.

The Ivory Coast produces three times the amount of cocoa produced in Ghana: $3x$

Chapter 2: First Degree Equations and Inequalities

Prep Test

1. $8 - 12 = -4$

2. $-9 + 3 = -6$

3. $\dfrac{-18}{-6} = 3$

4. $-\dfrac{3}{4}\left(-\dfrac{4}{3}\right) = \dfrac{3 \cdot 2 \cdot 2}{2 \cdot 2 \cdot 3} = 1$

5. $-\dfrac{5}{8}\left(\dfrac{4}{5}\right) = -\dfrac{5 \cdot 2 \cdot 2}{2 \cdot 2 \cdot 2 \cdot 5} = -\dfrac{1}{2}$

6. $3x - 5 + 7x = 10x - 5$

7. $6(x - 2) + 3 = 6x - 12 + 3 = 6x - 9$

8. $n + (n + 2) + (n + 4) = 3n + 6$

9. $0.08x + 0.05(400 - x)$
   $= 0.08x + 20 - 0.05x$
   $= 0.03x + 20$

10. The total weight of a snack mix of nuts and pretzels is twenty ounces.
    Let n represent the ounces of nuts in the mix.
    Ounces of peanuts in the mix: $20 - n$

Section 2.1

Objective A Exercises

1. An equation contains an equal sign while an expression does not contain an equal sign.

3. The Addition Property of Equations states that the same quantity can be added to each side of an equation without changing the solution of the equation.
   This property is used to remove a term from one side of an equation by adding the opposite of that term to each side of the equation.

5. $7 - 3(1) = 4$
   $7 - 3 = 4$
   $\qquad 4 = 4 \qquad$ Yes, 1 is a solution.

7. $6(-2) - 1 = 7(-2) + 1$
   $-12 - 1 = -14 + 1$
   $\qquad -13 = -13 \qquad$ Yes, -2 is a solution.

9. $\qquad x - 2 = 7$
   $x - 2 + 2 = 7 + 2$
   $\qquad\quad x = 9$
   The solution is 9.

11. $\qquad -7 = x + 8$
    $-7 - 8 = x + 8 - 8$
    $\quad -15 = x$
    The solution is -15.

13. $3x = 12$
    $\dfrac{3x}{3} = \dfrac{12}{3}$
    $\quad x = 4$
    The solution is 4.

15. $-3x = 2$
    $\dfrac{-3x}{-3} = \dfrac{2}{-3}$
    $\qquad x = -\dfrac{2}{3}$
    The solution is $-\dfrac{2}{3}$.

17. $\qquad -\dfrac{3}{2} + x = \dfrac{4}{3}$
    $-\dfrac{3}{2} + \dfrac{3}{2} + x = \dfrac{4}{3} + \dfrac{3}{2}$
    $\qquad\qquad x = \dfrac{8}{6} + \dfrac{9}{6}$
    $\qquad\qquad x = \dfrac{17}{6}$
    The solution is $\dfrac{17}{6}$.

19.
$$x + \frac{2}{3} = \frac{5}{6}$$
$$x + \frac{2}{3} - \frac{2}{3} = \frac{5}{6} - \frac{2}{3}$$
$$x = \frac{5}{6} - \frac{4}{6}$$
$$x = \frac{1}{6}$$
The solution is $\frac{1}{6}$.

21.
$$\frac{2}{3}y = 5$$
$$\frac{3}{2}\left(\frac{2}{3}y\right) = \frac{3}{2}(5)$$
$$y = \frac{15}{2}$$
The solution is $\frac{15}{2}$.

23.
$$\frac{4}{5} = -\frac{5}{8}x$$
$$-\frac{8}{5}\left(-\frac{5}{8}y\right) = -\frac{8}{5}\left(\frac{4}{5}\right)$$
$$y = -\frac{32}{25}$$
The solution is $-\frac{32}{25}$

25.
$$-12 = \frac{4x}{7}$$
$$\frac{7}{4}\left(\frac{4}{7}x\right) = \frac{7}{4}(-12)$$
$$x = -21$$
The solution is -21.

27.
$$-\frac{5y}{7} = \frac{10}{21}$$
$$-\frac{7}{5}\left(-\frac{5}{7}y\right) = -\frac{7}{5}\left(\frac{10}{21}\right)$$
$$y = -\frac{70}{105} = -\frac{2}{3}$$
The solution is $-\frac{2}{3}$.

29.
$$-\frac{3b}{5} = -\frac{3}{5}$$
$$-\frac{5}{3}\left(-\frac{3}{5}b\right) = -\frac{5}{3}\left(-\frac{3}{5}\right)$$
$$b = 1$$
The solution is 1.

31.
$$-\frac{2}{3}x = -\frac{5}{8}$$
$$-\frac{3}{2}\left(-\frac{2}{3}x\right) = -\frac{3}{2}\left(-\frac{5}{8}\right)$$
$$x = \frac{15}{16}$$
The solution is $\frac{15}{16}$.

33. $1.5x = 27$
$$\frac{1.5x}{1.5} = \frac{27}{1.5}$$
$$x = 18$$
The solution is 18.

35. $-0.015x = -12$
$$\frac{-0.015x}{-0.015} = \frac{-12}{-0.015}$$
$$x = 800$$
The solution is 800.

37. $3x + 5x = 12$
$$8x = 12$$
$$\frac{8x}{8} = \frac{12}{8}$$
$$x = \frac{3}{2}$$
The solution is $\frac{3}{2}$.

39. $3y - 5y = 0$
$$-2y = 0$$
$$\frac{-2y}{-2} = \frac{0}{-2}$$
$$y = 0$$
The solution is 0.

41. If a is a negative number less than -5, the solution to the equation $a = -5b$ is greater than 1.

Objective B Exercises

43. $2x - 4 = 12$
$$2x - 4 + 4 = 12 + 4$$
$$2x = 16$$
$$\frac{2x}{2} = \frac{16}{2}$$
$$x = 8$$
The solution is 8.

45. $4x - 6 = 3x$
$$4x + (-4x) - 6 = 3x + (-4x)$$
$$-6 = -x$$
$$(-1)(-6) = (-1)(-x)$$
$$6 = x$$
The solution is 6.

47. $7x + 12 = 9x$
$$7x + (-7x) + 12 = 9x + (-7x)$$
$$12 = 2x$$
$$\frac{12}{2} = \frac{2x}{2}$$
$$6 = x$$
The solution is 6.

49. $4x + 2 = 4x$
$$4x + (-4x) + 2 = 4x + 4x$$
$$2 = 0$$
There is no solution.

51. $2x + 2 = 3x + 5$
$$2x + (-2x) + 2 = 3x + (-2x) + 5$$
$$2 = x + 5$$
$$2 + (-5) = x + 5 + (-5)$$
$$-3 = x$$
The solution is -3.

53. $2 - 3t = 3t - 4$
$$2 - 3t + 3t = 3t + 3t - 4$$
$$2 = 6t - 4$$
$$2 + 4 = 6t - 4 + 4$$
$$6 = 6t$$
$$\frac{6}{6} = \frac{6t}{6}$$
$$1 = t$$
The solution is 1.

55. $3b - 2b = 4 - 2b$
$$b = 4 - 2b$$
$$b + 2b = 4 - 2b + 2b$$
$$3b = 4$$
$$\frac{3b}{3} = \frac{4}{3}$$
$$b = \frac{4}{3}$$
The solution is $\frac{4}{3}$.

57. $3x + 7 = 3 + 7x$
$$3x + (-3x) + 7 = 3 + 7x + (-3x)$$
$$7 = 3 + 4x$$
$$7 + (-3) = 3 + (-3) + 4x$$
$$4 = 4x$$
$$\frac{4}{4} = \frac{4x}{4}$$
$$1 = x$$
The solution is 1.

59.
$$\frac{1}{3} - 2b = 3$$
$$\frac{1}{3} + \left(-\frac{1}{3}\right) - 2b = 3 + \left(-\frac{1}{3}\right)$$
$$-2b = \frac{8}{3}$$
$$b = -\frac{4}{3}$$
The solution is $-\dfrac{4}{3}$.

61.
$$5.3y + 0.35 = 5.02y$$
$$5.3y + (-5.3y) + 0.35 = 5.02y + (-5.3y)$$
$$0.35 = -0.28y$$
$$-1.25 = y$$
The solution is -1.25.

63.
$$\frac{2x}{3} - \frac{1}{2} = \frac{5}{6}$$
$$\frac{2x}{3} - \frac{1}{2} + \frac{1}{2} = \frac{5}{6} + \frac{1}{2}$$
$$\frac{2x}{3} = \frac{5}{6} + \frac{3}{6}$$
$$\frac{2x}{3} = \frac{8}{6}$$
$$\frac{3}{2}\left(\frac{2}{3}x\right) = \frac{3}{2}\left(\frac{8}{6}\right)$$
$$x = 2$$
The solution is 2.

65.
$$\frac{3}{2} - \frac{10x}{9} = -\frac{1}{6}$$
$$\frac{3}{2} + \left(-\frac{3}{2}\right) - \frac{10x}{9} = -\frac{1}{6} + \left(-\frac{3}{2}\right)$$
$$-\frac{10x}{9} = -\frac{1}{6} + \left(-\frac{9}{6}\right)$$
$$-\frac{10x}{9} = -\frac{10}{6}$$
$$-\frac{9}{10}\left(-\frac{10}{9}x\right) = -\frac{10}{6}\left(-\frac{9}{10}\right)$$

$$x = \frac{3}{2}$$
The solution is $\dfrac{3}{2}$.

67. Solve $2y - 6 = 3y + 2$.
$$2y - 6 = 3y + 2$$
$$2y + (-2y) - 6 = 3y + (-2y) + 2$$
$$-6 = y + 2$$
$$-6 + (-2) = y + 2 + (-2)$$
$$-8 = y$$
Now evaluate $7y + 1$ for $y = -8$.
$$7(-8) + 1 = -56 + 1 = -55$$

69. If A is a negative number the solution of the equation $Ax - 2 = -5$ is positive.

Objective C Exercises

71. $2x + 3(x - 5) = 15$
$$2x + 3x - 15 = 15$$
$$5x - 15 = 15$$
$$5x = 30$$
$$\frac{5x}{5} = \frac{30}{5}$$
$$x = 6$$
The solution is 6.

73. $3(y - 5) - 5y = 2y + 9$
$$3y - 15 - 5y = 2y + 9$$
$$-2y - 15 = 2y + 9$$
$$-15 = 4y + 9$$
$$-24 = 4y$$
$$\frac{-24}{4} = \frac{4y}{4}$$
$$-6 = y$$
The solution is -6.

75.
$$4 - 3x = 7x - 2(3 - x)$$
$$4 - 3x = 7x - 6 + 2x$$
$$4 - 3x = 9x - 6$$
$$10 - 3x = 9x$$
$$10 = 12x$$
$$\frac{10}{12} = \frac{12x}{12}$$
$$x = \frac{5}{6}$$

The solution is $\dfrac{5}{6}$.

77.
$$-3x - 2(4 + 5x) = 14 - 3(2x - 3)$$
$$-3x - 8 - 10x = 14 - 6x + 9$$
$$-13x - 8 = 23 - 6x$$
$$-8 = 23 + 7x$$
$$-31 = 7x$$
$$-\frac{31}{7} = \frac{7x}{7}$$
$$-\frac{31}{7} = x$$

The solution is $-\dfrac{31}{7}$.

79.
$$3y = 2[5 - 3(2 - y)]$$
$$3y = 2[5 - 6 + 3y]$$
$$3y = 2[-1 + 3y]$$
$$3y = -2 + 6y$$
$$-3y = -2$$
$$\frac{-3y}{-3} = \frac{-2}{-3}$$
$$y = \frac{2}{3}$$

The solution is $\dfrac{2}{3}$.

81.
$$2[4 + 2(5 - x) - 2x] = 4x - 7$$
$$2[4 + 10 - 2x - 2x] = 4x - 7$$
$$2[14 - 4x] = 4x - 7$$
$$28 - 8x = 4x - 7$$
$$35 - 8x = 4x$$
$$35 = 12x$$
$$\frac{35}{12} = \frac{12x}{12}$$
$$\frac{35}{12} = x$$

The solution is $\dfrac{35}{12}$.

83.
$$2[3 - 2(z + 4)] = 3(4 - z)$$
$$2[3 - 2z - 8] = 12 - 3z$$
$$2[-2z - 5] = 12 - 3z$$
$$-4z - 10 = 12 - 3z$$
$$-4z - 22 = -3z$$
$$-22 = z$$

The solutions is -22.

85.
$$3[x - (2 - x) - 2x] = 3(4 - x)$$
$$3[x - 2 + x - 2x] = 12 - 3x$$
$$3[-2] = 12 - 3x$$
$$-6 = 12 - 3x$$
$$-18 = -3x$$
$$\frac{-18}{-3} = \frac{-3x}{-3}$$
$$6 = x$$

The solution is 6.

87.
$$\frac{2x - 5}{12} - \frac{3 - x}{6} = \frac{11}{12}$$
$$12\left(\frac{2x - 5}{12} - \frac{3 - x}{6}\right) = 12 \cdot \frac{11}{12}$$
$$\frac{12(2x - 5)}{12} - \frac{12(3 - x)}{6} = 11$$
$$(2x - 5) - 2(3 - x) = 11$$
$$2x - 5 - 6 + 2x = 11$$
$$4x - 11 = 11$$
$$4x = 22$$
$$\frac{4x}{4} = \frac{22}{4}$$
$$x = \frac{11}{2}$$

The solution is $\dfrac{11}{2}$.

89.
$$\frac{2x-1}{4} + \frac{3x+4}{8} = \frac{1-4x}{12}$$

$$24\left(\frac{2x-1}{4} + \frac{3x+4}{8}\right) = 24 \cdot \frac{1-4x}{12}$$

$$\frac{24(2x-1)}{4} + \frac{24(3x+4)}{8} = \frac{24(1-4x)}{12}$$

$$6(2x-1) + 3(3x+4) = 2(1-4x)$$

$$12x - 6 + 9x + 12 = 2 - 8x$$

$$21x + 6 = 2 - 8x$$

$$29x + 6 = 2$$

$$29x = -4$$

$$\frac{29x}{29} = \frac{-4}{29}$$

$$x = -\frac{4}{29}$$

The solution is $-\dfrac{4}{29}$.

91. $-1.6(b - 2.35) = -11.28$
$$-1.6b + 3.76 = -11.28$$
$$-1.6b = -15.04$$
$$b = 9.4$$
The solution is 9.4.

93. $0.05(300 - x) + 0.07x = 45$
$$15 - 0.05x + 0.07x = 45$$
$$15 + 0.02x = 45$$
$$0.02x = 30$$
$$x = 1500$$
The solution is 1500.

95. Solve $3(2x + 1) = 5 - 2(x - 2)$.
$$3(2x + 1) = 5 - 2(x - 2)$$
$$6x + 3 = 5 - 2x + 4$$
$$6x + 3 = 9 - 2x$$
$$8x + 3 = 9$$
$$8x = 6$$
$$\frac{8x}{8} = \frac{6}{8}$$
$$x = \frac{3}{4}$$

Now evaluate $2x^2 + 1$ for $x = \dfrac{3}{4}$.

$$2\left(\frac{3}{4}\right)^2 + 1 = 2\left(\frac{9}{16}\right) + 1 = \frac{18}{16} + 1$$

$$= \frac{9}{8} + \frac{8}{8} = \frac{17}{8}$$

97. Solve $5 - 2(4x - 1) = 3x + 7$.
$$5 - 2(4x - 1) = 3x + 7$$
$$5 - 8x + 2 = 3x + 7$$
$$7 - 8x = 3x + 7$$
$$7 - 11x = 7$$
$$-11x = 0$$
$$x = 0$$
Now evaluate $x^4 - x^2$.
$$(0)^4 - (0)^2 = 0$$

99. The equation $-15 + 12(x - 2) = 5x - 25$ is equivalent to the equation in Exercise 98.

Objective D Exercises

101.
$$C = 2\pi r$$
$$\frac{C}{2\pi} = \frac{2\pi r}{2\pi}$$
$$r = \frac{C}{2\pi}$$

103.
$$A = \frac{1}{2}bh$$
$$2 \cdot A = 2 \cdot \frac{1}{2}bh$$
$$2A = bh$$
$$\frac{2A}{b} = \frac{bh}{b}$$
$$h = \frac{2A}{b}$$

105. $I = \dfrac{100M}{C}$

$$C \cdot I = C \cdot \dfrac{100M}{C}$$

$$IC = 100M$$

$$\dfrac{IC}{100} = \dfrac{100M}{100}$$

$$M = \dfrac{IC}{100}$$

107. $A = P + Prt$

$$A - P = Prt$$

$$\dfrac{A - P}{Pt} = \dfrac{Prt}{Pt}$$

$$t = \dfrac{A - P}{Pt}$$

109. $s = \dfrac{1}{2}(a + b + c)$

$$2 \cdot s = 2 \cdot \dfrac{1}{2}(a + b + c)$$

$$2s = a + b + c$$

$$c = 2s - a - b$$

111. $S = 2\pi r^2 + 2\pi rh$

$$S - 2\pi r^2 = 2\pi rh$$

$$\dfrac{S - 2\pi r^2}{2\pi r} = \dfrac{2\pi rh}{2\pi r}$$

$$\dfrac{S - 2\pi r^2}{2\pi r} = h$$

113. $P = \dfrac{R - C}{n}$

$$n \cdot P = n \cdot \dfrac{R - C}{n}$$

$$nP = R - C$$

$$R = nP + C$$

115. $A = P(1 + i)$

$$\dfrac{A}{1 + i} = \dfrac{P(1 + i)}{1 + i}$$

$$\dfrac{A}{1 + i} = P$$

117. To solve the formula $P = 2L + 2W$ for L subtract $2W$ from each side of the equation.

Applying the Concepts

119. When the Multiplication Property of Equations is used, the quantity that multiplies each side of the equation must not be equal to zero otherwise the result would be the identity $0 = 0$ which does not help to solve the equation.

Section 2.2

Objective A Exercises

1. Strategy: Let x represent the unknown number.

$$\dfrac{3 + x}{10} = \dfrac{4}{5}$$

Solution: $\dfrac{3 + x}{10} = \dfrac{4}{5}$

$$10 \cdot \dfrac{3 + x}{10} = 10 \cdot \dfrac{4}{5}$$

$$3 + x = 8$$

$$x = 5$$

The number is 5.

3. Strategy: The sum of two integers is ten. Let x represent the smaller integer. The larger integer is $10 - x$. Three times the larger integer is three less than eight times the smaller integer: $3(10 - x) = 8x - 3$

Solution: $3(10 - x) = 8x - 3$

$$30 - 3x = 8x - 3$$

$$33 = 11x$$

$$3 = x$$

$10 - x = 10 - 3 = 7$

The integers are 3 and 7.

5. **Strategy:** Let x represent the larger integer.
The smaller integer is $x - 8$.
The sum of the two integers is 50.
$x + (x - 8) = 50$

Solution: $x + (x - 8) = 50$
$$2x - 8 = 50$$
$$2x = 58$$
$$x = 29$$
$x - 8 = 29 - 8 = 21$
The integers are 21 and 29.

7. **Strategy:** Let x represent the first number.
The second number is $2x + 2$.
The third number is $3x - 5$.
The sum of the numbers is 123.
$x + (2x + 2) + (3x - 5) = 123$

Solution: $x + (2x + 2) + (3x - 5) = 123$
$$6x - 3 = 123$$
$$6x = 126$$
$$x = 21$$
$2x + 2 = 2(21) + 2 = 42 + 2 = 44$
$3x - 5 = 3(21) - 5 = 63 - 5 = 58$
The numbers are 21, 44 and 58.

9. **Strategy:** Let x represent the first integer.
The second consecutive integer is $x + 1$.
The third consecutive integer is $x + 2$.
The sum of the integers is -57.
$x + (x + 1) + (x + 2) = -57$

Solution: $x + (x + 1) + (x + 2) = -57$
$$3x + 3 = -57$$
$$3x = -60$$
$$x = -20$$
$x + 1 = -20 + 1 = -19$
$x + 2 = -20 + 2 = -18$
The integers are -20, -19 and -18.

11. **Strategy:** Let x represent the first odd integer.
The second consecutive odd integer is $x + 2$.
The third consecutive odd integer is $x + 4$.
Five times the smallest is ten more than twice the largest.
$5x = 10 + 2(x + 4)$

Solution: $5x = 10 + 2(x + 4)$
$$5x = 10 + 2x + 8$$
$$3x = 18$$
$$x = 6$$
Since 6 is not an odd integer, there is no solution.

13. **Strategy:** Let x represent the first odd integer.
The second consecutive odd integer is $x + 2$.
The third consecutive odd integer is $x + 4$.
Three times the middle integer is seven more than the sum of the first and third integers:
$3(x + 2) = 7 + x + x + 4$

Solution: $3(x + 2) = 7 + x + x + 4$
$$3x + 6 = 11 + 2x$$
$$x = 5$$
$x + 2 = 5 + 2 = 7$
$x + 4 = 5 + 4 = 9$
The integers are 5, 7, and 9.

Objective B Exercises

15. a) $39n + 41m$
    b) $0.39n + 0.41m$

17. **Strategy:** The total number of coins in the coin bank is 22.
Let x represent the number of quarters.
The number of dimes is $4x$.
The number of nickels is $22 - 5x$.

Coin	Number	Value	Total Value
Quarter	x	25	$25x$
Dime	$4x$	10	$10(4x)$
Nickel	$22 - 5x$	5	$5(22 - 5x)$

The sum of the total values of each denomination of coin equals the total value of all of the coins (230 cents).
$25x + 10(4x) + 5(22 - 5x) = 230$

Solution: $25x + 10(4x) + 5(22 - 5x) = 230$
$$25x + 40x + 110 - 25x = 230$$
$$40x = 120$$
$$x = 3$$

$4x = 4(3) = 12$ dimes
There are 12 dimes in the collection.

19. Strategy: Let s represent the number of 20¢ stamps.
The number of 15¢ stamps is $3s - 8$.

Stamp	Number	Value	Total Value
20¢	s	20	$20s$
15¢	$3s - 8$	15	$15(3s - 8)$

The sum of the total value of each denomination of stamp equals the total value of all of the stamps (400 cents)
$20s + 15(3s - 8) = 400$.

Solution: $20s + 15(3s - 8) = 400$
$$20s + 45s - 120 = 400$$
$$65s = 520$$
$$s = 8$$

$3s - 8 = 3(8) - 8 = 24 - 8 = 16$
There are eight 20¢ stamps and sixteen 15¢ stamps.

21. Strategy: Let s represent the number of 3¢ stamps.
The number of 8¢ stamps is $2s - 3$.
The number of 13¢ stamps is $2(2s - 3)$.

Stamp	Number	Value	Total Value
3¢	s	3	$3s$
8¢	$2s - 3$	8	$8(2s - 3)$
13¢	$2(2s - 3)$	13	$13(2)(2s - 3)$

The sum of the total values of each denomination of stamp equals the total value of all of the stamps (253 cents).

$3s + 8(2s - 3) + 26(2s - 3) = 253$

Solution: $3s + 8(2s - 3) + 26(2s - 3) = 253$
$$3s + 16s - 24 + 52s - 78 = 253$$
$$71s - 102 = 252$$
$$71s = 355$$
$$s = 5$$

There are five 3¢ stamps in the collection.

23. Strategy: Let s represent the number of 18¢ stamps.
The number of 8¢ stamps is $2s$.
The number of 13¢ stamps is $s + 3$.

Stamp	Number	Value	Total Value
18¢	s	18	$18s$
8¢	$2s$	8	$8(2s)$
13¢	$s + 3$	13	$13(s + 3)$

The sum of the total values of each denomination of stamp equals the total value of all of the stamps (368 cents).
$18s + 8(2s) + 13(s + 3) = 368$

Solution: $18s + 8(2s) + 13(s + 3) = 368$
$$18s + 16s + 13s + 39 = 368$$
$$47s + 39 = 368$$
$$47s = 329$$
$$s = 7$$

There are seven 18¢ stamps in the collection.

Applying the Concepts

25. Strategy: Let x represent the first odd integer.
The second consecutive odd integer is $x + 2$.
The third consecutive odd integer is $x + 4$.
The product of the second and the third minus the product of the first and second is 42.
$(x + 2)(x + 4) - (x(x + 2) = 42$

Solution: $(x + 2)(x + 4) - x(x + 2) = 42$
$$x^2 + 6x + 8 - x^2 - 2x = 42$$
$$4x + 8 = 42$$
$$4x = 34$$
$$x = \frac{34}{4} = \frac{17}{2}$$

There is no solution because $\frac{17}{2}$ is not an integer.

Section 2.3

Objective A Exercises

1. It is not possible for the cost per pound of the mixture to be $11.00 or $6.50.

3. Strategy: Let x represent the cost of the mixture.

	Amount	Cost	Value
Cashews	40	9.20	40(9.20)
Peanuts	100	3.32	100(3.32)
Mixture	140	x	140x

The sum of the values before mixing equals the value after mixing.

Solution: $40(9.20) + 100(3.32) = 140x$
$$368 + 332 = 140x$$
$$700 = 140x$$
$$5 = x$$
The cost of the mixture is $5.00 per pound.

5. Strategy: Let x represent the number of adult tickets.
The number of children's tickets is $460 - x$.

	Amount	Cost	Value
Adult	x	10	10x
Children	$460 - x$	4	4(460 − x)

The total value of the tickets sold is $3760.

Solution: $10x + 4(460 - x) = 3760$
$$10x + 1840 - 4x = 3760$$
$$6x = 1920$$
$$x = 320$$
There were 320 adult tickets sold.

7. Strategy: Let x represent the number of liters of imitation maple syrup.

	Amount	Cost	Value
Imitation	x	4.00	4x
Maple	5	9.50	5(9.50)
Mixture	$5 + x$	5.00	5(5 + x)

The sum of the values before mixing equals the value after mixing.

Solution: $4x + 5(9.50) = 5(5 + x)$
$$4x + 47.5 = 25 + 5x$$
$$47.5 - 25 = x$$
$$22.5 = x$$
The mixture must contain 22.5 L of imitation maple syrup.

9. Strategy: Let x represent ounces of pure gold.
The ounces of gold alloy is $50 - x$.

	Amount	Cost	Value
Gold	x	890	890x
Alloy	$50 - x$	360	360(50 − x)
Mixture	50	519	519(50)

The sum of the values before mixing equals the value after mixing.

Solution: $890x + 360(50 - x) = 519(50)$
$$890x + 18,000 - 360x = 25,950$$
$$530x + 18,000 = 25,950$$
$$530x = 7950$$
$$x = 15$$
$50 - x = 50 - 15 = 35$
There were 15 oz of pure gold and 35 oz of gold alloy used in the mixture.

11. Strategy: Let x represent the cost per pound of the mixture.

	Amount	Cost	Value
$5.40 tea	40	5.40	5.4(40)
$3.25 tea	60	3.25	3.25(60)
Mixture	100	x	100x

The sum of the values before mixing equals the value after mixing.

Solution: $5.4(40) + 3.25(60) = 100x$
$$216 + 195 = 100x$$
$$411 = 100x$$
$$4.11 = x$$
The cost of the mixture is $4.11 per pound.

13. Strategy: Let x represent the number of gallons of cranberry juice.

	Amount	Cost	Value
Cranberry	x	28.50	$28.5(x)$
Apple	20	11.25	$11.25(20)$
Mixture	$20 + x$	17.00	$17(20 + x)$

The sum of the values before mixing equals the value after mixing.

Solution: $28.5x + 11.25(20) = 17(20 + x)$
$$28.5x + 225 = 340 + 17x$$
$$11.5x + 225 = 340$$
$$11.5x = 115$$
$$x = 10$$

The mixture must contain 10 gal of cranberry juice.

Objective B Exercises

15. A 30% salt solution is mixed with a 50% salt solution. It is not possible for the percent concentration of the resulting solution to be 30%, 25.8%, 80% or 50%.

17. Strategy: Let x represent the number of pounds of 15% aluminum alloy.

	Amount	Percent	Quantity
15%	x	0.15	$0.15x$
22%	500	0.22	$0.22(500)$
20%	$500 + x$	0.20	$0.20(500 + x)$

The sum of the quantities before mixing is equal to the quantity after mixing.

Solution: $0.15x + 0.22(500) = 0.20(500 + x)$
$$0.15x + 110 = 100 + 0.20x$$
$$110 = 100 + 0.05x$$
$$10 = 0.05x$$
$$200 = x$$

200 lb of the 15% aluminum alloy must be used.

19. Strategy: Let x represent the number of ounces of pure water.
The ounces of 70% alcohol is $3.5 - x$.

	Amount	Percent	Quantity
Water	x	0	$0x$
70% alcohol	$3.5 - x$	0.70	$0.7(3.5 - x)$
45% mixture	3.5	0.45	$0.45(3.5)$

The sum of the quantities before mixing is equal to the quantity after mixing.

Solution: $0x + 0.7(3.5 - x) = 0.45(3.5)$
$$2.45 - 0.7x = 1.575$$
$$-0.7x = -0.875$$
$$x = 1.25$$
$$3.5 - x = 3.5 - 1.25 = 2.25$$

The solution should contain 2.25 oz of rubbing alcohol and 1.25 oz of water.

21. Strategy: Let x represent the number of ounces of pure water.

	Amount	Percent	Quantity
Pure water	x	0	$0x$
8%	75	0.08	$0.08(75)$
5%	$75 + x$	0.05	$0.05(75 + x)$

The sum of the quantities before mixing is equal to the quantity after mixing.

Solution: $0x + 0.08(75) = 0.05(75 + x)$
$$6 = 3.75 + 0.05x$$
$$2.25 = 0.05x$$
$$45 = x$$

45 oz of pure water must be added.

23. Strategy: Let x represent the number of milliliters of alcohol.

	Amount	Percent	Quantity
Alcohol	x	0	$0x$
25% iodine	200	0.25	0.25(200)
10% iodine	$200 + x$	0.10	0.1(200 + x)

The sum of the quantities before mixing is equal to the quantity after mixing.

Solution: $0x + 0.25(200) = 0.1(200 + x)$
$$50 = 20 + 0.1x$$
$$30 = 0.1x$$
$$300 = x$$
300 ml of alcohol must be added.

25. Strategy: Let x represent the percent concentration of the remaining fruit drink.

	Amount	Percent	Quantity
5% fruit	12	0.05	0.05(12)
Water	2	0	0(2)
Mixture	10	x	$x(10)$

The sum of the quantities before mixing is equal to the quantity after mixing.

Solution: $0.05(12) + 0(2) = x(10)$
$$0.6 = 10x$$
$$0.06 = x$$
The remaining fruit drink is 6% fruit juice.

27. Strategy: Let x represent the number of quarts of 40% antifreeze replaced.
The quarts of pure antifreeze added is x.

	Amount	Percent	Quantity
40% antifreeze	12	0.40	0.40(12)
40% antifreeze	x	0.40	0.40(x)
Replaced by pure antifreeze	x	1.00	1.00(x)
Added 60% antifreeze	12	0.60	0.60(12)

The quantity in the radiator minus the quantity replaced plus the quantity added equals the quantity in the resulting solution.

Solution: $0.4(12) - 0.4(x) + x = 0.6(12)$
$$4.8 + 0.6x = 7.2$$
$$0.6x = 2.4$$
$$x = 4$$
4 qt will have to be replaced with pure antifreeze.

Objective C Exercises

29. a) The distance Ramona bikes is equal to the distance Eric bikes.
b) The time spent biking by Ramona is greater than the time spent biking by Eric.

31. Strategy: Let t represent policeman's time, in seconds.
The speeding car's time is $10 + t$ seconds.

	Rate	Time	Distance
Police officer	100	t	100(t)
Speeding car	80	$10 + t$	80(10 + t)

Police and the speeding car travel the same distance.

Solution $100t = 80(10 + t)$
$$100t = 800 + 80t$$
$$20t = 800$$
$$40 = t$$
The police officer will catch up to the speeding car in 2/3 minute.

33. Strategy: Let r represent the speed of the first car.
The speed of the second car is $r + 8$.

	Rate	Time	Distance
1st car	r	2.5	2.5(r)
2nd car	$r + 8$	2.5	2.5(r + 8)

The total distance traveled by the two cars is 310 mi.

Solution: $2.5r + 2.5(r + 8) = 310$
$$2.5r + 2.5r + 20 = 310$$
$$5r = 290$$
$$r = 58$$
$r + 8 = 58 + 8 = 66$
The speed of the first car is 58 mph.
The speed of the second car is 66 mph.

35. Strategy: Let t the time flying to the city. Time returning to the airport is $4 - t$.

	Rate	Time	Distance
Going	250	t	$250(t)$
Returning	150	$4 - t$	$150(4 - t)$

The distance to the city is the same as the distance returning to the airport.

Solution:
$$250t = 150(4 - t)$$
$$250t = 600 - 150t$$
$$400t = 600$$
$$t = 1.5$$
$$d = rt = 250(1.5) = 375$$
The distance between the airports is 375 mi.

37. Strategy: Let r represent the speed of the freight train.
The speed of the passenger train is $r + 18$.

	Rate	Time	Distance
Freight	r	4	$4(r)$
Passenger	$r + 18$	2.5	$2.5(r + 18)$

The distance traveled by the freight train is equal to the distance traveled by the passenger train.

Solution:
$$4r = 2.5(r + 18)$$
$$4r = 2.5r + 45$$
$$1.5r = 45$$
$$r = 30$$
$$r + 18 = 30 + 18 = 48$$
The speed of the freight train is 30 mph.
The speed of the passenger train is 48 mph.

39. Strategy: Let x represent the speed of the jogger.
The speed of the cyclist is $4x$.

	Rate	Time	Distance
Jogger	x	2	$2(x)$
Cyclist	$4x$	2	$2(4x)$

In two hours the cyclist is 33 mi ahead of the jogger.

Solution:
$$2(4x) - 2x = 33$$
$$8x - 2x = 33$$
$$6x = 33$$
$$x = 5.5$$
$$4x = 4(5.5) = 22$$
The speed of the cyclist is 22 mph.
$$d = rt = 22(2) = 44$$
The cyclist traveled 44 mi.

Applying the Concepts

41. Strategy: Let x represent the amount of 12-karat gold.

	Amount	Percent	Quantity
12-karat gold	x	$\dfrac{12}{24}$	$\dfrac{12}{24}x$
24-karat gold	3	$\dfrac{24}{24}$	$\dfrac{24}{24}(3)$
14-karat gold	$x + 3$	$\dfrac{14}{24}$	$\dfrac{14}{24}(x+3)$

The sum of the quantities before mixing is equal to the quantity after mixing.

Solution:
$$\frac{12}{24}x + \frac{24}{24}(3) = \frac{14}{24}(x + 3)$$
$$12x + 24(3) = 14(x + 3)$$
$$12x + 72 = 14x + 42$$
$$72 = 2x + 42$$
$$30 = 2x$$
$$15 = x$$
15 oz of the 12-karat gold should be used.

43. The distance between them 2 min before impact is equal to the sum of the distance each one can travel during 2 min.

$$2 \text{ minutes} \cdot \frac{1 \text{ hour}}{60 \text{ minutes}} = 0.03\overline{3} \text{ hour}$$

Distance between cars = rate of first car $\cdot 0.03\overline{3}$ + rate of second car $\cdot 0.03\overline{3}$.

Distance between cars
$$= 40(0.03\overline{3}) + 60(0.03\overline{3}) = 3.3\overline{3}$$

The cars are $3.3\overline{3}$ mi apart 2 min before impact.

Section 2.4
Objective A Exercises

1. The Addition Property of Inequalities states that the same number can be added to each side of an inequality without changing the solution set of the inequality.
    Examples will vary. For instance:

    $$8 > 6 \qquad\qquad -5 < -1$$
    $$8 + 7 > 6 + 7 \ \text{ and } \ -5 + (-2) < -1 + (-2)$$
    $$15 > 13 \qquad\qquad\quad -7 < -3$$

3. Replace x with each value to determine if the inequality holds.
    i) $-17 + 7 \le -3$; $-10 \le -3$; solution
    ii) $8 + 7 \le -3$; $15 \le -3$; not a solution
    iii) $-10 + 7 \le -3$; $-3 \le -3$; solution
    iv) $0 + 7 \le -3$; $7 \le -3$; not a solution

5. $x - 3 < 2$
    $x < 5$
    $\{x \mid x < 5\}$

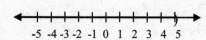

7. $4x \le 8$
    $\dfrac{4x}{4} \le \dfrac{8}{4}$
    $x \le 2$
    $\{x \mid x \le 2\}$

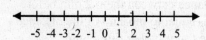

9. $-2x > 8$
    $\dfrac{-2x}{-2} < \dfrac{8}{-2}$
    $x < -4$
    $\{x \mid x < -4\}$

11. $3x - 1 > 2x + 2$
     $x - 1 > 2$
     $x > 3$
     $\{x \mid x > 3\}$

13. $2x - 1 > 7$
    $2x > 8$
    $\dfrac{2x}{2} > \dfrac{8}{2}$
    $x > 4$
    $\{x \mid x > 4\}$

15. $5x - 2 \le 8$
    $5x \le 10$
    $\dfrac{5x}{5} \le \dfrac{10}{5}$
    $x \le 2$
    $\{x \mid x \le 2\}$

17. $6x + 3 > 4x - 1$
    $6x > 4x - 4$
    $2x > -4$
    $\dfrac{2x}{2} > \dfrac{-4}{2}$
    $x > -2$
    $\{x \mid x > -2\}$

19. $8x + 1 \ge 2x + 13$
     $6x + 1 \ge 13$
     $6x \ge 12$
     $\dfrac{6x}{6} \ge \dfrac{12}{6}$
     $x \ge 2$
     $\{x \mid x \ge 2\}$

21. $4 - 3x < 10$
    $-3x < 6$
    $\dfrac{-3x}{-3} > \dfrac{6}{-3}$
    $x > -2$
    $\{x \mid x > -2\}$

23. $7 - 2x \geq 1$

$-2x \geq -6$

$\dfrac{-2x}{-2} \leq \dfrac{-6}{-2}$

$x \leq 3$

$\{x \mid x \leq 3\}$

25. $-3 - 4x > -11$

$-4x > -8$

$\dfrac{-4x}{-4} < \dfrac{-8}{-4}$

$x < 2$

$\{x \mid x < 2\}$

27. $4x - 2 < x - 11$

$3x - 2 < -11$

$3x < -9$

$\dfrac{3x}{3} < \dfrac{-9}{3}$

$x < -3$

$\{x \mid x < -3\}$

29. $x + 7 \geq 4x - 8$

$-3x + 7 \geq -8$

$-3x \geq -15$

$\dfrac{-3x}{-3} \leq \dfrac{-15}{-3}$

$x \leq 5$

$\{x \mid x \leq 5\}$

31. $3x + 2 \leq 7x + 4$

$-4x + 2 \leq 4$

$-4x \leq 2$

$\dfrac{-4x}{-4} \geq \dfrac{2}{-4}$

$x \geq -\dfrac{1}{2}$

$\{x \mid x \geq -\dfrac{1}{2}\}$

33. The solution set of the inequality $nx > a$, where both n and a are negative contains both positive and negative numbers.

35. The solution set of the inequality $x - n > -a$, where both n and a are positive and $n < a$ contains both positive and negative numbers.

37. $\dfrac{3}{5}x - 2 < \dfrac{3}{10} - x$

$10\left(\dfrac{3}{5}x - 2\right) < 10\left(\dfrac{3}{10} - x\right)$

$6x - 20 < 3 - 10x$

$16x - 20 < 3$

$16x < 23$

$\dfrac{16x}{16} < \dfrac{23}{16}$

$x < \dfrac{23}{16}$

$\left(-\infty, \dfrac{23}{16}\right)$

39. $\dfrac{2}{3}x - \dfrac{3}{2} < \dfrac{7}{6} - \dfrac{1}{3}x$

$6\left(\dfrac{2}{3}x - \dfrac{3}{2}\right) < 6\left(\dfrac{7}{6} - \dfrac{1}{3}x\right)$

$4x - 9 < 7 - 2x$

$6x - 9 < 7$

$6x < 16$

$\dfrac{6x}{6} < \dfrac{16}{6}$

$x < \dfrac{8}{3}$

$\left(-\infty, \dfrac{8}{3}\right)$

41. $\dfrac{1}{2}x - \dfrac{3}{4} < \dfrac{7}{4}x - 2$

$4\left(\dfrac{1}{2}x - \dfrac{3}{4}\right) < 4\left(\dfrac{7}{4}x - 2\right)$

$2x - 3 < 7x - 8$

$-5x - 3 < -8$

$-5x < -5$

$\dfrac{-5x}{-5} > \dfrac{-5}{-5}$

$x > 1$

$(1, \infty)$

43. $4(2x - 1) > 3x - 2(3x - 5)$

$8x - 4 > 3x - 6x + 10$

$8x - 4 > -3x + 10$

$11x - 4 > 10$

$11x > 14$

$\dfrac{11x}{11} > \dfrac{14}{11}$

$x > \dfrac{14}{11}$

$\left(\dfrac{14}{11}, \infty\right)$

45. $2 - 5(x + 1) \geq 3(x - 1) - 8$

$2 - 5x - 5 \geq 3x - 3 - 8$

$-3 - 5x \geq 3x - 11$

$-5x \geq 3x - 8$

$-8x \geq -8$

$\dfrac{-8x}{-8} \leq \dfrac{-8}{-8}$

$x \leq 1$

$(-\infty, 1]$

47. $3 + 2(x + 5) \geq x + 5(x + 1) + 1$

$3 + 2x + 10 \geq x + 5x + 5 + 1$

$2x + 13 \geq 6x + 6$

$-4x + 13 \geq 6$

$-4x \geq -7$

$\dfrac{-4x}{-4} \leq \dfrac{-7}{-4}$

$x \leq \dfrac{7}{4}$

$\left(-\infty, \dfrac{7}{4}\right]$

49. $3 - 4(x + 2) \leq 6 + 4(2x + 1)$

$3 - 4x - 8 \leq 6 + 8x + 4$

$-4x - 5 \leq 10 + 8x$

$-12x - 5 \leq 10$

$-12x \leq 15$

$\dfrac{-12x}{-12} \geq \dfrac{15}{-12}$

$x \geq -\dfrac{5}{4}$

$\left[-\dfrac{5}{4}, \infty\right)$

51. $12 - 2(3x - 2) \geq 5x - 2(5 - x)$

$12 - 6x + 4 \geq 5x - 10 + 2x$

$16 - 6x \geq 7x - 10$

$-6x \geq 7x - 26$

$-13x \geq -26$

$\dfrac{-13x}{-13} \leq \dfrac{-26}{-13}$

$x \leq 2$

$(-\infty, 2]$

Objective B Exercises

53. Writing $-3 > x > 4$ does not make senses because there is no number that is less than -3 *and* greater than 4.

55. $x - 3 \leq 1$ and $\quad 2x \geq -4$

$\qquad x \leq 4 \qquad\qquad\quad x \geq -2$

$\{x \mid x \leq 4\} \qquad\quad \{x \mid x \geq -2\}$

$\{x \mid x \leq 4\} \cap \{x \mid x \geq -2\} = [-2, 4]$

57. $2x < 6 \qquad$ or $\qquad x - 4 > 1$

$\quad x < 3 \qquad\qquad\qquad x > 5$

$\{x \mid x < 3\} \qquad\quad \{x \mid x > 5\}$

$\{x \mid x < 3\} \cup \{x \mid x > 5\} = (-\infty, 3) \cup (5, \infty)$

59. $\dfrac{1}{2}x > -2$ and $\quad 5x < 10$

$\quad x > -4 \qquad\qquad\quad x < 2$

$\{x \mid x > -4\} \qquad\quad \{x \mid x < 2\}$

$\{x \mid x > -4\} \cap \{x \mid x < 2\} = (-4, 2)$

61. $\dfrac{2}{3}x > 4 \qquad$ or $\qquad 2x < -8$

$\quad x > 6 \qquad\qquad\quad x < -4$

$\{x \mid x > 6\} \qquad\quad \{x \mid x < -4\}$

$\{x \mid x > 6\} \cup \{x \mid x < -4\} = (-\infty, -4) \cup (6, \infty)$

63. $3x < -9 \qquad$ and $\qquad x - 2 < 2$

$\quad x < -3 \qquad\qquad\quad x < 4$

$\{x \mid x < -3\} \qquad\quad \{x \mid x < 4\}$

$\{x \mid x < -3\} \cap \{x \mid x < 4\} = (-\infty, -3)$

65. $2x - 3 > 1$ and $3x - 1 < 2$
$\quad 2x > 4 \qquad\qquad 3x < 3$
$\quad x > 2 \qquad\qquad x < 1$
$\quad \{x \mid x > 2\} \qquad\quad \{x \mid x < 1\}$
$\quad \{x \mid x > 2\} \cap \{x \mid x < 1\} = \emptyset$

67. $4x + 1 < 5$ and $4x + 7 > -1$
$\quad 4x < 4 \qquad\qquad 4x > -8$
$\quad x < 1 \qquad\qquad x > -2$
$\quad \{x \mid x < 1\} \qquad\quad \{x \mid x > -2\}$
$\quad \{x \mid x < 1\} \cap \{x \mid x > -2\} = (-2, 1)$

69. The inequality $x > -3$ or $x < 2$ describes all real numbers.

71. The inequality $x < -3$ or $x > 2$ describes two intervals of real numbers.

73. $6x - 2 < -14$ or $5x + 1 > 11$
$\quad 6x < -12 \qquad\qquad 5x > 10$
$\quad x < -2 \qquad\qquad x > 2$
$\quad \{x \mid x < -2\} \qquad\quad \{x \mid x > 2\}$
$\quad \{x \mid x < -2\} \cup \{x \mid x > 2\} = \{x \mid x < -2 \text{ or } x > 2\}$

75. $5 < 4x - 3 < 21$
$\quad 5 + 3 < 4x - 3 + 3 < 21 + 3$
$\quad\quad 8 < 4x < 24$
$\quad\quad \dfrac{8}{4} < \dfrac{4x}{4} < \dfrac{24}{4}$
$\quad\quad 2 < x < 6$
$\quad \{x \mid 2 < x < 6\}$

77. $-2 < 3x + 7 < 1$
$\quad -2 + (-7) < 3x + 7 + (-7) < 1 + (-7)$
$\quad\quad -9 < 3x < -6$
$\quad\quad \dfrac{-9}{3} < \dfrac{3x}{3} < \dfrac{-6}{3}$
$\quad\quad -3 < x < -2$
$\quad \{x \mid -3 < x < -2\}$

79. $3x - 5 > 10$ or $3x - 5 < -10$
$\quad 3x > 15 \qquad\qquad 3x < -5$
$\quad x > 5 \qquad\qquad x < -\dfrac{5}{3}$
$\quad \{x \mid x > 5\} \qquad\quad \{x \mid x < -\dfrac{5}{3}\}$
$\quad \{x \mid x > 5\} \cup \{x \mid x < -\dfrac{5}{3}\} = \{x \mid x > 5 \text{ or } x < -\dfrac{5}{3}\}$

81. $8x + 2 \le -14$ and $4x - 2 > 10$
$\quad 8x \le -16 \qquad\qquad 4x > 12$
$\quad x \le -2 \qquad\qquad x > 3$
$\quad \{x \mid x \le -2\} \qquad\quad \{x \mid x > 3\}$
$\quad \{x \mid x \le -2\} \cap \{x \mid x > 3\} = \emptyset$

83. $5x + 12 \ge 2$ or $7x - 1 \le 13$
$\quad 5x \ge -10 \qquad\qquad 7x \le 14$
$\quad x \ge -2 \qquad\qquad x \le 2$
$\quad \{x \mid x \ge -2\} \qquad\quad \{x \mid x \le 2\}$
$\quad \{x \mid x \ge -2\} \cup \{x \mid x \le 2\} = \text{the set of real numbers}$

85. $3 \le 7x - 14 \le 31$
$\quad 3 + 14 \le 7x - 14 + 14 \le 31 + 14$
$\quad\quad 17 \le 7x \le 45$
$\quad\quad \dfrac{17}{7} \le \dfrac{7x}{7} \le \dfrac{45}{7}$
$\quad\quad \dfrac{17}{7} \le x \le \dfrac{45}{7}$
$\quad \{x \mid \dfrac{17}{7} \le x \le \dfrac{45}{7}\}$

87. $1 - 3x < 16$ and $1 - 3x > -16$
$\quad -3x < 15 \qquad\qquad -3x > -17$
$\quad x > -5 \qquad\qquad x < \dfrac{-17}{-3}$
$\quad \{x \mid x > -5\} \qquad\quad \{x \mid x < \dfrac{17}{3}\}$
$\quad \{x \mid x > -5\} \cap \{x \mid x < \dfrac{17}{3}\} = \{x \mid -5 < x < \dfrac{17}{3}\}$

89. $6x + 5 < -1$ or $1 - 2x < 7$
$\quad 6x < -6 \qquad\qquad -2x < 6$
$\quad x < -1 \qquad\qquad x > -3$
$\quad \{x \mid x < -1\} \qquad\quad \{x \mid x > -3\}$
$\quad \{x \mid x < -1\} \cup \{x \mid x > -3\} = \text{the set of real numbers.}$

91. $9 - x \ge 7$ and $9 - 2x < 3$
$\quad -x \ge -2 \qquad\qquad -2x < -6$
$\quad x \le 2 \qquad\qquad x > 3$
$\quad \{x \mid x \le 2\} \qquad\quad \{x \mid x > 3\}$
$\quad \{x \mid x \le 2\} \cap \{x \mid x > 3\} = \emptyset$

Objective C Exercises

93. "The temperature did not go above 42°F" can be written as $t \le 42$.

95. "The high temperature was 42°F." can be written as $t \leq 42$.

97. Strategy: Let x represent the smallest number.

Solution:

Two times the difference between a number and eight	$\leq$	five times the sum of the number and four.

$$2(x - 8) \leq 5(x + 4)$$
$$2x - 16 \leq 5x + 20$$
$$-3x - 16 \leq 20$$
$$-3x \leq 36$$
$$x \geq -12$$

The smallest number is -12.

99. Strategy: Let x represent the width of the rectangle.
The length of the rectangle is $2x - 5$.
To find the maximum width, solve the inequality $2L + 2W < 60$.

Solution:
$$2L + 2W < 60$$
$$2(2x - 5) + 2x < 60$$
$$4x - 10 + 2x < 60$$
$$6x - 10 < 60$$
$$6x < 70$$
$$x < \frac{70}{6} = 11\frac{2}{3}$$

The smallest width of the rectangle is 11 cm.

101. Strategy: Let d represent the number of days to run the advertisement.
To find the maximum number of days the advertisement can run on the website, solve the inequality $250 + 12d \leq 1500$.

Solution: $250 + 12d \leq 1500$
$$12d \leq 1250$$
$$d \leq \frac{1250}{12}$$
$$d \leq 104\frac{1}{6}$$

You can run the advertisement for 104 days.

103. Strategy: Let x represent the cost of a gallon of paint.
Since a gallon of paint covers 100 ft^2 and the room is 320 ft^2 the homeowner will need to buy 4 gal of paint.
To find the maximum cost per gallon, solve the inequality $24 + 4x \leq 100$.

Solution: $24 + 4x \leq 100$
$$4x \leq 76$$
$$x \leq 19$$

The maximum that the homeowner can pay for a gallon of paint is $19.

105. Strategy: To find the temperature range in Fahrenheit degrees, solve the compound inequality $0 < \frac{5(F - 32)}{9} < 30$.

Solution: $0 < \frac{5(F - 32)}{9} < 30$
$$\frac{9}{5}(0) < \frac{9}{5}\left(\frac{5(F - 32)}{9}\right) < \frac{9}{5}(30)$$
$$0 < F - 32 < 54$$
$$0 + 32 < F - 32 + 32 < 54 + 32$$
$$32° < F < 86°$$

107. Strategy: Let N represent the amount of sales.
To find the minimum amount of sales, solve the inequality $1000 + 0.05N \geq 3200$.

Solution: $1000 + 0.05N \geq 3200$
$$0.05N \geq 2200$$
$$N \geq 44{,}000$$

George's amount of sales must be $44,000 or more per month.

109. **Strategy:** Let N represent the number of checks.
To find the maximum number of checks solve the inequality
$8 + 0.12(N - 100) < 5 + 0.15(N - 100)$.

Solution:
$$8 + 0.12(N - 100) < 5 + 0.15(N - 100)$$
$$8 + 0.12N - 12 < 5 + 0.15N - 15$$
$$0.12N - 4 < 0.15N - 10$$
$$0.12N + 6 < 0.15N$$
$$6 < 0.03N$$
$$200 < N$$

The first account is less expensive when you write more than 200 checks.

111. **Strategy:** Let n represent the score on the last test.
To find the range of scores, solve the inequality
$$70 \le \frac{56 + 91 + 83 + 62 + n}{5} \le 79.$$

Solution:
$$70 \le \frac{56 + 91 + 83 + 62 + n}{5} \le 79$$
$$70 \le \frac{292 + n}{5} \le 79$$
$$5(70) \le 5 \cdot \frac{292 + n}{5} \le 5(79)$$
$$350 \le 292 + n \le 395$$
$$350 - 292 \le 292 - 292 + n \le 395 - 292$$
$$58 \le n \le 103$$

Since 100 is the maximum score, the range of scores needed to receive a C grade is
$58 \le n \le 100$.

113. **Strategy:** Let x represent the first even integer.
To find the three consecutive even integers, solve the inequality
$30 < x + (x + 2) + (x + 4) < 5$.

Solution:
$$30 < x + (x + 2) + (x + 4) < 51$$
$$30 < 3x + 6 < 51$$
$$30 - 6 < 3x + 6 - 6 < 51 - 6$$
$$24 < 3x < 45$$

$$\frac{24}{3} < \frac{3x}{3} < \frac{45}{3}$$
$$8 < x < 15$$

The four consecutive even integers are 10, 12 and 14; 12, 14 and 16; or 14, 16 and 18.

Applying the Concepts

115. a) Always true
   b) Sometimes true
   c) Sometimes true
   d) Sometimes true
   e) Always true

Section 2.5

Objective A Exercises

1. $|2 - 8| = 6$
   $|-6| = 6$
   $6 = 6$
   Yes, 2 is a solution.

3. $|3(-1) - 4| = 7$
   $|-3 - 4| = 7$
   $|-7| = 7$
   $7 = 7$
   Yes, -1 is a solution.

5. $|x| = 7$
   $x = 7$ or $x = -7$
   The solutions are 7 and -7.

7. $|b| = 4$
   $b = 4$ or $b = -4$
   The solutions are 4 and -4.

9. $|-y| = 6$
   $-y = 6$ or $-y = -6$
   $y = -6$ or $y = 6$
   The solutions are 6 and -6.

11. $|-a| = 7$
   $-a = 7$ or $-a = -7$
   $a = -7$ or $a = 7$
   The solutions are 7 and -7.

13. $|x| = -4$

There is no solution to this equation because the absolute value of a number must be nonnegative.

15. $|-t| = -3$

There is no solution to this equation because the absolute value of a number must be nonnegative.

17. $|x + 2| = 3$

$x + 2 = 3$ or $x + 2 = -3$

$x = 1$ $\qquad$ $x = -5$

The solutions are 1 and -5.

19. $|y - 5| = 3$

$y - 5 = 3$ or $y - 5 = -3$

$y = 8$ $\qquad$ $y = 2$

The solutions are 2 and 8.

21. $|a - 2| = 0$

$a - 2 = 0$

$a = 2$

The solution is 2.

23. $|x - 2| = -4$

There is no solution to this equation because the absolute value of a number must be nonnegative.

25. $|3 - 4x| = 9$

$3 - 4x = 9$ or $3 - 4x = -9$

$-4x = 6$ $\qquad$ $-4x = -12$

$x = -\dfrac{3}{2}$ $\qquad$ $x = 3$

The solutions are 3 and $-\dfrac{3}{2}$.

27. $|2x - 3| = 0$

$2x - 3 = 0$

$2x = 3$

$x = \dfrac{3}{2}$

The solution is $\dfrac{3}{2}$.

29. $|3x - 2| = -4$

There is no solution to this equation because the absolute value of a number must be nonnegative.

31. $|x - 2| - 2 = 3$

$|x - 2| = 5$

$x - 2 = 5$ or $x - 2 = -5$

$x = 7$ $\qquad$ $x = -3$

The solutions are 7 and -3.

33. $|3a + 2| - 4 = 4$

$|3a + 2| = 8$

$3a + 2 = 8$ or $3a + 2 = -8$

$3a = 6$ $\qquad$ $3a = -10$

$a = 2$ $\qquad$ $a = -\dfrac{10}{3}$

The solutions are 2 and $-\dfrac{10}{3}$.

35. $|2 - y| + 3 = 4$

$|2 - y| = 1$

$2 - y = 1$ or $2 - y = -1$

$-y = -1$ $\qquad$ $-y = -3$

$y = 1$ $\qquad$ $y = 3$

The solutions are 1 and 3.

37. $|2x - 3| + 3 = 3$

$|2x - 3| = 0$

$2x - 3 = 0$

$2x = 3$

$x = \dfrac{3}{2}$

The solution is $\dfrac{3}{2}$.

39. $|2x - 3| + 4 = -4$

$|2x - 3| = -8$

There is no solution to this equation because the absolute value of a number must be nonnegative.

41. $|6x - 5| - 2 = 4$

$|6x - 5| = 6$

$6x - 5 = 6$ or $6x - 5 = -6$

$6x = 11$ $\qquad$ $6x = -1$

$x = \dfrac{11}{6}$ $\qquad$ $x = -\dfrac{1}{6}$

The solutions are $\dfrac{11}{6}$ and $-\dfrac{1}{6}$.

43. $|3t + 2| + 3 = 4$
$|3t + 2| = 1$
$3t + 2 = 1$ or $3t + 2 = -1$
$3t = -1$ $3t = -3$
$x = -\dfrac{1}{3}$ $x = -1$

The solutions are $-\dfrac{1}{3}$ and -1.

45. $3 - |x - 4| = 5$
$-|x - 4| = 2$
$|x - 4| = -2$
There is no solution to this equation because the absolute value of a number must be nonnegative.

47. $8 - |2x - 3| = 5$
$-|2x - 3| = -3$
$|2x - 3| = 3$
$2x - 3 = 3$ or $2x - 3 = -3$
$2x = 6$ $2x = 0$
$x = 3$ $x = 0$
The solutions are 3 and 0.

49. $|2 - 3x| + 7 = 2$
$|2 - 3x| = -5$
There is no solution to this equation because the absolute value of a number must be nonnegative.

51. $|8 - 3x| - 3 = 2$
$|8 - 3x| = 5$
$8 - 3x = 5$ or $8 - 3x = -5$
$-3x = -3$ $-3x = -13$
$x = 1$ $x = \dfrac{13}{3}$

The solutions are 1 and $\dfrac{13}{3}$.

53. $|2x - 8| + 12 = 2$
$|2x - 8| = -10$
There is no solution to this equation because the absolute value of a number must be nonnegative.

55. $2 + |3x - 4| = 5$
$|3x - 4| = 3$
$3x - 4 = 3$ or $3x - 4 = -3$
$3x = 7$ $3x = 1$

$x = \dfrac{7}{3}$ $x = \dfrac{1}{3}$
The solutions are $\dfrac{7}{3}$ and $\dfrac{1}{3}$.

57. $5 - |2x + 1| = 5$
$-|2x + 1| = 0$
$2x + 1 = 0$
$2x = -1$
$x = -\dfrac{1}{2}$

The solution is $-\dfrac{1}{2}$.

59. $6 - |2x + 4| = 3$
$-|2x + 4| = -3$
$|2x + 4| = 3$
$2x + 4 = 3$ or $2x + 4 = -3$
$2x = -1$ $2x = -7$
$x = -\dfrac{1}{2}$ $x = -\dfrac{7}{2}$

The solutions are $-\dfrac{1}{2}$ and $-\dfrac{7}{2}$.

61. $8 - |1 - 3x| = -1$
$-|1 - 3x| = -9$
$|1 - 3x| = 9$
$1 - 3x = 9$ or $1 - 3x = -9$
$-3x = 8$ $-3x = -10$
$x = -\dfrac{8}{3}$ $x = \dfrac{10}{3}$

The solutions are $-\dfrac{8}{3}$ and $\dfrac{10}{3}$.

63. $5 + |2 - x| = 3$
$|2 - x| = -2$
There is no solution to this equation because the absolute value of a number must be nonnegative.

65. Two positive solutions

67. Two negative solutions

Objective B Exercises

69. $|x| > 3$

$\quad x > 3 \qquad$ or $\qquad x < -3$

$\quad \{x \mid x > 3\} \qquad\qquad \{x \mid x < -3\}$

$\quad \{x \mid x > 3\} \cup \{x \mid x < -3\} = \{x \mid x > 3 \text{ or } x < -3\}$

71. $|x + 1| > 2$

$\quad x + 1 > 2 \qquad$ or $\qquad x + 1 < -2$

$\quad x > 1 \qquad\qquad\qquad x < -3$

$\quad \{x \mid x > 1\} \qquad\qquad \{x \mid x < -3\}$

$\quad \{x \mid x > 1\} \cup \{x \mid x < -3\} = \{x \mid x > 1 \text{ or } x < -3\}$

73. $|x - 5| \le 1$

$\quad -1 \le x - 5 \le 1$

$\quad -1 + 5 \le x - 5 + 5 \le 1 + 5$

$\quad 4 \le x \le 6$

$\quad \{x \mid 4 \le x \le 6\}$

75. $|2 - x| \ge 3$

$\quad 2 - x \ge 3 \qquad$ or $\qquad 2 - x \le -3$

$\quad -x \ge 1 \qquad\qquad\qquad -x \le -5$

$\quad x \le -1 \qquad\qquad\qquad x \ge 5$

$\quad \{x \mid x \le -1\} \qquad\qquad \{x \mid x \ge 5\}$

$\quad \{x \mid x \le -1\} \cup \{x \mid x \ge 5\} = \{x \mid x \le -1 \text{ or } x \ge 5\}$

77. $|2x + 1| < 5$

$\quad -5 < 2x + 1 < 5$

$\quad -5 - 1 < 2x + 1 - 1 < 5 - 1$

$\quad -6 < 2x < 4$

$\quad -3 < x < 2$

$\quad \{x \mid -3 < x < 2\}$

79. $|5x + 2| > 12$

$\quad 5x + 2 > 12 \qquad$ or $\qquad 5x + 2 < -12$

$\quad 5x > 10 \qquad\qquad\qquad 5x < -14$

$\quad x > 2 \qquad\qquad\qquad x < -\dfrac{14}{5}$

$\quad \{x \mid x > 2\} \qquad\qquad \{x \mid x < -\dfrac{14}{5}\}$

$\quad \{x \mid x > 2\} \cup \{x \mid x < -\dfrac{14}{5}\} = \{x \mid x > 2 \text{ or } x < -\dfrac{14}{5}\}$

81. $|4x - 3| \le -2$

The absolute value of a number must be nonnegative. The solution set is the empty set $\varnothing$.

83. $|2x + 7| > -5$

$\quad 2x + 7 > -5 \qquad$ or $\qquad 2x + 7 < 5$

$\quad 2x > -12 \qquad\qquad\qquad 2x < -2$

$\quad x > -6 \qquad\qquad\qquad x < -1$

$\quad \{x \mid x > -6\} \qquad\qquad \{x \mid x < -1\}$

$\quad \{x \mid x > -6\} \cup \{x \mid x < -1\} = $ The set of all real numbers.

85. $|4 - 3x| \ge 5$

$\quad 4 - 3x \ge 5 \qquad$ or $\qquad 4 - 3x \le -5$

$\quad -3x \ge 1 \qquad\qquad\qquad -3x \le -9$

$\quad x \le -\dfrac{1}{3} \qquad\qquad\qquad x \ge 3$

$\quad \{x \mid x \le -\dfrac{1}{3}\} \qquad\qquad \{x \mid x \ge 3\}$

$\quad \{x \mid x \le -\dfrac{1}{3}\} \cup \{x \mid x \ge 3\} = \{x \mid x \le -\dfrac{1}{3} \text{ or } x \ge 3\}$

87. $|5 - 4x| \le 13$

$\quad -13 \le 5 - 4x \le 13$

$\quad -13 + (-5) \le 5 + (-5) - 4x \le 13 + (-5)$

$\quad -18 \le -4x \le 8$

$\quad \dfrac{18}{4} \ge x \ge -2$

$\quad \{x \mid -2 \le x \le \dfrac{9}{2}\}$

89. $|6 - 3x| \le 0$

$\quad 6 - 3x \le 0 \qquad$ or $\qquad 6 - 3x \ge 0$

$\quad -3x \le -6 \qquad\qquad\qquad -3x \ge -6$

$\quad x \ge 2 \qquad\qquad\qquad x \le 2$

$\quad \{x \mid x \ge 2\} \qquad\qquad \{x \mid x \le 2\}$

$\quad \{x \mid x \ge 2\} \cup \{x \mid x \le 2\} = \{x \mid x = 2\}$

91. $|2 - 9x| > 20$

$\quad 2 - 9x > 20 \qquad$ or $\qquad 2 - 9x < -20$

$\quad -9x > 18 \qquad\qquad\qquad -9x < -22$

$\quad x < -2 \qquad\qquad\qquad x > \dfrac{22}{9}$

$\quad \{x \mid x < -2\} \qquad\qquad \{x \mid x > \dfrac{22}{9}\}$

$\quad \{x \mid x < -2\} \cup \{x \mid x > \dfrac{22}{9}\} = \{x \mid x < -2 \text{ or } x > \dfrac{22}{9}\}$

93. $|2x - 3| + 2 < 8$

$|2x - 3| < 6$

$-6 < 2x - 3 < 6$

$-6 + 3 < 2x - 3 + 3 < 6 + 3$

$-3 < 2x < 9$

$-\dfrac{3}{2} < x < \dfrac{9}{2}$

$\{x \mid -\dfrac{3}{2} < x < \dfrac{9}{2}\}$

95. $|2 - 5x| - 4 > -2$

$|2 - 5x| > 2$

$2 - 5x > 2$ or $2 - 5x < -2$

$-5x > 0$ $\qquad\qquad$ $-5x < -4$

$x < 0$ $\qquad\qquad$ $x > \dfrac{4}{5}$

$\{x \mid x < 0\}$ $\qquad$ $\{x \mid x > \dfrac{4}{5}\}$

$\{x \mid x < 0\} \cup \{x \mid x > \dfrac{4}{5}\} = \{x \mid x < 0 \text{ or } x > \dfrac{4}{5}\}$

97. $8 - |2x - 5| < 3$

$-|2x - 5| < -5$

$|2x - 5| > 5$

$2x - 5 < -5$ or $2x - 5 > 5$

$2x < 0$ $\qquad\qquad$ $2x > 10$

$x < 0$ $\qquad\qquad$ $x > 5$

$\{x \mid x < 0\}$ $\qquad$ $\{x \mid x > 5\}$

$\{x \mid x < 0\} \cup \{x \mid x > 5\} = \{x \mid x < 0 \text{ or } x > 5\}$

99. All negative solutions

Objective C Exercises

101. The desired dosage is 3 ml. The tolerance is 0.2 ml.

103. **Strategy:** Let d represent the diameter of the bushing, T the tolerance and x the lower and upper limits of the diameter. Solve the absolute value inequality $|x - d| \le T$.

Solution: $|x - d| \le T$

$|x - 1.75| \le 0.008$

$-0.008 \le x - 1.75 \le 0.008$

$-0.008 + 1.75 \le x - 1.75 + 1.75 \le 0.008 + 1.75$

$1.742 \le x \le 1.758$

The lower and upper limits of the diameter of the bushing are 1.742 in. and 1.758 in.

105. **Strategy:** Let x represent the percent of American voters who felt the economy is an important issue. Solve the absolute value inequality $|x - 41| \le$.

Solution: $|x - 41| \le 3$

$-3 \le x - 41 \le 3$

$-3 + 41 \le x - 41 + 41 \le 3 + 41$

$38 \le x \le 44$

The lower and upper limits of American voters who felt the economy is an important issue 38% and 44%.

107. **Strategy:** Let L represent the length of the piston. Solve the absolute value inequality $|L - 9\dfrac{5}{8}| \le \dfrac{1}{32}$.

Solution: $|L - 9\dfrac{5}{8}| \le \dfrac{1}{32}$

$-\dfrac{1}{32} \le L - 9\dfrac{5}{8} \le \dfrac{1}{32}$

$-\dfrac{1}{32} + 9\dfrac{5}{8} \le L - 9\dfrac{5}{8} + 9\dfrac{5}{8} \le \dfrac{1}{32} + 9\dfrac{5}{8}$

$9\dfrac{19}{32} \le L \le 9\dfrac{21}{32}$

The lower and upper limits of the length of the piston are $9\dfrac{19}{32}$ in. and $9\dfrac{21}{32}$ in.

109. **Strategy:** Let M represent the range, in ohms, for a resistor.
Let T represent the tolerance of the resistor.
Solve the absolute value inequality $|M - 29{,}000| \le T$.

Solution: $T = (0.02)(29{,}000)$
$\qquad = 580$ ohms
$|M - 29{,}000| \le 580$
$-580 \le |M - 29{,}000| \le 580$
$-580 + 29{,}000 \le |M - 29{,}000 + 29{,}000|$
$\qquad\qquad\qquad \le 580 + 29{,}000$
$28{,}420 \le M \le 29{,}580$
The lower and upper limits of the resistor are 28,420 ohms and 29,580 ohms.

Applying the Concepts

111.
a) The equation $|x + 3| = x + 3$ is true for all x for which $x + 3 \ge 0$.
$x + 3 \ge 0$
$\quad x \ge -3$
$\{x \mid x \ge 3\}$

b) The equation $|a - 4| = 4 - a$ is true for all a for which $4 - a \ge 0$.
$4 - a \ge 0$
$\quad -a \ge -4$
$\quad a \le 4$
$\{a \mid a \le 4\}$

113.
a) $|x + y| \le |x| + |y|$
b) $|x - y| \ge |x| - |y|$
c) $||x| - |y|| \ge |x| - |y|$
d) $|xy| = |x||y|$

Chapter 2 Review Exercises

1. $3t - 3 + 2t = 7t - 15$
$\quad 5t - 3 = 7t - 15$
$\quad\quad -3 = 2t - 15$
$\quad -3 + 15 = 2t - 15 + 15$
$\quad\quad 12 = 2t$
$\quad\quad\; 6 = t$
The solution is 6.

2. $3x - 7 > -2$
$\quad 3x > 5$
$\quad \dfrac{3x}{3} > \dfrac{5}{3}$
$\quad x > \dfrac{5}{3}$
$\left(\dfrac{5}{3}, \infty\right)$

3. $\quad P = 2L + 2W$
$P - 2W = 2L$
$\dfrac{P - 2W}{2} = L$

4. $\quad x + 4 = -5$
$x + 4 - 4 = -5 - 4$
$\quad\quad x = -9$
The solution is -9.

5. $3x < 4 \quad$ and $\quad x + 2 > -1$
$\quad x < \dfrac{4}{3} \qquad\qquad x > -3$
$\left\{x \mid x < \dfrac{4}{3}\right\} \qquad \{x \mid x > -3\}$
$\left\{x \mid x < \dfrac{4}{3}\right\} \cap \{x \mid x > -3\} = \left\{x \mid -3 < x < \dfrac{4}{3}\right\}$

6. $\dfrac{3}{5}x - 3 = 2x + 5$
$5\left(\dfrac{3}{5}x - 3\right) = 5(2x + 5)$
$\quad 3x - 15 = 10x + 25$
$\quad\quad -15 = 7x + 25$
$\quad\quad -40 = 7x$
$\quad -\dfrac{40}{7} = x$
The solution is $-\dfrac{40}{7}$.

7.
$$-\frac{2}{3}x = \frac{4}{9}$$

$$-\frac{3}{2}\left(-\frac{2}{3}x\right) = -\frac{3}{2}\left(\frac{4}{9}\right)$$

$$x = -\frac{2}{3}$$

The solution is $-\frac{2}{3}$.

8. $|x - 4| - 8 = -3$

$|x - 4| = 5$

$x - 4 = 5$ or $x - 4 = -5$

$x = 9$ $x = -1$

The solutions are 9 and −1.

9. $|2x - 5| < 3$

$-3 < 2x - 5 < 3$

$-3 + 5 < 2x - 5 + 5 < 3 + 5$

$2 < 2x < 8$

$1 < x < 4$

$\{x \mid 1 < x < 4\}$

10.
$$\frac{2x - 3}{3} + 2 = \frac{2 - 3x}{5}$$

$$15\left(\frac{2x - 3}{3} + 2\right) = 15\left(\frac{2 - 3x}{5}\right)$$

$$5(2x - 3) + 15(2) = 3(2 - 3x)$$

$$10x - 15 + 30 = 6 - 9x$$

$$10x + 15 = 6 - 9x$$

$$19x + 15 = 6$$

$$19x = -9$$

$$x = -\frac{9}{19}$$

The solution is $-\frac{9}{19}$.

11. $2(a - 3) = 5(4 - 3a)$

$2a - 6 = 20 - 15a$

$17a - 6 = 20$

$17a = 26$

$a = \frac{26}{17}$

The solution is $\frac{26}{17}$.

12. $5x - 2 > 8$ or $3x + 2 < -4$

$5x > 10$ $3x < -6$

$x > 2$ $x < -2$

$\{x \mid x > 2\}$ $\{x \mid x < -2\}$

$\{x \mid x > 2\} \cup \{x \mid x < -2\} = \{x \mid x > 2 \text{ or } x < -2\}$

13. $|4x - 5| \geq 3$

$4x - 5 \geq 3$ or $4x - 5 \leq -3$

$4x \geq 8$ $4x \leq 2$

$x \geq 2$ $x \leq \frac{1}{2}$

$\{x \mid x \geq 2\}$ $\{x \mid x \leq \frac{1}{2}\}$

$\{x \mid x \geq 2\} \cup \{x \mid x \leq \frac{1}{2}\} = \{x \mid x \geq 2 \text{ or } x \leq \frac{1}{2}\}$

14.
$$P = \frac{R - C}{n}$$

$$n \cdot P = n\left(\frac{R - C}{n}\right)$$

$$nP = R - C$$

$$nP + C = R$$

$$C = R - nP$$

15.
$$\frac{1}{2}x - \frac{5}{8} = \frac{3}{4}x + \frac{3}{2}$$

$$8\left(\frac{1}{2}x - \frac{5}{8}\right) = 8\left(\frac{3}{4}x + \frac{3}{2}\right)$$

$$4x - 5 = 6x + 12$$

$$-5 = 2x + 12$$

$$-17 = 2x$$

$$-\frac{17}{2} = x$$

The solution is $-\frac{17}{2}$.

16. $6 + |3x - 3| = 2$

$|3x - 3| = -4$

There is no solution to this equation because the absolute value of a number must be nonnegative.

17. $3x - 2 > x - 4$ or $7x - 5 < 3x + 3$

$2x - 2 > -4$ $4x - 5 < 3$

$ 2x > -2$ $4x < 8$

$ x > -1$ $x < 2$

$\{x \mid x > -1\}$ $\{x \mid x < 2\}$

$\{x \mid x > -1\} \cup \{x \mid x < 2\} = \{x \mid x \text{ is any real number}\}$

$(-\infty, \infty)$

18. $2x - (3 - 2x) = 4 - 3(4 - 2x)$

$2x - 3 + 2x = 4 - 12 + 6x$

$4x - 3 = -8 + 6x$

$-3 = -8 + 2x$

$5 = 2x$

$\dfrac{5}{2} = x$

The solution is $\dfrac{5}{2}$.

19. $ x + 9 = -6$

$x + 9 - 9 = -6 - 9$

$ x = -15$

The solution is -15.

20. $ \dfrac{2}{3} = x + \dfrac{3}{4}$

$\dfrac{2}{3} - \dfrac{3}{4} = x + \dfrac{3}{4} - \dfrac{3}{4}$

$\dfrac{8}{12} - \dfrac{9}{12} = x$

$-\dfrac{1}{12} = x$

The solution is $-\dfrac{1}{12}$.

21. $ -3x = -21$

$\dfrac{-3x}{-3} = \dfrac{-21}{-3}$

$ x = 7$

The solution is 7.

22. $ \dfrac{2}{3}a = \dfrac{4}{9}$

$\dfrac{3}{2}\left(\dfrac{2}{3}a\right) = \dfrac{3}{2}\left(\dfrac{4}{9}\right)$

$a = \dfrac{2}{3}$

The solution is $\dfrac{2}{3}$.

23. $3y - 5 = 3 - 2y$

$5y - 5 = 3$

$ 5y = 8$

$ y = \dfrac{8}{5}$

The solution is $\dfrac{8}{5}$.

24. $4x - 5 + x = 6x - 8$

$5x - 5 = 6x - 8$

$ -5 = x - 8$

$ 3 = x$

The solution is 3.

25. $3(x - 4) = -5(6 - x)$

$3x - 12 = -30 + 5x$

$ -12 = -30 + 2x$

$ 18 = 2x$

$ 9 = x$

The solution is 9.

26. $ \dfrac{3x - 2}{4} + 1 = \dfrac{2x - 3}{2}$

$8\left(\dfrac{3x - 2}{4} + 1\right) = 8\left(\dfrac{2x - 3}{2}\right)$

$2(3x - 2) + 8(1) = 4(2x - 3)$

$6x - 4 + 8 = 8x - 12$

$6x + 4 = 8x - 12$

$4 = 2x - 12$

$16 = 2x$

$8 = x$

The solution is 8.

27. $ 5x - 8 < -3$

$5x - 8 + 8 < -3 + 8$

$ 5x < 5$

$ x < 1$

$(-\infty, 1)$

28. $2x - 9 \le 8x + 15$

$ -9 \le 6x + 15$

$ -24 \le 6x$

$ -4 \le x$

$[-4, \infty)$

29.
$$\frac{2}{3}x - \frac{5}{8} \geq \frac{3}{4}x + 1$$
$$24\left(\frac{2}{3}x - \frac{5}{8}\right) \geq 24\left(\frac{3}{4}x + 1\right)$$
$$16x - 15 \geq 18x + 24$$
$$-15 \geq 2x + 24$$
$$-39 \geq 2x$$
$$-\frac{39}{2} \geq x$$
$$\{x \mid x \leq -\frac{39}{2}\}$$

30. $2 - 3(2x - 4) \leq 4x - 2(1 - 3x)$
$$2 - 6x + 12 \leq 4x - 2 + 6x$$
$$14 - 6x \leq 10x - 2$$
$$14 \leq 16x - 2$$
$$16 \leq 16x$$
$$1 \leq x$$
$$\{x \mid x \geq 1\}$$

31. $-5 < 4x - 1 < 7$
$$-5 + 1 < 4x - 1 + 1 < 7 + 1$$
$$-4 < 4x < 8$$
$$-1 < x < 2$$
$$(-1, 2)$$

32. $|2x - 3| = 8$

$2x - 3 = 8$	$2x - 3 = -8$
$2x = 11$	$2x = -5$
$x = \dfrac{11}{2}$	$x = \dfrac{-5}{2}$

The solutions are $\dfrac{11}{2}$ and $\dfrac{-5}{2}$.

33. $|5x + 8| = 0$
$$5x + 8 = 0$$
$$5x = -8$$
$$x = \frac{-8}{5}$$
The solution is $-\dfrac{8}{5}$.

34. $|5x - 4| < -2$
$$\emptyset$$
The solution is the empty set because the absolute value of a number must be nonnegative.

35. Strategy: Let t be the time to travel to the island. Time returning to dock is $2\frac{1}{3} - t$.

	Rate	Time	Distance
To island	16	t	$16(t)$
Back to dock	12	$\frac{7}{3} - t$	$12(\frac{7}{3} - t)$

The distance to the island is the same as the distance back to the dock.
Determine t and then find the distance.

Solution: $16t = 12(\frac{7}{3} - t)$
$$16t = 28 - 12t$$
$$28t = 28$$
$$t = 1$$
$$d = rt = 16(1) = 16$$
The island is 16 min from the dock.

36. Strategy: Let x represent the number of gallons of apple juice.

	Amount	Cost	Value
Apple	x	12.50	$12.5(x)$
Cranberry	25	31.50	$25(31.5)$
Mixture	$25 + x$	25.00	$25(25 + x)$

The sum of the values before mixing equals the value after mixing.

Solution: $12.5x + 25(31.5) = 25(25 + x)$
$$12.5x + 787.5 = 625 + 25x$$
$$787.5 = 625 + 12.5x$$
$$162.5 = 12.5x$$
$$13 = x$$
The mixture must contain 13 gal of apple juice.

37. Strategy: Let N represent the amount of sales.
To find the minimum amount of sales solve the inequality $1200 + 0.08N \geq 5000$.

Solution: $1200 + 0.08N \geq 5000$
$$0.08N \geq 3200$$
$$N \geq 40,000$$
The executive's amount of sales must be $40,000 or more per month.

38. Strategy: Let x represent the number of nickels.
Number of dimes is $x + 3$.
Number of quarters is $30 - (2x + 3) = 27 - 2x$.

Coin	Number	Value	Total Value
Nickel	x	5	$5x$
Dime	$x + 3$	10	$10(x + 3)$
Quarter	$27 - 2x$	25	$25(27 - 2x)$

The sum of the total values of each denomination of coin equals the total value of all of the coins (355 cents)

Solution: $5x + 10(x + 3) + 25(27 - 2x) = 355$
$$5x + 10x + 30 + 675 - 50x = 355$$
$$-35x + 705 = 355$$
$$-35x = -350$$
$$x = 10$$

$27 - 2x = 27 - 2(10) = 27 - 20 = 7$
There are 7 quarters in the collection.

39. Strategy: Let d represent the diameter of the bushing, T the tolerance and x the lower and upper limits of the diameter.
Solve the absolute value inequality
$|x - d| \leq T$.

Solution: $|x - d| \leq T$
$|x - 2.75| \leq 0.003$
$$-0.003 \leq x - 2.75 \leq 0.003$$
$$-0.003 + 2.75 \leq x - 2.75 + 2.75 \leq 0.003 + 2.75$$
$$2.747 \leq x \leq 2.753$$
The lower and upper limits of the diameter of the bushing are 2.747 in. and 2.753 in.

40. Strategy: Let n represent the smaller integer.
The larger integer is $20 - n$.
Five times the smaller integer is two more than twice the larger integer.

Solution: $5n = 2 + 2(20 - n)$
$$5n = 2 + 40 - 2n$$
$$5n = 42 - 2n$$
$$7n = 42$$
$$n = 6$$
$20 - n = 20 - 6 = 14$
The integers are 6 and 14.

41. Strategy: Let N represent the score on the last test.
To find the range of scores, solve the inequality
$$80 \leq \frac{92 + 66 + 72 + 88 + N}{5} \leq 90$$

Solution:
$$80 \leq \frac{92 + 66 + 72 + 88 + N}{5} \leq 90$$
$$80 \leq \frac{318 + N}{5} \leq 90$$
$$5(80) \leq 5 \cdot \frac{318 + N}{5} \leq 5(90)$$
$$400 \leq 318 + N \leq 450$$
$$400 - 318 \leq 318 - 318 + N \leq 450 - 318$$
$$82 \leq N \leq 132$$
Since 100 is the maximum score, the range of scores needed to receive a C grade is
$82 \leq N \leq 100$.

42. Strategy: Let r be the speed of the first plane.
The speed of the second plane is $r + 80$.

	Rate	Time	Distance
1st plane	r	1.75	$1.75r$
2n plane	$r + 80$	1.75	$1.75(r + 80)$

The total distance traveled by the two planes is 1680 mi.
Solution: $1.75r + 1.75(r + 80) = 1680$
$$1.75r + 1.75r + 140 = 1680$$
$$3.5r + 140 = 1680$$
$$3.5r = 1540$$
$$r = 440$$

$r + 80 = 440 + 80 = 520$
The speed of the first plane is 440 mph and the speed of the second plane is 520 mph.

43. Strategy: Let x represent the number of pounds of 30% tin.
The pounds of 70% tin is $500 - x$.

	Amount	Percent	Quantity
30%	x	0.30	$0.30x$
70%	$500 - x$	0.70	$0.70(500 - x)$
40%	500	0.40	$0.40(500)$

The sum of the quantities before mixing is equal to the quantity after mixing.

Solution: $0.3x + 0.7(500 - x) = 0.4(500)$
$0.3x + 350 - 0.7x = 200$
$-0.4x + 350 = 200$
$-0.4x = -150$
$x = 375$

$500 - x = 500 - 375 = 125$
375 lb of the 30% tin alloy and 125 lb of the 70% tin alloy were used.

44. Strategy: Let L represent the range in length of the piston.
Solve the absolute value inequality
$$\left|L - 10\frac{3}{8}\right| \le \frac{1}{32}.$$

Solution: $\left|L - 10\frac{3}{8}\right| \le \frac{1}{32}$

$$-\frac{1}{32} \le L - 10\frac{3}{8} \le \frac{1}{32}$$

$$-\frac{1}{32} + 10\frac{3}{8} \le L - 10\frac{3}{8} + 10\frac{3}{8} \le \frac{1}{32} + 10\frac{3}{8}$$

$$10\frac{11}{32} \le L \le 10\frac{13}{32}$$

The lower and upper limits of the length of the piston are $10\frac{11}{32}$ in. and $10\frac{13}{32}$ in.

Chapter 2 Test

1. $x - 2 = -4$
$x - 2 + 2 = -4 + 2$
$x = -2$
The solution is -2.

2. $b + \frac{3}{4} = \frac{5}{8}$

$b + \frac{3}{4} - \frac{3}{4} = \frac{5}{8} - \frac{3}{4}$

$b = \frac{5}{8} - \frac{6}{8}$

$b = -\frac{1}{8}$

The solution is $-\frac{1}{8}$.

3. $-\frac{3}{4}y = -\frac{5}{8}$

$-\frac{4}{3}\left(-\frac{3}{4}\right)y = -\frac{4}{3}\left(-\frac{5}{8}\right)$

$y = \frac{5}{6}$

The solution is $\frac{5}{6}$.

4. $3x - 5 = 7$
$3x - 5 + 5 = 7 + 5$
$3x = 12$
$x = 4$
The solution is 4.

5. $\frac{3}{4}y - 2 = 6$

$\frac{3}{4}y - 2 + 2 = 6 + 2$

$\frac{3}{4}y = 8$

$\frac{4}{3}\left(\frac{3}{4}y\right) = 8\left(\frac{4}{3}\right)$

$y = \frac{32}{2}$

The solution is $\frac{32}{2}$.

6. $2x - 3 - 5x = 8 + 2x - 10$
$-3x - 3 = -2 + 2x$
$-3x - 1 = 2x$
$-1 = 5x$
$-\frac{1}{5} = x$

The solution is $-\frac{1}{5}$.

7. $2[a - (2 - 3a) - 4] = a - 5$
$2[a - 2 + 3a - 4] = a - 5$
$2[4a - 6] = a - 5$
$8a - 12 = a - 5$
$7a - 12 = -5$
$7a = 7$
$a = 1$
The solution is 1.

8. $E = IR + Ir$

$E - Ir = IR$

$$\dfrac{E - Ir}{I} = R$$

9.

$$\dfrac{2x+1}{3} - \dfrac{3x+4}{6} = \dfrac{5x-9}{9}$$

$$18\left(\dfrac{2x+1}{3} - \dfrac{3x+4}{6}\right) = 18\left(\dfrac{5x-9}{9}\right)$$

$6(2x + 1) - 3(3x + 4) = 2(5x - 9)$

$12x + 6 - 9x - 12 = 10x - 18$

$3x - 6 = 10x - 18$

$-6 = 7x - 18$

$12 = 7x$

$$\dfrac{12}{7} = x$$

The solution is $\dfrac{12}{7}$.

10. $3x - 2 \geq 6x + 7$

$-3x \geq 9$

$x \leq -3$

$\{x \mid x \leq -3\}$

11. $4 - 3(x + 2) < 2(2x + 3) - 1$

$4 - 3x - 6 < 4x + 6 - 1$

$-3x - 2 < 4x + 5$

$-2 < 7x + 5$

$-7 < 7x$

$-1 < x$

$(-1, \infty)$

12. $4x - 1 > 5$ or $2 - 3x < 8$

$4x > 6$ $-3x < 6$

$x > \dfrac{3}{2}$ $x > -2$

$\{x \mid x > \dfrac{3}{2}\} \cup \{x \mid x > -2\} = \{x \mid x > -2\}$

13. $4 - 3x \geq 7$ and $2x + 3 \geq 7$

$-3x \geq 3$ $2x \geq 4$

$x \leq -1$ $x \geq 2$

$\{x \mid x \leq -1\} \cap \{x \mid x \geq 2\} = \emptyset$

14. $|3 - 5x| = 12$

$3 - 5x = 12$ $3 - 5x = -12$

$-5x = 9$ $-5x = -15$

$x = -\dfrac{9}{5}$ $x = 3$

The solutions are $-\dfrac{9}{5}$ and 3.

15. $2 - |2x - 5| = -7$

$-|2x - 5| = -9$

$|2x - 5| = 9$

$2x - 5 = 9$ $2x - 5 = -9$

$2x = 14$ $2x = -4$

$x = 7$ $x = -2$

The solutions are 7 and -2.

16. $|3x - 5| \leq 4$

$-4 \leq 3x - 5 \leq 4$

$-4 + 5 \leq 3x - 5 + 5 \leq 4 + 5$

$1 \leq 3x \leq 9$

$\dfrac{1}{3} < x < 3$

$\{x \mid \dfrac{1}{3} < x < 3\}$

17. $|4x - 3| > 5$

$4x - 3 < -5$ or $4x - 3 > 5$

$4x < -2$ $4x > 8$

$x < -\dfrac{1}{2}$ $x > 2$

$\{x \mid x < -\dfrac{1}{2}\}$ $\{x \mid x > 2\}$

$\{x \mid x < -\dfrac{1}{2}\} \cup \{x \mid x > 2\} = \{x \mid x < -\dfrac{1}{2} \text{ or } x > 2\}$

18. Strategy: Let N represent the number of miles.

The cost of the Gambelli car < the cost of the McDougal car.

Solution: $40 + 0.25N < 58$

$0.25N < 18$

$N < 72$

It costs less to rent from Gambelli if the car is driven less than 72 mi.

19. Strategy: Let d represent the diameter of the bushing, T the tolerance and x the lower and upper limits of the diameter.

Solution: $|x - d| \leq T$
$|x - 2.65| \leq 0.002$
$-0.002 \leq x - 2.65 \leq 0.002$
$-0.002 + 2.65 \leq x - 2.65 + 2.65 \leq 0.002 + 2.65$
$2.648 \leq x \leq 2.652$

The lower and upper limits of the diameter of the bushing are 2.648 in. and 2.652 in.

20. Strategy: Let n represent the smaller integer
The larger integer is $15 - n$.
Eight times the smaller integer is one less than three times the larger integer.

Solution: $8n = 3(15 - n) - 1$
$8n = 45 - 3n - 1$
$8n = 44 - 3n$
$11n = 44$
$n = 4$
$15 - n = 15 - 4 = 11$
The integers are 4 and 11.

21. Strategy: Let s represent the number of 15¢ stamps.
The number of 11¢ stamps is $2s$.
The number of 24¢ stamps is $30 - 3s$.

Stamp	Number	Value	Total Value
15¢	s	15	$15s$
11¢	$2s$	11	$11(2s)$
24¢	$30 - 3s$	24	$24(30 - 3s)$

The sum of the total values of each denomination of stamp equals the total value of all of the stamps (440 cents)

Solution: $15s + 11(2s) + 24(30 - 3s) = 440$
$15s + 22s + 720 - 72s = 440$
$-35s + 720 = 440$
$-35s = -280$
$s = 8$
$30 - 3s = 30 - 3(8) = 30 - 24 = 6$
There are six 24¢ stamps.

22. Strategy: Let x represent the price of the hamburger mixture.

	Amount	Cost	Value
$3.10 hamburger	100	3.10	3.10(100)
$4.38 hamburger	60	4.38	4.38(60)
Mixture	160	x	$x(160)$

The sum of the values before mixing is equal to the quantity after mixing.

Solution: $3.1(100) + 4.38(60) = 160x$
$310 + 262.8 = 160x$
$572.8 = 160x$
$3.58 = x$

The price of the hamburger mixture is $3.58 per pound.

23. Strategy: Let t represent the time a jogger runs a distance.

Total running time is $1\dfrac{45}{60} = 1\dfrac{3}{4} = \dfrac{7}{4}$

	Rate	Time	Distance
Jogger runs a distance	8	t	$8t$
Jogger returns same distance	6	$\dfrac{7}{4} - t$	$6\left(\dfrac{7}{4} - t\right)$

A jogger runs a distance and returns the same distance.

Solution: $8t = 6\left(\dfrac{7}{4} - t\right)$
$8t = \dfrac{21}{2} - 6t$
$14t = \dfrac{21}{2}$
$t = \dfrac{3}{4}$

$8t = 8 \cdot \dfrac{3}{4} = 6$
The jogger ran 6 mi one way or a total distance of 12 mi.

24. Strategy: Let r represent the speed of the slower train.

	Rate	Time	Distance
Slower train	r	2	$2r$
Faster train	$r+5$	2	$2(r+5)$

The total distance traveled by the two trains is 250 mi.

Solution:
$$2r + 2(r+5) = 250$$
$$2r + 2r + 10 = 250$$
$$4r + 10 = 250$$
$$4r = 240$$
$$r = 60$$
$$r + 5 = 60 + 5 = 65$$
The rate of the slower train is 60 mph.
The rate of the faster train is 65 mph.

25. Strategy: Let x represent the number of ounces of pure water.

	Amount	Percent	Quantity
Pure water	x	0	$0x$
8% salt	60	0.08	$0.08(60)$
3% salt	$60+x$	0.03	$0.03(60+x)$

The sum of the quantities before mixing is equal to the quantity after mixing.

Solution:
$$0x + 0.08(60) = 0.03(60+x)$$
$$4.8 = 1.8 + 0.03x$$
$$3.0 = 0.3x$$
$$\mathbf{100 = x}$$
100 oz of pure water must be added.

Cumulative Review Exercises

1.
$$-4 - (-3) - 8 + (-2) = -4 + 3 + (-8) + (-2)$$
$$= -1 + (-8) + (-2)$$
$$= -9 + (-2)$$
$$= -11$$

2.
$$-2^2 \cdot 3^3 = -(2\cdot2)(3\cdot3\cdot3)$$
$$= -4\cdot27$$
$$= -108$$

3.
$$4 - (2-5)^2 \div 3 + 2 = 4 - (-3)^2 \div 3 + 2$$
$$= 4 - 9 \div 3 + 2$$

$$= 4 - 3 + 2$$
$$= 1 + 2$$
$$= 3$$

4.
$$4 \div \frac{\frac{3}{8}-1}{5} \cdot 2 = 4 \div \frac{-\frac{5}{8}}{5} \cdot 2$$
$$= 4 \div \left(-\frac{5}{8}\cdot\frac{1}{5}\right)\cdot 2$$
$$= 4 \div \left(-\frac{1}{8}\right)\cdot 2$$
$$= 4 \cdot (-8) \cdot 2$$
$$= -32\cdot 2$$
$$= -64$$

5.
$$2a^2 - (b-c)^2$$
$$2(2)^2 - (3 - (-1))^2$$
$$= 2(4) - (3+1)^2$$
$$= 2(4) - (4)^2$$
$$= 2(4) - 16$$
$$= 8 - 16$$
$$= -8$$

6.
$$\frac{a - b^2}{b-c}$$
$$\frac{2 - (-3)^2}{-3 - 4} = \frac{2 - 9}{-3 - 4}$$
$$= \frac{-7}{-7}$$
$$= 1$$

7. The Commutative Property of Addition

8. Let x represent the unknown number.
Three <u>times</u> the number <u>and</u> six is $3x + 6$.
$$3x + (3x + 6) = 6x + 6$$

9.
$$F = \frac{evB}{c}$$
$$Fc = evB$$
$$\frac{Fc}{ev} = B$$

10. $5[y - 2(3 - 2y) + 6] = 5[y - 6 + 4y + 6]$
$$= 5[5y]$$
$$= 25y$$

11. $\{-4, 0\}$

12. $\{x \mid x \le 3\} \cap \{x \mid x > -1\}$

13. $Ax + By + C = 0$
$$By = -Ax - C$$
$$y = \frac{-Ax - C}{B}$$

14. $-\dfrac{5}{6}b = -\dfrac{5}{12}$
$$-\frac{6}{5}\left(-\frac{5}{6}b\right) = -\frac{6}{5}\left(-\frac{5}{12}\right)$$
$$b = \frac{1}{2}$$
The solution is $\dfrac{1}{2}$.

15. $2x + 5 = 5x + 2$
$$5 = 3x + 2$$
$$3 = 3x$$
$$1 = x$$
The solution is 1.

16. $\dfrac{5}{12}x - 3 = 7$
$$\frac{5}{12}x = 10$$
$$\frac{12}{5}\left(\frac{5}{12}x\right) = \frac{12}{5}(10)$$
$$x = 24$$
The solution is 24.

17. $2[3 - 2(3 - 2x)] = 2(3 + x)$
$$2[3 - 6 + 4x] = 6 + 2x$$
$$2[-3 + 4x] = 6 + 2x$$
$$-6 + 8x = 6 + 2x$$
$$8x = 12 + 2x$$
$$6x = 12$$
$$x = 2$$
The solution is 2.

18. $3[2x - 3(4 - x)] = 2(1 - 2x)$
$$3[2x - 12 + 3x] = 2 - 4x$$
$$3[5x - 12] = 2 - 4x$$
$$15x - 36 = 2 - 4x$$
$$19x - 36 = 2$$
$$19x = 38$$
$$x = 2$$
The solution is 2.

19. $\dfrac{1}{2}y - \dfrac{2}{3}y + \dfrac{5}{12} = \dfrac{3}{4}y - \dfrac{1}{2}$
$$12\left(\frac{1}{2}y - \frac{2}{3}y + \frac{5}{12}\right) = 12\left(\frac{3}{4}y - \frac{1}{2}\right)$$
$$6y - 8y + 5 = 9y - 6$$
$$-2y + 5 = 9y - 6$$
$$-11y + 5 = -6$$
$$-11y = -11$$
$$y = 1$$
The solution is 1.

20. $\dfrac{3x - 1}{4} - \dfrac{4x - 1}{12} = \dfrac{3 + 5x}{8}$
$$24\left(\frac{3x - 1}{4} - \frac{4x - 1}{12}\right) = 24\left(\frac{3 + 5x}{8}\right)$$
$$6(3x - 1) - 2(4x - 1) = 3(3 + 5x)$$
$$18x - 6 - 8x + 2 = 9 + 15x$$
$$10x - 4 = 9 + 15x$$
$$-5x - 4 = 9$$
$$-5x = 13$$
$$x = -\frac{13}{5}$$
The solution is $-\dfrac{13}{5}$.

21. $3 - 2(2x - 1) \ge 3(2x - 2) + 1$
$$3 - 4x + 2 \ge 6x - 6 + 1$$
$$5 - 4x \ge 6x - 5$$
$$5 - 10x \ge -5$$
$$-10x \ge -10$$
$$x \le 1$$
$(-\infty, 1]$

22. $3x + 2 \le 5$ and $x + 5 > 1$
$$3x \le 3 \qquad \qquad x > -4$$
$$x \le 1$$
$$\{x \mid x \le 1\} \cap \{x \mid x > -4\} = \{x \mid -4 < x \le 1\}$$

23. $|3 - 2x| = 5$

$3 - 2x = 5$ or $3 - 2x = -5$

$-2x = 2$ $\qquad$ $-2x = -8$

$x = -1$ $\qquad$ $x = 4$

The solutions are -1 and 4.

24. $3 - |2x - 3| = -8$

$-|2x - 3| = -11$

$|2x - 3| = 11$

$2x - 3 = 11$ or $2x - 3 = -11$

$2x = 14$ $\qquad$ $2x = -8$

$x = 7$ $\qquad$ $x = -4$

The solutions are 7 and -4.

25. $|3x - 1| > 5$

$3x - 1 < -5$ or $3x - 1 > 5$

$3x < -4$ $\qquad$ $3x > 6$

$x < -\dfrac{4}{3}$ $\qquad$ $x > 2$

$\{x \mid x < -\dfrac{4}{3}\}$ $\qquad$ $\{x \mid x > 2\}$

$\{x \mid x < -\dfrac{4}{3}\} \cup \{x \mid x > 2\} = \{x \mid x < -\dfrac{4}{3} \text{ or } x > 2\}$

26. $|2x - 4| < 8$

$-8 < 2x - 4 < 8$

$-8 + 4 < 2x - 4 + 4 < 8 + 4$

$-4 < 2x < 12$

$-2 < x < 6$

$\{x \mid -2 < x < 6\}$

27. Strategy: Let N represent the number of checks.

To find the number of checks, solve the inequality

$5.00 + 0.04N > 2.00 + 0.10N$.

Solution:

$5.00 + 0.04N > 2.00 + 0.10N$

$5.00 > 2.00 + 0.06N$

$3.00 > 0.06N$

$50 > N$

The second account is cheaper if the customer writes less than 50 checks.

28. Strategy: Let n represent the first odd integer.

The second odd integer is $n + 2$.

The third odd integer is $n + 4$.

Four times the sum of the first and third integers is one less than seven times the second integer.

Solution: $4(n + n + 4) = 7(n + 2) - 1$

$4(2n + 4) = 7(n + 2) - 1$

$8n + 16 = 7n + 14 - 1$

$8n + 16 = 7n + 13$

$n + 16 = 13$

$n = -3$

The first integer is -3.

29. Strategy: Let x represent the number of quarters.

Number of dimes is $2x - 5$.

Coin	Number	Value	Total Value
Quarter	x	25	$25x$
Dime	$2x - 5$	10	$10(2x - 5)$

The sum of the total values of each denomination of coin equals the total value of all of the coins (400 cents).

Solution: $25x + 10(2x - 5) = 400$

$25x + 20x - 50 = 400$

$45x - 50 = 400$

$45x = 450$

$x = 10$

$2x - 5 = 2(10) - 5 = 20 - 5 = 15$

There are 15 dimes.

30. Strategy: Let x represent the number of ounces of pure silver.

	Amount	Cost	Value
Silver	x	15.78	$15.78x$
Alloy	100	8.26	$8.26(100)$
Mixture	$100 + x$	11.78	$11.78(100 + x)$

The sum of the values before mixing is equal to the quantity after mixing.

Solution:

$15.78x + 8.26(100) = 11.78(100 + x)$

$15.78x + 826 = 1178 + 11.78x$

$4x + 826 = 1178$

$4x = 352$

$x = 88$

88 oz of pure silver were used in the mixture.

31. Strategy: Let r represent the speed of the slower plane.

	Rate	Time	Distance
Slower plane	r	2.5	$2.5r$
Faster plane	$r + 120$	2.5	$2.5(r + 120)$

The total distance traveled by the two planes is 1400 mi.

Solution:
$$2.5r + 2.5(r + 120) = 1400$$
$$2.5r + 2.5r + 300 = 1400$$
$$5r + 300 = 1400$$
$$5r = 1100$$
$$r = 220$$

The speed of the slower plane is 220 mph.

32. Strategy: Let d represent the diameter of the bushing, T the tolerance and x the lower and upper limits of the diameter.

Solution: $|x - d| \leq T$
$$|x - 2.45| \leq 0.001$$
$$-0.001 \leq x - 2.45 \leq 0.001$$
$$-0.001 + 2.45 \leq x - 2.45 + 2.45 \leq 0.001 + 2.45$$
$$2.449 \leq x \leq 2.451$$

The lower and upper limits of the diameter of the bushing are 2.449 in. and 2.451 in.

33. Strategy: Let x represent the number of liters of 12% acid solution.

	Amount	Percent	Quantity
12% solution	x	0.12	$0.12x$
5% solution	4	0.05	$0.05(4)$
8% solution	$4 + x$	0.08	$0.08(4 + x)$

The sum of the quantities before mixing is equal to the quantity after mixing.

Solution:
$$0.12x + 0.05(4) = 0.08(4 + x)$$
$$0.12x + 0.2 = 0.32 + 0.08x$$
$$0.04x + 0.2 = 0.32$$
$$0.04x = 0.12$$
$$x = 3$$

3 L of 12% acid solution must be in the mixture.

Chapter 3: Linear Functions and Inequalities in Two Variables

Prep Test

1. $-4(x-3) = -4x + 12$

2. $\sqrt{(-6)^2 + (-8)^2} = \sqrt{36 + 64} = \sqrt{100} = 10$

3. $\dfrac{3 - (-5)}{2 - 6} = \dfrac{3 + 5}{2 - 6} = \dfrac{8}{-4} = -2$

4. $-2x + 5$
$-2(-3) + 5 = 6 + 5 = 11$

5. $\dfrac{2r}{r - 1}$

$\dfrac{2(5)}{5 - 1} = \dfrac{10}{4} = 2.5$

6. $2p^3 - 3p + 4$
$2(-1)^3 - 3(-1) + 4 = 2(-1) - 3(-1) + 4$
$\qquad\qquad\qquad = -2 + 3 + 4$
$\qquad\qquad\qquad = 5$

7. $\dfrac{x_1 + x_2}{2}$

$\dfrac{7 + (-5)}{2} = \dfrac{2}{2} = 1$

8. $\quad 3x - 4y = 12$
$\quad 3x - 4(0) = 12$
$\qquad\quad 3x = 12$
$\qquad\qquad x = 4$

9. $2x - y = 7$
$\quad -y = 7 - 2x$
$\qquad y = 2x - 7$

Section 3.1

Objective A Exercises

1. a) If the *x*-coordinate of an ordered pair is positive, the ordered pair will lie in quadrant I or IV.
b) If the ordinate of an ordered pair is negative, the ordered pair will lie in quadrant III or IV.

3.

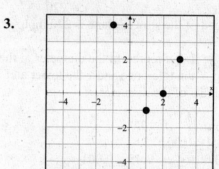

5. $A(0, 3),\ B(1, 1),\ C(3, -4),\ D(-4, 4)$

7.

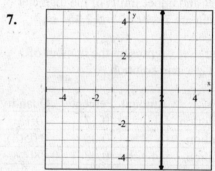

9.

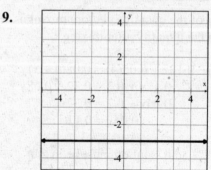

11. $y = x^2$

Ordered pairs: $(-2, -4)$, $(-1, 1)$, $(0, 0)$, $(1, 1)$, $(2, 4)$

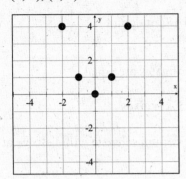

13. $y = |x + 1|$

Ordered pairs: $(-5, 4)$, $(-3, 2)$, $(0, 1)$, $(3, 4)$, $(5, 6)$

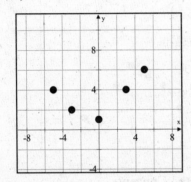

15. $y = -x^2 + 2$

Ordered pairs: $(-2, -2)$, $(-1, 1)$, $(0, 2)$, $(1, 1)$, $(2, -2)$

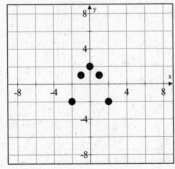

17. $y = x^3 - 2$

Ordered pairs: $(-1, -3)$, $(0, -2)$, $(1, -1)$, $(2, 6)$

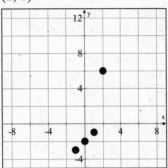

Objective B Exercises

19. $d = \sqrt{(x_1 - x_2)^2 + (y_1 - y_2)^2}$

$d = \sqrt{(3 - 5)^2 + (5 - 1)^2}$

$d = \sqrt{20} \approx 4.47$

$x_m = \dfrac{3 + 5}{2} = 4$

$y_m = \dfrac{5 + 1}{2} = 3$

Length: 4.47; midpoint (4, 3)

21. $d = \sqrt{(x_1 - x_2)^2 + (y_1 - y_2)^2}$

$d = \sqrt{(0 - (-2))^2 + (3 - 4)^2}$

$d = \sqrt{5} \approx 2.24$

$x_m = \dfrac{0 + (-2)}{2} = -1$

$y_m = \dfrac{3 + 4}{2} = \dfrac{7}{2}$

Length: 2.24; midpoint $\left(-1, \dfrac{7}{2}\right)$

23. $d = \sqrt{(x_1 - x_2)^2 + (y_1 - y_2)^2}$

$d = \sqrt{(-3 - 2)^2 + (-5 - (-4))^2}$

$d = \sqrt{26} \approx 5.10$

$x_m = \dfrac{-3 + 2}{2} = -\dfrac{1}{2}$

$y_m = \dfrac{-5 + (-4)}{2} = -\dfrac{9}{2}$

Length: 5.10; midpoint $\left(-\dfrac{1}{2}, -\dfrac{9}{2}\right)$

25. $d = \sqrt{(x_1 - x_2)^2 + (y_1 - y_2)^2}$

$d = \sqrt{(5 - (-2))^2 + (-2 - 5)^2}$

$d = \sqrt{98} \approx 9.90$

$x_m = \dfrac{5 + (-2)}{2} = \dfrac{3}{2}$

$y_m = \dfrac{-2 + 5}{2} = \dfrac{3}{2}$

Length: 9.90; midpoint $\left(\dfrac{3}{2}, \dfrac{3}{2}\right)$

27. $d = \sqrt{(x_1 - x_2)^2 + (y_1 - y_2)^2}$

$d = \sqrt{(5 - 2)^2 + (-5 - (-5))^2}$

$d = \sqrt{9} = 3$

$x_m = \dfrac{5 + 2}{2} = \dfrac{7}{2}$

$y_m = \dfrac{-5 + (-5)}{2} = -5$

Length: 3; midpoint $\left(\dfrac{7}{2}, -5\right)$

29. $d = \sqrt{(x_1 - x_2)^2 + (y_1 - y_2)^2}$

$d = \sqrt{\left(\dfrac{3}{2} - \left(-\dfrac{1}{2}\right)\right)^2 + \left(-\dfrac{4}{3} - \dfrac{7}{3}\right)^2}$

$d = \sqrt{\dfrac{157}{9}} \approx 4.18$

$x_m = \dfrac{\dfrac{3}{2} + \left(-\dfrac{1}{2}\right)}{2} = \dfrac{1}{2}$

$y_m = \dfrac{-\dfrac{4}{3} + \dfrac{7}{3}}{2} = \dfrac{1}{2}$

Length: 4.18; midpoint $\left(\dfrac{1}{2}, \dfrac{1}{2}\right)$

31. The x-coordinates are equal.

Objective C Exercises

33. a) 25 g of cerium selenate will dissolve at 50°C.

b) 5 g of cerium selenate will dissolve at 80°C.

35.

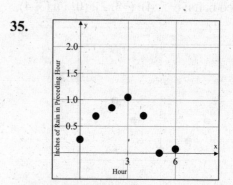

Applying the Concepts

37. Ordered pairs: $(-2, 4)$, $(-1, 1)$, $(0, 0)$, $(1, 1)$, $(2, 4)$

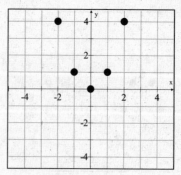

39. The graph of all ordered pairs (x, y) that are 5 units from the origin is a circle of radius 5 that has its center at $(0, 0)$.

41.

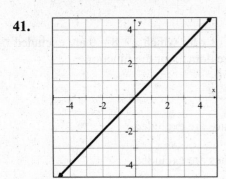

Section 3.2

Objective A Exercises

1. A function is a set of ordered pairs in which no two ordered pairs can have the same x-coordinate and different y-coordinates.

3. The diagram does represent a function because each number in the domain is paired with one number in the range.

5. The diagram does represent a function because each number in the domain is paired with one number in the range.

7. No, the diagram does not represent a function. The 6 in the domain is paired with two different values in the range.

9. This is a function because each x-coordinate is paired with only one y-coordinate.

11. This is a function because each x-coordinate is paired with only one y-coordinate.

13. This is a function because each x-coordinate is paired with only one y-coordinate.

15. This is not a function because there are x-coordinates paired with two different y-coordinates. They are $(1, 1)$, $(1, -1)$ and $(4, 2)$, $(4, -2)$.

17. a) Yes, this table defines a function because there is only one fine for any given number

of miles per hour over the speed limit.
　b) If $x = 16$, then $y = \$125$.

19. True

21. $f(x) = 5x - 4$
$f(3) = 5(3) - 4$
$f(3) = 15 - 4$
$f(3) = 11$

23. $f(x) = 5x - 4$
$f(0) = 5(0) - 4$
$f(0) = -4$

25. $G(t) = 4 - 3t$
$G(0) = 4 - 3(0)$
$G(0) = 4$

27. $G(t) = 4 - 3t$
$G(-2) = 4 - 3(-2)$
$G(-2) = 4 + 6$
$G(-2) = 10$

29. $q(r) = r^2 - 4$
$q(3) = 3^2 - 4$
$q(3) = 9 - 4$
$q(3) = 5$

31. $q(r) = r^2 - 4$
$q(-2) = (-2)^2 - 4$
$q(-2) = 4 - 4$
$q(-2) = 0$

33. $F(x) = x^2 + 3x - 4$
$F(4) = 4^2 + 3(4) - 4$
$F(4) = 16 + 12 - 4$
$F(4) = 24$

35. $F(x) = x^2 + 3x - 4$
$F(-3) = (-3)^2 + 3(-3) - 4$
$F(-3) = 9 - 9 - 4$
$F(-3) = -4$

37. $H(p) = \dfrac{3p}{p+2}$

$H(1) = \dfrac{3(1)}{1+2}$

$H(1) = \dfrac{3}{3}$

$H(1) = 1$

39. $H(p) = \dfrac{3p}{p+2}$

$H(t) = \dfrac{3t}{t+2}$

41. $s(t) = t^3 - 3t + 4$
$s(-1) = (-1)^3 - 3(-1) + 4$
$s(-1) = -1 + 3 + 4$
$s(-1) = 6$

43. $s(t) = t^3 - 3t + 4$
$s(a) = a^3 - 3a + 4$

45. $P(x) = 4x + 7$
$P(-2 + h) - P(-2)$
$= 4(-2 + h) + 7 - [4(-2) + 7]$
$= -8 + 4h + 7 + 8 - 7$
$= 4h$

47. $114.29

49. a) $4.75 per game
b) $4.00 per game

51. Domain = {1, 2, 3, 4, 5}
Range = {1, 4, 7, 10, 13}

53. Domain = {0, 2, 4, 6}
Range = {1, 2, 3, 4}

55. Domain = {1, 3, 5, 7, 9}
Range = {0}

57. Domain = {-2, -1, 0, 1, 2}
Range = {0, 1, 2}

59. Domain = {-2, -1, 0, 1, 2}
Range = {-3, 3, 6, 7, 9}

61. Values of x for which $x - 1 = 0$ are excluded from the domain of the function
$x - 1 = 0$
$x = 1$

63. Values of x for which $x + 8 = 0$ are excluded from the domain of the function
$x + 8 = 0$
$x = -8$

65. No values are excluded.

67. No values are excluded.

69. $x = 0$

71. No values are excluded.

73. No values are excluded.

75. Values of x for which $x + 2 = 0$ are excluded from the domain of the function
$x + 2 = 0$
$x = -2$

77. No values are excluded.

79. $f(x) = 4x - 3$
$f(0) = 4(0) - 3 = -3$
$f(1) = 4(1) - 3 = 1$
$f(2) = 4(2) - 3 = 5$
$f(3) = 4(3) - 3 = 9$
Range = {-3, 1, 5, 9}

81. $g(x) = 5x - 8$
$g(-3) = 5(-3) - 8 = -23$
$g(-1) = 5(-1) - 8 = -13$
$g(0) = 5(0) - 8 = -8$
$g(1) = 5(1) - 8 = -3$
$g(3) = 5(3) - 8 = 7$
Range = {-23, -13, -8, -3, 7}

83. $h(x) = x^2$
$h(-2) = (-2)^2 = 4$
$h(-1) = (-1)^2 = 1$
$h(0) = 0$
$h(1) = (1)^2 = 1$
$h(2) = (2)^2 = 4$
Range = {0, 1, 4}

85. $f(x) = 2x^2 - 2x + 2$

$f(-4) = 2(-4)^2 - 2(-4) + 2 = 42$

$f(-2) = 2(-2)^2 - 2(-2) + 2 = 14$

$f(0) = 2(0)^2 - 2(0) + 2 = 2$

$f(4) = 2(4)^2 - 2(4) + 2 = 26$

Range = $\{2, 14, 26, 42\}$

87. $H(x) = \dfrac{5}{1-x}$

$H(-2) = \dfrac{5}{1-(-2)} = \dfrac{5}{3}$

$H(0) = \dfrac{5}{1-0} = 5$

$H(2) = \dfrac{5}{1-2} = -5$

Range = $\{-5, \dfrac{5}{3}, 5\}$

89. $f(x) = \dfrac{2}{x-4}$

$f(-2) = \dfrac{2}{-2-4} = -\dfrac{1}{3}$

$f(0) = \dfrac{2}{0-4} = -\dfrac{1}{2}$

$f(2) = \dfrac{2}{2-4} = -1$

$f(6) = \dfrac{2}{6-4} = 1$

Range = $\{-1, -\dfrac{1}{2}, -\dfrac{1}{3}, 1\}$

91. $H(x) = 2 - 3x - x^2$

$H(-5) = 2 - 3(-5) - (-5)^2 = -8$

$H(0) = 2$

$H(5) = 2 - 3(5) - 5^2 = -38$

Range = $\{-38, -8, 2\}$

Applying the Concepts

93. A relation and a function are similar in that both are sets of ordered pairs. A function is a specific type of relation. A function is a relation in which there are no two ordered pairs with the same first element.

95. a) $\{(-2, -8), (-1, -1), (0, 0), (1, 1), (2, 8)\}$

b) Yes this set of ordered pairs defines a function because each member of the domain is assigned exactly one member of the range.

97. Evaluate the function $P = f(v) = 0.015v^3$ when $v = 15$.

$P = f(v) = 0.015(15)^3 = 50.625$

The power produced will be 50.625 watts.

99. a) The speed of the paratrooper 11.5 s after the beginning of the jump is 36.3 ft/s.

b) 30 ft/s

101. a) 275,000 malware attacks

b) 700,000 malware attacks

Section 3.3

Objective A Exercises

1. If the three ordered pairs appear to lie on the same line, then it is more likely that your calculations are accurate.

3. $y = x - 3$

x	y
−1	−4
0	−3
3	0

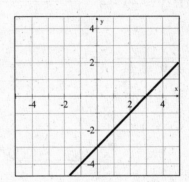

5. $y = -3x + 2$

x	y
0	2
1	−1
2	−4

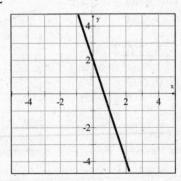

7. $f(x) = 3x - 4$

x	$f(x)$
0	−4
1	−1
2	2

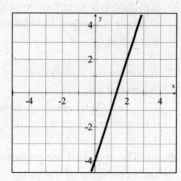

Objective B Exercises

15. $2x + y = -3$

$\qquad y = -2x - 3$

x	y
0	−3
1	−5
−1	−1

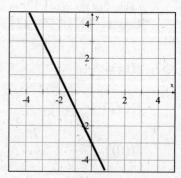

9. $f(x) = -\dfrac{2}{3}x$

x	$f(x)$
−3	2
0	0
3	−2

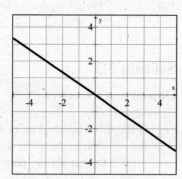

17. $x - 4y = 8$

$\qquad -4y = -x + 8$

$\qquad y = \dfrac{1}{4}x - 2$

x	y
0	−2
4	−1
8	0

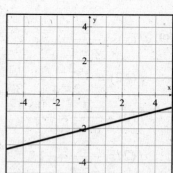

11. $y = \dfrac{2}{3}x - 4$

x	y
0	−4
3	−2
6	0

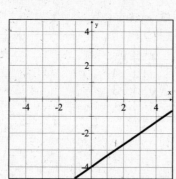

19. $4x + 3y = 12$

$\qquad 3y = -4x + 12$

$\qquad y = -\dfrac{4}{3}x + 4$

x	y
0	4
3	0
6	4

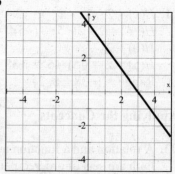

13. $f(x) = -\dfrac{1}{3}x + 2$

x	$f(x)$
−3	3
0	2
3	1

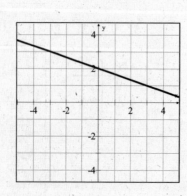

21. $x - 3y = 0$

$-3y = -x$

$y = \dfrac{1}{3}x$

x	y
0	0
3	1
-3	-1

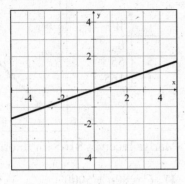

23. $y = -2$

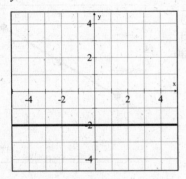

25. $x = -3$

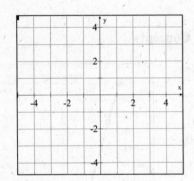

27. $3x - y = -2$

$-y = -3x - 2$

$y = 3x + 2$

x	y
0	2
1	5
-1	-1

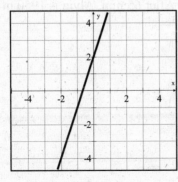

29. $3x - 2y = 8$

$-2y = -3x + 8$

$y = \dfrac{3}{2}x - 4$

x	y
0	-4
2	-1
4	2

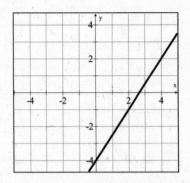

31. No. If $B = 0$, then it is not possible to solve $Ax + By = C$ for y.

Objective C Exercises

33. x-intercept:

$x - 2y = -4$

$x - 2(0) = -4$

$x = -4$

$(-4, 0)$

y-intercept:

$x - 2y = -4$

$0 - 2y = -4$

$-2y = -4$

$y = 2$

$(0, 2)$

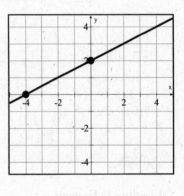

35. x-intercept:

$2x - 3y = 9$

$2x - 3(0) = 9$

$2x = 9$

$x = \dfrac{9}{2}$

$\left(\dfrac{9}{2}, 0\right)$

y-intercept:

$2x - 3y = 9$

$2(0) - 3y = 9$

$-3y = 9$

$y = -3$

$(0, -3)$

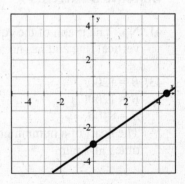

37. x-intercept:

$2x + y = 3$

$2x + 0 = 3$

$2x = 3$

$x = \dfrac{3}{2}$

$\left(\dfrac{3}{2}, 0\right)$

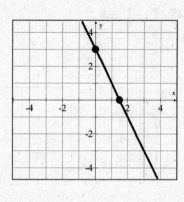

y-intercept:

$2x + y = 3$

$2(0) + y = 3$

$y = 3$

$(0, 3)$

39. x-intercept:

$3x + 2y = 4$

$3x + 2(0) = 4$

$3x = 4$

$x = \dfrac{4}{3}$

$\left(\dfrac{4}{3}, 0\right)$

y-intercept:

$3x + 2y = 4$

$3(0) + 2y = 4$

$2y = 4$

$y = 2$

$(0, 2)$

Objective D Exercises

41. No. The graph of the equation $x = a$ is a vertical line and has no y-intercept.

43. $B = 1200t$

$B = 1200(7)$

$B = 8400$

The heart of a hummingbird will beat 8400 times in 7 min.

45. $W = 11t$

t	W
0	0
5	55
15	165

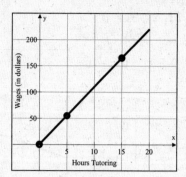

Marlys receives \$165 for tutoring 15 h.

47. $C = 80n + 5000$

n	C
0	5000
50	9000
100	13000

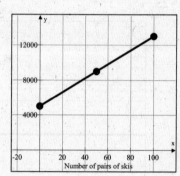

The cost of manufacturing 50 pairs of skis is \$9000.

Applying the Concepts

49. a) $D = -30t$

t	D
0	0
20	−600
65	−1950

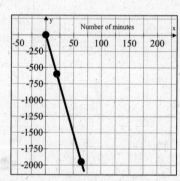

After 65 min, Alvin is 1950 m below sea level.

b) $D = -48t$

t	D
0	0
20	−960
65	−3120

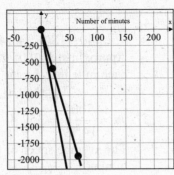

The point is below (65, −1950).

51. Yes. The points with y-coordinates less than −6500 represent depths that the replacement HOV cannot descend to.

53. To graph the equation of a straight line by plotting points, find three ordered pair solutions of the equation. Plot these ordered pairs in a rectangular coordinate system. Draw a straight line through the points.

55. The x and y intercepts of the graph of the equation $4x + 3y = 0$ are both (0,0). A straight line is determined by two points, so we need to find another point on the line in order to graph this equation.

57. $\dfrac{x}{2} + \dfrac{y}{3} = 1$

x-intercept (2, 0)
y-intercept (0, 3)

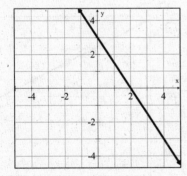

Section 3.4

Objective A Exercises

1. (1, 3), (3, 1)

$$m = \frac{y_2 - y_1}{x_2 - x_1} = \frac{1-3}{3-1} = \frac{-2}{2} = -1$$

The slope is −1.

3. (−1, 4), (2, 5)

$$m = \frac{y_2 - y_1}{x_2 - x_1} = \frac{5-4}{2-(-1)} = \frac{1}{3}$$

The slope is $\dfrac{1}{3}$.

5. (−1, 3), (−4, 5)

$$m = \frac{y_2 - y_1}{x_2 - x_1} = \frac{5-3}{-4-(-1)} = \frac{2}{-3} = -\frac{2}{3}$$

The slope is $-\dfrac{2}{3}$.

7. (0, 3), (4, 0)

$$m = \frac{y_2 - y_1}{x_2 - x_1} = \frac{3-0}{0-4} = \frac{3}{-4} = -\frac{3}{4}$$

The slope is $-\dfrac{3}{4}$.

9. (2, 4), (2, −2)

$$m = \frac{y_2 - y_1}{x_2 - x_1} = \frac{4-(-2)}{2-2} = \frac{6}{0}$$

The slope is undefined.

11. (2, 5), (−3, −2)

$$m = \frac{y_2 - y_1}{x_2 - x_1} = \frac{5-(-2)}{2-(-3)} = \frac{7}{5}$$

The slope is $\dfrac{7}{5}$.

13. (2, 3), (−1, 3)

$$m = \frac{y_2 - y_1}{x_2 - x_1} = \frac{3-3}{2-(-1)} = \frac{0}{3} = 0$$

The slope is 0.

15. $(0, 4), (-2, 5)$

$$m = \frac{y_2 - y_1}{x_2 - x_1} = \frac{5 - 4}{-2 - 0} = \frac{1}{-2} = -\frac{1}{2}$$

The slope is $-\frac{1}{2}$.

17. $(-3, -1), (-3, 4)$

$$m = \frac{y_2 - y_1}{x_2 - x_1} = \frac{4 - (-1)}{-3 - (-3)} = \frac{5}{0}$$

The slope is undefined.

19. If a and c are equal, the slope of l is undefined.

21. $m = \frac{240 - 80}{6 - 2} = \frac{160}{4} = 40$

The average speed of the motorist is 40 mph.

23. $m = \frac{275 - 125}{20 - 50} = \frac{150}{-30} = -5$

The temperature of the oven decreases $5°/\text{min}$.

25. $m = \frac{13 - 6}{40 - 180} = \frac{7}{-140} = -0.05$

Approximately 0.05 gal of fuel is used for each mile that the car is driven.

27. $m = \frac{5000}{14.19} = 352.4$

The average speed of the runner was 352.4 m/min.

29. a) $\frac{6\ in}{5\ ft} = \frac{6\ in}{60\ in} = \frac{1}{10} > \frac{1}{12}$

No it does not meet the requirements for ANSI.

b) $\frac{12}{170} = \frac{6}{85} < \frac{1}{12}$

Yes, it does meet the requirements for ANSI.

Objective B Exercises

	Equation	Value of m	Value of b	Slope	y-intercept
31.	$y = -3x + 5$	-3	5	-3	(0,5)
33.	$y = 4x$	4	0	4	(0,0)

35. $y = \frac{1}{2}x + 2$

$m = \frac{1}{2}$

y-intercept $(0, 2)$

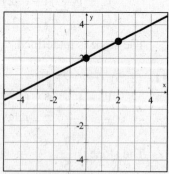

37. $y = -\frac{3}{2}x$

$m = -\frac{3}{2}$

y-intercept $(0, 0)$

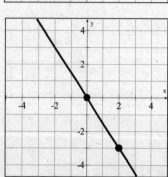

39. $y = -\frac{1}{2}x + 2$

$m = -\frac{1}{2}$

y-intercept $(0, 2)$

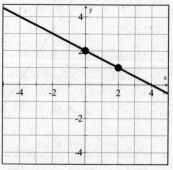

41. $y = 2x - 4$

$m = 2$

y-intercept $(0, -4)$

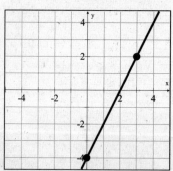

43. $4x - y = 1$
$-y = -4x + 1$
$y = 4x - 1$
$m = 4$
y-intercept
$(0, -1)$

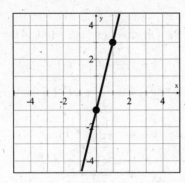

45. $x - 3y = 3$
$-3y = -x + 3$
$y = \dfrac{1}{3}x - 1$
$m = \dfrac{1}{3}$
y-intercept
$(0, -1)$

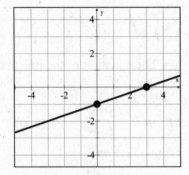

47.

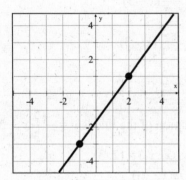

49.

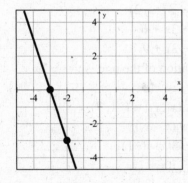

51.

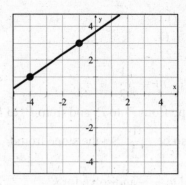

53. a) Below
b) Negative

Applying the Concepts

55. increases by 2

57. increases by 2

59. decreases by $\dfrac{2}{3}$

61. i. D   ii. C
iii. B   iv. F
v. E   vi. A

63. $P_1 = (3, 2)$
$P_2 = (4, 6)$
$P_3 = (5, k)$
P_1 to P_2: $m = 4$
The slope from P_2 to P_3 and from P_1 to P_3
must also be 4. Set the slope from P_1 to P_3
equal to 4 and solve for k.
$$\dfrac{2 - k}{-2} = 4$$
$$2 - k = -8$$
$$k = -10$$

65. $P_1 = (k, 1)$
$P_2 = (0, -1)$
$P_3 = (2, -2)$

P_2 to P_3: $m = -\dfrac{1}{2}$

The slope from P_1 to P_2 and from P_1 to P_3 must also be $-\dfrac{1}{2}$. Set the slope from P_1 to P_2 equal to $-\dfrac{1}{2}$ and solve for k.

$\dfrac{2}{k-0} = -\dfrac{1}{2}$

$k = -4$

Section 3.5

Objective A Exercises

1. When we know the slope and the y-intercept, we can find the equation of the line using the slope-intercept form, $y = mx + b$. The value of the slope is substituted for m, and the y-coordinate of the y-intercept is substituted for b.

3. Check that the coefficient of x is the given slope. Check that the coordinates of the given point are a solution of the equation you found.

5. $m = 2, b = 5$
$y = mx + b$
$y = 2x + 5$
The equation of the line is $y = 2x + 5$.

7. $m = \dfrac{1}{2}$, $(x_1, y_1) = (2, 3)$

$y - y_1 = m(x - x_1)$

$y - 3 = \dfrac{1}{2}(x - 2)$

$y - 3 = \dfrac{1}{2}x - 1$

$y = \dfrac{1}{2}x + 2$

The equation of the line is $y = \dfrac{1}{2}x + 2$.

9. $m = \dfrac{5}{4}$, $(x_1, y_1) = (-1, 4)$

$y - y_1 = m(x - x_1)$

$y - 4 = \dfrac{5}{4}[x - (-1)]$

$y - 4 = \dfrac{5}{4}(x + 1)$

$y - 4 = \dfrac{5}{4}x + \dfrac{5}{4}$

$y = \dfrac{5}{4}x + \dfrac{21}{4}$

The equation of the line is $y = \dfrac{5}{4}x + \dfrac{21}{4}$.

11. $m = -\dfrac{5}{3}$, $(x_1, y_1) = (3, 0)$

$y - y_1 = m(x - x_1)$

$y - 0 = -\dfrac{5}{3}(x - 3)$

$y = -\dfrac{5}{3}(x - 3)$

$y = -\dfrac{5}{3}x + 5$

The equation of the line is $y = -\dfrac{5}{3}x + 5$.

13. $m = -3$, $(x_1, y_1) = (2, 3)$
$y - y_1 = m(x - x_1)$
$y - 3 = -3(x - 2)$
$y - 3 = -3x + 6$
$y = -3x + 9$
The equation of the line is $y = -3x + 9$.

15. $m = -3$, $(x_1, y_1) = (-1, 7)$
$y - y_1 = m(x - x_1)$
$y - 7 = -3[x - (-1)]$
$y - 7 = -3(x + 1)$
$y - 7 = -3x - 3$
$y = -3x + 4$
The equation of the line is $y = -3x + 4$.

17. $m = \dfrac{2}{3}$, $(x_1, y_1) = (-1, -3)$

$y - y_1 = m(x - x_1)$

$y - (-3) = \dfrac{2}{3}[(x - (-1)]$

$y + 3 = \dfrac{2}{3}(x + 1)$

$y + 3 = \dfrac{2}{3}x + \dfrac{2}{3}$

$y = \dfrac{2}{3}x - \dfrac{7}{3}$

The equation of the line is $y = \dfrac{2}{3}x - \dfrac{7}{3}$.

19. $m = \dfrac{1}{2}$, $(x_1, y_1) = (0, 0)$

$y - y_1 = m(x - x_1)$

$y - 0 = \dfrac{1}{2}(x - 0)$

$y = \dfrac{1}{2}x$

The equation of the line is $y = \dfrac{1}{2}x$.

21. $m = 3$, $(x_1, y_1) = (2, -3)$

$y - y_1 = m(x - x_1)$

$y - (-3) = 3(x - 2)$

$y + 3 = 3x - 6$

$y = 3x - 9$

The equation of the line is $y = 3x - 9$.

23. $m = -\dfrac{2}{3}$, $(x_1, y_1) = (3, 5)$

$y - y_1 = m(x - x_1)$

$y - 5 = -\dfrac{2}{3}(x - 3)$

$y - 5 = -\dfrac{2}{3}x + 2$

$y = -\dfrac{2}{3}x + 7$

The equation of the line is $y = -\dfrac{2}{3}x + 7$.

25. $m = -1$, $b = -3$

$y = -x - 3$

The equation of the line is $y = -x - 3$.

27. $m = \dfrac{7}{5}$, $(x_1, y_1) = (1, -4)$

$y - y_1 = m(x - x_1)$

$y - (-4) = \dfrac{7}{5}(x - 1)$

$y + 4 = \dfrac{7}{5}x - \dfrac{7}{5}$

$y = \dfrac{7}{5}x - \dfrac{27}{5}$

The equation of the line is $y = \dfrac{7}{5}x - \dfrac{27}{5}$.

29. $m = -\dfrac{2}{5}$, $(x_1, y_1) = (4, -1)$

$y - y_1 = m(x - x_1)$

$y - (-1) = -\dfrac{2}{5}(x - 4)$

$y + 1 = -\dfrac{2}{5}x + \dfrac{8}{5}$

$y = -\dfrac{2}{5}x + \dfrac{3}{5}$

The equation of the line is $y = -\dfrac{2}{5}x + \dfrac{3}{5}$.

31. Slope is undefined, $(x_1, y_1) = (3, -4)$

The line is a vertical line. All points on the line have an abscissa of 3.

The equation of the line is $x = 3$.

33. $m = -\dfrac{5}{4}$, $(x_1, y_1) = (-2, -5)$

$y - y_1 = m(x - x_1)$

$y - (-5) = -\dfrac{5}{4}[x - (-2)]$

$y + 5 = -\dfrac{5}{4}(x + 2)$

$y + 5 = -\dfrac{5}{4}x - \dfrac{10}{4}$

$$y = -\frac{5}{4}x - \frac{15}{2}$$

The equation of the line is $y = -\frac{5}{4}x - \frac{15}{2}$.

35. $m = 0$, $(x_1, y_1) = (-2, -3)$

$y - y_1 = m(x - x_1)$

$y - (-3) = 0[x - (-2)]$

$y + 3 = 0$

$y = -3$

The equation of the line is $y = -3$.

37. $m = -2$, $(x_1, y_1) = (4, -5)$

$y - y_1 = m(x - x_1)$

$y - (-5) = -2(x - 4)$

$y + 5 = -2x + 8$

$y = -2x + 3$

The equation of the line is $y = -2x + 3$.

39. Slope is undefined, $(x_1, y_1) = (-5, -1)$

The line is a vertical line. All points on the line have an abscissa of -5.

The equation of the line is $x = -5$.

Objective B Exercises

41. Check that the coordinates of each given point are a solution of your equation.

43. $(0, 2)$, $(3, 5)$

$m = \dfrac{y_2 - y_1}{x_2 - x_1} = \dfrac{5 - 2}{3 - 0} = \dfrac{3}{3} = 1$

$y - y_1 = m(x - x_1)$

$y - 2 = 1(x - 0)$

$y - 2 = x$

$y = x + 2$

The equation of the line is $y = x + 2$.

45. $(0, -3)$, $(-4, 5)$

$m = \dfrac{y_2 - y_1}{x_2 - x_1} = \dfrac{5 - (-3)}{-4 - 0} = \dfrac{8}{-4} = -2$

$y - y_1 = m(x - x_1)$

$y - (-3) = -2(x - 0)$

$y + 3 = -2x$

$y = -2x - 3$

The equation of the line is $y = -2x - 3$.

47. $(2, 3)$, $(5, 5)$

$m = \dfrac{y_2 - y_1}{x_2 - x_1} = \dfrac{5 - 3}{5 - 2} = \dfrac{2}{3}$

$y - y_1 = m(x - x_1)$

$y - 3 = \dfrac{2}{3}(x - 2)$

$y - 3 = \dfrac{2}{3}x - \dfrac{4}{3}$

$y = \dfrac{2}{3}x + \dfrac{5}{3}$

The equation of the line is $y = \dfrac{2}{3}x + \dfrac{5}{3}$.

49. $(-1, 3)$, $(2, 4)$

$m = \dfrac{y_2 - y_1}{x_2 - x_1} = \dfrac{4 - 3}{2 - (-1)} = \dfrac{1}{3}$

$y - y_1 = m(x - x_1)$

$y - 3 = \dfrac{1}{3}[x - (-1)]$

$y - 3 = \dfrac{1}{3}(x + 1)$

$y - 3 = \dfrac{1}{3}x + \dfrac{1}{3}$

$y = \dfrac{1}{3}x + \dfrac{10}{3}$

The equation of the line is $y = \dfrac{1}{3}x + \dfrac{10}{3}$.

51. $(-1, -2)$, $(3, 4)$

$m = \dfrac{y_2 - y_1}{x_2 - x_1} = \dfrac{4 - (-2)}{3 - (-1)} = \dfrac{6}{4} = \dfrac{3}{2}$

$y - y_1 = m(x - x_1)$

$y - 4 = \dfrac{3}{2}(x - 3)$

$y - 4 = \dfrac{3}{2}x - \dfrac{9}{2}$

$y = \dfrac{3}{2}x - \dfrac{1}{2}$

The equation of the line is $y = \dfrac{3}{2}x - \dfrac{1}{2}$.

The equation of the line is $y = -2x - 3$.

53. $(0, 3), (2, 0)$

$$m = \frac{y_2 - y_1}{x_2 - x_1} = \frac{3 - 0}{0 - 2} = -\frac{3}{2}$$

$$y - y_1 = m(x - x_1)$$

$$y = -\frac{3}{2}(x - 2)$$

$$y = -\frac{3}{2}x + 3$$

The equation of the line is $y = -\frac{3}{2}x + 3$.

55. $(-3, -1), (2, -1)$

$$m = \frac{y_2 - y_1}{x_2 - x_1} = \frac{-1 - (-1)}{2 - (-3)} = \frac{0}{5} = 0$$

$$y - y_1 = m(x - x_1)$$
$$y - (-1) = 0[(x - (-3)]$$
$$y + 1 = 0$$
$$y = -1$$

The equation of the line is $y = -1$.

57. $(-2, -3), (-1, -2)$

$$m = \frac{y_2 - y_1}{x_2 - x_1} = \frac{-3 - (-2)}{-2 - (-1)} = \frac{-1}{-1} = 1$$

$$y - y_1 = m(x - x_1)$$
$$y - (-3) = 1[x - (-2)]$$
$$y + 3 = 1(x + 2)$$
$$y = x - 1$$

The equation of the line is $y = x - 1$.

59. $(-2, 3), (2, -1)$

$$m = \frac{y_2 - y_1}{x_2 - x_1} = \frac{3 - (-1)}{-2 - 2} = \frac{4}{-4} = -1$$

$$y - y_1 = m(x - x_1)$$
$$y - (-1) = -1(x - 2)$$
$$y + 1 = -x + 2$$
$$y = -x + 1$$

The equation of the line is $y = -x + 1$.

61. $(2, 3), (5, -5)$

$$m = \frac{y_2 - y_1}{x_2 - x_1} = \frac{3 - (-5)}{2 - 5} = \frac{8}{-3} = -\frac{8}{3}$$

$$y - y_1 = m(x - x_1)$$

$$y - 3 = -\frac{8}{3}(x - 2)$$

$$y - 3 = -\frac{8}{3}x + \frac{16}{3}$$

$$y = -\frac{8}{3}x + \frac{25}{3}$$

The equation of the line is $y = -\frac{8}{3}x + \frac{25}{3}$.

63. $(2, 0), (0, -1)$

$$m = \frac{y_2 - y_1}{x_2 - x_1} = \frac{0 - (-1)}{2 - 0} = \frac{1}{2}$$

$$y - y_1 = m(x - x_1)$$

$$y - 0 = \frac{1}{2}(x - 2)$$

$$y = \frac{1}{2}x - 1$$

The equation of the line is $y = \frac{1}{2}x - 1$.

65. $(3, -4), (-2, -4)$

$$m = \frac{y_2 - y_1}{x_2 - x_1} = \frac{-4 - (-4)}{3 - (-2)} = \frac{0}{5} = 0$$

$$y - y_1 = m(x - x_1)$$
$$y - (-4) = 0(x - 3)$$
$$y + 4 = 0$$
$$y = -4$$

The equation of the line is $y = -4$.

67. $(0, 0), (4, 3)$

$$m = \frac{y_2 - y_1}{x_2 - x_1} = \frac{3 - 0}{4 - 0} = \frac{3}{4}$$

$$y - y_1 = m(x - x_1)$$

$$y - 0 = \frac{3}{4}(x - 0)$$

$$y = \frac{3}{4}x$$

The equation of the line is $y = \frac{3}{4}x$.

69. $(2, -1), (-1, 3)$

$$m = \frac{y_2 - y_1}{x_2 - x_1} = \frac{3 - (-1)}{-1 - 2} = \frac{4}{-3} = -\frac{4}{3}$$

$$y - y_1 = m(x - x_1)$$

$$y - (-1) = -\frac{4}{3}(x - 2)$$

$$y + 1 = -\frac{4}{3}x + \frac{8}{3}$$

$$y = -\frac{4}{3}x + \frac{5}{3}$$

The equation of the line is $y = -\frac{4}{3}x + \frac{5}{3}$.

71. $(-2, 5), (-2, -5)$

$$m = \frac{y_2 - y_1}{x_2 - x_1} = \frac{5 - (-5)}{-2 - (-2)} = \frac{10}{0}$$

The slope is undefined. The line is a vertical line. All points on the line have an abscissa of -2. The equation of the line is $x = -2$.

73. $(2, 1), (-2, -3)$

$$m = \frac{y_2 - y_1}{x_2 - x_1} = \frac{1 - (-3)}{2 - (-2)} = \frac{4}{4} = 1$$

$$y - y_1 = m(x - x_1)$$

$$y - 1 = 1(x - 2)$$

$$y = x - 1$$

The equation of the line is $y = x - 1$.

75. $(-4, -3), (2, 5)$

$$m = \frac{y_2 - y_1}{x_2 - x_1} = \frac{5 - (-3)}{2 - (-4)} = \frac{8}{6} = \frac{4}{3}$$

$$y - y_1 = m(x - x_1)$$

$$y - 5 = \frac{4}{3}(x - 2)$$

$$y - 5 = \frac{4}{3}x - \frac{8}{3}$$

$$y = \frac{4}{3}x + \frac{7}{3}$$

The equation of the line is $y = \frac{4}{3}x + \frac{7}{3}$.

77. $(0, 3), (3, 0)$

$$m = \frac{y_2 - y_1}{x_2 - x_1} = \frac{3 - 0}{0 - 3} = \frac{3}{-3} = -1$$

$$y - y_1 = m(x - x_1)$$

$$y - 0 = -1(x - 3)$$

$$y = -x + 3$$

The equation of the line is $y = -x + 3$.

Objective C Exercises

79. Strategy: Let x represent the number of minutes after takeoff.
Let y represent the height of the plane in feet.
Use the slope-intercept form of an equation to find the equation of the line.

Solution:
a) y-intercept $(0, 0)$; slope is 1200

$y = mx + b$

$y = 1200x + 0$

The linear function is $f(x) = 1200x$.

$$0 \le x \le 26\frac{2}{3}$$

b) Find the height of the plane 11 min after takeoff.

$y = 1200(11) = 13{,}200$

Eleven minutes after takeoff, the height of the plane will be 13,200 ft.

81. Strategy: Let x represent the year. Let y represent the percent of trees that are hardwoods. Use the point-slope formula to find the equation of the line.

Solution:
a) $(1964, 57)$, $(2004, 82)$
$$m = \frac{82 - 57}{2004 - 1964} = \frac{25}{40} = 0.625$$
$$y - y_1 = m(x - x_1)$$
$$y - 57 = 0.625(x - 1964)$$
$$y - 57 = 0.625x - 1227.5$$
$$y = 0.625x - 1170.5$$
The linear equation is $f(x) = 0.625x - 1170.5$
b) Predict the percent of trees that will be hardwoods in 2012.
$y = 0.625(2012) - 1170.5 = 87$
In 2012 it is predicted that 87% of the trees will be hardwoods.

83. Strategy: Let x represent the number of miles driven. Let y represent the number of gallons of gas in the tank. Use the slope-intercept form of an equation to find the equation of the line.

Solution:
a) y-intercept $(0, 16)$; slope is -0.032
$y = -0.032x + 16$
Since $0 \le y \le 16$, we have
$$0 \le -0.032x + 16 \le 16$$
$$-16 \le -0.032x \le 0$$
$$500 \ge x \ge 0$$
The linear function is $f(x) = -0.032x + 16$, for $500 \ge x \ge 0$.
b) Find the number of gallons of gas left in the tank after driving 150 mi.
$y = -0.032(150) + 16 = 11.2$
After driving 150 mi, there are 11.2 gal of gas left in the tank.

85. Strategy: Let x represent the price of a motorcycle. Let y represent the number of motorcycles sold. Use the point-slope formula to find the equation of the line.

Solution:
a) $(9000, 50{,}000)$, $(8750, 55{,}000)$
$$m = \frac{55{,}000 - 50{,}000}{8750 - 9000} = \frac{5000}{-250} = -20$$
$$y - y_1 = m(x - x_1)$$
$$y - 50{,}000 = -20(x - 9000)$$
$$y - 50{,}000 = -20x + 180{,}000)$$
$$y = -20x + 230{,}000$$
The linear function is $f(x) = -20x + 230{,}000$
b) Find the number of motorcycles sold when the price is \$8500.
$y = -20(8500) + 230{,}000 = 60{,}000$

When the price of a motorcycle is \$8500, 60,000 will be sold.

87. Strategy: Let x represent the number of ounces of lean hamburger. Let y represent the number of calories. Use the point-slope formula to find the equation of the line.

Solution:
a) $(2, 126)$, $(3, 189)$
$$m = \frac{189 - 126}{3 - 2} = 63$$
$$y - y_1 = m(x - x_1)$$
$$y - 126 = 63(x - 2)$$
$$y - 126 = 63x - 126)$$
$$y = 63x$$
The linear function is $f(x) = 63x$
b) Find the number of calories in a 5-ounce serving.
$y = 63(5) + 315$
There are 315 calories in a 5-ounce serving of lean hamburger.

89. Substitute 15,000 for $f(x)$ and solve the equation for x.

91. $(2, 5), (0, 3)$

$$m = \frac{5-3}{2-0} = \frac{2}{2} = 1$$

$$y - y_1 = m(x - x_1)$$

$$y - 3 = 1(x - 0)$$

$$y = x + 3$$

$$f(x) = x + 3$$

93. $(1, 3), (-1, 5)$

$$m = \frac{5-3}{-1-1} = \frac{2}{-2} = -1$$

$$y - y_1 = m(x - x_1)$$

$$y - 3 = -1(x - 1)$$

$$y - 3 = -x + 1$$

$$y = -x + 4$$

$$f(x) = -x + 4$$

$$f(4) = -4 + 4 = 0$$

95. Given $m = \dfrac{4}{3}$ and a point $(3, 2)$

a) $y - y_1 = m(x - x_1)$

$$y - 2 = \frac{4}{3}(x - 3)$$

$$y - 2 = \frac{4}{3}x - 4$$

$$y = \frac{4}{3}x - 2$$

For $x = -6$ we have

$$y = \frac{4}{3}(-6) - 2 = -8 - 2 = -10$$

b) For $y = 6$ we have

$$6 = \frac{4}{3}x - 2$$

$$8 = \frac{4}{3}x$$

$$\frac{3}{4} \cdot 8 = \frac{3}{4} \cdot \frac{4}{3}x$$

$$6 = x$$

Applying the Concepts

97. The slope of any line parallel to the y-axis is undefined. In order to use the point-slope formula, we must be able to substitute the slope of the line for m. In other words, in order to use the point-slope formula, the slope of the line must be defined.

99. Student solutions will vary.

Find the equation of the line:

$$m = \frac{0-6}{6-(-3)} = \frac{-6}{9} = -\frac{2}{3}$$

$$y - 0 = -\frac{2}{3}(x - 6)$$

$$y = -\frac{2}{3}x + 4$$

Possible answers are:

If $x = 0$, $y = -\dfrac{2}{3}(0) + 4 = 4$

$x = 3$, $y = -\dfrac{2}{3}(3) + 4 = 2$

$x = 6$, $y = -\dfrac{2}{3}(6) + 4 = 0$

$(0,4), (3,2), (6,0)$

101. Find the x- and y-coordinates for the midpoint of the line segment:

$$x_m = \frac{2+(-4)}{2} = \frac{-2}{2} = -1$$

$$y_m = \frac{5+1}{2} = \frac{6}{2} = 3$$

The midpoint is $(-1, 3)$.

Use the point-slope formula to find the equation of the line.

$$y - y_1 = m(x - x_1)$$

$$y - 3 = -2[x - (-1)]$$

$$y - 3 = -2(x + 1)$$

$$y - 3 = -2x - 2$$

$$y = -2x + 1$$

Section 3.6

Objective A Exercises

1. Two lines are parallel if they have the same slope and different y-intercepts.

3. No. If two nonvertical lines are perpendicular, the product of their slopes is -1. Therefore one line must have a positive slope and the other line must have a negative slope.

5. $m = -5$

7. $m = -\dfrac{1}{4}$

9. Yes, the lines are perpendicular. $x = -2$ is a vertical line and $y = 3$ is a horizontal line.

11. No. the lines are not parallel. $x = -3$ is a vertical line and $y = \dfrac{1}{3}$ is a horizontal line.

13. No, the lines are not parallel because their slopes are not equal.

15. Yes, the lines are perpendicular. Their slopes are negative reciprocals of each other.

17. $2x + 3y = 2$
$$3y = -2x + 2$$
$$y = -\frac{2}{3}x + \frac{2}{3}$$
$$m_1 = -\frac{2}{3}$$
$$2x + 3y = -4$$
$$3y = -2x - 4$$
$$y = -\frac{2}{3}x - \frac{4}{3}$$
$$m_2 = -\frac{2}{3}$$
Since $m_1 = m_2 = -\dfrac{2}{3}$, the lines are parallel.

19. $x - 4y = 2$
$$-4y = -x + 2$$
$$y = \frac{1}{4}x - \frac{1}{2}$$
$$m_1 = \frac{1}{4}$$
$$4x + y = 8$$
$$y = -4x + 8$$
$$m_2 = -4$$
Since $m_1 \cdot m_2 = \dfrac{1}{4} \cdot (-4) = -1$, the lines are perpendicular.

21. $m_1 = \dfrac{6-2}{1-3} = \dfrac{4}{-2} = -2$
$$m_2 = \frac{-1-3}{-1-(-1)} = \frac{-4}{0}$$
$$m_1 \neq m_2$$
The lines are not parallel.

23. $m_1 = \dfrac{-1-2}{4-(-3)} = \dfrac{-3}{7} = -\dfrac{3}{7}$
$$m_2 = \frac{-4-3}{-2-1} = \frac{-7}{-3} = \frac{7}{3}$$
$$m_1 \cdot m_2 = -\frac{3}{7}\left(\frac{7}{3}\right) = -1$$
The lines are perpendicular.

25. Since the new line is parallel to $y = 2x + 1$, both lines will have the same slope. The slope of the new line is $m = 2$.

Use the point-slope formula to find the equation of the line.
$m = 2$ and $(3, -2)$
$$y - y_1 = m(x - x_1)$$
$$y - (-2) = 2(x - 3)$$
$$y + 2 = 2x - 6$$
$$y = 2x - 8$$
The equation of the line is $y = 2x - 8$.

27. Since the new line is perpendicular to $y = -\dfrac{2}{3}x - 2$, the slope of the new line must be the negative reciprocal of the slope of the given line. The slope of the new line is $m = \dfrac{3}{2}$.

Use the point-slope formula to find the equation of the line.

$m = \dfrac{3}{2}$ and $(-2, -1)$

$$y - y_1 = m(x - x_1)$$

$$y - (-1) = \dfrac{3}{2}[x - (-2)]$$

$$y + 1 = \dfrac{3}{2}(x + 2)$$

$$y + 1 = \dfrac{3}{2}x + 3$$

$$y = \dfrac{3}{2}x + 2$$

The equation of the line is $y = \dfrac{3}{2}x + 2$.

29. Since the new line is parallel to $2x - 3y = 2$, both lines will have the same slope.

$$2x - 3y = 2$$

$$-3y = -2x + 2$$

$$y = \dfrac{2}{3}x - \dfrac{2}{3}$$

The slope of the new line is $m = \dfrac{2}{3}$.

Use the point-slope formula to find the equation of the line.

$m = \dfrac{2}{3}$ and $(-2, -4)$

$$y - y_1 = m(x - x_1)$$

$$y - (-4) = \dfrac{2}{3}[x - (-2)]$$

$$y + 4 = \dfrac{2}{3}(x + 2)$$

$$y + 4 = \dfrac{2}{3}x + \dfrac{4}{3}$$

$$y = \dfrac{2}{3}x - \dfrac{8}{3}$$

The equation of the line is $y = \dfrac{2}{3}x - \dfrac{8}{3}$.

31. Since the new line is perpendicular to $y = -3x + 4$ the slope of the new line must be the negative reciprocal of the slope of the given line.

$$m_1 = -3$$

$$-3 \cdot m_2 = -1 \text{ therefore, } m_2 = \dfrac{1}{3}.$$

The slope of the new line is $m_2 = \dfrac{1}{3}$.

Use the point-slope formula to find the equation of the line.

$m_2 = \dfrac{1}{3}$ and $(4, 1)$

$$y - y_1 = m(x - x_1)$$

$$y - 1 = \dfrac{1}{3}(x - 4)$$

$$y - 1 = \dfrac{1}{3}x - \dfrac{4}{3}$$

$$y = \dfrac{1}{3}x - \dfrac{1}{3}$$

The equation of the line is $y = \dfrac{1}{3}x - \dfrac{1}{3}$.

33. Since the new line is perpendicular to $3x - 5y = 2$, the slope of the new line must be the negative reciprocal of the slope of the given line.

$$3x - 5y = 2$$

$$-5y = -3x + 2$$

$$y = \dfrac{3}{5}x - \dfrac{2}{5}$$

$$m_1 = \dfrac{3}{5}$$

$$\frac{3}{5} \cdot m_2 = -1$$

$$m_2 = -\frac{5}{3}$$

The slope of the new line is $m_2 = -\frac{5}{3}$.

Use the point-slope formula to find the equation of the line.

$$m_2 = -\frac{5}{3} \text{ and } (-1, -3)$$

$$y - y_1 = m(x - x_1)$$

$$y - (-3) = -\frac{5}{3}[x - (-1)]$$

$$y + 3 = -\frac{5}{3}(x + 1)$$

$$y + 3 = -\frac{5}{3}x - \frac{5}{3}$$

$$y = -\frac{5}{3}x - \frac{14}{3}$$

The equation of the line is $y = -\frac{5}{3}x - \frac{14}{3}$.

Applying the Concepts

35. Use the points $(0, 0)$ and $(6, 3)$.

$$m_1 = \frac{3 - 0}{6 - 0} = \frac{3}{6} = \frac{1}{2}$$

$$m_1 \cdot m_2 = -1$$

$$\frac{1}{2} \cdot m_2 = -1$$

$$m_2 = -2$$

Using the point $(6, 3)$,

$$y - 3 = -2(x - 6)$$

$$y - 3 = -2x + 12$$

$$y = -2x + 15$$

The equation of the line is $y = -2x + 15$.

37. Write the equations of the line in slope-intercept form.

$$A_1 x + B_1 y = C_1$$

$$B_1 y = -A_1 x + C_1$$

$$y = -\frac{A_1 x}{B_1} + \frac{C_1}{B_1}$$

$$A_2 x + B_2 y = C_2$$

$$B_2 y = -A_2 x + C_2$$

$$y = -\frac{A_2 x}{B_2} + \frac{C_2}{B_2}$$

If the two lines are parallel, their slopes must be equal.

$$-\frac{A_1}{B_1} = -\frac{A_2}{B_2} \text{ so } \frac{A_1}{B_1} = \frac{A_2}{B_2}.$$

Section 3.7

Objective A Exercises

1. A half-plane is the set of points on one side of a line in the plane.

3. $y > 2x - 7$
$0 > 2(0) - 7$
$0 > -7$
Yes, $(0, 0)$ is a solution.

5. $y \le -\frac{2}{3}x - 8$

$0 \le -\frac{2}{3}(0) - 8$

$0 \le -7$
No, $(0, 0)$ is not a solution.

7. $y \le \frac{3}{2}x - 3$

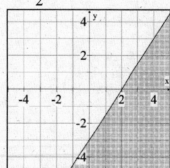

9. $y < -\dfrac{1}{3}x + 1$

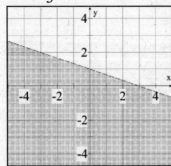

11. $4x - 5y > 10$

$-5y > -4x + 10$

$y < \dfrac{4}{5}x - 2$

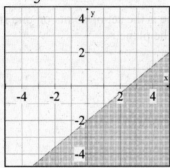

13. $x + 3y < 6$

$3y < -x + 6$

$y < -\dfrac{1}{3}x + 2$

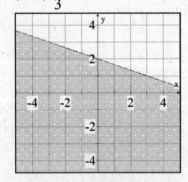

15. $2x + 3y \geq 6$

$3y \geq -2x + 6$

$y \geq -\dfrac{2}{3}x + 2$

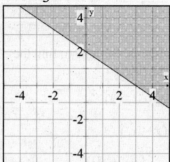

17. $-x + 2y > -8$

$2y > x - 8$

$y > \dfrac{1}{2}x - 4$

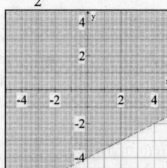

19. $y - 4 < 0$

$y < 4$

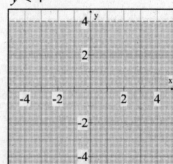

21. $6x + 5y < 15$

$5y < -6x + 15$

$y < -\dfrac{6}{5}x + 3$

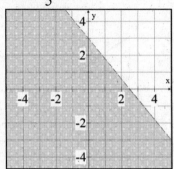

23. $-5x + 3y \geq -12$

$3y \geq 5x - 12$

$y \geq \dfrac{5}{3}x - 4$

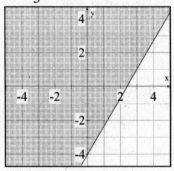

25. Quadrant I

Applying the Concepts

27. The inequality $y < 3x - 1$ is not a function because given a value of x there is more than one corresponding value of y. For example, both (3, 2) and (3, −1) are ordered pairs that satisfy the inequality. This contradicts the definition of a function because there are two ordered pairs with the same first coordinate and different second coordinates.

29. There are no points whose coordinates satisfy both $y \leq x - 1$ and $y \geq x + 2$. The solution set of $y \leq x - 1$ is all points on or below the line $y = x - 1$. The solution set

of $y \geq x + 2$ is all points on or above the line $y = x + 2$. Since the lines $y = x - 1$ and $y = x + 2$ are parallel lines and $y = x + 2$ is above the line $y = x - 1$, there are no points that lie both below $y = x - 1$ and above $y = x + 2$.

Chapter 3 Review Exercises

1. $y = \dfrac{x}{x - 2}$

$y = \dfrac{4}{4 - 2} = \dfrac{4}{2} = 2$

The ordered pair is (4, 2).

2. $P(x) = 3x + 4$

$P(-2) = 3(-2) + 4 = -6 + 4 = -2$

$P(a) = 3(a) + 4 = 3a + 4$

3. $y = 2x^2 - 5$

Ordered pairs: (−2, 3), (−1, −3), (0, −5), (1, −3) and (2, 3).

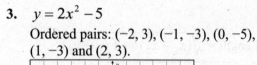
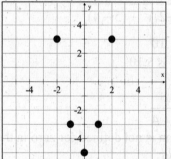

4.

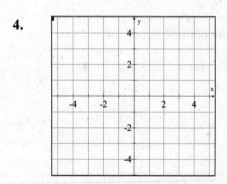

5. $f(x) = x^2 + x - 1$

$f(-2) = (-2)^2 + (-2) - 1 = 1$

$f(-1) = (-1)^2 + (-1) - 1 = -1$

$f(0) = (0)^2 + 0 - 1 = -1$

$f(1) = (1)^2 + 1 - 1 = 1$

$f(2) = (2)^2 + 2 - 1 = 5$

Range = $\{-1, 1, 5\}$

6. Domain = $\{-1, 0, 1, 5\}$

Range = $\{0, 2, 4\}$

7. $(-2, 4)$ and $(3, 5)$

$x_m = \dfrac{-2+3}{2} = \dfrac{1}{2}$

$y_m = \dfrac{4+5}{2} = \dfrac{9}{2}$

The midpoint is $\left(\dfrac{1}{2}, \dfrac{9}{2}\right)$.

Length = $\sqrt{(x_1 - x_2)^2 + (y_1 - y_2)^2}$

$= \sqrt{(3-(-2))^2 + (5-4)^2}$

$= \sqrt{26} \approx 5.10$

The length is 5.10.

8. $f(x) = \dfrac{x}{x+4}$

The function is not defined for zero in the denominator.

$x + 4 = 0$

$x = -4$

$f(x)$ is not defined for $x = -4$.

9. $y = -\dfrac{2}{3}x - 2$

x-intercept:

$0 = -\dfrac{2}{3}x - 2$

$2 = -\dfrac{2}{3}x$

$x = -3$

$(-3, 0)$

y-intercept:

$y = -\dfrac{2}{3}(0) - 2$

$y = -2$

$(0, -2)$

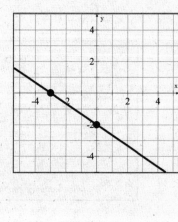

10. $3x + 2y = -6$

x-intercept:

$3x + 2(0) = -6$

$3x = -6$

$x = -2$

$(-2, 0)$

y-intercept:

$3(0) + 2y = -6$

$2y = -6$

$y = -3$

$(0, -3)$

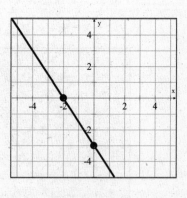

11. $y = -2x + 2$

x	y
0	2
1	0
-1	4

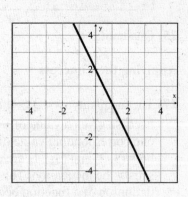

12. $4x - 3y = 12$

$-3y = -4x + 12$

$y = \dfrac{4}{3}x - 4$

x	y
0	-4
3	0
6	4

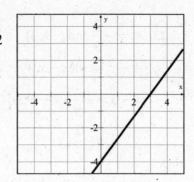

13. $(3, -2)$ and $(-1, 2)$

$m = \dfrac{y_2 - y_1}{x_2 - x_1}$

$m = \dfrac{-2 - 2}{3 - (-1)} = \dfrac{-4}{4} = -1$

14. Use the point-slope formula to find the equation of the line.

$m = \dfrac{5}{2}$ and $(-3, 4)$

$y - y_1 = m(x - x_1)$

$y - 4 = \dfrac{5}{2}[x - (-3)]$

$y - 4 = \dfrac{5}{2}(x + 3)$

$y - 4 = \dfrac{5}{2}x + \dfrac{15}{2}$

$y = \dfrac{5}{2}x + \dfrac{23}{2}$

15.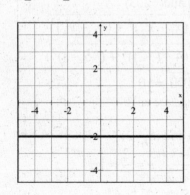

16. $m = -\dfrac{1}{4}$ and $(-2, 3)$

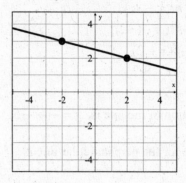

17. $f(x) = x^2 - 2$

$f(-2) = (-2)^2 - 2 = 2$

$f(-1) = (-1)^2 - 2 = -1$

$f(0) = (0)^2 - 2 = -2$

$f(1) = (1)^2 - 2 = -1$

$f(2) = (2)^2 - 2 = 2$

Range $= \{-2, -1, 2\}$

18. Strategy: Let x represent the room rate. Let y represent the number of rooms occupied.

Use the point – slope formula to find the equation of the line.

Solution:

a) $(95, 200)$, $(105, 190)$

$m = \dfrac{200 - 190}{95 - 105} = \dfrac{10}{-10} = -1$

$y - y_1 = m(x - x_1)$

$y - 200 = -1(x - 95)$

$y - 200 = -1x + 95$

$y = -x + 295$

The linear function is

$f(x) = -x + 295, \; 0 \le x \le 295$

b) Find the number of rooms occupied when the rate is \$120.

$f(120) = -1(120) + 295 = 175$

When the room rate is \$125, 175 rooms will be occupied.

19. The slope for the parallel line is $m = -4$.

$m = -4$ and $(-2, 3)$

$y - y_1 = m(x - x_1)$

$y - 3 = -4[(x - (-2)]$

$y - 3 = -4(x + 2)$

$y - 3 = -4x - 8$

$y = -4x - 5$

The equation of the line is $y = -4x - 5$.

20. The slope for the perpendicular line is

$m = \dfrac{5}{2}$.

$m = \dfrac{5}{2}$ and $(-2, 3)$

$y - y_1 = m(x - x_1)$

$y - 3 = \dfrac{5}{2}[(x - (-2)]$

$y - 3 = \dfrac{5}{2}(x + 2)$

$y - 3 = \dfrac{5}{2}x + 5$

$y = \dfrac{5}{2}x + 8$

The equation of the line is $y = \dfrac{5}{2}x + 8$.

21.

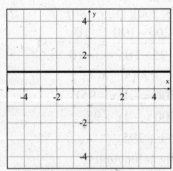

22.

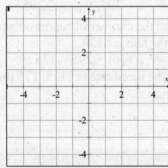

23. $m = -\dfrac{2}{3}$ and $(-3, 3)$

$y - y_1 = m(x - x_1)$

$y - 3 = -\dfrac{2}{3}[(x - (-3)]$

$y - 3 = -\dfrac{2}{3}(x + 3)$

$y - 3 = -\dfrac{2}{3}x - 2$

$y = -\dfrac{2}{3}x + 1$

The equation of the line is $y = -\dfrac{2}{3}x + 1$.

24. $(-8, 2)$ and $(4, 5)$

$m = \dfrac{5 - 2}{4 - (-8)} = \dfrac{3}{12} = \dfrac{1}{4}$

$y - y_1 = m(x - x_1)$

$y - 5 = \dfrac{1}{4}(x - 4)$

$y - 5 = \dfrac{1}{4}x - 1$

$y = \dfrac{1}{4}x + 4$

The equation of the line is $y = \dfrac{1}{4}x + 4$.

25. $(4, -5)$ and $(-2, 3)$

$d = \sqrt{(x_1 - x_2)^2 + (y_1 - y_2)^2}$

$= \sqrt{(4 - (-2))^2 + (-5 - 3)^2}$

$= \sqrt{100} = 10$

The distance is 10.

26. $(-3, 8)$ and $(5, -2)$

$x_m = \dfrac{-3 + 5}{2} = \dfrac{2}{2} = 1$

$y_m = \dfrac{8 + (-2)}{2} = \dfrac{6}{2} = 3$

The midpoint is $(1, 3)$.

27.

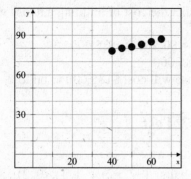

28. $y \geq 2x - 3$

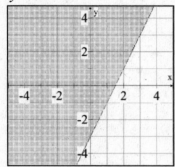

29. $3x - 2y < 6$

$-2y < -3x + 6$

$y > \dfrac{3}{2}x - 3$

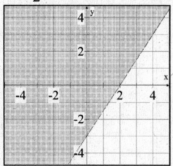

30. $(-2, 4)$ and $(4, -3)$

$m = \dfrac{4 - (-3)}{-2 - 4} = \dfrac{7}{-6} = -\dfrac{7}{6}$

$y - y_1 = m(x - x_1)$

$y - (-3) = -\dfrac{7}{6}(x - 4)$

$y + 3 = -\dfrac{7}{6}x + \dfrac{28}{6}$

$y = -\dfrac{7}{6}x + \dfrac{5}{3}$

The equation of the line is $y = -\dfrac{7}{6}x + \dfrac{5}{3}$.

31. $4x - 2y = 7$

$-2y = -4x + 7$

$y = 2x - \dfrac{7}{2}$

The slope for the parallel line is $m = 2$.

$m = 2$ and $(-2, -4)$

$y - y_1 = m(x - x_1)$

$y - (-4) = 2[(x - (-2)]$

$y + 4 = 2(x + 2)$

$y + 4 = 2x + 4$

$y = 2x$

The equation of the line is $y = 2x$.

32. The slope for the parallel line is $m = -3$.

$m = -3$ and $(3, -2)$

$y - y_1 = m(x - x_1)$

$y - (-2) = -3(x - 3)$

$y + 2 = -3x + 9$

$y = -3x + 7$

The equation of the line is $y = -3x + 7$.

33. $y = -\dfrac{2}{3}x + 6$

$m_1 = -\dfrac{2}{3}$

$m_1 \cdot m_2 = -1$

$-\dfrac{2}{3} \cdot m_2 = -1$

$m_2 = \dfrac{3}{2}$

The slope for the perpendicular line is

$m = \dfrac{3}{2}$.

$m = \dfrac{3}{2}$ and $(2, 5)$

$y - y_1 = m(x - x_1)$

$$y - 5 = \frac{3}{2}(x - 2)$$

$$y - 5 = \frac{3}{2}x - 3$$

$$y = \frac{3}{2}x + 2$$

The equation of the line is $y = \frac{3}{2}x + 2$.

34. $m = -\frac{1}{3}$ and $(-1, 4)$

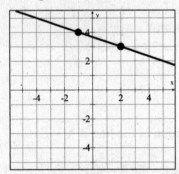

35.

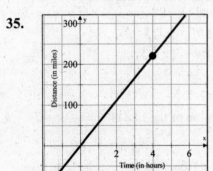

After 4 h, the car will travel 220 mi.

36. $(500, 12,000)$ and $(200, 6000)$

$$m = \frac{12,000 - 6000}{500 - 200} = \frac{6000}{300} = 20$$

The slope is 20. The manufacturing cost is $20 per calculator.

37. a) The y-intercept is $(0, 25,000)$.
The slope is 80.
$y = mx + b$
$y = 80x + 25,000$
The linear function is $f(x) = 80x + 25,000$

b) Predict the cost of building a house with 2000 ft^2.
$f(2000) = 80(2000) + 25,000$
$f(x) = 185,000$
The house will cost $185,000 to build.

Chapter 3 Test

1. $P(x) = 2 - x^2$
ordered pairs: $(-2, -2), (-1, 1), (0, 2),$
$(1, 1), (2, -2)$

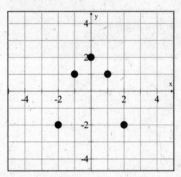

2. $y = 2x + 6$
$y = 2(-3) + 6$
$y = -6 + 6$
$y = 0$
The ordered pair is $(-3, 0)$.

3. $y = \frac{2}{3}x - 4$

x	y
0	−4
3	−2
6	0

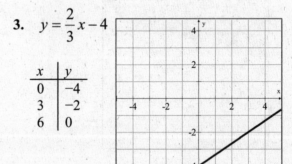

4. $2x + 3y = -3$

x	y
0	−1
3	−3
-3	1

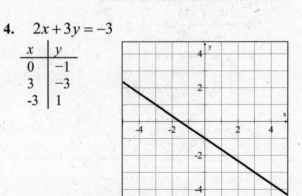

5. The equation of a vertical line that contains $(-2, 3)$ is $x = -2$.

6. $(4, 2)$ and $(-5, 8)$

Length $= \sqrt{(x_1 - x_2)^2 + (y_1 - y_2)^2}$

$= \sqrt{(4 - (-5))^2 + (2 - 8)^2}$

$= \sqrt{117} \approx 10.82$

The length is 10.82.

$x_m = \dfrac{x_1 + x_2}{2} = \dfrac{4 + (-5)}{2} = -\dfrac{1}{2}$

$y_m = \dfrac{y_1 + y_2}{2} = \dfrac{2 + 8}{2} = \dfrac{10}{2} = 5$

The midpoint is $\left(-\dfrac{1}{2}, 5\right)$.

7. $(-2, 3)$ and $(4, 2)$

$m = \dfrac{y_2 - y_1}{x_2 - x_1} = \dfrac{3 - 2}{-2 - 4} = \dfrac{1}{-6} = -\dfrac{1}{6}$

The slope of the line is $-\dfrac{1}{6}$.

8. $P(x) = 3x^2 - 2x + 1$

$P(2) = 3(2)^2 - 2(2) + 1$

$P(2) = 9$

9. $2x - 3y = 6$

x-intercept:

$2x - 3(0) = 6$

$2x = 6$

$x = 3$

$(3, 0)$

y-intercept:

$2(0) - 3y = 6$

$-3y = 6$

$y = -2$

$(0, -2)$

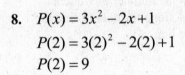

10. $(-2, 3)$ and $m = -\dfrac{3}{2}$

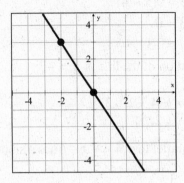

11. $m = \dfrac{2}{5}$ and $(-5, 2)$

$y - 2 = \dfrac{2}{5}[x - (-5)]$

$y - 2 = \dfrac{2}{5}(x + 5)$

$y - 2 = \dfrac{2}{5}x + 2$

$y = \dfrac{2}{5}x + 4$

The equation of the line is $y = \dfrac{2}{5}x + 4$.

12. $f(x) = \dfrac{2x + 1}{x}$

The function is not defined for zero in the denominator.

$x = 0$ is excluded from the domain of $f(x)$.

13. $(3, -4)$ and $(-2, 3)$

$m = \dfrac{3 - (-4)}{-2 - 3} = \dfrac{7}{-5} = -\dfrac{7}{5}$

$y - (-4) = -\dfrac{7}{5}(x - 3)$

$y + 4 = -\dfrac{7}{5}x + \dfrac{21}{5}$

$y = -\dfrac{7}{5}x + \dfrac{1}{5}$

The equation of the line is $y = -\dfrac{7}{5}x + \dfrac{1}{5}$.

14. A horizontal line has a slope of 0.

$m = 0$ and $(4, -3)$

$y - (-3) = 0(x - 4)$

$y + 3 = 0$

$y = -3$

The equation of the line is $y = -3$.

15. Domain = $\{-4, -2, 0, 3\}$

Range = $\{0, 2, 5\}$

16. A line parallel to $y = -\dfrac{3}{2}x - 6$ has a slope

of $m = -\dfrac{3}{2}$.

$m = -\dfrac{3}{2}$ and $(1, 2)$

$y - 2 = -\dfrac{3}{2}(x - 1)$

$y - 2 = -\dfrac{3}{2}x + \dfrac{3}{2}$

$y = -\dfrac{3}{2}x + \dfrac{7}{2}$

The equation of the line is $y = -\dfrac{3}{2}x + \dfrac{7}{2}$.

17. $y = -\dfrac{1}{2}x - 3$

$m_1 = -\dfrac{1}{2}$

$m_1 \cdot m_2 = -1$

$-\dfrac{1}{2} \cdot m_2 = -1$

$m_2 = 2$

The slope of the perpendicular line is
$m = 2$.

$m = 2$ and $(-2, -3)$

$y - (-3) = 2[x - (-2)]$

$y + 3 = 2(x + 2)$

$y + 3 = 2x + 4$

$y = 2x + 1$

The equation of the line is $y = 2x + 1$.

18. $3x - 4y > 8$

$-4y > -3x + 8$

$y < \dfrac{3}{4}x - 2$

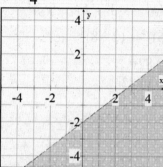

19. Strategy: Use two points on the graph to find the slope of the line.

Solution: $(3, 120{,}000)$ and $(12, 30{,}000)$

$m = \dfrac{120{,}000 - 30{,}000}{3 - 12} = \dfrac{90{,}000}{-9} = -10{,}000$

The value of the house decreases by $10,000 each year.

20. Strategy: Let x represent the tuition cost. Let y represent the number of students. Use the point – slope formula to find the equation of the line.

Solution:

a) $m = \dfrac{-6}{20} = -\dfrac{3}{10}$ and $(250, 100)$

$y - y_1 = m(x - x_1)$

$y - 100 = -\dfrac{3}{10}(x - 250)$

$y - 100 = -\dfrac{3}{10}x + 75$

$y = -\dfrac{3}{10}x + 175$

The linear function that will predict enrollment based on the cost of tuition is

$f(x) = -\dfrac{3}{10}x + 175$.

b) Find the number of students who enroll when tuition is $300.

$$f(300) = -\frac{3}{10}(300) + 175$$

$$f(300) = 85$$

When tuition is $300, 85 students will enroll.

Cumulative Review Exercises

1. Commutative Property of Multiplication

2.
$$3 - \frac{x}{2} = \frac{3}{4}$$

$$4 \cdot \left(3 - \frac{x}{2}\right) = \frac{3}{4}(4)$$

$$12 - 2x = 3$$

$$-2x = -9$$

$$x = \frac{9}{2}$$

The solution is $x = \frac{9}{2}$.

3. $2[y - 2(3 - y) + 4] = 4 - 3y$

$$2[y - 6 + 2y + 4] = 4 - 3y$$

$$2(3y - 2) = 4 - 3y$$

$$6y - 4 = 4 - 3y$$

$$9y - 4 = 4$$

$$9y = 8$$

$$y = \frac{8}{9}$$

The solution is $\frac{8}{9}$.

4.
$$\frac{1 - 3x}{2} + \frac{7x - 2}{6} = \frac{4x + 2}{9}$$

$$18 \cdot \left(\frac{1 - 3x}{2} + \frac{7x - 2}{6}\right) = 18\left(\frac{4x + 2}{9}\right)$$

$$9(1 - 3x) + 3(7x - 2) = 2(4x + 2)$$

$$9 - 27x + 21x - 6 = 8x + 4$$

$$3 - 6x = 8x + 4$$

$$3 = 14x + 4$$

$$-1 = 14x$$

$$x = -\frac{1}{14}$$

The solution is $-\frac{1}{14}$.

5. $x - 3 < -4$ or $2x + 2 > 3$

$x < -1$ $2x > 1$

$$x > \frac{1}{2}$$

$\{x|\, x < -1\}$ $\{x|\, x > \frac{1}{2}\}$

$\{x|\, x < -1\}$ or $\{x|\, x > \frac{1}{2}\}$

$\{x|\, x < -1 \text{ or } x > \frac{1}{2}\}$

6. $8 - |2x - 1| = 4$

$$-|2x - 1| = -4$$

$$|2x - 1| = 4$$

$2x - 1 = 4$ or $2x - 1 = -4$

$2x = 5$ $2x = -3$

$$x = \frac{5}{2} x = -\frac{3}{2}$$

The solutions are $\frac{5}{2}$ and $-\frac{3}{2}$.

7. $|3x - 5| < 5$

$$-5 < 3x - 5 < 5$$

$$-5 + 5 < 3x - 5 + 5 < 5 + 5$$

$$0 < 3x < 10$$

$$\frac{0}{3} < \frac{3x}{3} < \frac{10}{3}$$

$$0 < x < \frac{10}{3}$$

$$\{x \mid 0 < x < \frac{10}{3}\}$$

8. $4 - 2(4 - 5)^3 + 2$

$$= 4 - 2(-1)^3 + 2$$

$$= 4 - 2(-1) + 2$$

$$= 4 + 2 + 2$$

$$= 8$$

9. $(a-b)^2 \div (ab)$

$(4-(-2))^2 \div (4(-2))$

$= (6)^2 \div (-8) = 36 \div (-8)$

$= -4.5$

10. $\{x|\ x < -2\} \cup \{x|\ x > 0\}$

$$\begin{array}{c} \xleftarrow{\hspace{0.3cm}} \text{-5 -4 -3 -2 -1 0 1 2 3 4 5} \xrightarrow{\hspace{0.3cm}} \end{array}$$

11. $\quad P = \dfrac{R-C}{n}$

$P \cdot n = \dfrac{R-C}{n} \cdot n$

$P \cdot n = R - C$

$P \cdot n - R = -C$

$R - P \cdot n = C$

12. $2x + 3y = 6$

$2x = -3y + 6$

$x = -\dfrac{3}{2}y + 3$

13. $3x - 1 < 4$ and $x - 2 > 2$

$\quad 3x < 5 \qquad\qquad x > 4$

$\quad x < \dfrac{5}{3}$

$\{x|\ x < \dfrac{5}{3}\} \cap \{x|\ x > 4\} = \emptyset$

14. $P(x) = x^2 + 5$

$P(-3) = (-3)^2 + 5$

$P(-3) = 14$

15. $y = -\dfrac{5}{4}x + 3$

$y = -\dfrac{5}{4}(-8) + 3$

$y = 10 + 3$

$y = 13$

The ordered pair is $(-8, 13)$.

16. $(-1, 3)$ and $(3, -4)$

$m = \dfrac{y_2 - y_1}{x_2 - x_1} = \dfrac{3 - (-4)}{-1 - 3} = \dfrac{7}{-4} = -\dfrac{7}{4}$

17. $m = \dfrac{3}{2}$ and $(-1, 5)$

$y - y_1 = m(x - x_1)$

$y - 5 = \dfrac{3}{2}[(x - (-1)]$

$y - 5 = \dfrac{3}{2}(x + 1)$

$y - 5 = \dfrac{3}{2}x + \dfrac{3}{2}$

$y = \dfrac{3}{2}x + \dfrac{13}{2}$

The equation of the line is $y = \dfrac{3}{2}x + \dfrac{13}{2}$.

18. $(4, -2)$ and $(0, 3)$

$m = \dfrac{3 - (-2)}{0 - 4} = \dfrac{5}{-4} = -\dfrac{5}{4}$

$y - 3 = -\dfrac{5}{4}(x - 0)$

$y - 3 = -\dfrac{5}{4}x$

$y = -\dfrac{5}{4}x + 3$

The equation of the line is $y = -\dfrac{5}{4}x + 3$.

19. A line parallel to $y = -\dfrac{3}{2}x + 2$ has a slope

of $m = -\dfrac{3}{2}$.

$m = -\dfrac{3}{2}$ and $(2, 4)$

$y - 4 = -\dfrac{3}{2}(x - 2)$

$y - 4 = -\dfrac{3}{2}x + 3$

$$y = -\frac{3}{2}x + 7$$

The equation of the line is $y = -\frac{3}{2}x + 7$.

20. $3x - 2y = 5$

$-2y = -3x + 5$

$y = \frac{3}{2}x - \frac{5}{2}$

$m_1 = \frac{3}{2}$

$m_1 \cdot m_2 = -1$

$\frac{3}{2} \cdot m_2 = -1$

$m_2 = -\frac{2}{3}$

The slope of the perpendicular line is

$m = -\frac{2}{3}$.

$m = -\frac{2}{3}$ and $(4, 0)$

$y - 0 = -\frac{2}{3}(x - 4)$

$y = -\frac{2}{3}x + \frac{8}{3}$

The equation of the line is $y = -\frac{2}{3}x + \frac{8}{3}$.

21. Strategy: Let x represent the number of quarters.
The number of nickels is $4x$.
The number of dimes is $17 - 5x$.

Coin	Number	Value	Total Value
Quarter	x	25	$25x$
Nickels	$4x$	5	$5(4x)$
Dimes	$17 - 5x$	10	$10(17 - 5x)$

The sum of the total values of each denomination of coin equals the total value of all of the coins (160 cents)

Solution: $25x + 5(4x) + 10(17 - 5x) = 160$
$$25x + 20x + 170 - 50x = 160$$
$$-5x + 170 = 160$$
$$-5x = -10$$
$$x = 2$$

$17 - 5x = 17 - 5(2) = 17 - 10 = 7$
There are 7 dimes.

22. Strategy: Let r represent the speed of 1st plane.
The speed of the 2nd plane is $2r$.

	Rate	Time	Distance
1st plane	r	3	$3r$
2nd plane	$2r$	3	$3(2r)$

The total distance traveled by the two planes is 1800 mi.

Solution: $3r + 3(2r) = 1800$
$$3r + 6r = 1800$$
$$9r = 1800$$
$$r = 200$$

$2r = 2(200) = 400$
The first plane is traveling at 200 mph and the second plane is traveling at 400 mph.

23. Strategy: Let x represent the pounds of coffee costing $9.00.
Pounds of coffee costing $6.00: $60 - x$

	Amount	Cost	Value
$9 coffee	x	9	$9x$
$6 coffee	$60 - x$	6	$6(60 - x)$
Mixture	60	8	$8(60)$

The sum of the values before mixing is equal to the value after mixing.

Solution:
$9x + 6(60 - x) = 8(60)$
$9x + 360 - 6x = 480$
$3x + 360 = 480$
$3x = 120$
$x = 40$
$60 - x = 60 - 40 = 20$
The mixture contains 40 lb of $9.00 coffee and 20 lb of $6.00 coffee.

24. $3x - 5y = 15$
x-intercept:
$3x - 5(0) = 15$
$3x = 15$
$x = 5$
$(5, 0)$
y-intercept:
$3(0) - 5y = 15$
$-5y = 15$
$y = -3$
$(0, -3)$

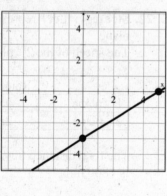

25.

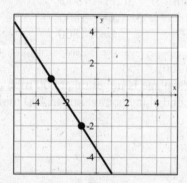

26. $3x - 2y \geq 6$
$-2y \geq -3x + 6$
$y \leq \dfrac{3}{2}x - 3$

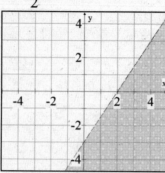

27. Strategy: Use two points on the graph to find the slope of the line.
Locate the y-intercept on the graph.
Use the slope-intercept form of an equation to write the equation of the line.

Solution: a) $(0, 30{,}000)$ and $(6, 0)$
$$m = \frac{30{,}000 - 0}{0 - 6} = \frac{30{,}000}{-6} = -5000$$
The y-intercept is $(0, 30{,}000)$.
The slope is -5000.
$y = mx + b$
$y = -5000x + 30{,}000$
The linear function is
$f(x) = -5000x + 30{,}000$
b) The value of the truck decreases by $5000 per year.

Chapter 4: Systems of Linear Equations and Inequalities

Prep Test

1. $10\left(\dfrac{3}{5}x + \dfrac{1}{2}y\right) = 10\left(\dfrac{3}{5}x\right) + 10\left(\dfrac{1}{2}y\right) = 6x + 5y$

2. $3x + 2y - z$
$3(-1) + 2(4) - (-2)$
$= -3 + 8 + 2$
$= 7$

3. $3x - 2z = 4$
$3x - 2(-2) = 4$
$3x + 4 = 4$
$3x = 0$
$x = 0$

4. $3x + 4(-2x - 5) = -5$
$3x - 8x - 20 = -5$
$-5x - 20 = -5$
$-5x = 15$
$x = -3$
The solution is -3.

5. $0.45x + 0.06(-x + 4000) = 630$
$0.45x - 0.06x + 240 = 630$
$0.39x + 240 = 630$
$0.39x = 390$
$x = 1000$
The solution is 1000.

6. $3x - 2y = 6$
$-2y = -3x + 6$

$y = \dfrac{3}{2}x - 3$

x	y
0	-3
2	0
4	3

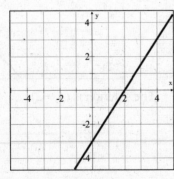

7. $y > -\dfrac{3}{5}x + 1$

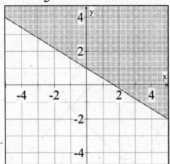

Section 4.1

Objective A Exercises

1. $3x - 2y = 2$
$3(0) - 2(-1) = 2$
$0 + 2 = 2$
$2 = 2$
$x + 2y = 6$
$0 + 2(-1) \neq 6$
$-2 \neq 6$
No, $(0, -1)$ is not a solution of the system of equations.

3. $x + y = -8$
$-3 + (-5) = -8$
$-8 = -8$
$2x + 5y = -31$
$2(-3) + 5(-5) = -31$
$-6 - 25 = -31$
$-31 = -31$
Yes, $(-3, -5)$ is a solution of the system of equations.

5. The graphs intersect at only one point. The system is an independent system of equations.

7. The graphs are parallel so the system of equations has no solution. The system is an inconsistent system of equations.

9. The system is an independent system of equations.

11. $x + y = 2$
$x - y = 4$

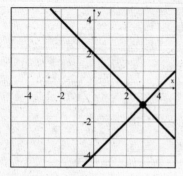

The solution is $(3, -1)$.

13. $x - y = -2$
$x + 2y = 10$

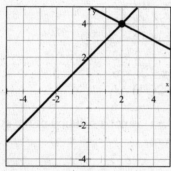

The solution is $(2, 4)$.

15. $3x - 2y = 6$
$y = 3$

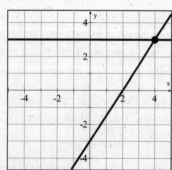

The solution is $(4, 3)$.

17. $x = 4$
$y = -1$

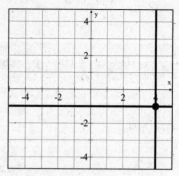

The solution is $(4, -1)$.

19. $2x + y = 3$
$x - 2 = 0$

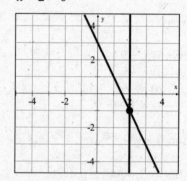

The solution is $(2, -1)$.

21. $x - y = 6$
$x + y = 2$

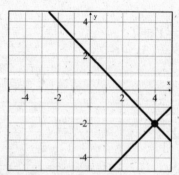

The solution is $(4, -2)$.

23. $y = x - 5$
$2x + y = 4$

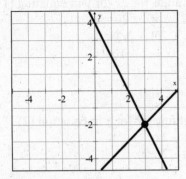

The solution is $(3, -2)$.

25. $y = \dfrac{1}{2}x - 2$
$x - 2y = 8$

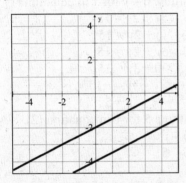

The lines are parallel and do not intersect so there is no solution.

27. $2x - 5y = 10$
$y = \dfrac{2}{5}x - 2$

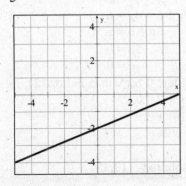

The two equations represent the same line. The system of equations is dependent.

The solutions are the ordered pairs $\left(x, \dfrac{2}{5}x - 2 \right)$.

Objective B Exercises

29. (1) $y = -x + 1$
(2) $2x - y = 5$
Substitute $-x + 1$ for y in equation (2).
$2x - y = 5$
$2x - (-x + 1) = 5$
$2x + x - 1 = 5$
$3x - 1 = 5$
$3x = 6$
$x = 2$
Substitute 2 for x in equation (1).
$y = -x + 1$
$y = -2 + 1$
$y = -1$
The solution is $(2, -1)$.

31. (1) $x = 2y - 3$
(2) $3x + y = 5$
Substitute $2y - 3$ for x in equation (2).
$3x + y = 5$
$3(2y - 3) + y = 5$
$6y - 9 + y = 5$
$7y - 9 = 5$
$7y = 14$
$y = 2$
Substitute 2 for y in equation (1).
$x = 2y - 3$
$x = 2(2) - 3$
$x = 4 - 3$
$x = 1$
The solution is $(1, 2)$.

33. (1) $3x + 5y = -1$
(2) $y = 2x - 8$
Substitute $2x - 8$ for y in equation (1).
$3x + 5y = -1$
$3x + 5(2x - 8) = -1$
$3x + 10x - 40 = -1$
$13x - 40 = -1$
$13x = 39$
$x = 3$

Substitute 3 for x in equation (2).
$$y = 2x - 8$$
$$y = 2(3) - 8$$
$$y = 6 - 8$$
$$y = -2$$
The solution is $(3, -2)$.

35. (1) $\qquad 4x - 3y = 2$
(2) $\qquad y = 2x + 1$
Substitute $2x + 1$ for y in equation (1).
$$4x - 3y = 2$$
$$4x - 3(2x + 1) = 2$$
$$4x - 6x - 3 = 2$$
$$-2x - 3 = 2$$
$$-2x = 5$$
$$x = -\frac{5}{2}$$
Substitute $-\dfrac{5}{2}$ for x in equation (2).
$$y = 2x + 1$$
$$y = 2\left(-\frac{5}{2}\right) + 1$$
$$y = -5 + 1$$
$$y = -4$$
The solution is $\left(-\dfrac{5}{2}, -4\right)$.

37. (1) $\qquad 3x - 2y = -11$
(2) $\qquad x = 2y - 9$
Substitute $2y - 9$ for x in equation (1).
$$3x - 2y = -11$$
$$3(2y - 9) - 2y = -11$$
$$6y - 27 - 2y = -11$$
$$4y - 27 = -11$$
$$4y = 16$$
$$y = 4$$
Substitute 4 for y in equation (2).
$$x = 2y - 9$$
$$x = 2(4) - 9$$
$$x = 8 - 9$$
$$x = -1$$
The solution is $(-1, 4)$.

39. (1) $3x + 2y = 4$
(2) $\qquad y = 1 - 2x$
Substitute $1 - 2x$ for y in equation (1).
$$3x + 2y = 4$$
$$3x + 2(1 - 2x) = 4$$
$$3x + 2 - 4x = 4$$
$$-x + 2 = 4$$
$$-x = 2$$
$$x = -2$$
Substitute -2 for x in equation (2).
$$y = 1 - 2x$$
$$y = 1 - 2(-2)$$
$$y = 1 + 4$$
$$y = 5$$
The solution is $(-2, 5)$.

41. (1) $\qquad 5x + 2y = 15$
(2) $\qquad x = 6 - y$
Substitute $6 - y$ for x in equation (1).
$$5x + 2y = 15$$
$$5(6 - y) + 2y = 15$$
$$30 - 5y + 2y = 15$$
$$30 - 3y = 15$$
$$-3y = -15$$
$$y = 5$$
Substitute 5 for y in equation (2).
$$x = 6 - y$$
$$x = 6 - 5$$
$$x = 1$$
The solution is $(1, 5)$.

43. (1) $\qquad 3x - 4y = 6$
(2) $\qquad x = 3y + 2$
Substitute $3y + 2$ for x in equation (1).
$$3x - 4y = 6$$
$$3(3y + 2) - 4y = 6$$
$$9y + 6 - 4y = 6$$
$$5y + 6 = 6$$
$$5y = 0$$
$$y = 0$$
Substitute 0 for y in equation (2).
$$x = 3y + 2$$
$$x = 2(0) + 2$$
$$x = 0 + 2$$
$$x = 2$$
The solution is $(2, 0)$.

45. (1) $3x + 7y = -5$
      (2) $y = 6x - 5$
Substitute $6x - 5$ for y in equation (1).
$$3x + 7y = -5$$
$$3x + 7(6x - 5) = -5$$
$$3x + 42x - 35 = -5$$
$$45x - 35 = -5$$
$$45x = 30$$
$$x = \frac{2}{3}$$

Substitute $\frac{2}{3}$ for x in equation (2).

$$y = 6x - 5$$
$$y = 6 \cdot \frac{2}{3} - 5$$
$$y = 4 - 5$$
$$y = -1$$

The solution is $\left(\frac{2}{3}, -1\right)$.

47. (1) $3x - y = 10$
      (2) $6x - 2y = 5$
Solve equation (1) for y.
$$3x - y = 10$$
$$-y = -3x + 10$$
$$y = 3x - 10$$
Substitute $3x - 10$ for y in equation (2).
$$6x - 2y = 5$$
$$6x - 2(3x - 10) = 5$$
$$6x - 6x + 20 = 5$$
$$20 = 5$$
No solution. This is not a true equation.
The lines are parallel and the system is inconsistent.

49. (1) $3x + 4y = 14$
      (2) $2x + y = 1$
Solve equation (2) for y.
$$2x + y = 1$$
$$y = -2x + 1$$
Substitute $-2x + 1$ for y in equation (1).
$$3x + 4y = 14$$
$$3x + 4(-2x + 1) = 14$$
$$3x - 8x + 4 = 14$$
$$-5x + 4 = 14$$
$$-5x = 10$$
$$x = -2$$

Substitute -2 for x in equation (2).
$$2x + y = 1$$
$$2(-2) + y = 1$$
$$-4 + y = 1$$
$$y = 5$$
The solution is $(-2, 5)$.

51. (1) $3x + 5y = 0$
      (2) $x - 4y = 0$
Solve equation (2) for x.
$$x - 4y = 0$$
$$x = 4y$$
Substitute $4y$ for x in equation (1).
$$3x + 5y = 0$$
$$3(4y) + 5y = 0$$
$$12y + 5y = 0$$
$$17y = 0$$
$$y = 0$$
Substitute 0 for y in equation (2).
$$x - 4y = 0$$
$$x - 4(0) = 0$$
$$x - 0 = 0$$
$$x = 0$$
The solution is $(0, 0)$.

53. (1) $2x - 4y = 16$
      (2) $-x + 2y = -8$
Solve equation (2) for x.
$$-x + 2y = -8$$
$$x = 2y + 8$$
Substitute $2y + 8$ for x in equation (1).
$$2x - 4y = 16$$
$$2(2y + 8) - 4y = 16$$
$$4y + 16 - 4y = 16$$
$$16 = 16$$
This is a true equation. The equations are dependent. The solutions are the ordered pairs $\left(x, \frac{1}{2}x - 4\right)$.

55. (1) $\quad\quad y = 3x + 2$
(2) $\quad\quad y = 2x + 3$
Substitute $2x + 3$ for y in equation (1).
$y = 3x + 2$
$2x + 3 = 3x + 2$
$3 = x + 2$
$x = 1$
Substitute 1 for x in equation (2).
$y = 2x + 3$
$y = 2(1) + 3$
$y = 5$
The solution is $(1, 5)$.

57. (1) $\quad\quad y = 3x + 1$
(2) $\quad\quad y = 6x - 1$
Substitute $6x - 1$ for y in equation (1).
$y = 3x + 1$
$6x - 1 = 3x + 1$
$3x - 1 = 1$
$3x = 2$
$x = \dfrac{2}{3}$

Substitute $\dfrac{2}{3}$ for x in equation (2).

$y = 6x - 1$
$y = 6 \cdot \dfrac{2}{3} - 1$
$y = 3$
The solution is $\left(\dfrac{2}{3}, 3\right)$.

59. The value of $\dfrac{a}{b}$ is $\dfrac{2}{3}$.

Objective C Exercises

61. The interest rates on the two accounts are 5.5% and 7.2%.

63. Strategy: Let x represent the amount invested at 4.2%.
$2800 is invested at 3.5%.

	Principal	Rate	Interest
Amount at 3.5%	2800	0.035	0.035(2800)
Amount at 4.2%	x	0.042	0.042x

The sum of the interest earned is $329.

Solution: $0.035(2800) + 0.042x = 329$
$98 + 0.042x = 329$
$0.042x = 231$
$x = 5500$
$5500 is invested at 4.2%.

65. Strategy: Let x represent the amount invested at 6.5%.
Let y represent the total amount invested at 5%.
$6000 is invested at 4%.

	Principal	Rate	Interest
Amount at 4%	6000	0.04	0.04(6000)
Amount at 6.5%	x	0.065	0.065x
Total invested	y	0.05	0.05y

The total amount invested is y.
$y = 6000 + x$
The total interest earned is equal to 5% of the total investment.
$0.04(6000) + 0.065x = 0.05y$

Solution: (1) $\quad y = 6000 + x$
(2) $\quad 0.04(6000) + 0.065x = 0.05y$
Substitute $6000 + x$ for y in equation (2).
$0.04(6000) + 0.065x = 0.05(6000 + x)$
$240 + 0.065x = 300 + 0.05x$
$240 + 0.015x = 300$
$0.015x = 60$
$x = 4000$
$4000 must be invested at 6.5%.

67. Strategy: Let x represent the amount invested at 3.5%.
Let y represent the amount invested at 4.5%.

	Principal	Rate	Interest
Amount at 3.5%	x	0.035	0.035x
Amount at 4.5%	y	0.045	0.045y

The total amount invested is $42,000.
$x + y = 42,000$
The interest earned from the 3.5% investment is equal to the interest earned from the 4.5% investment.
$0.035x = 0.045y$

Solution: (1) $x + y = 42,000$
(2) $0.035x = 0.045y$
Solve equation (1) for y and substitute for y in equation (2).
$y = 42,000 - x$
$0.035x = 0.045(42,000 - x)$
$0.035x = 1890 - 0.045x$
$0.080x = 1890$
$x = 23,625$
$y = 42,000 - x = 42,000 - 23,625 = 18,375$

$23,625 is invested at 3.5% and $18,375 is invested at 4.5%.

69. Strategy: Let x represent the amount invested at 4.5%.
Let y represent the amount invested at 8%.

	Principal	Rate	Interest
Amount at 4.5%	x	0.045	0.045x
Amount at 8%	y	0.08	0.08y

The total amount invested is $16,000.
$x + y = 16,000$
The total interest earned is $1070.
$0.045x + 0.08y = 1070$

Solution: (1) $x + y = 16,000$
(2) $0.045x + 0.08y = 1070$
Solve equation (1) for y and substitute for y in equation (2).
$y = 16,000 - x$
$0.045x + 0.08(16,000 - x) = 1070$
$0.045x + 1280 - 0.08x = 1070$
$-0.035x + 1280 = 1070$
$-0.035x = -210$
$x = 6000$
$6000 is invested at 4.5%.

Applying the Concepts

71.
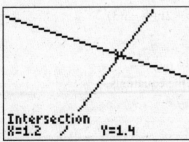
The solution is (1.2, 1.4).

73.
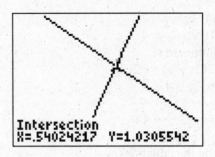
The solution is (0.54, 1.03).

Section 4.2

Objective A Exercises

1. Student answers may vary. Possible answers are 6 and -5.

3. (1) $x - y = 5$
(2) $x + y = 7$
Eliminate y. Add the two equations.
$2x = 12$
$x = 6$
Replace x with 6 in equation (1).
$x - y = 5$
$6 - y = 5$
$-y = -1$
$y = 1$
The solution is (6, 1).

5. (1) $3x + y = 4$
(2) $x + y = 2$
Eliminate y.
$3x + y = 4$
$-1(x + y) = -1(2)$

$3x + y = 4$
$-x - y = -2$
Add the equations.
$2x = 2$
$x = 1$
Replace x with 1 in equation (2).
$x + y = 2$
$1 + y = 2$
$y = 1$
The solution is (1, 1).

7. (1) $3x + y = 7$
   (2) $x + 2y = 4$
   Eliminate y.
   $-2(3x + y) = -2(7)$
   $x + 2y = 4$

   $-6x - 2y = -14$
   $x + 2y = 4$
   Add the equations.
   $-5x = -10$
   $x = 2$
   Replace x with 2 in equation (2).
   $x + 2y = 4$
   $2 + 2y = 4$
   $2y = 2$
   $y = 1$
   The solution is $(2, 1)$.

9. (1) $2x + 3y = -1$
   (2) $x + 5y = 3$
   Eliminate x.
   $2x + 3y = -1$
   $-2(x + 5y) = -2(3)$

   $2x + 3y = -1$
   $-2x - 10y = -6$
   Add the equations.
   $-7y = -7$
   $y = 1$
   Replace y with 1 in equation (2).
   $x + 5y = 3$
   $x + 5(1) = 3$
   $x + 5 = 3$
   $x = -2$
   The solution is $(-2, 1)$.

11. (1) $3x - y = 4$
    (2) $6x - 2y = 8$
    Eliminate y.
    $-2(3x - y) = -2(4)$
    $-6x + 2y = -8$

    $6x - 2y = 8$
    $6x - 2y = 8$
    Add the equations.
    $0 = 0$
    This is a true equation. The equations are dependent. The solutions are the ordered pairs $(x, 3x - 4)$.

13. (1) $2x + 5y = 9$
    (2) $4x - 7y = -16$
    Eliminate x.
    $-2(2x + 5y) = -2(9)$
    $4x - 7y = -16$

    $-4x - 10y = -18$
    $4x - 7y = -16$
    Add the equations.
    $-17y = -34$
    $y = 2$

    Replace y with 2 in equation (1).
    $2x + 5y = 9$
    $2x + 5(2) = 9$
    $2x + 10 = 9$
    $2x = -1$
    $x = -\dfrac{1}{2}$
    The solution is $\left(-\dfrac{1}{2}, 2\right)$.

15. (1) $4x - 6y = 5$
    (2) $2x - 3y = 7$
    Eliminate y.
    $4x - 6y = 5$
    $-2(2x - 3y) = -2(7)$

    $4x - 6y = 5$
    $-4x + 6y = -14$
    Add the equations.
    $0 = -9$
    This is not a true equation. The system of equations is inconsistent and therefore has no solution.

17. (1) $3x - 5y = 7$
    (2) $x - 2y = 3$
    Eliminate x.
    $3x - 5y = 7$
    $-3(x - 2y) = -3(3)$

    $3x - 5y = 7$
    $-3x + 6y = -9$
    Add the equations.
    $y = -2$

Replace y with -2 in equation (2).
$x - 2y = 3$
$x - 2(-2) = 3$
$x + 4 = 3$
$x = -1$
The solution is $(-1, -2)$.

19. (1) $x + 3y = 7$
    (2) $-2x + 3y = 22$
    Eliminate x.
    $2(x + 3y) = 2(7)$
    $-2x + 3y = 22$

    $2x + 6y = 14$
    $-2x + 3y = 22$
    Add the equations.
    $9y = 36$
    $y = 4$
    Replace y with 4 in equation (1).
    $x + 3y = 7$
    $x + 3(4) = 7$
    $x + 12 = 7$
    $x = -5$
    The solution is $(-5, 4)$.

21. (1) $3x + 2y = 16$
    (2) $2x - 3y = -11$
    Eliminate x.
    $-2(3x + 2y) = -2(16)$
    $3(2x - 3y) = 3(-11)$

    $-6x - 4y = -32$
    $6x - 9y = -33$
    Add the equations,
    $-13y = -65$
    $y = 5$
    Replace y with 5 in equation (1).
    $3x + 2y = 16$
    $3x + 2(5) = 16$
    $3x + 10 = 16$
    $3x = 6$
    $x = 2$
    The solution is $(2, 5)$.

23. (1) $4x + 4y = 5$
    (2) $2x - 8y = -5$
    Eliminate x.
    $4x + 4y = 5$
    $-2(2x - 8y) = -2(-5)$

$4x + 4y = 5$
$-4x + 16y = 10$
Add the equations.
$20y = 15$
$y = \dfrac{3}{4}$

Replace y with $\dfrac{3}{4}$ in equation (1).

$4x + 4y = 5$

$4x + 4\left(\dfrac{3}{4}\right) = 5$

$4x + 3 = 5$
$4x = 2$
$x = \dfrac{1}{2}$

The solution is $\left(\dfrac{1}{2}, \dfrac{3}{4}\right)$.

25. (1) $5x + 4y = 0$
    (2) $3x + 7y = 0$
    Eliminate x.
    $-3(5x + 4y) = -3(0)$
    $5(3x + 7y) = 5(0)$
    $-15x - 12y = 0$
    $15x + 35y = 0$
    Add the equations.
    $23y = 0$
    $y = 0$
    Replace y with 0 in equation (1).
    $5x + 4y = 0$
    $5x + 4(0) = 0$
    $3x + 0 = 0$
    $3x = 0$
    $x = 0$
    The solution is $(0, 0)$.

27. (1) $5x + 2y = 1$
    (2) $2x + 3y = 7$
    Eliminate x.
    $-2(5x + 2y) = -2(1)$
    $5(2x + 3y) = 5(7)$

    $-10x - 4y = -2$
    $10x + 15y = 35$
    Add the equations.
    $11y = 33$
    $y = 3$
    Replace y with 3 in equation (1).

$5x + 2y = 1$
$5x + 2(3) = 1$
$5x + 6 = 1$
$5x = -5$
$x = -1$
The solution is $(-1, 3)$.

29. (1) $3x - 6y = 6$
(2) $9x - 3y = 8$
Eliminate y.
$3x - 6y = 6$
$-2(9x - 3y) = -2(8)$

$3x - 6y = 6$
$-18x + 6y = -16$
Add the equations.
$-15x = -10$
$x = \dfrac{2}{3}$

Replace x with $\dfrac{2}{3}$ in equation (1).
$3x - 6y = 6$
$3\left(\dfrac{2}{3}\right) - 6y = 6$
$2 - 6y = 6$
$-6y = 4$
$y = -\dfrac{2}{3}$

The solution is $\left(\dfrac{2}{3}, -\dfrac{2}{3}\right)$.

31. (1) $\dfrac{3}{4}x + \dfrac{1}{3}y = -\dfrac{1}{2}$

(2) $\dfrac{1}{2}x - \dfrac{5}{6}y = -\dfrac{7}{2}$

Clear the fractions.

$12\left(\dfrac{3}{4}x + \dfrac{1}{3}y\right) = 12\left(-\dfrac{1}{2}\right)$

$6\left(\dfrac{1}{2}x - \dfrac{5}{6}y\right) = 6\left(-\dfrac{7}{2}\right)$

$9x + 4y = -6$
$3x - 5y = -21$
Eliminate x.
$9x + 4y = -6$
$-3(3x - 5y) = -3(-21)$

$9x + 4y = -6$
$-9x + 15y = 63$
Add the equations.
$19y = 57$
$y = 3$
Replace y with 3 in equation (1).
$\dfrac{3}{4}x + \dfrac{1}{3}y = -\dfrac{1}{2}$

$\dfrac{3}{4}x + \dfrac{1}{3}(3) = -\dfrac{1}{2}$

$\dfrac{3}{4}x + 1 = -\dfrac{1}{2}$

$\dfrac{3}{4}x = -\dfrac{3}{2}$

$x = -2$
The solution is $(-2, 3)$.

33. (1) $\dfrac{5x}{6} + \dfrac{y}{3} = \dfrac{4}{3}$

(2) $\dfrac{2x}{3} - \dfrac{y}{2} = \dfrac{11}{6}$

Clear the fractions.

$6\left(\dfrac{5x}{6} + \dfrac{y}{3}\right) = 6\left(\dfrac{4}{3}\right)$

$6\left(\dfrac{2x}{3} - \dfrac{y}{2}\right) = 6\left(\dfrac{11}{6}\right)$

$5x + 2y = 8$
$4x - 3y = 11$
Eliminate y.
$3(5x + 2y) = 3(8)$
$2(4x - 3y) = 2(11)$

$15x + 6y = 24$
$8x - 6y = 22$
Add the equations.
$23x = 46$
$x = 2$
Replace x with 2 in equation (1).
$\dfrac{5x}{6} + \dfrac{y}{3} = \dfrac{4}{3}$

$\dfrac{5(2)}{6} + \dfrac{y}{3} = \dfrac{4}{3}$

$\dfrac{5}{3} + \dfrac{y}{3} = \dfrac{4}{3}$

$$\frac{y}{3} = -\frac{1}{3}$$

$$y = -1$$

The solution is $(2, -1)$.

35. (1) $\dfrac{2x}{5} - \dfrac{y}{2} = \dfrac{13}{2}$

(2) $\dfrac{3x}{4} - \dfrac{y}{5} = \dfrac{17}{2}$

Clear the fractions.

$$10\left(\frac{2x}{5} - \frac{y}{2}\right) = 10\left(\frac{13}{2}\right)$$

$$20\left(\frac{3x}{4} - \frac{y}{5}\right) = 20\left(\frac{17}{2}\right)$$

$4x - 5y = 65$

$15x - 4y = 170$

Eliminate y.

$4(4x - 5y) = 4(65)$

$-5(15x - 4y) = -5(170)$

$16x - 20y = 260$

$-75x + 20y = -850$

Add the equations.

$-59x = -590$

$x = 10$

Replace x with 10 in equation (1).

$$\frac{2x}{5} - \frac{y}{2} = \frac{13}{2}$$

$$\frac{2(10)}{5} - \frac{y}{2} = \frac{13}{2}$$

$$4 - \frac{y}{2} = \frac{13}{2}$$

$$-\frac{y}{2} = \frac{5}{2}$$

$$y = -5$$

The solution is $(10, -5)$.

37. (1) $\dfrac{3x}{2} - \dfrac{y}{4} = -\dfrac{11}{12}$

(2) $\dfrac{x}{3} - y = -\dfrac{5}{6}$

Clear the fractions.

$$12\left(\frac{3x}{2} - \frac{y}{4}\right) = 12\left(-\frac{11}{12}\right)$$

$$6\left(\frac{x}{3} - y\right) = 6\left(-\frac{5}{6}\right)$$

$18x - 3y = -11$

$2x - 6y = -5$

Eliminate y.

$-2(18x - 3y) = -2(-11)$

$2x - 6y = -5$

$-36x + 6y = 22$

$2x - 6y = -5$

Add the equations,

$-34x = 17$

$$x = -\frac{1}{2}$$

Replace x with $-\dfrac{1}{2}$ in equation (1).

$$\frac{3x}{2} - \frac{y}{4} = -\frac{11}{12}$$

$$\frac{3}{2}\left(-\frac{1}{2}\right) - \frac{y}{4} = -\frac{11}{12}$$

$$-\frac{3}{4} - \frac{y}{4} = -\frac{11}{12}$$

$$-\frac{y}{4} = -\frac{1}{6}$$

$$y = \frac{2}{3}$$

The solution is $\left(-\dfrac{1}{2}, \dfrac{2}{3}\right)$.

39. (1) $4x - 5y = 3y + 4$

(2) $2x + 3y = 2x + 1$

Write the equations in the form $Ax + By = C$.

Solve the system of equations.

(1) $4x - 8y = 4$

(2) $3y = 1$

Solve equation (2) for y.

$3y = 1$

$$y = \frac{1}{3}$$

Replace y with $\dfrac{1}{3}$ in equation (1).

$4x - 5y = 3y + 4$

$$4x - 5 \cdot \frac{1}{3} = 3 \cdot \frac{1}{3} + 4$$

$$4x - \frac{5}{3} = 1 + 4$$

$$4x - \frac{5}{3} = 5$$

$$4x = \frac{20}{3}$$

$$x = \frac{5}{3}$$

The solution is $\left(\frac{5}{3}, \frac{1}{3}\right)$.

41. (1) $2x + 5y = 5x + 1$
(2) $3x - 2y = 3y + 3$
Write the equations in the form $Ax + By = C$.
Solve the system of equations.
(1) $-3x + 5y = 1$
(2) $3x - 5y = 3$

Add the equations.
$0 = 4$
This is not a true equation. The system of equations is inconsistent and therefore has no solution.

43. (1) $5x + 2y = 2x + 1$
(2) $2x - 3y = 3x + 2$
Write the equations in the form $Ax + By = C$.
Solve the system of equations.
(1) $3x + 2y = 1$
(2) $-x - 3y = 2$

Eliminate x.
$3x + 2y = 1$
$3(-x - 3y) = 3(2)$

$3x + 2y = 1$
$-3x - 9y = 6$
Add the equations.
$-7y = 7$
$y = -1$
Replace y with -1 in equation (1).
$5x + 2y = 2x + 1$
$5x + 2(-1) = 2x + 1$
$5x - 2 = 2x + 1$
$3x = 3$
$x = 1$
The solution is $(1, -1)$.

Objective B Exercises

45. (1) $x + 2y - z = 1$
(2) $2x - y + z = 6$
(3) $x + 3y - z = 2$
Eliminate z. Add equations (1) and (2).
$x + 2y - z = 1$
$2x - y + z = 6$

(4) $3x + y = 7$

Add equations (2) and (3).
$2x - y + z = 6$
$x + 3y - z = 2$

(5) $3x + 2y = 8$

Use equations (4) and (5) to solve for x and y.
$3x + y = 7$
$3x + 2y = 8$
Eliminate x.
$-1(3x + y) = -1(7)$
$3x + 2y = 8$

$-3x - y = -7$
$3x + 2y = 8$
$y = 1$
Replace y with 1 in equation (4).
$3x + y = 7$
$3x + 1 = 7$
$3x = 6$
$x = 2$
Replace x with 2 and y with 1 in equation (1).
$x + 2y - z = 1$
$2 + 2(1) - z = 1$
$2 + 2 - z = 1$
$4 - z = 1$
$-z = -3$
$z = 3$
The solution is $(2, 1, 3)$.

47. (1) $2x - y + 2z = 7$
(2) $x + y + z = 2$
(3) $3x - y + z = 6$
Eliminate y. Add equations (1) and (2).
$2x - y + 2z = 7$
$x + y + z = 2$

(4) $3x + 3z = 9$

Add equations (2) and (3).
$x + y + z = 2$
$3x - y + z = 6$

(5) $4x + 2z = 8$

Use equations (4) and (5) to solve for x and z.
$3x + 3z = 9$
$4x + 2z = 8$
Eliminate z.
$-2(3x + 3z) = -2(9)$
$3(4x + 2z) = 3(8)$

$-6x - 6z = -18$
$12x + 6z = 24$
$6x = 6$
$x = 1$
Replace x with 1 in equation (4).
$3x + 3z = 9$
$3(1) + 3z = 9$
$3 + 3z = 9$
$3z = 6$
$z = 2$
Replace x with 1 and z with 2 in equation (1).
$2x - y + 2z = 7$
$2(1) - y + 2(2) = 7$
$2 - y + 4 = 7$
$6 - y = 7$
$-y = 1$
$y = -1$
The solution is $(1, -1, 2)$.

49. (1) $3x + y = 5$
    (2) $3y - z = 2$
    (3) $x + z = 5$
    Eliminate z. Add equations (2) and (3).
    $3y - z = 2$
    $x + z = 5$

    (4) $x + 3y = 7$

    Use equations (1) and (4). Solve for x and y.
    $3x + y = 5$
    $x + 3y = 7$

    Eliminate y.
    $-3(3x + y) = -3(5)$
    $x + 3y = 7$

$-9x - 3y = -15$
$x + 3y = 7$
$-8x = -8$
$x = 1$
Replace x with 1 in equation (1).
$3x + y = 5$
$3(1) + y = 5$
$3 + y = 5$
$y = 2$
Replace y with 2 in equation (2).
$3y - z = 2$
$3(2) - z = 2$
$6 - z = 2$
$-z = -4$
$z = 4$
The solution is $(1, 2, 4)$.

51. (1) $x - y + z = 1$
    (2) $2x + 3y - z = 3$
    (3) $-x + 2y - 4z = 4$
    Eliminate z. Add equations (1) and (2).
    $x - y + z = 1$
    $2x + 3y - z = 3$

    (4) $3x + 2y = 4$

    Multiply equation (1) by 4 and add to equation (3).
    $4(x - y + z) = 4(1)$
    $-x + 2y - 4z = 4$

    $4x - 4y + 4z = 4$
    $-x + 2y - 4z = 4$

    (5) $3x - 2y = 8$
    Use equations (4) and (5) to solve for x and y.
    Add equation (4) and (5) to eliminate x.
    $3x + 2y = 4$
    $3x - 2y = 8$
    $6x = 12$
    $x = 2$
    Replace x with 2 in equation (4).
    $3x + 2y = 4$
    $3(2) + 2y = 4$
    $6 + 2y = 4$
    $2y = -2$
    $y = -1$

    Replace x with 2 and y with -1 in equation (1).

$x - y + z = 1$
$2 - (-1) + z = 1$
$2 + 1 + z = 1$
$3 + z = 1$
$z = -2$
The solution is $(2, -1, -2)$

53. (1) $2x + 3z = 5$
(2) $3y + 2z = 3$
(3) $3x + 4y = -10$
Eliminate z. Use equations (1) and (2).
$2x + 3z = 5$
$3y + 2z = 3$

$-2(2x + 3z) = -2(5)$
$3(3y + 2z) = 3(3)$

$-4x - 6z = -10$
$9y + 6z = 9$

(4) $-4x + 9y = -1$

Use equations (3) and (4) to solve for x and y.
$3x + 4y = -10$
$-4x + 9y = -1$
Eliminate x.
$4(3x + 4y) = 4(-10)$
$3(-4x + 9y) = 3(-1)$

$12x + 16y = -40$
$-12x + 27y = -3$
$43y = -43$
$y = -1$
Replace y with -1 in equation (2).
$3y + 2z = 3$
$3(-1) + 2z = 3$
$-3 + 2z = 3$
$2z = 6$
$z = 3$
Replace z with 3 in equation (1).
$2x + 3z = 5$
$2x + 3(3) = 5$
$2x + 9 = 5$
$2x = -4$
$x = -2$
The solution is $(-2, -1, 3)$.

55. (1) $2x + 4y - 2z = 3$
(2) $x + 3y + 4z = 1$
(3) $x + 2y - z = 4$

Eliminate x. Use equations (1) and (2).
$2x + 4y - 2z = 3$
$x + 3y + 4z = 1$

$2x + 4y - 2z = 3$
$-2(x + 3y + 4z) = -2(1)$

$2x + 4y - 2z = 3$
$-2x - 6y - 8z = -2$

(4) $-2y - 10z = 1$

Use equations (2) and (3).
$x + 3y + 4z = 1$
$x + 2y - z = 4$

$x + 3y + 4z = 1$
$-1(x + 2y - z) = -1(4)$

$x + 3y + 4z = 1$
$-x - 2y + z = -4$

(5) $y + 5z = -3$

Use equations (4) and (5) to solve for y and z.
$-2y - 10z = 1$
$y + 5z = -3$
Eliminate y.
$-2y - 10z = 1$
$2(y + 5z) = 2(-3)$

$-2y - 10z = 1$
$2y + 10z = -6$
$0 = -6$
This is not a true equation. The system of equations is inconsistent and therefore has no solution.

57. (1) $2x + y - z = 5$
(2) $x + 3y + z = 14$
(3) $3x - y + 2z = 1$
Eliminate z. Add equations (1) and (2).
$2x + y - z = 5$
$x + 3y + z = 14$

(4) $3x + 4y = 19$

Use equations (2) and (3).
$x + 3y + z = 14$
$3x - y + 2z = 1$

$-2(x + 3y + z) = -2(14)$
$3x - y + 2z = 1$

$-2x - 6y - 2z = -28$
$3x - y + 2z = 1$

(5) $x - 7y = -27$

Use equations (4) and (5) to solve for x and y.
$3x + 4y = 19$
$x - 7y = -27$
Eliminate x.
$3x + 4y = 19$
$-3(x - 7y) = -3(-27)$

$3x + 4y = 19$
$-3x + 21y = 81$
$25y = 100$
$y = 4$
Replace y with 4 in equation (4).
$3x + 4y = 19$
$3x + 4(4) = 19$
$3x + 16 = 19$
$3x = 3$
$x = 1$
Replace x with 1 and y with 4 in equation (1).
$2x + y - z = 5$
$2(1) + 4 - z = 5$
$2 + 4 - z = 5$
$6 - z = 5$
$-z = -1$
$z = 1$
The solution is $(1, 4, 1)$.

59. (1) $3x + y - 2z = 2$
(2) $x + 2y + 3z = 13$
(3) $2x - 2y + 5z = 6$
Eliminate x. Add equations (1) and (2).
$3x + y - 2z = 2$
$x + 2y + 3z = 13$

$3x + y - 2z = 2$
$-3(x + 2y + 3z) = -3(13)$

$3x + y - 2z = 2$
$-3x - 6y - 9z = -39$

(4) $-5y - 11z = -37$

Use equations (2) and (3).
$x + 2y + 3z = 13$
$2x - 2y + 5z = 6$

$-2(x + 2y + 3z) = -2(13)$
$2x - 2y + 5z = 6$

$-2x - 4y - 6z = -26$
$2x - 2y + 5z = 6$

(5) $-6y - z = -20$

Use equations (4) and (5) to solve for y and z.
$-5y - 11z = -37$
$-6y - z = -20$
Eliminate z.
$-5y - 11z = -37$
$-11(-6y - z) = -11(-20)$

$-5y - 11z = -37$
$66y + 11z = 220$
$61y = 183$
$y = 3$
Replace y with 3 in equation (4).
$-5y - 11z = -37$
$-5(3) - 11z = -37$
$-15 - 11z = -37$
$-11z = -22$
$z = 2$
Replace y with 3 and z with 2 in equation (1).
$3x + y - 2z = 2$
$3x + 3 - 2(2) = 2$
$3x + 3 - 4 = 2$
$3x - 1 = 2$
$3x = 3$
$x = 1$
The solution is $(1, 3, 2.)$

61. (1) $2x - y + z = 6$
(2) $3x + 2y + z = 4$
(3) $x - 2y + 3z = 12$
Eliminate y. Use equations (1) and (2).
$2x - y + z = 6$
$3x + 2y + z = 4$

$2(2x - y + z) = 2(6)$
$3x + 2y + z = 4$

$$4x - 2y + 2z = 12$$
$$3x + 2y + z = 4$$

(4) $7x + 3z = 16$

Add equations (2) and (3).
$$3x + 2y + z = 4$$
$$x - 2y + 3z = 12$$

(5) $4x + 4z = 16$

Use equations (4) and (5) to solve for x and z.
$$7x + 3z = 16$$
$$4x + 4z = 16$$
Eliminate z.
$$4(7x + 3z) = 4(16)$$
$$-3(4x + 4z) = -3(16)$$

$$28x + 12z = 64$$
$$-12x - 12z = -48$$
$$16x = 16$$
$$x = 1$$
Replace x with 1 in equation (4).
$$7x + 3z = 16$$
$$7(1) + 3z = 16$$
$$7 + 3z = 16$$
$$3z = 9$$
$$z = 3$$
Replace x with 1 and z with 3 in equation (1).
$$2x - y + z = 6$$
$$2(1) - y + 3 = 6$$
$$-y + 5 = 6$$
$$-y = 1$$
$$y = -1$$
The solution is $(1, -1, 3)$.

63. (1) $3x - 2y + 3z = -4$
(2) $2x + y - 3z = 2$
(3) $3x + 4y + 5z = 8$
Eliminate y. Use equations (1) and (2).
$$3x - 2y + 3z = -4$$
$$2x + y - 3z = 2$$

$$3x - 2y + 3z = -4$$
$$2(2x + y - 3z) = 2(2)$$

$$3x - 2y + 3z = -4$$
$$4x + 2y - 6z = 4$$

(4) $7x - 3z = 0$

Use equations (2) and (3).
$$2x + y - 3z = 2$$
$$3x + 4y + 5z = 8$$

$$-4(2x + y - 3z) = -4(2)$$
$$3x + 4y + 5z = 8$$

$$-8x - 4y + 12z = -8$$
$$3x + 4y + 5z = 8$$

(5) $-5x + 17z = 0$

Use equations (4) and (5) solve for x and z.
$$7x - 3z = 0$$
$$-5x + 17z = 0$$
Eliminate x.
$$5(7x - 3z) = 5(0)$$
$$7(-5x + 17z) = 7(0)$$

$$35x - 15z = 0$$
$$-35x + 119z = 0$$
$$104z = 0$$
$$z = 0$$

Replace z with 0 in equation (4).
$$7x - 3z = 0$$
$$7x - 3(0) = 0$$
$$7x = 0$$
$$x = 0$$
Replace x with 0 and z with 0 in equation (1).
$$3x - 2y + 3z = -4$$
$$3(0) - 2y + 3(0) = -4$$
$$0 - 2y + 0 = -4$$
$$-2y = -4$$
$$y = 2$$
The solution is $(0, 2, 0)$.

65. (1) $3x - y + 2z = 2$
(2) $4x + 2y - 7z = 0$
(3) $2x + 3y - 5z = 7$
Eliminate y. Use equations (1) and (2).
$$3x - y + 2z = 2$$
$$4x + 2y - 7z = 0$$

$$2(3x - y + 2z) = 2(2)$$
$$4x + 2y - 7z = 0$$

$6x - 2y + 4z = 4$
$4x + 2y - 7z = 0$

(4) $10x - 3z = 4$

Use equations (1) and (3).
$3x - y + 2z = 2$
$2x + 3y - 5z = 7$

$3(3x - y + 2z) = 3(2)$
$2x + 3y - 5z = 7$

$9x - 3y + 6z = 6$
$2x + 3y - 5z = 7$

(5) $11x + z = 13$

Use equations (4) and (5) to solve for x and z.
$10x - 3z = 4$
$11x + z = 13$
Eliminate z.
$10x - 3z = 4$
$3(11x + z) = 3(13)$

$10x - 3z = 4$
$33x + 3z = 39$
$43x = 43$
$x = 1$

Replace x with 1 in equation (4).
$10x - 3z = 4$
$10(1) - 3z = 4$
$-3z = -6$
$z = 2$
Replace x with 1 and z with 2 in equation (1).
$3x - y + 2z = 2$
$3(1) - y + 2(2) = 2$
$3 - y + 4 = 2$
$7 - y = 2$
$-y = -5$
$y = 5$
The solution is (1, 5, 2).

67. (1) $2x - 3y + 7z = 0$
(2) $x + 4y - 4z = -2$
(3) $3x + 2y + 5z = 1$
Eliminate x. Use equations (1) and (2).
$2x - 3y + 7z = 0$

$x + 4y - 4z = -2$

$2x - 3y + 7z = 0$
$-2(x + 4y - 4z) = -2(-2)$

$2x - 3y + 7z = 0$
$-2x - 8y + 8z = 4$

(4) $-11y + 15z = 4$

Use equations (2) and (3).
$x + 4y - 4z = -2$
$3x + 2y + 5z = 1$

$-3(x + 4y - 4z) = -3(-2)$
$3x + 2y + 5z = 1$

$-3x - 12y + 12z = 6$
$3x + 2y + 5z = 1$

(5) $-10y + 17z = 7$

Use equations (4) and (5) to solve for y and z.
$-11y + 15z = 4$
$-10y + 17z = 7$
Eliminate y.
$10(-11y + 15z) = 10(4)$
$-11(-10y + 17z) = -11(7)$

$-110y + 150z = 40$
$110y - 187z = -77$
$-37z = -37$
$z = 1$

Replace z with 1 in equation (4).
$-11y + 15z = 4$
$-11y + 15(1) = 4$
$-11y = -11$
$y = 1$
Replace y with 1 and z with 1 in equation (1).
$2x - 3y + 7z = 0$
$2x - 3(1) + 7(1) = 0$
$2x - 3 + 7 = 0$
$2x + 4 = 0$
$2x = -4$
$x = -2$
The solution is (-2, 1, 1).

69. a) (iii) no points
b) (ii) more than one point
c) (i) exactly one point

Applying the Concepts

71. (1) $\dfrac{1}{x} - \dfrac{2}{y} = 3$

(2) $\dfrac{2}{x} + \dfrac{3}{y} = -1$

Clear the fractions.

$$xy\left(\dfrac{1}{x} - \dfrac{2}{y}\right) = xy(3)$$

$$xy\left(\dfrac{2}{x} + \dfrac{3}{y}\right) = xy(-1)$$

$y - 2x = 3xy$
$2y + 3x = -xy$
Eliminate y.
$-2(y - 2x) = -2(3xy)$
$2y + 3x = -xy$

$-2y + 4x = -6xy$
$2y + 3x = -xy$
$7x = -7xy$
$y = -1$
Replace y with -1 in equation (1).
$$\dfrac{1}{x} - \dfrac{2}{y} = 3$$

$$\dfrac{1}{x} - \dfrac{2}{-1} = 3$$

$$\dfrac{1}{x} + 2 = 3$$

$$\dfrac{1}{x} = 1$$

$x = 1$
The solution is $(1, -1)$.

73. (1) $\dfrac{3}{x} + \dfrac{2}{y} = 1$

(2) $\dfrac{2}{x} + \dfrac{4}{y} = -2$

Clear the fractions.

$$xy\left(\dfrac{3}{x} + \dfrac{2}{y}\right) = xy(1)$$

$$xy\left(\dfrac{2}{x} + \dfrac{4}{y}\right) = xy(-2)$$

$3y + 2x = xy$
$2y + 4x = -2xy$
Eliminate x.
$-2(3y + 2x) = -2(xy)$
$2y + 4x = -2xy$

$-6y - 4x = -2xy$
$2y + 4x = -2xy$
$-4y = -4xy$
$x = 1$
Replace x with 1 in equation (1).
$$\dfrac{3}{x} + \dfrac{2}{y} = 1$$

$$\dfrac{3}{1} + \dfrac{2}{y} = 1$$

$$\dfrac{2}{y} = -2$$

$y = -1$
The solution is $(1, -1)$.

Section 4.3

Objective A Exercises

1. The determinant associated with the 2 X 2

matrix $\begin{bmatrix} a & b \\ c & d \end{bmatrix} = \begin{vmatrix} a & b \\ c & d \end{vmatrix}$.

Its value is $ad - bc$.

3. $\begin{vmatrix} 2 & -1 \\ 3 & 4 \end{vmatrix} = 2(4) - 3(-1) = 8 + 3 = 11$

5. $\begin{vmatrix} 6 & -2 \\ -3 & 4 \end{vmatrix} = 6(4) - (-3)(-2) = 24 - 6 = 18$

7. $\begin{vmatrix} 3 & 6 \\ 2 & 4 \end{vmatrix} = 3(4) - 2(6) = 12 - 12 = 0$

9. $\begin{vmatrix} 1 & -1 & 2 \\ 3 & 2 & 1 \\ 1 & 0 & 4 \end{vmatrix} = 1\begin{vmatrix} 2 & 1 \\ 0 & 4 \end{vmatrix} + 1\begin{vmatrix} 3 & 1 \\ 1 & 4 \end{vmatrix} + 2\begin{vmatrix} 3 & 2 \\ 1 & 0 \end{vmatrix}$

$$= 1(8 - 0) + 1(12 - 1) + 2(0 - 2)$$
$$= 8 + 11 - 4$$
$$= 15$$

11. $\begin{vmatrix} 3 & -1 & 2 \\ 0 & 1 & 2 \\ 3 & 2 & -2 \end{vmatrix} = 3\begin{vmatrix} 1 & 2 \\ 2 & -2 \end{vmatrix} + 1\begin{vmatrix} 0 & 2 \\ 3 & -2 \end{vmatrix} + 2\begin{vmatrix} 0 & 1 \\ 3 & 2 \end{vmatrix}$

$$= 3(-2 - 4) + 1(0 - 6) + 2(0 - 3)$$
$$= -18 - 6 - 6$$
$$= -30$$

13. $\begin{vmatrix} 4 & 2 & 6 \\ -2 & 1 & 1 \\ 2 & 1 & 3 \end{vmatrix} = 4\begin{vmatrix} 1 & 1 \\ 1 & 3 \end{vmatrix} - 2\begin{vmatrix} -2 & 1 \\ 2 & 3 \end{vmatrix} + 6\begin{vmatrix} -2 & 1 \\ 2 & 1 \end{vmatrix}$

$$= 4(3 - 1) - 2(-6 - 2) + 6(-2 - 2)$$
$$= 8 + 16 - 24$$
$$= 0$$

15. If one row of a matrix is all zeros, the value of the determinant will be 0.

Objective B Exercises

17. $2x - 5y = 26$
$5x + 3y = 3$

$$D = \begin{vmatrix} 2 & -5 \\ 5 & 3 \end{vmatrix} = 31$$

$$D_x = \begin{vmatrix} 26 & -5 \\ 3 & 3 \end{vmatrix} = 93, \, D_y = \begin{vmatrix} 2 & 26 \\ 5 & 3 \end{vmatrix} = -124$$

$$x = \frac{D_x}{D} = \frac{93}{31} = 3, \, y = \frac{D_y}{D} = \frac{-124}{31} = -4$$

The solution is $(3, -4)$.

19. $x - 4y = 8$
$3x + 7y = 5$

$$D = \begin{vmatrix} 1 & -4 \\ 3 & 7 \end{vmatrix} = 19$$

$$D_x = \begin{vmatrix} 8 & -4 \\ 5 & 7 \end{vmatrix} = 76, \, D_y = \begin{vmatrix} 1 & 8 \\ 3 & 5 \end{vmatrix} = -19$$

$$x = \frac{D_x}{D} = \frac{76}{19} = 4, \, y = \frac{D_y}{D} = \frac{-19}{19} = -1$$

The solution is $(4, -1)$.

21. $2x + 3y = 4$
$6x - 12y = -5$

$$D = \begin{vmatrix} 2 & 3 \\ 6 & -12 \end{vmatrix} = -42$$

$$D_x = \begin{vmatrix} 4 & 3 \\ -5 & -12 \end{vmatrix} = -33,$$

$$D_y = \begin{vmatrix} 2 & 4 \\ 6 & -5 \end{vmatrix} = -34$$

$$x = \frac{D_x}{D} = \frac{-33}{-42} = \frac{11}{14},$$

$$y = \frac{D_y}{D} = \frac{-34}{-42} = \frac{17}{21}$$

The solution is $\left(\frac{11}{14}, \frac{17}{21} \right)$.

23. $2x + 5y = 6$
$6x - 2y = 1$

$$D = \begin{vmatrix} 2 & 5 \\ 6 & -2 \end{vmatrix} = -34$$

$$D_x = \begin{vmatrix} 6 & 5 \\ 1 & -2 \end{vmatrix} = -17, \, D_y = \begin{vmatrix} 2 & 6 \\ 6 & 1 \end{vmatrix} = -34$$

$$x = \frac{D_x}{D} = \frac{-17}{-34} = \frac{1}{2}, \, y = \frac{D_y}{D} = \frac{-34}{-34} = 1$$

The solution is $\left(\frac{1}{2}, 1 \right)$.

25. $-2x + 3y = 7$
$4x - 6y = -5$

$$D = \begin{vmatrix} -2 & 3 \\ 4 & -6 \end{vmatrix} = 0$$

Since $D = 0$, $\dfrac{D_x}{D}$ is undefined. Therefore,

the system of equations does not have a unique solution. The equations are not independent.

27. $2x - 5y = -2$
$3x - 7y = -3$

$$D = \begin{vmatrix} 2 & -5 \\ 3 & -7 \end{vmatrix} = 1$$

$$D_x = \begin{vmatrix} -2 & -5 \\ -3 & -7 \end{vmatrix} = -1, D_y = \begin{vmatrix} 2 & -2 \\ 3 & -3 \end{vmatrix} = 0$$

$$x = \frac{D_x}{D} = \frac{-1}{1} = -1, \; y = \frac{D_y}{D} = \frac{0}{1} = 0$$

The solution is $(-1, 0)$.

29. $2x - y + 3z = 9$
$x + 4y + 4z = 5$
$3x + 2y + 2z = 5$

$$D = \begin{vmatrix} 2 & -1 & 3 \\ 1 & 4 & 4 \\ 3 & 2 & 2 \end{vmatrix} = -40$$

$$D_x = \begin{vmatrix} 9 & -1 & 3 \\ 5 & 4 & 4 \\ 5 & 2 & 2 \end{vmatrix} = -40,$$

$$D_y = \begin{vmatrix} 2 & 9 & 3 \\ 1 & 5 & 4 \\ 3 & 5 & 2 \end{vmatrix} = 40$$

$$D_z = \begin{vmatrix} 2 & -1 & 9 \\ 1 & 4 & 5 \\ 3 & 2 & 5 \end{vmatrix} = -80$$

$$x = \frac{D_x}{D} = \frac{-40}{-40} = 1, \; y = \frac{D_y}{D} = \frac{40}{-40} = -1$$

$$z = \frac{D_z}{D} = \frac{-80}{-40} = 2$$

The solution is $(1, -1, 2)$.

31. $3x - y + z = 11$
$x + 4y - 2z = -12$
$2x + 2y - z = -3$

$$D = \begin{vmatrix} 3 & -1 & 1 \\ 1 & 4 & -2 \\ 2 & 2 & -1 \end{vmatrix} = -3$$

$$D_x = \begin{vmatrix} 11 & -1 & 1 \\ -12 & 4 & -2 \\ -3 & 2 & -1 \end{vmatrix} = -6,$$

$$D_y = \begin{vmatrix} 3 & 11 & 1 \\ 1 & -12 & -2 \\ 2 & -3 & -1 \end{vmatrix} = 6$$

$$D_z = \begin{vmatrix} 3 & -1 & 11 \\ 1 & 4 & -12 \\ 2 & 2 & -3 \end{vmatrix} = -9$$

$$x = \frac{D_x}{D} = \frac{-6}{-3} = 2, \; y = \frac{D_y}{D} = \frac{6}{-3} = -2$$

$$z = \frac{D_z}{D} = \frac{-9}{-3} = 3$$

The solution is $(2, -2, 3)$.

33. $4x - 2y + 6z = 1$
$3x + 4y + 2z = 1$
$2x - y + 3z = 2$

$$D = \begin{vmatrix} 4 & -2 & 6 \\ 3 & 4 & 2 \\ 2 & -1 & 3 \end{vmatrix} = 0$$

Since $D = 0$, $\dfrac{D_x}{D}$ is undefined. Therefore,

the system of equations does not have a unique solution. The equations are not independent.

35. No, Cramer's Rule cannot be used to solve a dependent system of equations.

Applying the Concepts

37. a) Sometimes true
b) Always true
c) Sometimes true

39. $A = \dfrac{1}{2}\left\{ \begin{vmatrix} 9 & 26 \\ -3 & 6 \end{vmatrix} + \begin{vmatrix} 26 & 18 \\ 6 & 21 \end{vmatrix} + \begin{vmatrix} 18 & 16 \\ 21 & 10 \end{vmatrix} + \begin{vmatrix} 16 & 1 \\ 10 & 11 \end{vmatrix} + \begin{vmatrix} 1 & 9 \\ 11 & -3 \end{vmatrix} \right\}$

$A = \dfrac{1}{2}(132 + 438 - 156 + 166 - 102)$

$A = \dfrac{1}{2}(478)$

$ = 239 \text{ ft}^2$

Section 4.4

Objective A Exercises

1. n is less than m.

3. **Strategy:** Let x represent the rate of the motorboat in calm water.
The rate of the current is y.

	Rate	Time	Distance
with current	$x + y$	2	$2(x + y)$
against current	$x - y$	3	$3(x - y)$

The distance traveled with the current is 36 mi. The distance traveled against the current is 36 mi.
$2(x + y) = 36$
$3(x - y) = 36$

Solution: $2(x + y) = 36$
$ 3(x - y) = 36$

$\dfrac{1}{2} \cdot 2(x + y) = \dfrac{1}{2} \cdot 36$

$\dfrac{1}{3} \cdot 3(x - y) = \dfrac{1}{3} \cdot 36$

$x + y = 18$
$x - y = 12$
$2x = 30$
$x = 15$

$x + y = 18$
$15 + y = 18$
$y = 3$

The rate of the motorboat in calm water is 15 mph. The rate of the current is 3 mph.

5. **Strategy:** Let p represent the rate of the plane in calm air.
The rate of the wind is w.

	Rate	Time	Distance
with wind	$p + w$	4	$4(p + w)$
against wind	$p - w$	4	$4(p - w)$

The distance traveled with the wind is 2200 mi. The distance traveled against the wind is 1820 mi.
$4(p + w) = 2200$
$4(p - w) = 1820$

Solution: $4(p + w) = 2200$
$ 4(p - w) = 1820$

$\dfrac{1}{4} \cdot 4(p + w) = \dfrac{1}{4} \cdot 2200$

$\dfrac{1}{4} \cdot 4(p - w) = \dfrac{1}{4} \cdot 1820$

$p + w = 550$
$p - w = 455$
$2p = 1005$
$p = 502.5$

$p + w = 550$
$502.5 + w = 550$
$w = 47.5$

The rate of the plane in calm air is 502.5 mph. The rate of the wind is 47.5 mph.

7. **Strategy:** Let x represent the rate of the team in calm water.
The rate of the current is y.

	Rate	Time	Distance
with current	$x + y$	2	$2(x + y)$
against current	$x - y$	2	$2(x - y)$

The distance traveled with the current is 20 km. The distance traveled against the current is 12 km.
$2(x + y) = 20$
$2(x - y) = 12$

Solution: $2(x + y) = 20$
$\qquad\qquad 2(x - y) = 12$

$$\frac{1}{2} \cdot 2(x + y) = \frac{1}{2} \cdot 20$$

$$\frac{1}{2} \cdot 2(x - y) = \frac{1}{2} \cdot 12$$

$x + y = 10$
$x - y = 6$
$2x = 16$
$x = 8$

$x + y = 10$
$8 + y = 10$
$y = 2$

The rate of the team in calm water is 8 km/h.
The rate of the current is 2 km/h.

9. **Strategy:** Let x represent the rate of the plane in calm air.
The rate of the wind is y.

	Rate	Time	Distance
with wind	$x + y$	4	$4(x + y)$
against wind	$x - y$	5	$5(x - y)$

The distance traveled with the wind is 800 mi. The distance traveled against the wind is 800 mi.
$4(x + y) = 800$
$5(x - y) = 800$

Solution: $4(x + y) = 800$
$\qquad\qquad 5(x - y) = 800$

$$\frac{1}{4} \cdot 4(x + y) = \frac{1}{4} \cdot 800$$

$$\frac{1}{5} \cdot 5(x - y) = \frac{1}{5} \cdot 800$$

$x + y = 200$
$x - y = 160$
$2x = 360$
$x = 180$

$x + y = 200$
$180 + y = 200$
$y = 20$

The rate of the plane in calm air is 180 mph.
The rate of the wind is 20 mph.

11. **Strategy:** Let x represent the rate of the plane in calm air.
The rate of the wind is y.

	Rate	Time	Distance
with wind	$x + y$	5	$5(x + y)$
against wind	$x - y$	6	$6(x - y)$

The distance traveled with the wind is 600 mi. The distance traveled against the wind is 600 mi.
$5(x + y) = 800$
$6(x - y) = 800$

Solution: $5(x + y) = 600$
$\qquad\qquad 6(x - y) = 600$

$$\frac{1}{5} \cdot 5(x + y) = \frac{1}{5} \cdot 600$$

$$\frac{1}{6} \cdot 6(x - y) = \frac{1}{6} \cdot 600$$

$x + y = 120$
$x - y = 100$
$2x = 220$
$x = 110$

$x + y = 120$
$110 + y = 120$
$y = 10$

The rate of the plane in calm air is 110 mph.
The rate of the wind is 10 mph.

Objective B Exercises

13. The cost per pound of dark roast coffee is greater than the cost per pound of light roast coffee.

15. Strategy: Let n represent the number of nickels.
The number of dimes is d.

Coin	Number	Value	Total Value
Nickels	n	5	$5n$
Dimes	d	10	$10d$

Coins in the bank if the nickels were dimes and the dimes were nickels.
The value of the nickels and dimes in the bank is $2.50.

Coin	Number	Value	Total Value
Nickels	d	5	$5d$
Dimes	n	10	$10n$

The value of the nickels and dimes in the bank would be $3.50.

$5n + 10d = 250$
$10n + 5d = 350$

Solution: $5n + 10d = 250$
$\qquad\qquad 10n + 5d = 350$

$-2(5n + 10d) = -2(250)$
$\quad 10n + 5d = 350$

$-10n - 20d = -500$
$\quad 10n + 5d = 350$

$-15d = -150$
$\quad d = 10$

$5n + 10d = 250$
$5n + 10(10) = 250$
$5n = 150$
$n = 30$

There are 30 nickels in the bank.

17. Strategy: Let x represent the cost of redwood.
The cost of pine is y.
First purchase:

	Amount	Rate	Total Value
Redwood	60	x	$60x$
Pine	80	y	$80y$

Second purchase:

	Amount	Rate	Total Value
Redwood	100	x	$100x$
Pine	60	y	$60y$

The first purchase costs $286. The second purchase costs $396.

$60x + 80y = 286$
$100x + 60y = 396$

Solution: $60x + 80y = 286$
$\qquad\qquad 100x + 60y = 396$

$3(60x + 80y) = 3(286)$
$-4(100x + 60y) = -4(396)$

$180x + 240y = 858$
$-400x - 240y = -1584$

$-220x = -726$
$x = 3.3$

$60x + 80y = 286$
$60(3.3) + 80y = 286$
$198 + 80y = 286$
$80y = 88$
$y = 1.1$
The cost of the pine is $1.10/ft. The cost of the redwood is $3.30/ft.

19. Strategy: Let x represent the cost of nylon carpet.
The cost of wool carpet is y.
First purchase:

	Amount	Rate	Total Cost
Nylon	16	x	$16x$
Wood	20	y	$20y$

Second purchase:

	Amount	Rate	Total Cost
Nylon	18	x	$18x$
Wool	25	y	$25y$

The first purchase costs $1840. The second purchase costs $2200.

$$16x + 20y = 1840$$
$$18x + 25y = 2200$$

Solution:
$$16x + 20y = 1840$$
$$18x + 25y = 2200$$

$$5(16x + 20y) = 5(1840)$$
$$-4(18x + 25y) = -4(2200)$$

$$80x + 100y = 9200$$
$$-72x - 100y = -8800$$

$$8x = 400$$
$$x = 50$$

$$16x + 20y = 1840$$
$$16(50) + 20y = 1840$$
$$800 + 20y = 1840$$
$$20y = 1040$$
$$y = 52$$

The cost of the wool carpet is $52/yd.

21. Strategy: Let m represent the number of mountain bikes to be manufactured.
The number of trail bikes to be manufactured is t.

Cost of materials:

Type	Number	Cost	Total Cost
Mountain	m	70	$70m$
Trail	t	50	$50t$

Cost of labor:

Type	Number	Cost	Total Cost
Mountain	m	80	$80m$
Trail	t	40	$40t$

The company has budgeted $2500 for materials and $2600 for labor.

$$70m + 50t = 2500$$
$$80m + 40t = 2600$$

Solution:
$$70m + 50t = 2500$$
$$80m + 40t = 2600$$

$$4(70m + 50t) = 4(2500)$$
$$-5(80m + 40t) = -5(2600)$$

$$280m + 200t = 10,000$$
$$-400m - 200t = -13,000$$

$$-120m = -3000$$
$$m = 25$$

The company plans to manufacture 25 mountain bikes during the week.

23. Strategy: Let x represent the number of miles driven in the city.
The number of miles driven on the highway is $394 - x$.
Cost of hybrid driving:

	Number	Cost	Total Cost
City	x	0.09	$0.09x$
Highway	$394 - x$	0.08	$0.08(394 - x)$

The total amount spent on gasoline was $34.74.

Solution:
$$0.09x + 0.08(394 - x) = 34.74$$
$$0.09x + 31.52 - 0.08x = 34.74$$
$$0.01x = 3.22$$
$$x = 322$$

$$394 - x = 394 - 322 = 72$$

The owner drives 322 miles in the city and 72 on the highway.

25. Strategy: Let x represent the amount of the first alloy.
The amount of the second alloy is y.
Gold:

	Amount	Percent	Quantity
1st alloy	x	0.10	$0.10x$
2nd alloy	y	0.30	$0.30y$

Lead:

	Amount	Percent	Quantity
1st alloy	x	0.15	$0.15x$
2nd alloy	y	0.40	$0.40y$

The resulting alloy contains 60 g of gold and 88 g of lead.
$$0.10x + 0.30y = 60$$
$$0.15x + 0.40y = 88$$

Solution:
$$0.10x + 0.30y = 60$$
$$0.15x + 0.40y = 88$$
$$3(0.10x + 0.30y) = 3(60)$$
$$-2(0.15x + 0.40y) = -2(88)$$

$$0.30x + 0.90y = 180$$
$$-0.30x - 0.80y = -176$$
$$0.10y = 4$$
$$y = 40$$

$$0.10x + 0.30y = 60$$
$$0.10x + 0.30(40) = 60$$
$$0.10x + 12 = 60$$
$$0.10x = 48$$
$$x = 480$$

The chemist should use 480 g of gold and 40 g of lead.

27. Strategy: Let x represent the cost of the Model II computer.
The cost of the Model IV computer is y.
The cost of the Model IX computer is z.

First shipment:

	Number	Unit Cost	Value
Model II	4	x	$4x$
Model IV	6	y	$6y$
Model IX	10	z	$10z$

Second shipment:

	Number	Unit Cost	Value
Model II	8	x	$8x$
Model IV	3	y	$3y$
Model IX	5	z	$5z$

Third shipment:

	Number	Unit Cost	Value
Model II	2	x	$2x$
Model IV	9	y	$9y$
Model IX	5	z	$5z$

The value of the first shipment was $114,000. The value of the second shipment was $72,000. The value of the third shipment was $81,000.

Solution:
$$(1)\ 4x + 6y + 10z = 114,000$$
$$(2)\ 8x + 3y + 5z = 72,000$$
$$(3)\ 2x + 9y + 5z = 81,000$$

Multiply equation (2) by -2 and add to equation (1).
$$4x + 6y + 10z = 114,000$$
$$-16x - 6y - 10z = -144,000$$
$$-12x = -30,000$$
$$x = 2500$$

Multiply equation (3) by -1 and add to equation (2).

$8x + 3y + 5z = 72,000$
$-2x - 9y - 5z = -81,000$
$6x - 6y = -9000$

$6(2500) - 6y = 9000$
$-6y = -24,000$
$y = 4000$

The Model IV computer costs $4000.

29. Strategy: Let x represent the amount
deposited at 8%.
The amount deposited at 6% is y.
The amount deposited at 4% is z.

	Principal	Rate	Interest
8%	x	0.08	$0.08x$
6%	y	0.06	$0.06y$
4%	z	0.04	$0.04z$

The amount deposited in the 8% account is
twice the amount deposited in the 6% account.
The total amount invested is $25,000. The
total interest earned is $1520.

Solution: (1) $\qquad\qquad x = 2y$
(2) $\qquad\quad x + y + z = 25,000$
(3) $\ 0.08x + 0.06y + 0.04z = 1520$

Substitute $2y$ for x in equation (2) and
equation (3)
$2y + y + z = 25000$
(4) $3y + z = 25000$
$0.08(2y) + 0.06y + 0.04z = 1520$
(5) $0.22y + 0.04z = 1520$

Solve equation (4) for z and substitute into
equation (5).
$z = 25,000 - 3y$
$0.22y + 0.04(25,000 - 3y) = 1520$
$0.22y + 1000 - 0.12y = 1520$
$0.10y = 520$
$y = 5200$

$z = 25,000 - 3(5200)$
$z = 9400$

$x = 2(5200)$
$x = 10,400$

The investor placed $10,400 in the 8%

account, $5200 in the 6% account and $9400
in the 4% account.

Applying the Concepts

31. Strategy: Let n represent the measure of the
smaller angle.
The measure of the larger angle is m.
First relationship: $m + n = 180$
Second relationship: $m = 3n + 40$

Solution: Solve for n by substitution:
$(3n + 40) + n = 180$
$3n + 40 + n = 180$
$4n = 140$
$n = 35$
$m + n = 180$
$m + 35 = 180$
$m = 145$
The angles have measures of 35° and 145°.

33. Strategy: Let x represent the age of the oil
painting.
The age of the watercolor is y.
First relationship: $x - y = 35$
Second relationship:
$x + 5 = 2(y - 5)$
$x + 5 = 2y - 10$
$x = 2y - 15$

$x - y = 35$
$x = 2y - 15$

Solution: Solve for y by substitution:
$(2y - 15) - y = 35$
$y - 15 = 35$
$y = 50$
$x - y = 35$
$x - 50 = 35$
$x = 85$
The age of the oil painting is 85 years and
the age of the watercolor is 50 years.

Section 4.5

Objective A Exercises

1. Solve each inequality for y.

$x - y \geq 3$
$-y \geq -x + 3$
$y \leq x - 3$

$x + y \leq 5$
$y \leq 5 - x$

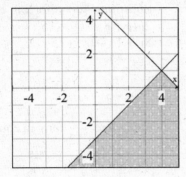

3. Solve each inequality for y.

$3x - y < 3$
$-y < -3x + 3$
$y > 3x - 3$

$2x + y \geq 2$
$y \geq 2 - 2x$

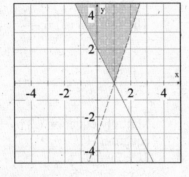

5. Solve each inequality for y.

$2x + y \geq -2$
$y \geq -2x - 2$

$6x + 3y \leq 6$
$3y \leq -6x + 6$
$y \leq -2x + 2$

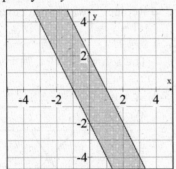

7. Solve the first inequality for y.

$3x - 2y < 6$
$-2y < -3x + 6$

$y > \dfrac{3}{2}x - 3$

$y \leq 3$

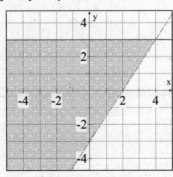

9. Solve the second inequality for y.

$y > 2x + 6$

$x + y < 0$
$y < -x$

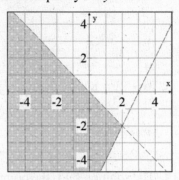

11. Solve each inequality for the variable.

$x + 1 \geq 0$
$x \geq -1$

$y - 3 \leq 0$
$y \leq 3$

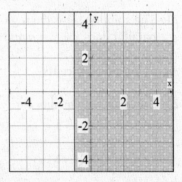

13. Solve each inequality for y.

$2x + y \geq 4$
$y \geq -2x + 4$

$3x - 2y < 6$
$-2y < -3x + 6$

$y > \dfrac{3}{2}x - 3$

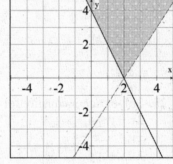

15. Solve each inequality for y.

$x - 2y \leq 6$
$-2y \leq -x + 6$

$y \geq \dfrac{1}{2}x - 3$

$2x + 3y \leq 6$
$3y \leq -2x + 6$

$y \geq -\dfrac{2}{3}x + 2$

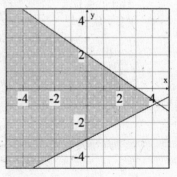

17. Solve each inequality for y.

$x - 2y \leq 4$

$-2y \leq -x + 4$

$y \geq \dfrac{1}{2}x - 2$

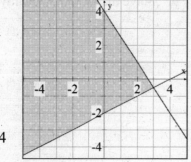

$3x + 2y \leq 8$

$2y \leq -3x + 8$

$y \leq -\dfrac{3}{2}x + 4$

19. The solution set is the points below the line $x + y = b$.

21. The solution set is the region between the parallel lines $x + y = a$ and $x + y = b$.

Applying the Concepts

23. Solve each inequality for y.

$2x + 3y \leq 15$

$3y \leq -2x + 15$

$y \leq -\dfrac{2}{3}x + 5$

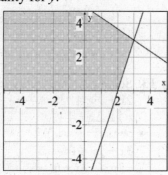

$3x - y \leq 6$

$-y \leq -3x + 6$

$y \geq 3x - 6$

$y \geq 0$

25. Solve each inequality for y.

$2x - y \leq 4$

$-y \leq -2x + 4$

$y \geq 2x - 4$

$3x + y < 1$

$y < -3x + 1$

$y \leq 0$

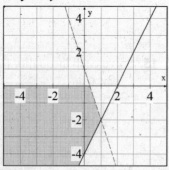

Chapter 4 Review Exercises

1. (1) $2x - 6y = 15$

   (2) $\qquad x = 4y + 8$

Substitute $4y + 8$ for x in equation (1).

$2(4y + 8) - 6y = 15$

$8y + 16 - 6y = 15$

$2y = -1$

$y = -\dfrac{1}{2}$

Substitute $-\dfrac{1}{2}$ for y in equation (2).

$x = 4\left(-\dfrac{1}{2}\right) + 8$

$x = 6$

The solution is $\left(6, -\dfrac{1}{2}\right)$.

2. (1) $3x + 2y = 2$

   (2) $\quad x + y = 3$

Multiply equation (2) by -2 and add to equation (1).

$3x + 2y = 2$

$-2x - 2y = -6$

$x = -4$

Replace x with -4 in equation (2).

$-4 + y = 3$

$y = 7$

The solution is $(-4, 7)$.

3. $x + y = 3$

$3x - 2y = -6$

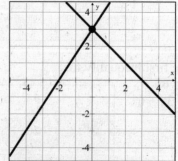

The solution is $(0, 3)$.

4. $2x - y = 4$
    $y = 2x - 4$

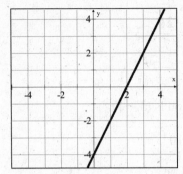

The two equations represent the same line. The system of equations is dependent. The solutions are the ordered pairs $(x, 2x - 4)$.

5. (1) $3x + 12y = 18$
    (2) $x + 4y = 6$

Solve equation (2) for x.
$x + 4y = 6$
$x = -4y + 6$
Substitute $-4y + 6$ for x in equation (1).
$3(-4y + 6) + 12y = 18$
$-12y + 18 + 12y = 18$
$18 = 18$
This is a true equation. The equations are dependent. The solutions are the ordered pairs $\left(x, -\dfrac{1}{4}x + \dfrac{3}{2} \right)$.

6. (1) $5x - 15y = 30$
    (2) $x - 3y = 6$

Multiply equation (2) by -5 and add to equation (1).
$5x - 15y = 30$
$-5x + 15y = -30$
$0 = 0$
This is a true equation. The equations are dependent. The solutions are the ordered pairs $\left(x, \dfrac{1}{3}x - 2 \right)$.

7. 1) $3x - 4y - 2z = 17$
    (2) $4x - 3y + 5z = 5$
    (3) $5x - 5y + 3z = 14$

Eliminate z. Multiply equation (1) by 3 and equation (3) by 2. Then add the equations.
$3(3x - 4y - 2z) = 3(17)$
$2(5x - 5y + 3z) = 2(14)$

$9x - 12y - 6z = 51$
$10x - 10y + 6z = 28$

(4) $19x - 22y = 79$

Multiply equation (1) by 5 and equation (2) by 2. Then add the equations.
$5(3x - 4y - 2z) = 5(17)$
$2(4x - 3y + 5z) = 2(5)$

$15x - 20y - 10z = 85$
$8x - 6y + 10z = 10$

(5) $23x - 26y = 95$

Multiply equation (4) by 23 and equation (5) by -19. Then add the equations.

$23(19x - 22y) = 23(79)$
$-19(23x - 26y) = -19(95)$

$437x - 506y = 1817$
$-437x + 494y = -1805$
$-12y = 12$
$y = -1$

Replace y with -1 in equation (4).
$19x - 22(-1) = 79$
$19x + 22 = 79$
$19x = 57$
$x = 3$
Replace x with 3 and y with -1 in equation (1).
$3(3) - 4(-1) - 2z = 17$
$9 + 4 - 2z = 17$
$-2z = 4$
$z = -2$
The solution is $(3, -1, -2)$.

8. 1) $3x + y = 13$
    (2) $2y + 3z = 5$
    (3) $x + 2z = 11$

Eliminate y. Multiply equation (1) by -2 then add to equation (2).
$-2(3x + y) = -2(13)$

$2y + 3z = 5$

$-6x - 2y = -26$
$ 2y + 3z = 5$

(4) $-6x + 3z = -21$

Multiply equation (3) by 6 then add to equation (4).
$6(x + 2z) = 6(11)$
$-6x + 3z = -21$

$6x + 12z = 66$
$-6x + 3z = -21$
$15z = 45$
$z = 3$

Replace z with 3 in equation (3).
$x + 2(3) = 11$
$x = 5$
Replace x with 5 in equation (1).
$3(5) + y = 13$
$y = -2$
The solution is $(5, -2, 3)$.

9. $\begin{vmatrix} 6 & 1 \\ 2 & 5 \end{vmatrix} = 6(5) - 2(1) = 30 - 2 = 28$

10. $\begin{vmatrix} 1 & 5 & -2 \\ -2 & 1 & 4 \\ 4 & 3 & -8 \end{vmatrix}$

$= 1\begin{vmatrix} 1 & 4 \\ 3 & -8 \end{vmatrix} - 5\begin{vmatrix} -2 & 4 \\ 4 & -8 \end{vmatrix} - 2\begin{vmatrix} -2 & 1 \\ 4 & 3 \end{vmatrix}$
$= 1(-8 - 12) - 5(16 - 16) - 2(-6 - 4)$
$= 1(-20) - 5(0) - 2(-10)$
$= -20 + 20$
$= 0$

11. $2x - y = 7$
$3x + 2y = 7$
$D = \begin{vmatrix} 2 & -1 \\ 3 & 2 \end{vmatrix} = 7$

$D_x = \begin{vmatrix} 7 & -1 \\ 7 & 2 \end{vmatrix} = 21$

$D_y = \begin{vmatrix} 2 & 7 \\ 3 & 7 \end{vmatrix} = -7$

$x = \dfrac{D_x}{D} = \dfrac{21}{7} = 3$

$y = \dfrac{D_y}{D} = \dfrac{-7}{7} = -1$

The solution is $(3, -1)$.

12. $3x - 4y = 10$
$2x + 5y = 15$
$D = \begin{vmatrix} 3 & -4 \\ 2 & 5 \end{vmatrix} = 23$

$D_x = \begin{vmatrix} 10 & -4 \\ 15 & 5 \end{vmatrix} = 110$

$D_y = \begin{vmatrix} 3 & 10 \\ 2 & 15 \end{vmatrix} = 25$

$x = \dfrac{D_x}{D} = \dfrac{110}{23}$

$y = \dfrac{D_y}{D} = \dfrac{25}{23}$

The solution is $\left(\dfrac{110}{23}, \dfrac{25}{23} \right)$.

13. $x + y + z = 0$
$x + 2y + 3z = 5$
$2x + y + 2z = 3$

$D = \begin{vmatrix} 1 & 1 & 1 \\ 1 & 2 & 3 \\ 2 & 1 & 2 \end{vmatrix} = 2$

$D_x = \begin{vmatrix} 0 & 1 & 1 \\ 5 & 2 & 3 \\ 3 & 1 & 2 \end{vmatrix} = -2$

$D_y = \begin{vmatrix} 1 & 0 & 1 \\ 1 & 5 & 3 \\ 2 & 3 & 2 \end{vmatrix} = -6$

$$D_z = \begin{vmatrix} 1 & 1 & 0 \\ 1 & 2 & 5 \\ 2 & 1 & 3 \end{vmatrix} = 8$$

$$x = \frac{D_x}{D} = \frac{-2}{2} = -1$$

$$y = \frac{D_y}{D} = \frac{-6}{2} = -3$$

$$z = \frac{D_z}{D} = \frac{8}{2} = 4$$

The solution is $(-1, -3, 4)$.

14. $x + 3y + z = 6$
$2x + y - z = 12$
$x + 2y - z = 13$

$$D = \begin{vmatrix} 1 & 3 & 1 \\ 2 & 1 & -1 \\ 1 & 2 & -1 \end{vmatrix} = 7$$

$$D_x = \begin{vmatrix} 6 & 3 & 1 \\ 12 & 1 & -1 \\ 13 & 2 & -1 \end{vmatrix} = 14$$

$$D_y = \begin{vmatrix} 1 & 6 & 1 \\ 2 & 12 & -1 \\ 1 & 13 & -1 \end{vmatrix} = 21$$

$$D_z = \begin{vmatrix} 1 & 3 & 6 \\ 2 & 1 & 12 \\ 1 & 2 & 13 \end{vmatrix} = -35$$

$$x = \frac{D_x}{D} = \frac{14}{7} = 2$$

$$y = \frac{D_y}{D} = \frac{21}{7} = 3$$

$$z = \frac{D_z}{D} = \frac{-35}{7} = -5$$

The solution is $(2, 3, -5)$.

15. Solve each inequality for y.

$x + 3y \le 6$
$3y \le -x + 6$
$y \le -\dfrac{1}{3}x + 2$

$2x - y \ge 4$
$-y \ge -2x + 4$
$y \le 2x - 4$

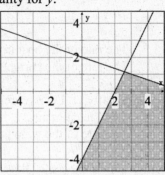

16. Solve each inequality for y.

$2x + 4y \ge 8$
$4y \ge -2x + 8$
$y \ge -\dfrac{1}{2}x + 2$

$x + y \le 3$
$y \le -x + 3$

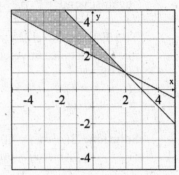

17. Strategy: Let x represent the rate of the cabin cruiser in calm water.
The rate of the current is y.

	Rate	Time	Distance
with current	$x+y$	3	$3(x+y)$
against current	$x-y$	5	$5(x-y)$

The distance traveled with the current is 60 mi. The distance traveled against the current is 60 mi.

$3(x + y) = 60$
$5(x - y) = 60$

Solution: $3(x + y) = 60$
$\quad\quad\quad\quad 5(x - y) = 60$

$\dfrac{1}{3} \cdot 3(x + y) = \dfrac{1}{3} \cdot 60$

$\dfrac{1}{5} \cdot 5(x - y) = \dfrac{1}{5} \cdot 60$

$x + y = 20$
$x - y = 12$
$2x = 32$
$x = 16$

$x + y = 20$
$16 + y = 20$
$y = 4$
The rate of the boat in calm water is 16 mph.
The rate of the current is 4 mph.

18. Strategy: Let p represent the rate of the plane in calm air.
The rate of the wind is w.

	Rate	Time	Distance
with wind	$p + w$	3	$3(p + w)$
against wind	$p - w$	4	$4(p - w)$

The distance traveled with the wind is 600 mi. The distance traveled against the wind is 600 mi.
$3(p + w) = 600$
$4(p - w) = 600$

Solution: $3(p + w) = 600$
$\quad\quad\quad\quad 4(p - w) = 600$

$\dfrac{1}{3} \cdot 3(p + w) = \dfrac{1}{3} \cdot 600$

$\dfrac{1}{4} \cdot 4(p - w) = \dfrac{1}{4} \cdot 600$

$p + w = 200$
$p - w = 150$
$2p = 350$
$p = 175$

$p + w = 200$
$175 + w = 200$
$w = 25$
The rate of the plane in calm air is 175 mph.
The rate of the wind is 25 mph.

19. Strategy: Let x represent the number of children's tickets sold.
The number of adult tickets sold is y.

Friday:

	Amount	Rate	Quantity
Children	x	5	$5x$
Adult	y	8	$8y$

Saturday:

	Amount	Rate	Quantity
Children	$3x$	5	$5(3x)$
Adult	$½ y$	8	$8(½)y$

The total receipts for Friday were $2500.
The total receipts for Saturday were $2500.
$5x + 8y = 2500$
$15x + 4y = 2500$

Solution: (1) $5x + 8y = 2500$
$\quad\quad\quad\quad$ (2) $15x + 4y = 2500$
Multiply equation (2) by -2 then add to equation (1).
$5x + 8y = 2500$
$-2(15x + 4y) = -2(2500)$
$5x + 8y = 2500$
$-30x - 8y = -5000$
$-25x = -2500$
$x = 100$
On Friday, 100 children attended.

20. Strategy: Let x represent the amount invested at 3%.
The amount invested at 7% is y.

	Amount	Rate	Quantity
Amount at 3%	x	0.03	$0.03x$
Amount at 7%	y	0.07	$0.07y$

The total amount invested is $20,000.
$x + y = 20,000$
The total annual interest earned is $1200.
$0.03x + 0.07y = 1200$

Solution: (1) $x + y = 20,000$
$\quad\quad\quad\quad$ (2) $0.03x + 0.07y = 1200$
Multiply equation (1) by -0.07 then add to equation (2).

$-0.07x - 0.07y = -1400$
$0.03x + 0.07y = 1200$
$-0.04x = -200$
$x = 5000$
Substitute 5000 for x in equation (1).
$x + y = 20,000$
$5000 + y = 20,000$
$y = 15,000$
The amount invested at 3% is $5000.
The amount invested at 7% is $15,000.

Chapter 4 Test

1. $2x - 3y = -6$
$2x - y = 2$

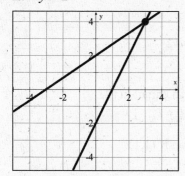

The solution is $(3, 4)$.

2. $x - 2y = -6$
$y = \dfrac{1}{2}x - 4$

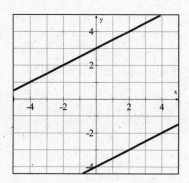

There is no solution.

3. Solve each inequality for y.
$2x - y < 3$
$-y < -2x + 3$
$y > 2x - 3$

$4x + 3y < 11$
$3y < -4x + 11$
$y < -\dfrac{4}{3}x + \dfrac{11}{3}$

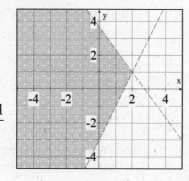

4. Solve each inequality for y.
$x + y > 2$
$y > -x + 2$

$2x - y < -1$
$-y < -2x - 1$
$y > 2x + 1$

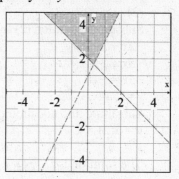

5. (1) $3x + 2y = 4$
(2) $x = 2y - 1$

Substitute $2y - 1$ for x in equation (1).
$3(2y - 1) + 2y = 4$
$6y - 3 + 2y = 4$
$8y = 7$
$y = \dfrac{7}{8}$

Substitute into equation (2).
$x = 2\left(\dfrac{7}{8}\right) - 1 = \dfrac{7}{4} - 1 = \dfrac{3}{4}$

The solution is $\left(\dfrac{3}{4}, \dfrac{7}{8}\right)$.

6. (1) $5x + 2y = -23$
(2) $2x + y = -10$

Solve equation (2) for y.
$y = -2x - 10$
Substitute $-2x - 10$ for y in equation (1).
$5x + 2(-2x - 10) = -23$
$5x - 4x - 20 = -23$
$x = -3$
Substitute in equation (2).

$2(-3) + y = -10$
$-6 + y = -10$
$y = -4$
The solution is $(-3, -4)$.

7. (1) $y = 3x - 7$
(2) $y = -2x + 3$
Substitute equation (2) into equation (1).
$-2x + 3 = 3x - 7$
$-5x + 3 = -7$
$-5x = -10$
$x = 2$
Substitute into equation (1).
$y = 3(2) - 7$
$y = -1$
The solution is $(2, -1)$.

8. (1) $3x + 4y = -2$
(2) $2x + 5y = 1$

Multiply equation (1) by 2 and equation (2)
by -3 then add the new equations.
$2(3x + 4y) = 2(-2)$
$-3(2x + 5y) = -3(1)$

$6x + 8y = -4$
$-6x - 15y = -3$
$-7y = -7$
$y = 1$
Substitute into equation (1).
$3x + 4(1) = -2$
$3x = -6$
$x = -2$
The solution is $(-2, 1)$.

9. (1) $4x - 6y = 5$
(2) $6x - 9y = 4$

Multiply equation (1) by 3 and equation (2)
by -2 then add the new equations.
$3(4x - 6y) = 3(5)$
$-2(6x - 9y) = -2(4)$

$12x - 18y = 15$
$-12x + 18y = -8$
$0 = 7$
This is not a true equation. The system of
equations is inconsistent and therefore has
no solution.

10. (1) $3x - y = 2x + y - 1$
(2) $5x + 2y = y + 6$
Write the equations in the form $Ax + By = C$.
(3) $x - 2y = -1$
(4) $5x + y = 6$

Multiply equation (4) by 2 then add to
equation (3).
$x - 2y = -1$
$2(5x + y) = 2(6)$

$x - 2y = -1$
$10x - 2y = 12$
$11x = 11$
$x = 1$
Substitute in equation (4).
$5(1) + y = 6$
$y = 1$
The solution is $(1,1)$.

11. (1) $2x + 4y - z = 3$
(2) $x + 2y + z = 5$
(3) $4x + 8y - 2z = 7$

Eliminate z. Add equations (1) and (2).
$2x + 4y - z = 3$
$x + 2y + z = 5$

(4) $3x + 6y = 8$

Multiply equation (2) by 2 and add to
equation (3).

$2(x + 2y + z) = 2(5)$
$4x + 8y - 2z = 7$

$2x + 4y + 2z = 10$
$4x + 8y - 2z = 7$

(5) $6x + 12y = 17$

Multiply equation (4) by -2 then add to
equation (5).

$-2(3x + 6y) = -2(8)$
$6x + 12y = 17$
$-6x - 12y = -16$
$6x + 12y = 17$
$0 = 1$
This is not a true equation. The system of
equations is inconsistent and therefore has
no solution.

12. (1) $x - y - z = 5$
(2) $2x + z = 2$
(3) $3y - 2z = 1$

Multiply equation (1) by 3 and then add to equation (3).
$3(x - y - z) = 3(5)$
$3x - 3y - 3z = 15$
$3y - 2z = 1$

(4) $3x - 5z = 16$

Multiply equation (2) by 5 and add to equation (4).
$5(2x + z) = 5(2)$
$10x + 5z = 10$
$3x - 5z = 16$
$13x = 26$
$x = 2$

Substitute 2 in for x in equation (2).
$2(2) + z = 2$
$z = -2$

Substitute 2 in for x and -2 in for z in equation (1).
$2 - y - (-2) = 5$
$4 - y = 5$
$-y = 1$
$y = -1$

The solution is $(2, -1, -2)$.

13. $\begin{vmatrix} 3 & -1 \\ -2 & 4 \end{vmatrix} = 3(4) - (-2)(-1) = 12 - 2 = 10$

14. $\begin{vmatrix} 1 & -2 & 3 \\ 3 & 1 & 1 \\ 2 & -1 & -2 \end{vmatrix} = 1\begin{vmatrix} 1 & 1 \\ -1 & -2 \end{vmatrix} - (-2)\begin{vmatrix} 3 & 1 \\ 2 & -2 \end{vmatrix} + 3\begin{vmatrix} 3 & 1 \\ 2 & -1 \end{vmatrix}$

$= 1(-2 - (-1)) + 2(-6 - 2) + 3(-3 - 2)$
$= 1(-1) + 2(-8) + 3(-5)$
$= -1 - 16 - 15$
$= -32$

15. $x - y = 3$
$2x + y = -4$

$D = \begin{vmatrix} 1 & -1 \\ 2 & 1 \end{vmatrix} = 3$

$D_x = \begin{vmatrix} 3 & -1 \\ -4 & 1 \end{vmatrix} = -1$

$D_y = \begin{vmatrix} 1 & 3 \\ 2 & -4 \end{vmatrix} = -10$

$x = \dfrac{D_x}{D} = \dfrac{-1}{3} = -\dfrac{1}{3}$

$y = \dfrac{D_y}{D} = \dfrac{-10}{3} = -\dfrac{10}{3}$

The solution is $\left(-\dfrac{1}{3}, -\dfrac{10}{3}\right)$.

16. $5x + 2y = 9$
$3x + 5y = -7$

$D = \begin{vmatrix} 5 & 2 \\ 3 & 5 \end{vmatrix} = 19$

$D_x = \begin{vmatrix} 9 & 2 \\ -7 & 5 \end{vmatrix} = 59$

$D_y = \begin{vmatrix} 5 & 9 \\ 3 & -7 \end{vmatrix} = -62$

$x = \dfrac{D_x}{D} = \dfrac{59}{19}$

$y = \dfrac{D_y}{D} = \dfrac{-62}{19}$

The solution is $\left(\dfrac{59}{19}, -\dfrac{62}{19}\right)$.

17. $x - y + z = 2$
$2x - y - z = 1$
$x + 2y - 3z = -4$

$$D = \begin{vmatrix} 1 & -1 & 1 \\ 2 & -1 & -1 \\ 1 & 2 & -3 \end{vmatrix} = 5$$

$$D_x = \begin{vmatrix} 2 & -1 & 1 \\ 1 & -1 & -1 \\ -4 & 2 & -3 \end{vmatrix} = 1$$

$$D_y = \begin{vmatrix} 1 & 2 & 1 \\ 2 & 1 & -1 \\ 1 & -4 & -3 \end{vmatrix} = -6$$

$$D_z = \begin{vmatrix} 1 & -1 & 2 \\ 2 & -1 & 1 \\ 1 & 2 & -4 \end{vmatrix} = 3$$

$$x = \frac{D_x}{D} = \frac{1}{5}$$

$$y = \frac{D_y}{D} = \frac{-6}{5}$$

$$z = \frac{D_z}{D} = \frac{3}{5}$$

The solution is $\left(\dfrac{1}{5}, -\dfrac{6}{5}, \dfrac{3}{5} \right)$.

18. Strategy: Let x represent the rate of the plane in calm air.
The rate of the wind is y.

	Rate	Time	Distance
with wind	$x + y$	2	$2(x + y)$
against wind	$x - y$	2.8	$2.8(x - y)$

The distance traveled with the wind is 350 mi. The distance traveled against the wind is 350 mi.
$2(x + y) = 350$
$2.8(x - y) = 350$

Solution: $2(x + y) = 350$
$2.8(x - y) = 350$

$$\frac{1}{2} \cdot 2(x + y) = \frac{1}{2} \cdot 350$$

$$\frac{1}{2.8} \cdot 2.8(x - y) = \frac{1}{2.8} \cdot 350$$

$x + y = 175$
$x - y = 125$
$2x = 300$
$x = 150$

$x + y = 175$
$150 + y = 175$
$y = 25$

The rate of the plane in calm air is 150 mph.
The rate of the wind is 25 mph.

19. Strategy: Let x represent the cost per yard of cotton.
The cost per yard of wool is y.
First purchase:

	Amount	Rate	Total Value
Cotton	60	x	$60x$
Wool	90	y	$90y$

Second purchase:

	Amount	Rate	Total Value
Cotton	80	x	$80x$
Wool	20	y	$20y$

The total cost of the first purchase was $1800. The total cost of the second purchase was $1000.
$60x + 90y = 1800$
$80x + 20y = 1000$

Solution: $-4(60x + 90y) = -4(1800)$
$3(80x + 20y) = 3(1000)$

$-240x - 360y = -7200$
$240x + 60y = 3000$
$-300y = -4200$
$y = 14$

$60x + 90(14) = 1800$
$60x + 1260 = 1800$
$60x = 540$
$x = 9$

The cost of cotton is $9.00/yd.
The cost of wool is $14.00/yd.

20. Strategy: Let x represent the amount invested at 2.7%.
The amount invested at 5.1% is y.

	Amount	Rate	Quantity
Amount at 2.7%	x	0.027	$0.027x$
Amount at 5.1%	y	0.051	$0.051y$

The total amount invested is \$15,000.
$x + y = 15,000$
The total annual interest earned is \$549.
$0.027x + 0.051y = 549$

Solution: (1) $x + y = 15,000$
(2) $0.027x + 0.051y = 549$
Multiply equation (1) by -0.051 then add to equation (2).

$-0.051x - 0.051y = -765$
$0.027x + 0.051y = 549$
$-0.024x = -216$
$x = 9000$
Substitute 9000 for x in equation (1).
$x + y = 15,000$
$9000 + y = 15,000$
$y = 6000$
The amount invested at 2.7% is \$9000.
The amount invested at 5.1% is \$6000.

Cumulative Review Exercises

1.
$$\frac{3}{2}x - \frac{3}{8} + \frac{1}{4}x = \frac{7}{12}x - \frac{5}{6}$$
$$24\left(\frac{3}{2}x - \frac{3}{8} + \frac{1}{4}x\right) = 24\left(\frac{7}{12}x - \frac{5}{6}\right)$$
$$36x - 9 + 6x = 14x - 20$$
$$42x - 9 = 14x - 20$$
$$28x - 9 = -20$$
$$28x = -11$$
$$x = -\frac{11}{28}$$
The solution is $-\dfrac{11}{28}$.

2. $(2, -1)\ (3, 4)$
$$m = \frac{y_2 - y_1}{x_2 - x_1} = \frac{4 - (-1)}{3 - 2} = \frac{5}{1} = 5$$
$$y - y_1 = m(x - x_1)$$

$$y - 4 = 5(x - 3)$$
$$y - 4 = 5x - 15$$
$$y = 5x - 11$$
The equation of the line is $y = 5x - 11$.

3. $3[x - 2(5 - 2x) - 4x] + 6$
$= 3(x - 10 + 4x - 4x) + 6$
$= 3(x - 10) + 6$
$= 3x - 30 + 6$
$= 3x - 24$

4. $a + bc \div 2$
$4 + 8(-2) \div 2 = 4 - 16 \div 2 = 4 - 8 = -4$

5. $2x - 3 < 9$ or $5x - 1 < 4$
Solve each inequality.

$2x - 3 < 9$ ‎ ‎ ‎ ‎ $5x - 1 < 4$
$2x < 12$ ‎ ‎ ‎ ‎ ‎ ‎ $5x < 5$
$x < 6$ ‎ or ‎ ‎ ‎ $x < 1$
$(-\infty, 6) \cup (-\infty, 1) = (-\infty, 6)$

6. $|x - 2| - 4 < 2$
$|x - 2| < 6$
$-6 < x - 2 < 6$
$-6 + 2 < x - 2 + 2 < 6 + 2$
$-4 < x < 8$
$\{x \mid -4 < x < 8\}$

7. $|2x - 3| > 5$
Solve each inequality.
$2x - 3 < -5$ ‎ or ‎ $2x - 3 > 5$
$2x < -2$ ‎ or ‎ ‎ ‎ ‎ $2x > 8$
$x < -1$ ‎ or ‎ ‎ ‎ ‎ $x > 4$
$\{x \mid x < -1\} \cup \{x \mid x > 4\}$
$\{x \mid x < -1 \text{ or } x > 4\}$

8. $f(x) = 3x^3 - 2x^2 + 1$
$f(-3) = 3(-3)^3 - 2(-3)^2 + 1$
$f(-3) = 3(-27) - 2(9) + 1$
$f(-3) = -98$

9. $f(x) = 3x^2 - 2x$
$f(-2) = 3(-2)^2 - 2(-2) = 16$
$f(-1) = 3(-1)^2 - 2(-1) = 5$
$f(0) = 3(0)^2 - 2(0) = 0$
$f(1) = 3(1)^2 - 2(1) = 1$
$f(2) = 3(2)^2 - 2(2) = 8$

The range is $\{0, 1, 5, 8, 16\}$.

10. $F(x) = x^2 - 3$

$F(2) = (2)^2 - 3 = 1$

11. $f(x) = 3x - 4$

$f(2 + h) = 3(2 + h) - 4 = 6 + 3h - 4 = 2 + 3h$

$f(2) = 3(2) - 4 = 6 - 4 = 2$

$f(2 + h) - f(2) = 2 + 3h - 2 = 3h$

12. $\{x \mid x \le 2\} \cap \{x \mid x > -3\}$

![number line from -5 to 5 with open circle at -3 and closed circle at 2, shaded between]

$$-5\ -4\ -3\ -2\ -1\ 0\ 1\ 2\ 3\ 4\ 5$$

13. $(-2, 3)$, $m = -\dfrac{2}{3}$

$y - y_1 = m(x - x_1)$

$y - 3 = -\dfrac{2}{3}(x - (-2))$

$y - 3 = -\dfrac{2}{3}x - \dfrac{4}{3}$

$y = -\dfrac{2}{3}x - \dfrac{4}{3} + 3$

$y = -\dfrac{2}{3}x + \dfrac{5}{3}$

14. The slope of the line $2x - 3y = 7$ is found by solving the equation for y.

$-3y = -2x + 7$

$y = \dfrac{2}{3}x - \dfrac{7}{3}$

The slope is $\dfrac{2}{3}$.

Use $(-1, 2)$ and the perpendicular slope $-\dfrac{3}{2}$

$y - y_1 = m(x - x_1)$

$y - 2 = -\dfrac{3}{2}(x - (-1))$

$y - 2 = -\dfrac{3}{2}x - \dfrac{3}{2}$

$y = -\dfrac{3}{2}x - \dfrac{3}{2} + 2$

$y = -\dfrac{3}{2}x + \dfrac{1}{2}$

15. The distance between two points is

$$d = \sqrt{(x_1 - x_2)^2 + (y_1 - y_2)^2}$$

$(-4, 2)$ and $(2, 0)$

$$d = \sqrt{(2 - (-4))^2 + (0 - 2)^2} = \sqrt{6^2 + (-2)^2}$$
$$= \sqrt{36 + 4} = \sqrt{40} \approx 6.32$$

16. $\left(\dfrac{x_1 + x_2}{2}, \dfrac{y_1 + y_2}{2} \right)$

$(-4, 3)$ and $(3, 5)$

$$\left(\dfrac{-4 + 3}{2}, \dfrac{3 + 5}{2} \right) = \left(-\dfrac{1}{2}, 4 \right)$$

17. $2x - 5y = 10$

$-5y = -2x + 10$

$y = \dfrac{2}{5}x - 2$

slope is $\dfrac{2}{5}$; y-intercept is -2

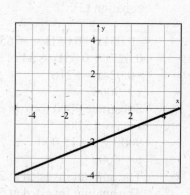

18. $3x - 4y \ge 8$

$-4y \ge -3x + 8$

$y \le \dfrac{3}{4}x - 2$

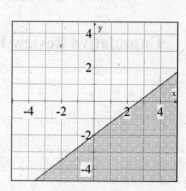

19. $5x - 2y = 10$
$3x + 2y = 6$

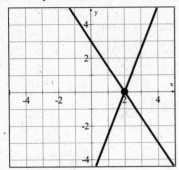

The solution is $(2, 0)$.

20. Solve each inequality for y.
$3x - 2y \geq 4$
$-2y \geq -3x + 4$
$y \leq \dfrac{3}{2}x - 2$

$x + y < 3$
$y < -x + 3$

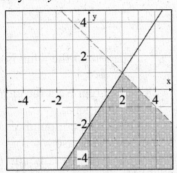

21. (1) $3x + 2z = 1$
(2) $2y - z = 1$
(3) $x + 2y = 1$

Multiply equation (2) by -1 and add to equation (3).
$-1(2y - z) = -1(1)$
$\quad x + 2y = 1$
$-2y + z = -1$
$x + 2y = 1$

(4) $x + z = 0$

Multiply equation (4) by -2 and add to equation (1).

$-2(x + z) = -2(0)$
$3x + 2z = 1$
$-2x - 2z = 0$
$3x + 2z = 1$
$x = 1$
Substitute 1 for x in equation (3).
$1 + 2y = 1$

$2y = 0$
$y = 0$
Substitute 0 for y in equation (2).
$2(0) - z = 1$
$-z = 1$
$z = -1$
The solution is $(1, 0, -1)$.

22. $\begin{vmatrix} 2 & -5 & 1 \\ 3 & 1 & 2 \\ 6 & -1 & 4 \end{vmatrix} = 2\begin{vmatrix} 1 & 2 \\ -1 & 4 \end{vmatrix} - 3\begin{vmatrix} -5 & 1 \\ -1 & 4 \end{vmatrix} + 6\begin{vmatrix} -5 & 1 \\ 1 & 2 \end{vmatrix}$

$= 2(4 + 2) - 3(-20 + 1) + 6(-10 - 1)$
$= 2(6) - 3(-19) + 6(-11)$
$= 12 + 57 - 66$
$= 3$

23. $4x - 3y = 17$
$3x - 2y = 12$

$D = \begin{vmatrix} 4 & -3 \\ 3 & -2 \end{vmatrix} = 1$

$D_x = \begin{vmatrix} 17 & -3 \\ 12 & -2 \end{vmatrix} = 2$

$D_y = \begin{vmatrix} 4 & 17 \\ 3 & 12 \end{vmatrix} = -3$

$x = \dfrac{D_x}{D} = \dfrac{2}{1} = 2$

$y = \dfrac{D_y}{D} = \dfrac{-3}{1} = -3$

The solution is $(2, -3)$.

24. (1) $3x - 2y = 7$
(2) $y = 2x - 1$

Solve by substitution
$3x - 2(2x - 1) = 7$
$3x - 4x + 2 = 7$
$-x + 2 = 7$
$-x = 5$
$x = -5$
Substitute -5 for x in equation (2).
$y = 2(-5) - 1$
$y = -10 - 1$
$y = -11$
The solution is $(-5, -11)$.

25. Let x represent the number of quarters.
The number of dimes is $3x$.
The number of nickels is $40 - (x + 3x)$.

Coin	Number	Value	Total Value
Quarters	x	25	$25x$
Dimes	$3x$	10	$10(3x)$
Nickels	$40 - 4x$	5	$5(40 - 4x)$

The sum of the total values of all the coins is $4.10 (410 cents).
$25x + 10(3x) + 5(40 - 4x) = 410$

Solution:
$$25x + 10(3x) + 5(40 - 4x) = 410$$
$$25x + 30x + 200 - 20x = 410$$
$$35x + 200 = 410$$
$$35x = 210$$
$$x = 6$$
$$40 - 4x = 40 - 4(6) = 40 - 24 = 16$$
There are 16 nickels in the purse.

26. Let x represent the amount of pure water.

	Amount	Percent	Quantity
Water	x	0	$0x$
4%	100	0.04	$0.04(100)$
2.5%	$100 + x$	0.025	$0.025(100 + x)$

The sum of the quantities before mixing is equal to the quantity after mixing.
$0x + 0.04(100) = 0.025(100 + x)$

Solution:
$$0x + 0.04(100) = 0.025(100 + x)$$
$$0 + 4 = 2.5 + 0.025x$$
$$1.5 = 0.025x$$
$$x = 60$$
The amount of water that should be added is 60 ml.

27. Strategy: Let x represent the rate of the plane in calm air.
The rate of the wind is y.

	Rate	Time	Distance
with wind	$x + y$	2	$2(x + y)$
against wind	$x - y$	3	$3(x - y)$

The distance traveled with the wind is 150 mi. The distance traveled against the wind is 150 mi.

$$2(x + y) = 150$$
$$3(x - y) = 150$$

Solution:
$$2(x + y) = 150$$
$$3(x - y) = 150$$
$$\frac{1}{2} \cdot 2(x + y) = \frac{1}{2} \cdot 150$$
$$\frac{1}{3} \cdot 3(x - y) = \frac{1}{3} \cdot 150$$
$$x + y = 75$$
$$x - y = 50$$
$$2x = 125$$
$$x = 62.5$$

$$x + y = 75$$
$$62.5 + y = 75$$
$$y = 12.5$$

The rate of the wind is 12.5 mph.

28. Let x represent the cost per pound of hamburger.
The cost per pound of steak is y.
First purchase:

	Amount	Cost	Quantity
Hamburger	100	x	$100x$
Steak	50	y	$50y$

Second purchase:

	Amount	Cost	Quantity
Hamburger	150	x	$150x$
Steak	100	y	$100y$

The total cost of the first purchase is $541. The total cost of the second purchase is $960.
$$100x + 50y = 540$$
$$150x + 100y = 960$$

Solution:
$$100x + 50y = 540$$
$$150x + 100y = 960$$

$$-2(100x + 50y) = -2(540)$$
$$150x + 100y = 960$$

$$-200x - 100y = -1080$$
$$150x + 100y = 960$$
$$-50x = -120$$
$$x = 2.4$$

$100(2.4) + 50y = 540$
$240 + 50y = 540$
$50y = 300$
$y = 6$
The cost of steak is $6.00/lb.

29. **Strategy:** Let M represent the number of
ohms, T the tolerance and r the value of the
resistor. Find the tolerance and solve
$|M - 12,000| \leq T$ for M.

Solution: $T = 0.15 \cdot 12,000 = 1800$ ohms
$|M - 12,000| \leq 1800$
$-1800 \leq M - 12,000 \leq 1800$
$-1800 + 12,000 \leq M - 12,000 + 12,000$
$\leq 1800 + 12,000$
$10,200 \leq M \leq 13,800$

The lower and upper limits of the resistor
are 10,200 ohms and 13,800 ohms.

30. The slope of the line is
$$\frac{5000 - 1000}{100 - 0} = \frac{4000}{100} = 40 .$$
The commission rate of the executive is $40
for every $1000 in sales.

Chapter 5: Polynomials

Prep Test

1. $-4(3y) = -12y$

2. $(-2)^3 = (-2)(-2)(-2) = -8$

3. $-4a - 8b + 7a = 3a - 8b$

4. $3x - 2[y - 4(x + 1) + 5]$
$= 3x - 2(y - 4x - 4 + 5)$
$= 3x - 2(y - 4x + 1)$
$= 3x - 2y + 8x - 2$
$= 11x - 2y - 2$

5. $-(x + y) = -x - y$

6. $40 = 2 \cdot 20$
$\quad = 2 \cdot 4 \cdot 5$
$\quad = 2 \cdot 2 \cdot 2 \cdot 5$

7. $16 = \underline{2 \cdot 2} \cdot 2 \cdot 2$
$20 = \underline{2 \cdot 2} \cdot 5$
$24 = \underline{2 \cdot 2} \cdot 2 \cdot 3$
The GCF is $2 \cdot 2$ or 4.

8. $x^3 - 2x^2 + x + 5$
$(-2)^3 - 2(-2)^2 + (-2) + 5$
$= -8 - 2(4) - 2 + 5$
$= -8 - 8 - 2 + 5$
$= -13$

9. $3x + 1 = 0$
$\quad 3x = -1$
$\quad x = -\dfrac{1}{3}$

The solution is $-\dfrac{1}{3}$.

Section 5.1

Objective A Exercises

1. $(ab^3)(a^3b) = a^4b^4$

3. $(9xy^2)(-2x^2y^2) = -18x^3y^4$

5. $(x^2y^4)^4 = x^8y^{16}$

7. $(-3x^2y^3)^4 = (-3)^4 x^8 y^{12} = 81x^4y^{12}$

9. $(27a^5b^3)^2 = (27)^2 a^{10}b^6 = 729a^{10}b^6$

11. $[(2a^4b^3)^3]^2 = (2a^4b^3)^6 = (2)^6 a^{24}b^{18} = 64a^{24}b^{18}$

13. $(x^2y^2)(xy^3)^3 = (x^2y^2)(x^3y^9) = x^5y^{11}$

15. $(-5ab)(3a^3b^2)^2 = (-5ab)(3^2a^6b^4)$
$\qquad\qquad\qquad = (-5ab)(9a^6b^4)$
$\qquad\qquad\qquad = -45a^7b^5$

17. $(3x^5y)(-4x^3)^3 = (3x^5y)((-4)^3x^9)$
$\qquad\qquad\qquad = (3x^5y)(-64x^9)$
$\qquad\qquad\qquad = -192x^{14}y$

19. $(-6a^4b^2)(-7a^2c^5) = 42a^6b^2c^5$

21. $(-2ab^2)(-3a^4b^5)^3 = (-2ab^2)((-3)^3a^{12}b^{15})$
$\qquad\qquad\qquad = (-2ab^2)(-27a^{12}b^{15})$
$\qquad\qquad\qquad = 54a^{13}b^{17}$

23. $(-3ab^3)^3(-2^2a^2b)^2 = ((-3)^3a^3b^9)(2^4a^4b^2)$
$\qquad\qquad\qquad = (-27a^3b^9)(16a^4b^2)$
$\qquad\qquad\qquad = -432a^7b^{11}$

25. $(-2x^2y^3z)(3x^2yz^4) = -6x^4y^4z^5$

27. $(2xy)(-3x^2yz)(x^2y^3z^3) = -6x^5y^5z^4$

29. $(3b^5)(2ab^2)(-2ab^2c^2) = -12a^2b^9c^2$

30. $x^n \cdot x^{n+1} = x^{n+n+1} = x^{2n+1}$

31. $y^{3n} \cdot y^{3n-2} = y^{3n+3-2} = y^{6n-2}$

35. $(a^{n-3})^{2n} = a^{2n^2-6n}$

37. $(x^{3n+2})^5 = x^{15n+10}$

39. The value of n is 33.

Objective B Exercises

41. $\dfrac{1}{3^{-5}} = 3^5 = 243$

43. $\dfrac{1}{y^{-3}} = y^3$

45. $\dfrac{a^3}{4b^{-2}} = \dfrac{a^3 b^2}{4}$

47. $xy^{-4} = \dfrac{x}{y^4}$

49. $\dfrac{1}{2x^0} = \dfrac{1}{2}$

51. $\dfrac{-3^{-2}}{(2y)^0} = \dfrac{-1}{3^2} = -\dfrac{1}{9}$

53. $(x^3 y^5)^{-2} = x^{-6} y^{-10} = \dfrac{1}{x^6 y^{10}}$

55. $(-3a^{-4} b^{-5})(-5a^{-2} b^4) = 15a^{-6} b^{-1} = \dfrac{15}{a^6 b}$

57. $(4y^{-3} z^{-4})(-3y^3 z^{-3})^{-2}$

$= (4y^{-3} z^{-4})((-3)^{-2} y^{-6} z^6)$

$= (4)(-3)^{-2} y^{-9} z^2 = \dfrac{4z^2}{(-3)^2 y^9}$

$= \dfrac{4z^2}{9y^9}$

59. $(4x^{-3} y^2)^{-3}(2xy^{-3})^4$

$= (4^{-3} x^9 y^{-6})(2^4 x^4 y^{-12})$

$= (4^{-3})(2)^4 x^{13} y^{-18} = \dfrac{2^4 x^{13}}{(4)^3 y^{18}}$

$= \dfrac{16x^{13}}{64y^{18}} = \dfrac{x^{13}}{4y^{18}}$

61. $\dfrac{9x^5}{12x^8} = \dfrac{3}{4x^3}$

63. $\dfrac{-6x^2 y}{12x^4 y} = -\dfrac{1}{2x^2}$

65. $\dfrac{y^{-2}}{y^6} = y^{-2-(6)} = y^{-8} = \dfrac{1}{y^8}$

67. $\dfrac{x^4 y^3}{x^{-1} y^{-2}} = x^5 y^5$

69. $\dfrac{a^{61} b^{-4}}{a^{-2} b^5} = a^8 b^{-9} = \dfrac{a^8}{b^9}$

71. $\dfrac{-3ab^2}{(9a^2 b^4)^3} = \dfrac{-3ab^2}{9^3 a^6 b^{12}} = \dfrac{-3ab^2}{729a^6 b^{12}}$

$= -\dfrac{1}{243a^5 b^{10}}$

73. $\dfrac{(3a^2 b)^3}{(-6ab^3)^2} = \dfrac{3^3 a^6 b^3}{(-6)^2 a^2 b^6} = \dfrac{27a^6 b^3}{36a^2 b^6} = \dfrac{3a^4}{4b^3}$

75. $\dfrac{(-8x^2 y^2)^4}{(16x^3 y^7)^2} = \dfrac{(-8)^4 x^4 y^8}{(16)^2 x^6 y^{14}} = \dfrac{4096x^4 y^8}{256x^6 y^{14}}$

$= \dfrac{16x^2}{y^6}$

77. $\dfrac{(3a^4 b^{-2})^{-2}}{(2a^{-3} b)^3} = \dfrac{(3)^{-2} a^{-8} b^4}{(2)^3 a^{-9} b^3} = \dfrac{ab}{(2)^3 (3)^2} = \dfrac{ab}{72}$

79. $\dfrac{(-2x^{-5} y^2)^{-3}}{(4xy^{-2})^{-4}} = \dfrac{(-2)^{-3} x^{15} y^{-6}}{(4)^{-4} x^{-4} y^8} = \dfrac{(4)^{-4} x^{19}}{(-2)^3 y^{14}}$

$= -\dfrac{32x^{19}}{y^{14}}$

81. $\dfrac{b^{6n}}{b^{10n}} = b^{6n-10n} = b^{-4n} = \dfrac{1}{b^{4n}}$

83. $\dfrac{y^{2n}}{-y^{8n}} = -y^{2n-8n} = -y^{-6n} = -\dfrac{1}{y^{6n}}$

85. $\dfrac{y^{3n+2}}{y^{2n+4}} = y^{3n+2-(2n+4)} = y^{n-2}$

87. $\dfrac{x^{2n-1}y^{n-3}}{x^{n+4}y^{n+3}} = \dfrac{x^{2n-1-(n+4)}}{y^{n+3-(n-3)}} = \dfrac{x^{n-5}}{y^{6}}$

89. $\left(\dfrac{9ab^{-2}}{8a^{-2}b}\right)^{-2}\left(\dfrac{3a^{-2}b}{2a^{2}b^{-2}}\right)^{3}$

$= \left(\dfrac{9^{-2}a^{-2}b^{4}}{8^{-2}a^{4}b^{-2}}\right)\left(\dfrac{3^{3}a^{-6}b^{3}}{2^{3}a^{6}b^{-6}}\right)$

$= \left(\dfrac{9^{-2}b^{6}}{8^{-2}a^{6}}\right)\left(\dfrac{3^{3}b^{9}}{2^{3}a^{12}}\right)$

$= \left(\dfrac{8^{2}b^{6}}{9^{2}a^{6}}\right)\left(\dfrac{3^{3}b^{9}}{2^{3}a^{12}}\right) = \dfrac{8b^{15}}{3a^{18}}$

91. The value of $p - q$ is 0.

Objective C Exercises

93. 4.67×10^{-6}

95. 1.7×10^{-10}

97. 2×10^{11}

99. 0.000000123

101. 8,200,000,000,000,000

103. 0.039

105. $(3 \times 10^{-12})(5 \times 10^{16})$
$= (3)(5) \times 10^{-12+16}$
$= 15 \times 10^{4}$
$= 150,000$

107. $(0.0000065)(3,200,000,000,000)$
$= (6.5 \times 10^{-6})(3.2 \times 10^{12})$
$= (6.5)(3.2) \times 10^{-6+12}$

$= 20.8 \times 10^{6}$
$= 20,800,000$

109. $\dfrac{9 \times 10^{-3}}{6 \times 10^{5}} = 1.5 \times 10^{-3-5}$
$= 1.5 \times 10^{-8} = 0.000000015$

111. $\dfrac{0.0089}{500,000,000} = \dfrac{8.9 \times 10^{-3}}{5 \times 10^{8}}$
$= 1.78 \times 10^{-3-8} = 1.78 \times 10^{-11}$
$= 0.0000000000178$

113. $\dfrac{(3.3 \times 10^{-11})(2.7 \times 10^{15})}{8.1 \times 10^{-3}}$
$= \dfrac{(3.3)(2.7) \times 10^{-11+15-(-3)}}{8.1}$
$= 1.1 \times 10^{7} = 11,000,000$

115. $\dfrac{(0.00000004)(84,000)}{(0.0003)(1,400,000)}$
$= \dfrac{4 \times 10^{-8} \times 8.4 \times 10^{4}}{3 \times 10^{-4} \times 1.4 \times 10^{6}}$
$= \dfrac{4(8.4) \times 10^{-8+4-(-4)-6}}{3(1.4)}$
$= 8 \times 10^{-6} = 0.000008$

117. Greater than zero

Objective D Exercises

119. **Strategy:** To find the number of years needed to cross the galaxy, divide the width of the galaxy by the product of the rate of the space ship and the number of hours in a year.
Solution:
$\dfrac{5.6 \times 10^{19}}{2.5 \times 10^{4} \times 8.76 \times 10^{3}} \approx 2.6 \times 10^{11}$
It would takes a space ship 2.6×10^{11} years to cross the galaxy.

121. **Strategy:** To find the number of times larger the mass of the proton is, divide the mass of the proton by the mass of an electron.

Solution: $\dfrac{1.673 \times 10^{-27}}{9.109 \times 10^{-31}} \approx 1.837 \times 10^{3}$

The mass of a proton is approximately 1.837×10^{3} times larger than an electron.

123. **Strategy:** To find the number of times larger the mass of the sun is, divide the mass of the sun by the mass of Earth.

Solution: $\dfrac{2 \times 10^{30}}{5.9 \times 10^{24}} \approx 3.39 \times 10^{5}$

The sun is approximately 3.39×10^{5} times larger than Earth.

125. **Strategy:** To find the rate of the signals divide the distance by the time.

Solution:

$\dfrac{119,000,000}{11} = \dfrac{1.19 \times 10^{8}}{1.1 \times 10^{1}} = 1.08\overline{1} \times 10^{7}$

The signal travels $1.08\overline{1} \times 10^{7}$ mi/min.

127. **Strategy:** To find the number of seeds produced, divide the number of seeds by the number of pine seedlings growing.

Solution:

$\dfrac{2,000,000}{12,000} = \dfrac{2 \times 10^{6}}{1.2 \times 10^{4}} = 1.\overline{6} \times 10^{2}$

$1.\overline{6} \times 10^{2}$ seeds are produced.

Applying the Concepts

129. a) $x^{0} = 0$ is incorrect.
   $x^{0} = 1$ is correct.
   The Definition of Zero as an Exponent was used incorrectly.

   b) $(x^{4})^{5} = x^{9}$ is incorrect.
   $(x^{4})^{5} = x^{20}$ is correct.
   The Rule for Simplifying the Power of an Exponential Expression was used incorrectly.

   c) $x^{2} \cdot x^{3} = x^{6}$ is incorrect.
   $x^{2} \cdot x^{3} = x^{5}$ is correct.

The Rule for Multiplying Exponential Expressions was used incorrectly.

Section 5.2

Objective A Exercises

1. $P(x) = 3x^{2} - 2x - 8$
$P(3) = 3(3)^{2} - 2(3) - 8$
$P(3) = 13$

3. $R(x) = 2x^{3} - 3x^{2} + 4x - 2$
$R(2) = 2(2)^{3} - 3(2)^{2} + 4(2) - 2$
$R(2) = 10$

5. $f(x) = x^{4} - 2x^{2} - 10$
$f(-1) = (-1)^{4} - 2(-1)^{2} - 10$
$f(-1) = -11$

7. Polynomial: (a) -1 (b) 8 (c) 2

9. This is not a polynomial. The terms are not monomials.

11. This is not a polynomial. The terms are not monomials.

13. Polynomial: (a) 3 (b) π (c) 5

15. Polynomial: (a) -5 (b) 2 (c) 3

17. Polynomial: (a) 14 (b) 14 (c) 0

19. $P(x) = x^{2} - 1$

x	y
-2	3
-1	0
0	-1
1	0
2	3

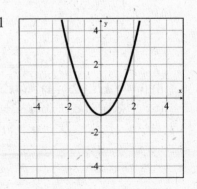

21. $R(x) = x^3 + 2$

x	y
-2	-6
-1	1
0	2
1	3
2	10

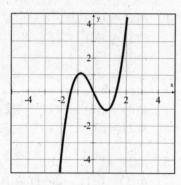

23. $f(x) = x^3 - 2x$

x	y
-2	-4
-1	1
0	0
1	-1
2	4

25. (a) $f(c) - g(c) > 0$

    (b) $f(c) - g(c) > 0$

Objective B Exercises

27. $\begin{aligned}5x^2 + 2x - 7 \\ \underline{x^2 - 8x + 12} \\ 6x^2 - 6x + 5\end{aligned}$

Applying the Concepts

41. a) $(2x^3 + 3x^2 + kx + 5) - (x^3 + 2x^2 + 3x + 7) = x^3 + x^2 + 5x - 2$

$(2x^3 - x^3) + (3x^2 - 2x^2) + (kx - 3x) + (5 - 7) = x^3 + x^2 + 5x - 2$

$x^3 + x^2 + (k-3)x - 2 = x^3 + x^2 + 5x - 2$

$(k-3)x = 5x$

$k - 3 = 5$

$k = 8$

b) $(6x^3 + kx^2 - 2x - 1) - (4x^3 - 3x^2 + 1) = 2x^3 - x^2 - 2x - 2$

$(6x^3 - 4x^3) + (kx^2 + 3x^2) - 2x + (-1 - 1) = 2x^3 - x^2 - 2x - 2$

$2x^3 + (k+3)x^2 - 2x - 6 = 2x^3 - x^2 - 2x - 2$

$(k+3)x^2 = -x^2$

$k + 3 = -1$

$k = -4$

29. $\begin{aligned}x^2 - 3x + 8 \\ -2x^2 + 3x - 7 \\ \underline{-x^2 \qquad\ \ + 1}\end{aligned}$

31. $(3y^2 - 7y) + (2y^2 - 8y + 2)$
$= (3y^2 + 2y^2) + (-7y - 8y) + 2$
$= 5y^2 - 15y + 2$

33. $(2a^2 - 3a - 7) - (-5a^2 - 2a - 9)$
$= (2a^2 + 5a^2) + (-3a + 2a) + (-7 + 9)$
$= 7a^2 - a + 2$

35. $P(x) + R(x) = (x^2 - 3xy + y^2) + (2x^2 - 3y^2)$
$= (x^2 + 2x^2) - 3xy + (y^2 - 3y^2)$
$= 3x^2 - 3xy - 2y^2$

37. $P(x) - R(x)$
$= (3x^2 + 2y^2) - (-5x^2 + 2xy - 3y^2)$
$= (3x^2 + 5x^2) - 2xy + (2y^2 + 3y^2)$
$= 8x^2 - 2xy + 5y^2$

39. $S(x) = P(x) + R(x)$
$= (3x^4 - 3x^3 - x^2) + (3x^3 - 7x^2 + 2x)$
$= 3x^4 + (-3x^3 + 3x^3) + (-x^2 - 7x^2) + 2x$
$S(x) = 3x^4 - 8x^2 + 2x$
$S(2) = 3(2)^4 - 8(2)^2 + 2(2)$
$S(2) = 20$

43. If $P(x)$ is a fifth degree polynomial and $Q(x)$ is a fourth degree polynomial, then $P(x) -$ $Q(x)$ is a fifth degree polynomial.

Example: $P(x) = 3x^5 - 5x - 8$

$\quad\quad Q(x) = 3x^4 - 2x + 1$

$P(x) - Q(x) = 3x^5 - 3x^4 - 3x - 9$

Section 5.3

Objective A Exercises

1. The Distributive Property

3. $2x(x - 3) = 2x^2 - 6x$

5. $3x^2(2x^2 - x) = 6x^4 - 3x^3$

7. $3xy(2x - 3y) = 6x^2y - 9xy^2$

9. $x^n(x + 1) = x^{n+1} + x^n$

11. $x^n(x^n + y^n) = x^{2n} + x^ny^n$

13. $2b + 4b(2 - b) = 2b + 8b - 4b^2 = -4b^2 + 10b$

15. $-2a^2(3a^2 - 2a + 3) = -6a^4 + 4a^3 - 6a^2$

17. $(-3y^2 - 4y + 2)(y^2) = -3y^4 - 4y^3 + 2y^2$

19. $-5x^2(4 - 3x + 3x^2 + 4x^3)$
$\quad = -20x^2 + 15x^3 - 15x^4 - 20x^5$

21. $-2x^2y(x^2 - 3xy + 2y^2) = -2x^4y + 6x^3y^2 - 4x^2y^3$

23. $5x^3 - 4x(2x^2 + 3x - 7)$
$\quad = 5x^3 - 8x^3 - 12x^2 + 28$
$\quad = -3x^3 - 12x^2 + 28$

25. $2y^2 - y[3 - 2(y - 4) - y]$
$\quad = 2y^2 - y(3 - 2y + 8 - y)$
$\quad = 2y^2 - y(11 - 3y)$
$\quad = 2y^2 - 11y + 3y^2$
$\quad = 5y^2 - 11y$

27. $2y - 3[y - 2y(y - 3) + 4y]$
$\quad = 2y - 3(y - 2y^2 + 6y + 4y)$
$\quad = 2y - 3(-2y^2 + 11y)$
$\quad = 2y + 6y^2 - 33y$
$\quad = 6y^2 - 31y$

29. $P(b) = 3b$ and $Q(b) = 3b^4 - 3b^2 + 8$
$\quad P(b){\cdot}Q(b) = 3b(3b^4 - 3b^2 + 8)$
$\quad\quad = 9b^5 - 9b^3 + 24b$

Objective B Exercises

31. $(x - 2)(x + 7) = x^2 + 7x - 2x - 14$
$\quad\quad = x^2 + 5x - 14$

33. $(2y - 3)(4y + 7) = 8y^2 + 14y - 12y - 21$
$\quad\quad = 8y^2 + 2y - 21$

35. $(a + 3c)(4a - 5c) = 4a^2 - 5ac + 12ac - 15c^2$
$\quad\quad = 4a^2 + 7ac - 15c^2$

37. $(5x - 7)(5x - 7) = 25x^2 - 35x - 35x + 49$
$\quad\quad = 25x^2 - 70x + 49$

39. $2(2x - 3y)(2x + 5y)$
$\quad = 2(4x^2 + 10xy - 6xy - 15y^2)$
$\quad = 2(4x^2 + 4xy - 15y^2)$
$\quad = 8x^2 + 8xy - 30y^2$

41. $(xy + 4)(xy - 3) = x^2y^2 - 3xy + 4xy - 12$
$\quad\quad = x^2y^2 + xy - 12$

43. $(2x^2 - 5)(x^2 - 5) = 2x^4 - 10x^2 - 5x^2 + 25$
$\quad\quad = 2x^4 - 15x^2 + 25$

45. $(5x^2 - 5y)(2x^2 - y)$
$\quad = 10x^4 - 5x^2y - 10x^2y + 5y^2$
$\quad = 10x^4 - 15x^2y + 5y^2$

47. $(x + 5)(x^2 - 3x + 4)$
$\quad = x(x^2 - 3x + 4) + 5(x^2 - 3x + 4)$
$\quad = x^3 - 3x^2 + 4x + 5x^2 - 15x + 20$
$\quad = x^3 + 2x^2 - 11x + 20$

49. $(2a - 3b)(5a^2 - 6ab + 4b^2)$

$= 2a(5a^2 - 6ab + 4b^2) - 3b(5a^2 - 6ab + 4b^2)$

$= 10a^3 - 12a^2b + 8ab^2 - 15a^2b + 18ab^2 - 12b^3$

$= 10a^3 - 27a^2b + 26ab^2 - 12b^3$

51. $(2x^3 + 3x^2 - 2x + 5)(2x - 3)$

$= (2x^3 + 3x^2 - 2x + 5)2x + (2x^3 + 3x^2 - 2x + 5)(-3)$

$= 4x^4 + 6x^3 - 4x^2 + 10x - 6x^3 - 9x^2 + 6x - 15$

$= 4x^4 - 13x^2 + 16x - 15$

53. $(2x - 5)(2^{x4} - 3^{x3} - 2x + 9)$

$= 2x(2^{x4} - 3^{x3} - 2x + 9) - 5(2^{x4} - 3^{x3} - 2x + 9)$

$= 4^{x5} - 6^{x4} - 4^{x2} + 18x - 10^{x4} + 15^{x3} + 10x - 45$

$= 4^{x5} - 16^{x4} + 15^{x3} - 4^{x2} + 28x - 45$

55. $(x^2 + 2x - 3)(x^2 - 5x + 7)$

$= x^2(x^2 - 5x + 7) + 2x(x^2 - 5x + 7) - 3(x^2 - 5x + 7)$

$= x^4 - 5x^3 + 7x^2 + 2x^3 - 10x^2 + 14x - 3x^2 + 15x - 21$

$= x^4 - 3x^3 - 6x^2 + 29x - 21$

57. $(a - 2)(2a - 3)(a + 7)$

$= (2a^2 - 3a - 4a + 6)(a + 7)$

$= (2a^2 - 7a + 6)(a + 7)$

$= (2a^2 - 7a + 6)a + (2a^2 - 7a + 6)7$

$= 2a^3 - 7a^2 + 6a + 14a^2 - 49a + 42$

$= 2a^3 + 7a^2 - 43a + 42$

59. $(2x + 3)(x - 4)(3x + 5)$

$= (2x^2 - 8x + 3x - 12)(3x + 5)$

$= (2x^2 - 5x - 12)(3x + 5)$

$= (2x^2 - 5x - 12)3x + (2x^2 - 5x - 12)5$

$= 6x^3 - 15x^2 - 36x + 10x^2 - 25x - 60$

$= 6x^3 - 5x^2 - 61x - 60$

61. $P(y) = 2y^2 - 1$ and $Q(y) = y^3 - 5y^2 - 3$

$P(y) \cdot Q(y) = (2y^2 - 1)(y^3 - 5y^2 - 3)$

$= 2y^2(y^3 - 5y^2 - 3) - 1(y^3 - 5y^2 - 3)$

$= 2y^5 - 10y^4 - 6y^2 - y^3 + 5y^2 + 3$

$= 2y^5 - 10y^4 - y^3 - y^2 + 3$

63. 5

Objective C Exercises

65. $(3x - 2)(3x + 2) = 9x^2 - 4$

67. $(6 - x)(6 + x) = 36 - x^2$

69. $(2a - 3b)(2a + 3b) = 4a^2 - 9b^2$

71. $(3ab + 4)(3ab - 4) = 9a^2b^2 - 16$

73. $(x^2 + 1)(x^2 - 1) = x^4 - 1$

75. $(x - 5)^2 = x^2 - 10x + 25$

77. $(3a + 5b)^2 = 9a^2 + 30ab + 25b^2$

79. $(x^2 - 3)^2 = x^4 - 6x^2 + 9$

81. $(2x^2 - 3y^2)^2 = 4x^4 + 12x^2y^2 + 9y^4$

83. $(3mn - 5)^2 = 9m^2n^2 - 30mn + 25$

85. $y^2 - (x - y)^2 = y^2 - (x^2 - 2xy + y^2)$

$= y^2 - x^2 + 2xy - y^2$

$= -x^2 + 2xy$

87. $(x - y)^2 - (x + y)^2$

$= x^2 - 2xy + y^2 - (x^2 + 2xy + y^2)$

$= x^2 - 2xy + y^2 - x^2 - 2xy - y^2$

$= -4xy$

89. False

Objective D Exercises

91. ft^2

93. Strategy: To find the area, substitute the given values for L and W in the equation $A = L \cdot W$ and solve for A.

Solution: $A = L \cdot W$
$A = (3x - 2)(x + 4)$
$A = 3x^2 + 12x - 2x - 8$
$A = 3x^2 + 10x - 8$
The area is $(3x^2 + 10x - 8)$ ft^2.

95. Strategy: To find the area, add the area of the small rectangle to the area of the large rectangle.
Larger rectangle:
Length $= L_1 = x + 5$
Width $= W_1 = x - 2$
Smaller rectangle:
Length $= L_2 = 5$
Width $= W_2 = 2$

Solution: $A = L_1 \cdot W_1 + L_2 \cdot W_2$
$A = (x + 5)(x - 2) + (5)(2)$
$A = x^2 - 2x + 5x - 10 + 10$
$A = x^2 + 3x$
The area is $(x^2 + 3x)$ ft^2.

97. Strategy: To find the volume, substitute the given value for s in the equation $V = s^3$ and solve for V.

Solution: $V = s^3$
$V = (x + 3)^3$
$V = (x + 3)(x + 3)(x + 3)$
$V = (x^2 + 6x + 9)(x + 3)$
$V = x^3 + 9x^2 + 27x + 27$
The volume is $(x^3 + 9x^2 + 27x + 27)$ cm^3.

99. Strategy: To find the volume, subtract the volume of the small rectangular solid from the volume of the large rectangular solid.
Large rectangular solid:
Length $= L_1 = x + 2$
Width $= W_1 = 2x$
Height $= h_1 = x$

Small rectangular solid:
Length $= L_2 = x$
Width $= W_2 = 2x$
Height $= h_2 = 2$

Solution: $V = (L_1 \cdot W_1 \cdot h_1) - (L_2 \cdot W_2 \cdot h_2)$
$V = (x + 2)(2x)(x) - (x)(2x)(2)$
$V = (2x^2 + 4x)(x) - (2x^2)(2)$
$V = 2x^3 + 4x^2 - 4x^2$
$V = 2x^3$
The volume is $(2x^3)$ in^3.

101. Strategy: To find the area, substitute the given value for r into the equation $A = \pi r^2$ and solve for A.

Solution: $A = \pi r^2$
$A = 3.14(5x + 4)^2$
$A = 3.14(25x^2 + 40x + 16)$
$A = 78.5x^2 + 125.6x + 50.24$
The area is $(78.5x^2 + 125.6x + 50.24)$ in^2.

Applying the Concepts

103.   a) $(a - b)(a^2 + ab + b^2)$
$= a(a^2 + ab + b^2) - b(a^2 + ab + b^2)$
$= a^3 + a^2b + ab^2 - a^2b - ab^2 - b^3$
$= a^3 - b^3$

   b) $(x + y)(x^2 - xy + y^2)$
$= x(x^2 - xy + y^2) + y(x^2 - xy + y^2)$
$= x^3 - x^2y + xy^2 + x^2y - xy^2 + y^3$
$= x^3 + y^3$

105.   a) $(3x - k)(2x + k) = 6x^2 + 5x + k^2$
$6x^2 + xk - k^2 = 6x^2 + 5x + k^2$
$xk = 5x$
$k = 5$

   b) $(4x + k)^2 = 16x^2 + 8x + k^2$
$16x^2 + 8xk + k^2 = 16x^2 + 8x + k^2$
$8xk = 8x$
$k = 1$

107.   The product of $(4a + b)$ and $(2a - b)$:
$(4a + b)(2a - b) = 8a^2 - 2ab - b^2$.
Subtract the product from $9a^2 - 2ab$.
$9a^2 - 2ab - (8a^2 - 2ab - b^2)$
$= 9a^2 - 2ab - 8a^2 + 2ab + b^2$
$= a^2 + b^2$

Section 5.4

Objective A Exercises

1. $\dfrac{3x^2 - 6x}{3x} = \dfrac{3x^2}{3x} - \dfrac{6x}{3x} = x - 2$

3. $\dfrac{5x^2 - 10x}{-5x} = \dfrac{5x^2}{-5x} - \dfrac{10x}{-5x} = -x + 2$

5. $\dfrac{5x^2 y^2 + 10xy}{5xy} = \dfrac{5x^2 y^2}{5xy} + \dfrac{10xy}{5xy} = xy + 2$

7. $\dfrac{x^3 + 3x^2 - 5x}{x} = \dfrac{x^3}{x} + \dfrac{3x^2}{x} - \dfrac{5x}{x} = x^2 + 3x - 5$

9. $\dfrac{9b^5 + 12b^4 + 6b^3}{3b^2} = \dfrac{9b^5}{3b^2} + \dfrac{12b^4}{3b^2} + \dfrac{6b^3}{3b^2}$
$= 3b^3 + 4b^2 + 2b$

11. $\dfrac{a^5 b - 6a^3 b + ab}{ab} = \dfrac{a^5 b}{ab} - \dfrac{6a^3 b}{ab} + \dfrac{ab}{ab}$
$= a^4 - 6a^2 + 1$

13. $P(x) = 6x^3 + 21x^2 - 15x$

Objective B Exercises

15.
$$
\begin{array}{r}
x + 8 \\
x - 5 \overline{)\, x^2 + 3x - 40} \\
\underline{x^2 - 5x} \\
8x - 40 \\
\underline{8x - 40} \\
0
\end{array}
$$

$(x^2 + 3x - 40) \div (x - 5) = x + 8$

17.
$$
\begin{array}{r}
x^2 + 3x + 6 \\
x - 3 \overline{)\, x^3 + 0x^2 - 3x + 2} \\
\underline{x^3 - 3x^2} \\
3x^2 - 3x \\
\underline{3x^2 - 9x} \\
6x + 2 \\
\underline{6x - 18} \\
20
\end{array}
$$

$(x^3 - 3x + 2) \div (x - 3) = x^2 + 3x + 6 + \dfrac{20}{x - 3}$

19.
$$
\begin{array}{r}
3x + 5 \\
2x + 1 \overline{)\, 6x^2 + 13x + 8} \\
\underline{6x^2 + 3x} \\
10x + 8 \\
\underline{10x + 5} \\
3
\end{array}
$$

$(6x^2 + 13x + 8) \div (2x + 1) = 3x + 5 + \dfrac{3}{2x + 1}$

21.
$$
\begin{array}{r}
5x + 7 \\
2x - 1 \overline{)\, 10x^2 + 9x - 5} \\
\underline{10x^2 - 5x} \\
14x - 5 \\
\underline{14x - 7} \\
2
\end{array}
$$

$(10x^2 + 9x - 5) \div (2x - 1) = 5x + 7 + \dfrac{2}{2x - 1}$

23.

$$
2x-3 \overline{\smash{\big)}\, 8x^3 + 0x^2 + 0x - 9} \quad \overset{\textstyle 4x^2 + 6x + 9}{}
$$

$$
\underline{8x^3 - 12x^2}
$$
$$
12x^2 + 0x
$$
$$
\underline{12x^2 - 18x}
$$
$$
18x - 9
$$
$$
\underline{18x - 27}
$$
$$
18
$$

$$(8x^3 - 9) \div (2x - 3) = 4x^2 + 6x + 9 + \frac{18}{2x-3}$$

25.

$$
2x^2 - 5 \overline{\smash{\big)}\, 6x^4 + 0x^3 - 13x^2 + 0x - 4} \quad \overset{\textstyle 3x^2 + 1}{}
$$

$$
\underline{6x^4 \qquad -15x^2}
$$
$$
2x^2 + 0x - 4
$$
$$
\underline{2x^2 \qquad -5}
$$
$$
1
$$

$$(6x^4 - 13x^2 - 4) \div (2x^2 - 5)$$
$$= 3x^2 + 1 + \frac{1}{2x^2 - 5}$$

27.

$$
3x + 1 \overline{\smash{\big)}\, 3x^3 - 8x^2 - 33x - 10} \quad \overset{\textstyle x^2 - 3x - 10}{}
$$

$$
\underline{3x^3 + x^2}
$$
$$
-9x^2 - 33x
$$
$$
\underline{-9x^2 - 3x}
$$
$$
-30x - 10
$$
$$
\underline{-30x - 10}
$$
$$
0
$$

$$\frac{3x^3 - 8x^2 - 33x - 10}{3x + 1} = x^2 - 3x - 10$$

29.

$$
x - 3 \overline{\smash{\big)}\, x^3 - 5x^2 + 7x - 4} \quad \overset{\textstyle x^2 - 2x + 1}{}
$$

$$
\underline{x^3 - 3x^2}
$$
$$
-2x^2 + 7x
$$
$$
\underline{-2x^2 + 6x}
$$
$$
x - 4
$$
$$
\underline{x - 3}
$$
$$
-1
$$

$$\frac{x^3 - 5x^2 + 7x - 4}{x - 3} = x^2 - 2x + 1 - \frac{1}{x-3}$$

31.

$$
x - 5 \overline{\smash{\big)}\, 2x^4 - 13x^3 + 16x^2 - 9x + 20} \quad \overset{\textstyle 2x^3 - 3x^2 + x - 4}{}
$$

$$
\underline{2x^4 - 10x^3}
$$
$$
-3x^3 + 16x^2
$$
$$
\underline{-3x^3 + 15x^2}
$$
$$
x^2 - 9x
$$
$$
\underline{x^2 - 5x}
$$
$$
-4x + 20
$$
$$
\underline{-4x + 20}
$$
$$
0
$$

$$\frac{2x^4 - 13x^3 + 16x^2 - 9x + 20}{x - 5} = 2x^3 - 3x^2 + x - 4$$

33.

$$
x^2 + 2x - 1 \overline{\smash{\big)}\, 2x^3 + 4x^2 - x + 2} \quad \overset{\textstyle 2x}{}
$$

$$
\underline{2x^3 + 4x^2 - 2x}
$$
$$
x + 2
$$

$$\frac{2x^3 + 4x^2 - x + 2}{x^2 + 2x - 1} = 2x + \frac{x + 2}{x^2 + 2x - 1}$$

35.

$$x^2 - 2x - 1 \overline{\smash{\big)}\ x^4 + 2x^3 - 3x^2 - 6x + 2}$$ with quotient $x^2 + 4x + 6$

$$\underline{x^4 - 2x^3 - x^2}$$
$$4x^3 - 2x^2 - 6x$$
$$\underline{4x^3 - 8x^2 - 4x}$$
$$6x^2 - 2x + 2$$
$$\underline{6x^2 - 12x - 6}$$
$$10x + 8$$

$$\frac{x^4 + 2x^3 - 3x^2 - 6x + 2}{x^2 - 2x - 1} = x^2 + 4x + 6 + \frac{10x + 8}{x^2 - 2x - 1}$$

37. $\dfrac{P(x)}{Q(x)} = \dfrac{2x^3 + x^2 + 8x + 7}{2x + 1}$

$$2x + 1 \overline{\smash{\big)}\ 2x^3 + x^2 + 8x + 7}$$ with quotient $x^2 + 4$

$$\underline{2x^3 + x^2}$$
$$8x + 7$$
$$\underline{8x + 4}$$
$$3$$

$$\frac{2x^3 + x^2 + 8x + 7}{2x + 1} = x^2 + 4 + \frac{3}{2x + 1}$$

39. False

Objective C Exercise

41. 3

43.

-1	2	-6	-8
		-2	8
	2	-8	0

$$(2x^2 - 6x - 8) \div (x + 1) = 2x - 8$$

45.

-1	3	3	-1	3	2
		-3	0	1	-4
	3	0	-1	4	-2

$$(3x^2 - 14x + 16) \div (x - 2) = 3x - 8$$

47.

1	3	0	-4
		3	3
	3	3	-1

$$(4x^2 - 23x + 28) \div (x - 4) = 4x - 7$$

49.

-1	2	-1	6	9
		-2	3	-9
	2	-3	9	0

$$(4x^2 - 8) \div (x - 2) = 4x + 8 + \frac{8}{x - 2}$$

51.

2	4	0	-1	-18
		8	16	30
	4	8	15	12

$$(4x^3 - x - 18) \div (x - 2) = 4x^2 + 8x + 15 + \frac{12}{x - 2}$$

53.

-4	2	5	-5	20
		-8	12	-28
	2	-3	7	-8

$$(2x^3 + 5x^2 - 5x + 20) \div (x + 4) = 2x^2 - 3x + 7 - \frac{8}{x + 4}$$

55.

2	3	-4	8	-5	-5
		6	4	24	38
	3	2	12	19	33

$$(3x^4 - 4x^3 + 8x^2 - 5x - 5) \div (x - 2)$$
$$= 3x^3 + 2x^2 + 12x + 19 + \frac{33}{x - 2}$$

57.

2	3	-14	16
		6	-16
	3	-8	0

$$(3x^4 + 3x^3 - x^2 + 3x + 2) \div (x + 1)$$
$$= 3x^3 - x + 4 - \frac{2}{x + 1}$$

59. $\dfrac{P(x)}{Q(x)} = \dfrac{3x^2 - 5x + 6}{x - 2}$

$$
\begin{array}{r|rrr}
2 & 3 & -5 & 6 \\
  &   & 6 & 2 \\
\hline
  & 3 & 1 & 8
\end{array}
$$

$$(3x^2 - 5x + 6) \div (x - 2) = 3x + 1 + \dfrac{8}{x - 2}$$

Objective D Exercises

61. $x - 3$

63.
$$
\begin{array}{r|rrr}
3 & 2 & -3 & -1 \\
  &   & 6 & 9 \\
\hline
  & 2 & 3 & 8
\end{array}
$$

$P(3) = 8$

65.
$$
\begin{array}{r|rrrr}
4 & 1 & -2 & 3 & -1 \\
  &   & 4 & 8 & 44 \\
\hline
  & 1 & 2 & 11 & 43
\end{array}
$$

$R(4) = 43$

67.
$$
\begin{array}{r|rrrr}
-2 & 2 & -4 & 3 & -1 \\
   &   & -4 & 16 & -38 \\
\hline
   & 2 & -8 & 19 & -39
\end{array}
$$

$P(-2) = -39$

69.
$$
\begin{array}{r|rrrr}
-3 & 2 & -1 & 0 & 3 \\
   &   & -6 & 21 & -63 \\
\hline
   & 2 & -7 & 21 & -60
\end{array}
$$

$Z(-3) = -60$

71.
$$
\begin{array}{r|rrrrr}
2 & 1 & 3 & -2 & 4 & -9 \\
  &   & 2 & 10 & 16 & 40 \\
\hline
  & 1 & 5 & 8 & 20 & 31
\end{array}
$$

$Q(2) = 31$

73.
$$
\begin{array}{r|rrrrr}
-3 & 2 & -1 & 0 & 2 & -5 \\
   &   & -6 & 21 & -63 & 183 \\
\hline
   & 2 & -7 & 21 & -61 & 178
\end{array}
$$

$F(-3) = 178$

75.
$$
\begin{array}{r|rrrr}
5 & 1 & 0 & 0 & -3 \\
  &   & 5 & 25 & 125 \\
\hline
  & 1 & 5 & 25 & 122
\end{array}
$$

$P(5) = 122$

77.
$$
\begin{array}{r|rrrrr}
-3 & 4 & 0 & -3 & 0 & 5 \\
   &   & -12 & 36 & -99 & 297 \\
\hline
   & 4 & -12 & 33 & -99 & 302
\end{array}
$$

$R(-3) = 302$

79.
$$
\begin{array}{r|rrrrr}
3 & 1 & -2 & -3 & -1 & 7 \\
  &   & 3 & 3 & 0 & -3 \\
\hline
  & 1 & 1 & 0 & -1 & 4
\end{array}
$$

Applying the Concepts

81. a)

$$
\begin{array}{r}
a^2 - ab + b^2 \\
a + b\overline{)a^3 \qquad\qquad\quad + b^3} \\
\underline{a^3 + a^2 b} \\
-a^2 b \\
\underline{-a^2 b - ab^2} \\
ab^2 + b^3 \\
\underline{ab^2 + b^3} \\
0
\end{array}
$$

$$\dfrac{a^3 + b^3}{a + b} = a^2 - ab + b^2$$

b)

$$x + y \overline{)\,x^5 \qquad\qquad\qquad + y^5\,}$$

$$x^4 - x^3 y + x^2 y^2 - xy^3 + y^4$$

$$\underline{x^5 + x^4 y}$$
$$-x^4 y$$
$$\underline{-x^4 y - x^3 y^2}$$
$$x^3 y^2$$
$$\underline{x^3 y^2 + x^2 y^3}$$
$$-x^2 y^3$$
$$\underline{-x^2 y^3 - xy^4}$$
$$xy^4 + y^5$$
$$\underline{xy^4 + y^5}$$
$$0$$

c)

$$x + y \overline{)\,x^6 \qquad\qquad\qquad\qquad - y^6\,}$$

$$x^5 - x^4 y + x^3 y^2 - x^2 y^3 + xy^4 - y^5$$

$$\underline{x^6 + x^5 y}$$
$$-x^5 y$$
$$\underline{-x^5 y - x^4 y^2}$$
$$x^4 y^2$$
$$\underline{x^4 y^2 + x^3 y^3}$$
$$-x^3 y^3$$
$$\underline{-x^3 y^3 - x^2 y^4}$$
$$x^2 y^4$$
$$\underline{x^2 y^4 + xy^5}$$
$$-xy^5 - y^6$$
$$\underline{-xy^5 - y^6}$$

Divide both the dividend and the divisor by a. The divisor is now in the form $x + \dfrac{b}{a}$ and the expression $\dfrac{-b}{a}$ can be used for a in the $x - a$ of synthetic division.

Section 5.5

Objective A Exercises

1. The GCF of $6a^2$ and $15a$ is $3a$.
   $6a^2 - 15a = 3a(2a - 5)$

3. The GCF of $4x^3$ and $3x^2$ is x^2.
   $4x^3 - 3x^2 = x^2(4x - 3)$

5. There is no common factor.

7. The GCF of x^5, x^3 and x is x.
   $x^5 - x^3 - x = x(x^4 - x^2 - 1)$

9. The GCF of $16x^2$, $12x$ and 24 is 4.
   $16x^2 - 12x + 24 = 4(4x^2 - 3x + 6)$

11. The GCF of $25b^4$, $10b^3$ and $5b^2$ is $5b^2$.
    $5b^2 - 10b^3 + 25b^4 = 5b^2(1 - 2b + 5b^2)$

13. The GCF of x^{2n}, and x^n is x^n.
    $x^{2n} - x^n = x^n(x^n - 1)$

15. The GCF of x^{2n}, and x^n is x^n.
    $x^{2n} - x^n = x^n(x^n - 1)$

17. The GCF of a^{2n+2} and a^2 is a^2.
    $a^{2n+2} + a^2 = a^2(a^{2n} + 1)$

19. The GCF of $12x^2 y^2$, $18x^3 y$ and $24x^2 y$ is $6x^2 y$.
    $12x^2 y^2 - 18x^3 y + 24x^2 y = 6x^2 y(2y - 3x + 4)$

21. The GCF of $24a^3 b^2$, $4a^2 b^2$ and $16a^2 b^4$ is $4a^2 b^2$.
    $24a^3 b^2 - 4a^2 b^2 - 16a^2 b^4$
    $\quad = 4a^2 b^2(6a - 1 - 4b^2)$

23. $3x, 2x + 1$

83. Synthetic division can be modified so that the divisor may be of the form $ax + b$.

Objective B Exercises

25. $x(a + 2) - 2(a + 2) = (a + 2)(x - 2)$

27. $a(x - 2) - b(2 - x) = a(x - 2) + b(x - 2)$
$= (x - 2)(a + b)$

29. $x(a - 2b) + y(2b - a) = x(a - 2b) - y(a - 2b)$
$= (a - 2b)(x - y)$

31. $xy + 4y - 2x - 8 = (xy + 4y) - (2x + 8)$
$= y(x + 4) - 2(x + 4)$
$= (x + 4)(y - 2)$

33. $ax + bx - ay - by = (ax + bx) - (ay + by)$
$= x(a + b) - y(a + b)$
$= (a + b)(x - y)$

35. $x^2y - 3x^2 - 2y + 6$
$= (x^2y - 3x^2) - (2y - 6)$
$= x^2(y - 3) - 2(y - 3)$
$= (y - 3)(x^2 - 2)$

37. $6 + 2y + 3x^2 + x^2y$
$= (6 + 2y) + (3x^2 + x^2y)$
$= 2(3 + y) + x^2(3 + y)$
$= (3 + y)(2 + x^2)$

39. $2ax^2 + bx^2 - 4ay - 2by$
$= (2ax^2 + bx^2) - (4ay + 2by)$
$= x^2(2a + b) - 2y(2a + b)$
$= (2a + b)(x^2 - 2y)$

41. $6xb + 3ax - 4by - 2ay$
$= (6xb + 3ax) - (4by + 2ay)$
$= 3x(2b + a) - 2y(2b + a)$
$= (2b + a)(3 - 2y)$

42. $x^ny - 5x^n + y - 5 = (x^ny - 5x^n) + (y - 5)$
$= x^n(y - 5) + (y - 5)$
$= (y - 5)(x^n + 1)$

45. $2x^3 - x^2 + 4x - 2 = (2x^3 - x^2) + (4x - 2)$
$= x^2(2x - 1) + 2(2x - 1)$
$= (2x - 1)(x^2 + 2)$

47. a) All three expressions are equivalent to
$x^2 - x - 6$.
b) (ii) $x^2 - 3x + 2x - 6$

Objective C Exercises

49. $-11, -4, -1, 1, 4, 11$

51. $x^2 + 12x + 20 = (x + 10)(x + 2)$

53. $a^2 + a - 72 = (a + 9)(a - 8)$

55. $a^2 + 7a + 6 = (a + 1)(a + 6)$

57. $y^2 - 18y + 72 = (y - 6)(y - 12)$

59. $x^2 + x - 132 = (x + 12)(x - 11)$

61. $x^2 + 15x + 50 = (x + 10)(x + 5)$

63. $b^2 - 6b - 16 = (b - 8)(b + 2)$

65. $a^2 - 3ab + 2b^2 = (a - b)(a - 2b)$

67. $a^2 + 8ab - 33b^2 = (a + 11b)(a - 3b)$

69. $x^2 + 5xy + 6y^2 = (x + 2y)(x + 3y)$

71. $2 + x - x^2 = (1 + x)(2 - x) = -(x - 2)(x + 1)$

73. $5 + 4x - x^2 = (1 + x)(5 - x) = -(x - 5)(x + 1)$

75. $x^2 - 5x + 6 = (x - 2)(x - 3)$

Objective D Exercises

77. $2x^2 + 7x + 3 = (2x + 1)(x + 3)$

79. $6y^2 + 5y - 6 = (2y + 3)(3y - 2)$

81. $6b^2 - b - 35 = (3b + 7)(2b - 5)$

83. $3y^2 - 22y + 39 = (3y - 13)(y - 3)$

85. Not factorable

87. $4a^2 - a - 5 = (a + 1)(4a - 5)$

89. $10x^2 - 29x + 10 = (5x - 2)(2x - 5)$

91. Not factorable

93. $6x^2 + 41xy - 7y^2 = (6x - y)(x + 7y)$

95. $7a^2 + 46ab - 21b^2 = (7a - 3b)(a + 7b)$

97. $18x^2 + 27xy + 10y^2 = (6x + 5y)(3x + 2y)$

99. $6 - 7x - 5x^2 = (2 + x)(3 - 5x)$
$\qquad = -(x + 2)(5x - 3)$

101. Not factorable

103. $35 - 6b - 8b^2 = (5 + 2b)(7 - 4b)$
$\qquad = -(2b + 5)(4b - 7)$

105. $5y^4 - 29y^3 + 20y^2 = y^2(5y^2 - 29y + 20)$
$\qquad = y^2(5y - 4)(y - 5)$

109. $20x^2 - 38x^3 - 30x^4$
$\qquad = 2x^2(10 - 19x - 15x^2)$
$\qquad = 2x^2(5 + 3x)(2 - 5x)$
$\qquad = -2x^2(3x + 5)(5x - 2)$

110. $3y - 16y^2 + 16y^3 = y(3 - 16y + 16y^2)$
$\qquad = y(4y - 3)(4y - 1)$

111. Answers will vary. One possible answer
is $2x^2 + 3x + 5$.

113. $2x^2 - 5x + 2 = (2x - 1)(x - 2)$
$\qquad f(x) = 2x - 1;\ h(x) = x - 2$

115. $3a^2 + 11a - 4 = (3a - 1)(a + 4)$
$\qquad f(a) = 3a - 1;\ g(a) = a + 4$

117. $2x^2 + 13x - 24 = (2x - 3)(x + 8)$
$\qquad f(x) = 2x - 3;\ g(x) = x + 8$

119. $6x^2 + 7x - 5 = (2x - 1)(3x + 5)$
$\qquad g(x) = 2x - 1;\ h(x) = 3x + 5$

121. $6t^2 - 17t - 3 = (6t + 1)(t - 3)$
$\qquad f(t) = 6t + 1;\ g(t) = t - 3$

Applying the Concepts

123.   a) $x^2 + kx + 8$.
Find two positive or two negative factors of
8. Their sum is the value of k.

Factors	Sum
8, 1	9
2, 4	6
−8, −1	−9
−2, −4	−6

Values of k are 9, 6, −9, −6.

b) $x^2 + kx - 6$.
Find the factors of 6 with opposite signs.
Their sum is the value of k.

Factors	Sum
−2, 3	1
−3, 2	−1
−6, 1	−5
−1, 6	5

Values of k are 1, −1, −5, 5.

c) $2x^2 + kx + 3$
Find two positive or two negative factors of
6 (2·3). Their sum is the value of k.

Factors	Sum
2, 3	5
−3, −2	−5
1, 6	7
−1, −6	−7

Values of k are 5, −5, 7, −7.

d) $2x^2 + kx - 5$
Find factors of 10 with opposite signs.
Their sum is the value of k.

Factors	Sum
2, −5	−3
−2, 5	3
−10, 1	−9
−1, 10	9

The values of k are −3, 3, −9, 9.

e) $3x^2 + kx + 5$
Find two positive or two negative factors of
15. Their sum is the value of k.

Factors	Sum
3, 5	8
−3, −5	−8
15, 1	16
−15, −1	−16

The values of k are −8, 8, −16, 16.

f) $2x^2 + kx - 3$.

Find the factors of 6 with opposite signs.
Their sum is the value of k.

Factors	Sum
−2, 3	1
−3, 2	−1
−6, 1	−5
−1, 6	5

Values of k are 1, −1, −5, 5.

Section 5.6

Objective A Exercises

1. perfect squares: 4; $25x^6$; $100x^4y^4$

3. $4z^4$

5. $9a^2b^3$

7. (iv) $m^4 - n^2$

9. $x^2 - 16 = x^2 - 4^2 = (x + 4)(x - 4)$

11. $4x^2 - 1 = (2x)^2 - 1^2 = (2x + 1)(2x - 1)$

13. $16x^2 - 121 = (4x)^2 - 11^2 = (4x + 11)(4x - 11)$

15. $1 - 9a^2 = 1^2 - (3a)^2 = (1 + 3a)(1 - 3a)$

17. $x^2y^2 - 100 = (xy)^2 - 10^2 = (xy + 10)(xy - 10)$

19. Not factorable

21. $25 - a^2b^2 = 5^2 - (ab)^2 = (5 + ab)(5 - ab)$

22. $a^{2n} - 1 = (a^n)^2 - 1^2 = (a^n + 1)(a^n - 1)$

25. $x^2 - 12x + 36 = (x - 6)^2$

27. $b^2 - 2b + 1 = (b - 1)^2$

29. $16x^2 - 40x + 25 = (4x - 5)^2$

31. Not factorable

33. Not factorable

35. $x^2 + 6xy + 9y^2 = (x + 3y)^2$

37. $25a^2 - 40ab + 16b^2 = (5a - 4b)^2$

39. $x^{2n} + 6x^n + 9 = (x^n + 3)^2$

41. $(x - 4)^2 - 9 = [(x - 4) - 3][(x - 4) + 3]$
$= (x - 7)(x - 1)$

43. $(x - y)^2 - (a + b)^2$
$= [(x - y) - (a + b)][(x - y) + (a + b)]$
$= (x - y - a - b)(x - y + a + b)$

Objective B Exercises

45. 8; x^9; $27c^{15}d^{18}$

47. $2x^3$

49. $4a^2b^6$

51. Yes

53. Yes

55. $x^3 - 27 = x^3 - 3^3$
$= (x - 3)(x^2 + 3x + 9)$

57. $8x^3 - 1 = (2x)^3 - 1^3$
$= (2x - 1)(4x^2 + 2x + 1)$

59. $x^3 - y^3 = (x - y)(x^2 + xy + y^2)$

61. $m^3 + n^3 = (m + n)(m^2 - mn + n^2)$

63. $64x^3 + 1 = (4x)^3 - 1^3$
$= (4x + 1)(16x^2 - 4x + 1)$

65. $27x^3 - 8y^3 = (3x)^3 - (2y)^3$
$= (3x - 2y)(9x^2 + 6xy + 4y^2)$

67. $x^3y^3 + 64 = (xy)^3 + (4)^3$
$= (xy + 4)(x^2y^2 - 4xy + 16)$

69. Not factorable

71. Not factorable

73. $(a - b)^3 - b^3$
$= [(a - b) - b][(a - b)^2 + b(a - b) + b^2]$
$= (a - 2b)(a^2 - 2ab + b^2 + ab - b^2 + b^2)$
$= (a - 2b)(a^2 - ab + b^2)$

75. $x^{6n} + y^{3n} = (x^{2n})^3 + (y^n)^3$
$= (x^{2n} + y^n)(x^{4n} - x^{2n}y^n + y^{2n})$

77. $x^{3n} + 8 = (x^n)^3 + (2)^3$
$= (x^n + 2)(x^{2n} - 2x^n + 4)$

Objective C Exercises

79. No. Polynomials cannot have square roots as variable terms.

81. Let $u = xy$
$x^2y^2 - 8xy + 15 = u^2 - 8u + 15$
$= (u - 3)(u - 5)$
$= (xy - 3)(xy - 5)$

83. Let $u = xy$
$x^2y^2 - 17xy + 60 = u^2 - 17u + 60$
$= (u - 12)(u - 5)$
$= (xy - 12)(xy - 5)$

85. Let $u = x^2$
$x^4 - 9x^2 + 18 = u^2 - 9u + 18$
$= (u - 3)(u - 6)$
$= (x^2 - 3)(x^2 - 6)$

87. Let $u = b^2$
$b^4 - 13b^2 - 90 = u^2 - 13u - 90$
$= (u + 5)(u - 18)$
$= (b^2 + 5)(b^2 - 18)$

89. Let $u = x^2y^2$
$x^4y^4 - 8x^2y^2 + 12 = u^2 - 8u + 12$
$= (u - 2)(u - 6)$
$= (x^2y^2 - 2)(x^2y^2 - 6)$

91. Let $u = x^n$.
$x^{2n} + 3x^n + 2 = u^2 + 3u + 2$
$= (u + 1)(u + 2)$
$= (x^n + 1)(x^n + 2)$

93. Let $u = xy$
$3x^2y^2 - 14xy + 15 = 3u^2 - 14u + 15$
$= (3u - 5)(u - 3)$
$= (3xy - 5)(xy - 3)$

95. Let $u = ab$
$6a^2b^2 - 23ab + 21 = 6u^2 - 23u + 21$
$= (2u - 3)(3u - 7)$
$= (2ab - 3)(3ab - 7)$

97. Let $u = x^2$
$2x^4 - 13x^2 - 15 = 2u^2 - 13u - 15$
$= (2u - 15)(u + 1)$
$= (2x^2 - 15)(x^2 + 1)$

99. Let $u = x^n$
$2x^{2n} - 7x^n + 3 = 2u^2 - 7u + 3$
$= (2u - 1)(u - 3)$
$= (2x^n - 1)(x^n - 3)$

101. Let $u = a^n$
$6a^{2n} + 19a^n + 10 = 6u^2 + 19u + 10$
$= (2u + 5)(3u + 2)$
$= (2a^n + 5)(3a^n + 2)$

Objective D Exercises

103. $12x^2 - 36x + 27 = 3(4x^2 - 12x + 9)$
$= 3(2x - 3)^2$

105. $27a^4 - a = a(27a^3 - 1)$
$= a(3a - 1)(9a^2 + 3a + 1)$

107. $20x^2 - 5 = 5(4x^2 - 1)$
$= 5(2x + 1)(2x - 1)$

109. $y^5 + 6y^4 - 55y^3 = y^3(y^2 + 6y - 55)$
$= y^3(y + 11)(y - 5)$

111. $16x^4 - 81 = (4x^2 + 9)(4x^2 - 9)$
$= (4x^2 + 9)(2x + 3)(2x - 3)$

113. $16a - 2a^4 = 2a(8 - a^3)$
$= 2a(2 - a)(4 + 2a + a^2)$

115. $a^3b^6 - b^3 = b^3(a^3b^3 - 1)$
$= b^3(ab - 1)(a^2b^2 + ab + 1)$

117. $8x^4 - 40x^3 + 50x^2 = 2x^2(4x^2 - 20x + 25)$
$= 2x^2(2x - 5)^2$

119. $x^4 - y^4 = (x^2 + y^2)(x^2 - y^2)$
$= (x^2 + y^2)(x + y)(x - y)$

121. $x^6 + y^6 = (x^2 + y^2)(x^4 - x^2y^2 + y^4)$

123. Not factorable

125. $16a^4 - 2a = 2a(8a^3 - 1)$
$= 2a(2a - 1)(4a^2 + 2a + 1)$

127. $a^4b^2 - 8a^3b^3 - 48a^2b^4$
$= a^2b^2(a^2 - 8ab - 48b^2)$
$= a^2b^2(a + 4b)(a - 12b)$

129. $x^3 - 2x^2 - 4x + 8 = x^2(x - 2) - 4(x - 2)$
$= (x - 2)(x^2 - 4)$
$= (x - 2)(x + 2)(x - 2)$
$= (x - 2)^2(x + 2)$

131. $2x^3 + x^2 - 32x - 16$
$= x^2(2x + 1) - 16(2x + 1)$
$= (2x + 1)(x^2 - 16)$
$= (2x + 1)(x + 4)(x - 4)$

133. $4x^4 - x^2 - 4x^2y^2 + y^2$
$= x^2(4x^2 - 1) - y^2(4x^2 - 1)$
$= (4x^2 - 1)(x^2 - y^2)$
$= (2x + 1)(2x - 1)(x + y)(x - y)$

135. $3b^{n+2} + 4b^{n+1} - 4b^n$
$= b^n(3b^2 + 4b - 4)$
$= b^n(b + 2)(3b - 2)$

137. The coefficient is 8.

Applying the Concepts

139. If $x - 3$ and $x + 4$ are factors of
$x^3 + 6x^2 - 7x - 60$ then $x^3 + 6x^2 - 7x - 60$ is
divisible by $x - 3$ and $x + 4$.
Divide $x^3 + 6x^2 - 7x - 60$ by $x - 3$. The
quotient is $x^2 + 9x + 20$. Divide this quotient
by $x + 4$. The quotient is $x + 5$.
Therefore $x + 5$ is a third first-degree factor
of $x^3 + 6x^2 - 7x - 60$.

Section 5.7

Objective A Exercises

1. No. If $a = 0$, then b can be any number.

3. $(x - 5)(x + 3) = 0$
$x - 5 = 0 \quad x + 3 = 0$
$x = 5 \qquad x = -3$
The solutions are -3 and 5.

5. $(x + 7)(x - 8) = 0$
$x + 7 = 0 \quad x - 8 = 0$
$x = -7 \qquad x = 8$
The solutions are -7 and 8.

7. $2x(3x - 2)(x + 4) = 0$
$2x = 0 \quad 3x - 2 = 0 \quad x + 4 = 0$
$x = 0 \qquad 3x = 2 \qquad x = -4$
$$x = \frac{2}{3}$$
The solutions are -4, 0, and $\frac{2}{3}$.

9. $x^2 + 2x - 15 = 0$
$(x + 5)(x - 3) = 0$
$x + 5 = 0 \quad x - 3 = 0$
$x = -5 \qquad x = 3$
The solutions are -5 and 3.

11. $z^2 - 4z + 3 = 0$
$(z - 1)(z - 3) = 0$
$z - 1 = 0 \quad z - 3 = 0$
$z = 1 \qquad z = 3$
The solutions are 1 and 3.

13. $r^2 - 10 = 3r$

$r^2 - 3r - 10 = 0$

$(r - 5)(r + 2) = 0$

$r - 5 = 0 \quad r + 2 = 0$

$\quad r = 5 \qquad r = -2$

The solutions are -2 and 5.

15. $4t^2 = 4t + 3$

$4t^2 - 4t - 3 = 0$

$(2t + 1)(2t - 3) = 0$

$2t + 1 = 0 \quad 2t - 3 = 0$

$\quad 2t = -1 \qquad 2t = 3$

$\quad t = -\dfrac{1}{2} \qquad t = \dfrac{3}{2}$

The solutions are $-\dfrac{1}{2}$ and $\dfrac{3}{2}$.

17. $4v^2 - 4v + 1 = 0$

$(2v - 1)(2v - 1) = 0$

$2v - 1 = 0 \quad 2v - 1 = 0$

$\quad 2v = 1 \qquad 2v = 1$

$\quad v = \dfrac{1}{2} \qquad v = \dfrac{1}{2}$

The solution is $\dfrac{1}{2}$.

19. $x^2 - 9 = 0$

$(x - 3)(x + 3) = 0$

$x - 3 = 0 \quad x + 3 = 0$

$\quad x = 3 \qquad x = -3$

The solutions are -3 and 3.

21. $4y^2 - 1 = 0$

$(2y - 1)(2y + 1) = 0$

$2y - 1 = 0 \quad 2y + 1 = 0$

$\quad 2y = 1 \qquad 2y = -1$

$\quad y = \dfrac{1}{2} \qquad y = -\dfrac{1}{2}$

The solutions are $-\dfrac{1}{2}$ and $\dfrac{1}{2}$.

23. $x(x - 1) = x + 15$

$x^2 - x = x + 15$

$x^2 - 2x - 15 = 0$

$(x - 5)(x + 3) = 0$

$x - 5 = 0 \quad x + 3 = 0$

$\quad x = 5 \qquad x = -3$

The solutions are -3 and 5.

25. $2x^2 - 3x - 8 = x^2 + 20$

$x^2 - 3x - 28 = 0$

$(x - 7)(x + 4) = 0$

$x - 7 = 0 \quad x + 4 = 0$

$\quad x = 7 \qquad x = -4$

The solutions are -4 and 7.

27. $(3v + 2)(v - 4) = v^2 + v + 5$

$3v^2 - 10v - 8 = v^2 + v + 5$

$2v^2 - 11x - 13 = 0$

$(2v - 13)(v + 1) = 0$

$2v - 13 = 0 \quad v + 1 = 0$

$\quad 2v = 13 \qquad v = -1$

$\quad v = \dfrac{13}{2}$

The solutions are -1 and $\dfrac{13}{2}$.

29. $4x^2 + x - 10 = (x - 2)(x + 1)$

$4x^2 + x - 10 = x^2 - x - 2$

$3x^2 + 2x - 8 = 0$

$(3x - 4)(x + 2) = 0$

$3x - 4 = 0 \quad x + 2 = 0$

$\quad 3x = 4 \qquad x = -2$

$\quad x = \dfrac{4}{3}$

The solutions are -2 and $\dfrac{4}{3}$.

31. $c^3 + 3c^2 - 10c = 0$

$c(c^2 + 3c - 10) = 0$

$c(c + 5)(c - 2) = 0$

$c = 0 \quad c + 5 = 0 \quad c - 2 = 0$

$\qquad\qquad c = -5 \qquad c = 2$

The solutions are -5, 0 and 2.

33. $a^3 + a^2 - 9a - 9 = 0$

$a^2(a + 1) - 9(a + 1) = 0$

$(a + 1)(a^2 - 9) = 0$

$(a + 1)(a + 3)(a - 3) = 0$

$a + 1 = 0 \quad a + 3 = 0 \quad a - 3 = 0$

$\quad a = -1 \qquad a = -3 \qquad a = 3$

The solutions are $-3, -1$ and 3.

35. $3x^3 + 2x^2 - 12x - 8 = 0$

$x^2(3x + 2) - 4(3x + 2) = 0$

$(3x + 2)(x^2 - 4) = 0$

$(3x + 2)(x + 2)(x - 2) = 0$

$3x + 2 = 0 \quad x + 2 = 0 \quad x - 2 = 0$

$\quad 3x = -2 \qquad x = -2 \qquad x = 2$

$\quad x = -\dfrac{2}{3}$

The solutions are $-2, -\dfrac{2}{3}$ and 2.

37. $5x^3 + 2x^2 - 20x - 8 = 0$

$x^2(5x + 2) - 4(5x + 2) = 0$

$(5x + 2)(x^2 - 4) = 0$

$(5x + 2)(x + 2)(x - 2) = 0$

$5x + 2 = 0 \quad x + 2 = 0 \quad x - 2 = 0$

$\quad 5x = -2 \qquad x = -2 \qquad x = 2$

$\quad x = -\dfrac{2}{5}$

The solutions are $-2, -\dfrac{2}{5}$ and 2.

Objective B Exercises

39. Strategy: Let x represent the number.
The square of the number is x^2.
The sum of the number and its square is 210.

Solution: $x + x^2 = 210$

$x^2 + x - 210 = 0$

$(x - 14)(x + 15) = 0$

$x = 14 \quad x = -15$

The number is 14 or -15.

41. Strategy: Let b represent the base of the triangle.
The height of the triangle is $b + 8$.

The area of a triangle is

$\dfrac{1}{2} \cdot$ base $\cdot$ height $= \dfrac{1}{2} b \cdot (b + 8)$

The area of the rectangle is 64 cm^2.

Solution: $\dfrac{1}{2} b(b + 8) = 64$

$b^2 + 8b = 128$

$b^2 + 8b - 128 = 0$

$(b - 8)(b + 16) = 0$

$b = 8 \quad b = -16$

The base cannot be a negative number so we can eliminate -16 from consideration.
The base is 8 cm.
The height is $b + 8 = 8 + 8 = 16$ cm.

43. Strategy: Let x represent the width of the rectangle.
The length of the rectangle is $3x - 3$.
The area of a rectangle is
length $\cdot$ width $= x \cdot (3x - 3)$
The area of the rectangle is 60 cm^2.

Solution: $x(3x - 3) = 60$

$3x^2 - 3x - 60 = 0$

$(3x + 12)(x - 5) = 0$

$x = -4 \quad x = 5$

The width cannot be a negative number so we can eliminate -4 from consideration.
The width is 5 cm.
The length is $3x - 3 = 3(5) - 3 = 12$ cm.

45. Larry incorrectly assumed that if $ab = 15$ then $a = 3$ and $b = 5$.

Applying the Concepts

47. $(x - 3)(x - 7) = 0$

$x^2 - 10x + 21 = 0$

49. $(x - (-1))(x - (-3)) = 0$

$(x + 1)(x + 3) = 0$

$x^2 + 4x + 3 = 0$

51. Strategy: The height of the box is 2 in.
The width of the box is $w - 4$.
The length of the box is
$(w + 10) - 4 = w + 6$.
The volume of a box is
length $\cdot$ width $\cdot$ height and equals 112 cm³.

Solution: $2(w - 4)(w + 6) = 112$
$w^2 + 2w - 24 = 56$
$w^2 + 2w - 80 = 0$
$(w - 8)(w + 10) = 0$
$w = 8 \qquad w = -10$
The width cannot be a negative number so
we can eliminate -10 from consideration.
The width is 8 cm.
The length is $w + 10 = 8 + 10 = 18$ cm.

Chapter 5 Review Exercises

1. The GCF of $18a^5b^2 - 12a^3b^3 + 30a^2b$ is $6a^2b$.
$6a^2b(3a^3b - 2ab^2 + 5)$

2.

$$
\begin{array}{r}
5x + 4 \\
3x - 2 \overline{)15x^2 + 2x - 2} \\
\underline{15x^2 - 10x} \\
12x - 2 \\
\underline{12x - 8} \\
6
\end{array}
$$

$\dfrac{15x^2 + 2x - 2}{3x - 2} = 5x + 4 + \dfrac{6}{3x - 2}$

3. $(2x^{-1}y^2z^5)^4(-3x^3yz^{-3})^2$
$= (16x^{-4}y^8z^{20})(9x^6y^2z^{-6})$
$= 144x^2y^{10}z^{14}$

4. $2ax + 4bx - 3ay - 6by$
$= 2x(a + 2b) - 3y(a + 2b)$
$= (a + 2b)(2x - 3y)$

5. $12 + x - x^2 = (4 - x)(3 + x)$
$= -(x - 4)(x + 3)$

6.

$$
\begin{array}{r|rrrr}
2 & 1 & -2 & 3 & -5 \\
 & & 2 & 0 & 6 \\
\hline
 & 1 & 0 & 3 & 1
\end{array}
$$

$P(2) = 1$

7. $(5x^2 - 8xy + 2y^2) - (x^2 - 3y^2)$
$= (5x^2 - x^2) - 8xy + (2y^2 + 3y^2)$
$= 4x^2 - 8xy + 5y^2$

8. $24x^2 + 38x + 15 = (6x + 5)(4x + 3)$

9. $4x^2 + 12xy + 9y^2 = (2x + 3y)^2$

10. $(-2a^2b^4)(3ab^2) = -6a^3b^6$

11. $64a^3 - 27b^3 = (4a)^3 - (3b)^3$
$= (4a - 3b)(16a^2 + 12ab + 9b^2)$

12.

$$
\begin{array}{r|rrrr}
-6 & 4 & 27 & 10 & 2 \\
 & & -24 & -18 & 48 \\
\hline
 & 4 & 3 & -8 & 50
\end{array}
$$

$\dfrac{4x^3 + 27x^2 + 10x + 2}{x + 6} = 4x^2 + 3x - 8 + \dfrac{50}{x + 6}$

13. $P(x) = 2x^3 - x + 7$
$P(-2) = 2(-2)^3 - (-2) + 7$
$P(-2) = -16 + 2 + 7$
$P(-2) = -7$

14. $x^2 - 3x - 40 = (x - 8)(x + 5)$

15. Let $u = xy$
$x^2y^2 - 9 = u^2 - 9 = (u + 3)(u - 3)$
$= (xy + 3)(xy - 3)$

16. $4x^2y(3x^3y^2 + 2xy - 7y^3)$
$= 12x^5y^3 + 8x^3y^2 - 28x^2y^4$

17. Let $u = x^n$
$x^{2n} - 12x^n + 36$
$= u^2 - 12u + 36$
$= (u - 6)^2$
$= (x^n - 6)^2$

18. $6x^2 + 60 = 39x$

$6x^2 - 39x + 60 = 0$

$3(2x^2 - 13x + 20) = 0$

$3(2x - 5)(x - 4) = 0$

$2x - 5 = 0 \qquad x - 4 = 0$

$2x = 5 \qquad\quad x = 4$

$x = \dfrac{5}{2}$

The solutions are $\dfrac{5}{2}$ and 4.

19. $5x^2 - 4x[x - 3(3x + 2) + x]$

$= 5x^2 - 4x(x - 9x - 6 + x)$

$= 5x^2 - 4x(-7x - 6)$

$= 5x^2 + 28x^2 + 24x$

$= 33x^2 + 24x$

20. $3a^6 - 15a^4 - 18a^2$

$= 3a^2(a^4 - 5a^2 - 6)$

$= 3a^2(a^2 - 6)(a^2 + 1)$

21. $(4x - 3y)^2 = 16x^2 - 24xy + 9y^2$

22.

$$\begin{array}{r|rrrrr} 4 & 1 & 0 & 0 & 0 & -4 \\ & & 4 & 16 & 64 & 256 \\ \hline & 1 & 4 & 16 & 64 & 252 \end{array}$$

$\dfrac{x^4 - 4}{x - 4} = x^3 + 4x^2 + 16x + 64 + \dfrac{252}{x - 4}$

23. $(3x^2 - 2x - 6) + (-x^2 - 3x + 4)$

$= 2x^2 - 5x - 2$

24. $(5x^2yz^4)(2xy^3z^{-1})(7x^{-2}y^{-2}z^3)$

$= (10x^3y^4z^3)(7x^{-2}y^{-2}z^3)$

$= 70xy^2z^6$

25. $\dfrac{3x^4yz^{-1}}{-12xy^3z^2} = -\dfrac{x^3}{4y^2z^3}$

26. 9.48×10^8

27. $\dfrac{3 \times 10^{-3}}{15 \times 10^2} = 0.2 \times 10^{-5} = 2 \times 10^{-6}$

28.

$$\begin{array}{r|rrrr} -3 & -2 & 2 & 0 & -4 \\ & & 6 & -24 & 72 \\ \hline & -2 & 8 & -24 & 68 \end{array}$$

$P(-3) = 68$

29. $\dfrac{16x^5 - 8x^3 + 20x}{4x} = \dfrac{16x^5}{4x} - \dfrac{8x^3}{4x} + \dfrac{20x}{4x}$

$= 4x^4 - 2x^2 + 5$

30.

$$\begin{array}{r} 2x - 3 \\ 6x + 1 \overline{) 12x^2 - 16x - 7} \\ \underline{12x^2 + 2x} \\ -18x - 7 \\ \underline{-18x - 3} \\ -4 \end{array}$$

$\dfrac{12x^2 - 16x - 7}{6x + 1} = 2x - 3 - \dfrac{4}{6x + 1}$

31. $a^3(a^4 - 5a + 2) = a^7 - 5a^4 + 2a^3$

32. $(x + 6)(x^3 - 3x^2 - 5x + 1)$

$= x(x^3 - 3x^2 - 5x + 1) + 6(x^3 - 3x^2 - 5x + 1)$

$= x^4 - 3x^3 - 5x^2 + x + 6x^3 - 18x^2 - 30x + 6$

$= x^4 + 3x^3 - 23x^2 - 29x + 6$

33. $10a^3b^3 - 20a^2b^4 + 35ab^2$

$= 5ab^2(2a^2b - 4ab^2 + 7)$

34. $5x^5 + x^3 + 4x^2 = x^2(5x^3 + x + 4)$

35. $x(y - 3) + 4(3 - y) = x(y - 3) - 4(y - 3)$

$= (y - 3)(x - 4)$

36. $x^2 - 16x + 63 = (x - 7)(x - 9)$

37. $24x^2 + 61x - 8 = (8x - 1)(3x + 8)$

38. We are looking for two polynomials that
when multiplied together give us
$5x^2 + 3x - 2$.
$5x^2 + 3x - 2 = (5x - 2)(x + 1)$
$f(x) = 5x - 2$ and $g(x) = x + 1$

39. $36 - a^{2n} = (6 + a^n)(6 - a^n)$

40. $8 - y^{3n} = 2^3 - (y^n)^3$
$= (2 - y^n)(4 + 2y^n + y^{2n})$

41. Let $u = x^4$
$36x^8 - 36x^4 + 5$
$= 36u^2 - 36u + 5$
$= (6u - 5)(6u - 1)$
$= (6x^4 - 5)(6x^4 - 1)$

42. $3a^4b - 3ab^4 = 3ab(a^3 - b^3)$
$= 3ab(a - b)(a^2 + ab + b^2)$

43. Let $u = x^2$.
$x^4 - 8x^2 + 16 = u^2 - 8u + 16$
$= (u - 4)^2 = (x^2 - 4)^2$
$= (x - 2)^2(x + 2)^2$

44. $x^3 - x^2 - 6x = 0$
$x(x^2 - x - 6) = 0$
$x(x - 3)(x + 2) = 0$
$x = 0 \quad x - 3 = 0 \quad x + 2 = 0$
$\qquad\qquad x = 3 \qquad x = -2$
The solutions are -2, 0 and 3.

45. $x^3 - 16x = 0$
$x(x^2 - 16) = 0$
$x(x - 4)(x + 4) = 0$
$x = 0 \quad x - 4 = 0 \quad x + 4 = 0$
$\qquad\qquad x = 4 \qquad x = -4$
The solutions are -4, 0 and 4.

46. $y^3 + y^2 - 36y - 36 = 0$
$y^2(y + 1) - 36(y + 1) = 0$
$(y + 1)(y^2 - 36) = 0$
$(y + 1)(y - 6)(y + 6) = 0$
$y + 1 = 0 \quad y - 6 = 0 \quad y + 6 = 0$
$\quad y = -1 \qquad y = 6 \qquad y = -6$
The solutions are -6, -1 and 6.

47. Let $u = x^2$.
$15x^4 + x^2 - 6 = 15u^2 + u - 6$
$= (3u + 2)(5u - 3)$
$= (3x^2 + 2)(5x^2 - 3)$

48. $\dfrac{(2a^4b^{-3}c^2)^3}{(2a^3b^2c^{-1})^4} = \dfrac{8a^{12}b^{-9}c^6}{16a^{12}b^8c^{-4}}$

$= \dfrac{c^{10}}{2b^{17}}$

49. $(x - 4)(3x + 2)(2x - 3)$
$= (x - 4)(6x^2 - 5x - 6)$
$= x(6x^2 - 5x - 6) - 4(6x^2 - 5x - 6)$
$= 6x^3 - 5x^2 - 6x - 24x^2 + 20x + 24$
$= 6x^3 - 29x^2 + 14x + 24$

50. Let $u = x^2y^2$.
$21x^4y^4 + 23x^2y^2 + 6 = 21u^2 + 23u + 6$
$= (7u + 3)(3u + 2)$
$= (7x^2y^2 + 3)(3x^2y^2 + 2)$

51. $x^3 + 16 = x(x + 16)$
$x^3 + 16 = x^2 + 16x$
$x^3 - x^2 - 16x + 16 = 0$
$x^2(x - 1) - 16(x - 1) = 0$
$(x - 1)(x^2 - 16) = 0$
$(x - 1)(x - 4)(x + 4) = 0$
$x - 1 = 0 \quad x - 4 = 0 \quad x + 4 = 0$
$\quad x = 1 \qquad x = 4 \qquad x = -4$
The solutions are -4, 1 and 4.

52. $(5a + 2b)(5a - 2b) = 25a^2 - 4b^2$

53. 0.00254

54. $6x^2 - 31x + 18 = (3x - 2)(2x - 9)$

55. $y = x^2 + 1$

x	x
-2	5
-1	2
0	1
1	2
2	5

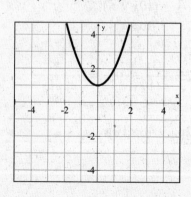

56. a) 3

b) 8

c) 5

57. Strategy: To find the mass of the moon, multiply the mass of the sun by 3.7×10^{-8}.

Solution:
$$(2.19 \times 10^{27})(3.7 \times 10^{-8}) = 8.103 \times 10^{19}$$
The mass of the moon is 8.103×10^{19} tons.

58. Strategy: Let x represent the number.
The square of the number is x^2.
The sum of the number and its square is 56.

Solution: $x + x^2 = 56$
$x^2 + x - 56 = 0$
$(x - 7)(x + 8) = 0$
$x = 7 \qquad x = -8$
The number is 7 or -8.

59. Strategy: To find how far Earth is from the Great Galaxy of Andromeda, use the equation $d = rt$, where $r = 6.7 \times 10^8$ mph and $t = 2.2 \times 10^6$ years.
$2.2 \times 10^6 \times 24 \times 365 = 1.9272 \times 10^{10}$ hours.

Solution: $d = rt$
$d = (6.7 \times 10^8)(1.9272 \times 10^{10})$
$= 1.291224 \times 10^{19}$
The distance from Earth to the Great Galaxy of Andromeda is 1.291224×10^{19} mi.

60. Strategy: To find the area substitute the given values for L and W in the equation $A = LW$ and solve for A.

Solution: $A = LW$
$A = (5x + 3)(2x - 7) = 10x^2 - 29x - 21$
The area is $(10x^2 - 29x - 21)$ cm^2.

Chapter 5 Test

1. $16t^2 + 24t + 9 = (4t + 3)^2$

2. $-6rs^2(3r - 2s - 3) = -18r^2s^2 + 12rs^3 + 18rs^2$

3. $P(x) = 3x^2 - 8x + 1$
$P(2) = 3(2)^2 - 8(2) + 1$
$P(2) = 12 - 16 + 1$
$P(2) = -3$

4. $27x^3 - 8 = (3x)^3 - (2)^3$
$= (3x - 2)(9x^2 + 6x + 4)$

5. $16x^2 - 25 = (4x + 5)(4x - 5)$

6. $(3t^3 - 4t^2 + 1)(2t^2 - 5)$
$= 2t^2(3t^3 - 4t^2 + 1) - 5(3t^3 - 4t^2 + 1)$
$= 6t^5 - 8t^4 + 2t^2 - 15t^3 + 20t^2 - 5$
$= 6t^5 - 8t^4 - 15t^3 + 22t^2 - 5$

7. $-5x[3 - 2(2x - 4) - 3x]$
$= -5x(3 - 4x + 8 - 3x)$
$= -5x(-7x + 11)$
$= 35x^2 - 55x$

8. $12x^3 + 12x^2 - 45x = 3x(4x^2 + 4x - 15)$
$= 3x(2x - 3)(2x + 5)$

9. $6x^3 + x^2 - 6x - 1 = 0$
$x^2(6x + 1) - 1(6x + 1) = 0$
$(6x + 1)(x^2 - 1) = 0$
$(6x + 1)(x + 1)(x - 1) = 0$

$6x + 1 = 0 \quad x + 1 = 0 \quad x - 1 = 0$
$6x = -1 \qquad x = -1 \qquad x = 1$
$x = -\dfrac{1}{6}$

The solutions are $-1, -\dfrac{1}{6}, 1$.

10. $(6x^3 - 7x^2 + 6x - 7) - (4x^3 - 3x^2 + 7)$
$= (6x^3 - 4x^3) + (-7x^2 + 3x^2) + 6x + (-7 - 7)$
$= 2x^3 - 4x^2 + 6x - 14$

11. 5.01×10^{-7}

12.

$$\begin{array}{r} 2x+1 \\ 7x-3{\overline{\smash{\big)}\,14x^2+x+1}} \\ \underline{14x^2-6x} \\ 7x+1 \\ \underline{7x-3} \\ 4 \end{array}$$

$$\frac{14x^2+x+1}{7x-3}=2x+1+\frac{4}{7x-3}$$

13. $(7-5x)(7+5x)=49-25x^2$

14. Let $u=a^2$.
$$6a^4-13a^2-5=6u^2-13u-5$$
$$=(2u-5)(3u+1)$$
$$=(2a^2-5)(3a^2+1)$$

15. $(3a+4b)(2a-7b)=6a^2-13ab-28b^2$

16. $3x^4-23x^2-36=(3x^2+4)(x^2-9)$
$$=(3x^2+4)(x+3)(x-3)$$

17. $(-4a^2b)^3(-ab^4)=(-64a^6b^3)(-ab^4)=64a^7b^7$

18. $6x^2=x+1$
$$6x^2-x-1=0$$
$$(3x+1)(2x-1)=0$$
$$3x+1=0 \qquad 2x-1=0$$
$$3x=-1 \qquad 2x=1$$
$$x=-\frac{1}{3} \qquad x=\frac{1}{2}$$

The solutions are $-\dfrac{1}{3}$ and $\dfrac{1}{2}$.

19.

$$\begin{array}{r|rrrr} -2 & -1 & 0 & 4 & -8 \\ & & 2 & -4 & 0 \\ \hline & -1 & 2 & 0 & -8 \end{array}$$

$P(-2)=-8$

20. $\dfrac{(2a^{-4}b^2)^3}{4a^{-2}b^{-1}}=\dfrac{8a^{-12}b^6}{4a^{-2}b^{-1}}=\dfrac{2b^7}{a^{10}}$

21.

$$\begin{array}{r} x^2-5x+10 \\ x+3{\overline{\smash{\big)}\,x^3-2x^2-5x+7}} \\ \underline{x^3+3x^2} \\ -5x^2-5x \\ \underline{-5x^2-15x} \\ 10x+7 \\ \underline{10x+30} \\ -23 \end{array}$$

$$\frac{x^3-2x^2-5x+7}{x+3}=x^2-5x+10-\frac{23}{x+3}$$

22. $12-17x+6x^2=(3x-4)(2x-3)$

23. $6x^2-4x-3xa+2a$
$$=2x(3x-2)-a(3x-2)$$
$$=(3x-2)(2x-a)$$

24. Strategy: To find the number of seconds in one week:
Multiply the number of seconds in a minute (60) by the number of minutes in an hour (60) by the number of hours in a day (24) by the number of days in a week (7).
Convert your answer to scientific notation.

Solution:
$(60)(60)(24)(7)=604,800$
$=6.048\times10^5\,\text{s}$
There are $6.048\times10^5\,\text{s}$ in one week.

25. Strategy: The distance is $h=64$ ft.

Solution: $h=32+48t-16t^2$
$$64=32+48t-16t^2$$
$$0=-32+48t-16t^2$$
$$16t^2-48t+32=0$$
$$t^2-3t+2=0$$
$$(t-2)(t-1)=0$$
$$t-2=0 \qquad t-1=0$$
$$t=2 \qquad t=1$$

The arrow will be 64 ft above the ground at 1 s and at 2 s after the arrow is released.

26. Strategy: To find the area substitute the given values for L and W in the equation $A = LW$ and solve for A.

Solution: $A = LW$

$A = (5x + 1)(2x - 1) = 10x^2 - 3x - 1$

$$
\begin{array}{c|cccc}
3 & 1 & 0 & 0 & -3 \\
3x & & 3 & 9 & 27 \\
\hline
& 1 & 3 & 9 & 24
\end{array}
$$

The area is $(10x^2 - 1)$ ft².

Cumulative Review Exercises

1. $8 - 2[-3 - (-1)]^2 + 4$

$= 8 - 2(-3 + 1)^2 + 4$

$= 8 - 2(-2)^2 + 4$

$= 8 - 2(4) + 4$

$= 4$

2. $\dfrac{2a - b}{b - c}$

$\dfrac{2(4) - (-2)}{(-2) - 6} = \dfrac{8 + 2}{-8} = \dfrac{10}{-8} = -\dfrac{5}{4}$

3. Inverse Property of Addition

4. $2x - 4[x - 2(3 - 2x) + 4]$

$= 2x - 4(x - 6 + 4x + 4)$

$= 2x - 4(5x - 2)$

$= 2x - 20x + 8$

$= -18x + 8$

5. $\dfrac{2}{3} - y = \dfrac{5}{6}$

$\dfrac{2}{3} - y - \dfrac{2}{3} = \dfrac{5}{6} - \dfrac{2}{3}$

$-y = \dfrac{1}{6}$

$(-1)(-y) = -\dfrac{1}{6}(-1)$

$y = -\dfrac{1}{6}$

The solution is $-\dfrac{1}{6}$.

6. $8x - 3 - x = -6 + 3x - 8$

$7x - 3 = 3x - 14$

$4x - 3 = -14$

$4x = -11$

$x = -\dfrac{11}{4}$

The solution is $-\dfrac{11}{4}$.

7.

$$
\begin{array}{c|cccc}
3 & 1 & 0 & 0 & -3 \\
& & 3 & 9 & 27 \\
\hline
& 1 & 3 & 9 & 24
\end{array}
$$

$\dfrac{x^3 - 3}{x - 3} = x^2 + 3x + 9 + \dfrac{24}{x - 3}$

8. $3 - |2 - 3x| = -2$

$-|2 - 3x| = -5$

$|2 - 3x| = 5$

$2 - 3x = 5 \qquad 2 - 3x = -5$

$-3x = 3 \qquad\ \ -3x = -7$

$x = -1 \qquad\ \ x = \dfrac{7}{3}$

The solutions are -1 and $\dfrac{7}{3}$.

9. $P(x) = 3x^2 - 2x + 2$

$P(-2) = 3(-2)^2 - 2(-2) + 2$

$P(-2) = 3(4) + 4 + 2$

$P(-2) = 18$

10. $x = -2$ is excluded from the domain of $f(x)$.

11. $F(x) = 3x^2 - 4$

$F(-2) = 3(-2)^2 - 4 = 8$

$F(-1) = 3(-1)^2 - 4 = -1$

$F(0) = 3(0)^2 - 4 = -4$

$F(1) = 3(1)^2 - 4 = -1$

$F(2) = 3(2)^2 - 4 = 8$

The range = $\{-4, -1, 8\}$

12. $m = \dfrac{y_2 - y_1}{x_2 - x_1} = \dfrac{2 - 3}{4 - (-2)} = -\dfrac{1}{6}$

13. Use the point-slope formula.
$$y - y_1 = m(x - x_1)$$
$$y - 2 = -\frac{3}{2}[x - (-1)]$$
$$y - 2 = -\frac{3}{2}x - \frac{3}{2}$$
$$y = -\frac{3}{2}x + \frac{1}{2}$$

14. Solve the equation $3x + 2y = 4$ for y to find the slope of this line.
$$3x + 2y = 4$$
$$2y = -3x + 4$$
$$y = -\frac{3}{2}x + 2$$
$$m = -\frac{3}{2}$$
The perpendicular line will have a slope that is the negative reciprocal of $-\frac{3}{2}$.
$$m = \frac{2}{3} \text{ and } (-2, 4)$$
$$y - y_1 = m(x - x_1)$$
$$y - 4 = \frac{2}{3}[x - (-2)]$$
$$y - 4 = \frac{2}{3}x + \frac{4}{3}$$
$$y = \frac{2}{3}x + \frac{16}{3}$$
The equation of the perpendicular line is
$$y = \frac{2}{3}x + \frac{16}{3}.$$

15. $2x - 3y = 2$
$\quad x + y = -3$
$$D = \begin{vmatrix} 2 & -3 \\ 1 & 1 \end{vmatrix} = 5$$
$$D_x = \begin{vmatrix} 2 & -3 \\ -3 & 1 \end{vmatrix} = -7$$
$$D_y = \begin{vmatrix} 2 & 2 \\ 1 & -3 \end{vmatrix} = -8$$
$$x = \frac{D_x}{D} = \frac{-7}{5} = -\frac{7}{5}$$

$$y = \frac{D_y}{D} = \frac{-8}{5} = -\frac{8}{5}$$
The solution is $(-\frac{7}{5}, -\frac{8}{5})$.

16. (1) $\quad x - y + z = 0$
$\quad$ (2) $\quad 2x + y - 3z = -7$
$\quad$ (3) $\quad -x + 2y + 2z = 5$
Add equations (1) and (3) to eliminate x.
$$x - y + z = 0$$
$$-x + 2y + 2z = 5$$

(4) $y + 3z = 5$
Add -2 times equation (1) and equation (2) to eliminate x.
$$-2x + 2y - 2z = 0$$
$$2x + y - 3z = -7$$

(5) $3y - 5z = -7$
Add -3 times equation (4) to equation (5) to eliminate y.
$$-3y - 9z = -15$$
$$3y - 5z = -7$$
$$-14z = -22$$
$$z = \frac{11}{7}$$

Substitute $\frac{11}{7}$ for z in equation (4).
$$y + 3(\frac{11}{7}) = 5$$
$$y = \frac{2}{7}$$

Substitute in values for y and z.
$$x - y + z = 0$$
$$x - \frac{2}{7} + \frac{11}{7} = 0$$
$$x = -\frac{9}{7}$$

The solution is $\left(-\frac{9}{7}, \frac{2}{7}, \frac{11}{7}\right)$.

17. $3x - 4y = 12$

x-intercept:

$3x - 4(0) = 12$

$3x = 12$

$x = 4$

$(4, 0)$

y-intercept:

$3(0) - 4y = 12$

$-4y = 12$

$y = -3$

$(0, -3)$

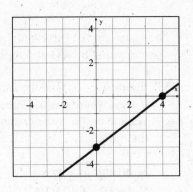

18. $-3x + 2y < 6$

$2y < 3x + 6$

$y < \dfrac{3}{2}x + 3$

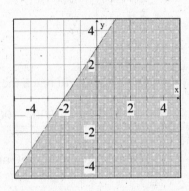

19. $x - 2y = 3$

$-2y = -x + 3$

$y = \dfrac{1}{2}x - \dfrac{3}{2}$

$-2x + y = -3$

$y = 2x - 3$

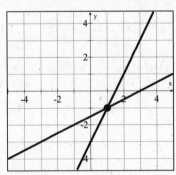

The solutions is $(1, -1)$.

20. Solve each inequality for y.

$2x + y < 3$

$y < -2x + 3$

$-2x + y \geq 1$

$y \geq 2x + 1$

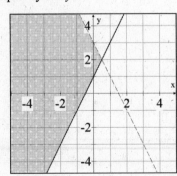

21. $(4a^{-2}b^3)(2a^3b^{-1})^{-2} = 4a^{-2}b^3(2^{-2}a^{-6}b^2)$

$= 4(2^{-2})a^{-8}b^5$

$= \dfrac{b^5}{a^8}$

22. $\dfrac{(5x^3y^{-3}z)^{-2}}{y^4z^{-2}} = \dfrac{5^{-2}x^{-6}y^6z^{-2}}{y^4z^{-2}} = \dfrac{y^2}{25x^6}$

23. $3 - (3 - 3^{-1})^{-1} = 3 - \left(3 - \dfrac{1}{3}\right)^{-1}$

$= 3 - \left(\dfrac{8}{3}\right)^{-1} = 3 - \dfrac{3}{8}$

$= \dfrac{21}{8}$

24. $(2x + 3)(2x^2 - 3x + 1)$

$= 2x(2x^2 - 3x + 1) + 3(2x^2 - 3x + 1)$

$= 4x^3 - 6x^2 + 2x + 6x^2 - 9x + 3$

$= 4x^3 - 7x + 3$

25. $-4x^3 + 14x^2 - 12x = -2x(2x^2 - 7x + 6)$

$= -2x(2x - 3)(x - 2)$

26. $a(x - y) - b(y - x) = a(x - y) + b(x - y)$

$= (x - y)(a + b)$

27. $x^4 - 16 = (x^2 + 4)(x^2 - 4)$

$= (x^2 + 4)(x - 2)(x + 2)$

28. $2x^3 - 16 = 2(x^3 - 8)$

$= 2(x - 2)(x^2 + 2x + 4)$

29. Strategy Let x represent the smaller integer. The larger integer is $24 - x$.

The difference between four times the smaller and nine is 3 less than twice the larger.

$4x - 9 = 2(24 - x) - 3$

30. Solution: $4x - 9 = 2(24 - x) - 3$
$4x - 9 = 48 - 2x - 3$
$4x - 9 = 45 - 2x$
$6x - 9 = 45$
$6x = 54$
$x = 9$
$24 - x = 24 - 9 = 15$
The integers are 9 and 15.

31. Strategy: Let x represent the number of ounces of pure gold.

	Amount	Cost	Value
Pure gold	x	360	$360x$
Alloy	80	120	80(120)
Mixture	$x + 80$	200	$200(x + 80)$

The sum of values before mixing is equal to the value after mixing.

Solution: $360x + 80(120) = 200(x + 80)$
$360x + 9600 = 200x + 16,000$
$160x + 9600 = 16,000$
$160x = 6400$

$x = 40$

40 oz of pure gold must be mixed with the alloy.

32. Strategy: Let x represent the speed of the faster cyclist.

The speed of the slower cyclist is $\frac{2}{3}x$.

	Rate	Time	Distance
Faster cyclist	x	2	$2x$
Slower cyclist	$\frac{2}{3}x$	2	$2\left(\frac{2}{3}x\right)$

The sum of the distances is 25 mi.

Solution: $2x + 2\left(\frac{2}{3}x\right) = 25$

$2x + \frac{4}{3}x = 25$

$\frac{10}{3}x = 25$

$x = 7.5$

$\frac{2}{3}x = \frac{2}{3}(7.5) = 5$

The faster cyclist travels at 7.5 mph and the slower cyclist travels at 5 mph.

33. Strategy: To find the time, use the equation $d = rt$, where r is the speed of the space vehicle and d is the distance from Earth to the moon.

Solution: $d = rt$
$2.4 \times 10^5 = (2 \times 10^4)t$
$\dfrac{2.4 \times 10^5}{2 \times 10^4} = t$
$1.2 \times 10^1 = t$
The vehicle will reach the moon in 12 h.

34. $m = \dfrac{y_2 - y_1}{x_2 - x_1} = \dfrac{300 - 100}{6 - 2} = \dfrac{200}{4} = 50$
The average speed is 50 mph.

Chapter 6: Rational Expressions

Prep Test

1. Multiples of 10: 10, 20, 30, 40, 50, 60, …
Multiples of 25: 25, 50, 75, …
LCM is 50.

2. $-\dfrac{3}{8} \cdot \dfrac{4}{9} = -\dfrac{3 \cdot 2 \cdot 2}{2 \cdot 2 \cdot 2 \cdot 3 \cdot 3} = -\dfrac{1}{6}$

3. $-\dfrac{4}{5} \div \dfrac{8}{15} = -\dfrac{4}{5} \cdot \dfrac{15}{8} = -\dfrac{2 \cdot 2 \cdot 3 \cdot 5}{2 \cdot 2 \cdot 2 \cdot 5} = -\dfrac{3}{2}$

4. $-\dfrac{5}{6} + \dfrac{7}{8} = -\dfrac{20}{24} + \dfrac{21}{24} = \dfrac{1}{24}$

5. $-\dfrac{3}{8} - \left(-\dfrac{7}{12}\right) = -\dfrac{3}{8} + \dfrac{7}{12} = -\dfrac{9}{24} + \dfrac{14}{24} = \dfrac{5}{24}$

6. $\dfrac{\dfrac{2}{3} - \dfrac{1}{4}}{\dfrac{1}{8} - \dfrac{2}{1}} = \dfrac{\dfrac{8}{12} - \dfrac{3}{12}}{\dfrac{1}{8} - \dfrac{16}{8}} = \dfrac{\dfrac{5}{12}}{-\dfrac{15}{8}} = \dfrac{5}{12} \div \dfrac{-15}{8}$

$= \dfrac{5}{12} \cdot \left(-\dfrac{8}{15}\right) = -\dfrac{2 \cdot 2 \cdot 2 \cdot 5}{2 \cdot 2 \cdot 3 \cdot 3 \cdot 5} = -\dfrac{2}{9}$

7. $\dfrac{2x - 3}{x^2 - x + 1}$

$\dfrac{2(2) - 3}{2^2 - 2 + 1} = \dfrac{4 - 3}{4 - 2 + 1} = \dfrac{1}{3}$

8. $4(2x + 1) = 3(x - 2)$

$8x + 4 = 3x - 6$

$5x = -10$

$x = -2$

The solution is −2.

9. $10\left(\dfrac{t}{2} + \dfrac{t}{5}\right) = 10(1)$

$\dfrac{10t}{2} + \dfrac{10t}{5} = 10$

$5t + 2t = 10$

$7t = 10$

$t = \dfrac{10}{7}$

The solution is $\dfrac{10}{7}$.

10. Strategy: Let x represent the rate of the second plane.
The rate of the first plane is $x - 20$.

	Rate	Time	Distance
First plane	$x - 20$	2	$2(x - 20)$
Second plane	x	2	$2x$

The planes are 480 mi apart. The total distance traveled by the two planes is $2(x - 20) + 2x = 480$.
Solution:
$2(x - 20) + 2x = 480$

$2x - 40 + 2x = 480$

$4x - 40 = 480$

$4x = 520$

$x = 130$

$x - 20 = 130 - 20 = 110$

The rate of the first plane is 110 mph and the rate of the second plane is 130 mph.

Section 6.1

Objective A Exercises

1. A rational function is a function that is written as an expression in which the numerator and denominator are polynomials.

Example: $\dfrac{5x - 2}{x^2 + 2x + 6}$

3. $f(x) = \dfrac{2}{x-3}$

$f(4) = \dfrac{2}{4-3} = \dfrac{2}{1} = 2$

5. $f(x) = \dfrac{x-2}{x+4}$

$f(-2) = \dfrac{-2-2}{-2+4} = \dfrac{-4}{2} = -2$

7. $f(x) = \dfrac{1}{x^2 - 2x + 1}$

$f(-2) = \dfrac{1}{(-2)^2 - 2(-2) + 1} = \dfrac{1}{9}$

9. $f(x) = \dfrac{x-2}{2x^2 + 3x + 8}$

$f(3) = \dfrac{3-2}{2(3)^2 + 3(3) + 8} = \dfrac{1}{35}$

11. $f(x) = \dfrac{x^2 - 2x}{x^3 - x + 4}$

$f(-1) = \dfrac{(-1)^2 - 2(-1)}{(-1)^3 - (-1) + 4} = \dfrac{3}{4}$

13. $f(x) = \dfrac{4}{x-3}$

$x - 3 = 0$

$x = 3$

The domain is $\{x \mid x \neq 3\}$.

15. $H(x) = \dfrac{x}{x+4}$

$x + 4 = 0$

$x = -4$

The domain is $\{x \mid x \neq -4\}$.

17. $h(x) = \dfrac{5x}{3x+9}$

$3x + 9 = 0$

$3x = -9$

$x = -3$

The domain is $\{x \mid x \neq -3\}$.

19. $q(x) = \dfrac{4-x}{(x-4)(3x-2)}$

$(x-4)(3x-2) = 0$

$x - 4 = 0 \quad 3x - 2 = 0$

$x = 4 \qquad 3x = 2$

$\qquad\qquad x = \dfrac{2}{3}$

The domain is $\left\{x \mid x \neq \dfrac{2}{3}, 4\right\}$.

21. $f(x) = \dfrac{2x-1}{x^2 + x - 6}$

$f(x) = \dfrac{2x-1}{(x+3)(x-2)}$

$(x+3)(x-2) = 0$

$x + 3 = 0 \quad x - 2 = 0$

$x = -3 \qquad x = 2$

The domain is $\{x \mid x \neq -3, 2\}$.

23. $f(x) = \dfrac{x+1}{x^2 + 1}$

The domain is $\{x \mid x \in \text{real numbers}\}$..

25. (i), (ii) and (iii) represent rational functions.

Objective B Exercises

27. A rational function is in simplest form when the numerator and denominator have no common factors other than 1.

29. $\dfrac{4-8x}{4} = \dfrac{4(1-2x)}{4} = 1 - 2x$

31. $\dfrac{6x^2-2x}{2x}=\dfrac{2x(3x-1)}{2x}=3x-1$

33. $\dfrac{3x^3y^3-12x^2y^2+15xy}{3xy}=\dfrac{3xy(x^2y^2-4xy+5)}{3xy}$

$=x^2y^2-4xy+5$

35. $\dfrac{8x^2(x-3)}{4x(x-3)}=\dfrac{8x^2}{4x}=2x$

37. $\dfrac{-36a^2-48a}{18a^3+24a^2}=\dfrac{-12a(3a+4)}{6a^2(3a+4)}$

$=\dfrac{-12a}{6a^2}=-\dfrac{2}{a}$

39. $\dfrac{3x-6}{x^2+2x}$

This expression is in simplest form.

41. $\dfrac{x^2-7x+12}{x^2-9x+20}=\dfrac{(x-3)(x-4)}{(x-5)(x-4)}=\dfrac{x-3}{x-5}$

43. $\dfrac{2x^2-5x-3}{2x^2-3x-9}=\dfrac{(2x+1)(x-3)}{(2x+3)(x-3)}=\dfrac{2x+1}{2x+3}$

45. $\dfrac{3x^2+10x-8}{8-14x+3x^2}=\dfrac{(3x-2)(x+4)}{(2-3x)(4-x)}$

$=\dfrac{(3x-2)(x+4)}{-(3x-2)(4-x)}=-\dfrac{x+4}{4-x}=\dfrac{x+4}{x-4}$

47. $\dfrac{a^2-b^2}{a^2+b^3}=\dfrac{(a+b)(a-b)}{(a+b)(a^2-ab+b^2)}$

$=\dfrac{a-b}{a^2-ab+b^2}$

49. $\dfrac{8x^3-y^3}{4x^2-y^2}=\dfrac{(2x-y)(4x^2+2xy+y^2)}{(2x-y)(2x+y)}$

$=\dfrac{4x^2+2xy+y^2}{2x+y}$

51. $\dfrac{x^2(a-2)-a+2}{ax^2-ax}=\dfrac{x^2(a-2)-(a-2)}{ax(x-1)}$

$\dfrac{(a-2)(x^2-1)}{ax(x-1)}=\dfrac{(a-2)(x-1)(x+1)}{ax(x-1)}$

$=\dfrac{(a-2)(x+1)}{ax}$

53. $\dfrac{x^4-2x^2-3}{x^4+2x^2+1}=\dfrac{(x^2+1)(x^2-3)}{(x^2+1)(x^2+1)}=\dfrac{x^2-3}{x^2+1}$

55. $\dfrac{6x^2y^2+11xy+4}{9x^2y^2+9xy-4}=\dfrac{(3xy+4)(2xy+1)}{(3xy+4)(3xy-1)}$

$=\dfrac{2xy+1}{3xy-1}$

57. $\dfrac{a^{2n}+a^n-12}{a^{2n}-2a^n-3}=\dfrac{(a^n-3)(a^n+4)}{(a^n-3)(a^n+1)}=\dfrac{a^n+4}{a^n+1}$

59. $\dfrac{a^{2n}+2a^nb^n+b^{2n}}{a^{2n}-b^{2n}}=\dfrac{(a^n+b^n)(a^n+b^n)}{(a^n+b^n)(a^n-b^n)}$

$=\dfrac{a^n+b^n}{a^n-b^n}$

61. $\dfrac{x^2(a+b)+a+b}{x^4-1}=\dfrac{x^2(a+b)+1(a+b)}{(x^2-1)(x^2+1)}$

$\dfrac{(a+b)(x^2+1)}{(x+1)(x-1)(x^2+1)}=\dfrac{a+b}{(x+1)(x-1)}$

63. $k=4$

Objective B Exercises

65. $\dfrac{15x^2y^4}{24ab^3}\cdot\dfrac{28a^2b^4}{35xy^4}=\dfrac{15\cdot28x^2y^4a^2b^4}{24\cdot35ab^3xy^4}$

$=\dfrac{abx}{2}$

67. $\dfrac{2x^2+4x}{8x^2-40x}\cdot\dfrac{6x^3-30x^2}{3x^2+6x}$

$=\dfrac{2x(x+2)}{8x(x-5)}\cdot\dfrac{6x^2(x-5)}{3x(x+2)}$

$=\dfrac{2x(x+2)\cdot 6x^2(x-5)}{8x(x-5)\cdot 3x(x+2)}$

$=\dfrac{2x\cdot 6x^2}{8x\cdot 3x}=\dfrac{x}{2}$

69. $\dfrac{2x^2-5x+3}{x^6y^3}\cdot\dfrac{x^4y^4}{2x^2-x-3}$

$=\dfrac{(2x-3)(x-1)}{x^6y^3}\cdot\dfrac{x^4y^4}{(2x-3)(x+1)}$

$=\dfrac{(2x-3)(x-1)\cdot x^4y^4}{x^6y^3\cdot(2x-3)(x+1)}$

$=\dfrac{y(x-1)}{x^2(x+1)}$

71. $\dfrac{x^2+x-6}{12+x-x^2}\cdot\dfrac{x^2+x-20}{x^2-4x+4}$

$=\dfrac{(x-2)(x+3)}{(3+x)(4-x)}\cdot\dfrac{(x-4)(x+5)}{(x-2)(x-2)}$

$=\dfrac{(x-2)(x+3)\cdot(x-4)(x+5)}{(3+x)(4-x)\cdot(x-2)(x-2)}$

$=-\dfrac{x+5}{x-2}$

73. $\dfrac{x^4-5x^2+4}{3x^2-4x-4}\cdot\dfrac{3x^2-10x-8}{x^2-4}$

$=\dfrac{(x^2-4)(x^2-1)}{(x-2)(3x+2)}\cdot\dfrac{(3x+2)(x-4)}{(x^2-4)}$

$=\dfrac{(x^2-4)(x+1)(x-1)\cdot(x-4)(3x+2)}{(x-2)(3x+2)\cdot(x^2-4)}$

$=\dfrac{(x+1)(x-1)(x-4)}{x-2}$

75. $\dfrac{x^2-y^2}{x^2+xy+y^2}\cdot\dfrac{x^2-xy}{3x^2-3xy}\cdot\dfrac{x^3-y^3}{x^2-2xy+y^2}$

$=\dfrac{(x-y)(x+y)}{x^2+xy+y^2}\cdot\dfrac{x(x-y)}{3x(x-y)}\cdot\dfrac{(x-y)(x^2+xy+y^2)}{(x-y)(x-y)}$

$=\dfrac{(x-y)(x+y)\cdot x(x-y)\cdot(x-y)(x^2+xy+y^2)}{(x^2+xy+y^2)\cdot 3x(x-y)\cdot(x-y)(x-y)}$

$=\dfrac{x+y}{3}$

77. $\dfrac{x^{2n}+3x^n+2}{x^{2n}-x^n-6}\cdot\dfrac{x^{2n}+x^n-12}{x^{2n}-1}$

$=\dfrac{(x^n+1)(x^n+2)}{(x^n-3)(x^n+2)}\cdot\dfrac{(x^n+4)(x^n-3)}{(x^n-1))(x^n+1)}$

$=\dfrac{(x^n+1)(x^n+2)\cdot(x^n+4)(x^n-3)}{(x^n-3)(x^n+2)\cdot(x^n-1)(x^n+1)}$

$=\dfrac{x^n+4}{x^n-1}$

79. $p(x)=x^2+2x-15$

Objective D Exercises

81. $\dfrac{12a^4b^7}{13x^2y^2}\div\dfrac{18a^5b^6}{26xy^3}=\dfrac{12a^4b^7}{13x^2y^2}\cdot\dfrac{26xy^3}{18a^5b^6}$

$=\dfrac{12\cdot 26a^4b^7xy^3}{13\cdot 18a^5b^6x^2y^2}=\dfrac{4by}{3ax}$

83. $\dfrac{4x^2 - 4y^2}{6x^2 y^2} \div \dfrac{3x^2 + 3xy}{2x^2 y - 2xy^2}$

$= \dfrac{4x^2 - 4y^2}{6x^2 y^2} \cdot \dfrac{2x^2 y - 2xy^2}{3x^2 + 3xy}$

$= \dfrac{4(x-y)(x+y)}{6x^2 y^2} \cdot \dfrac{2xy(x-y)}{3x(x+y)}$

$= \dfrac{4(x-y)(x+y) \cdot 2xy(x-y)}{6x^2 y^2 \cdot 3x(x+y)}$

$= \dfrac{8xy(x-y)(x-y)}{18x^3 y^2}$

$= \dfrac{4(x-y)^2}{9x^2 y}$

85. $\dfrac{8x^3 + 12x^2 y}{4x^2 - 9y^2} \div \dfrac{16x^2 y^2}{4x^2 - 12xy + 9y^2}$

$= \dfrac{8x^3 + 12x^2 y}{4x^2 - 9y^2} \cdot \dfrac{4x^2 - 12xy + 9y^2}{16x^2 y^2}$

$= \dfrac{4x^2(2x+3y)}{(2x+3y)(2x-3y)} \cdot \dfrac{(2x-3y)(2x-3y)}{16x^2 y^2}$

$= \dfrac{4x^2(2x+3y) \cdot (2x-3y)(2x-3y)}{(2x+3y)(2x-3y) \cdot 16x^2 y^2}$

$= \dfrac{4x^2(2x-3y)}{16x^2 y^2}$

$= \dfrac{2x-3y}{4y^2}$

87. $\dfrac{2x^2 + 13x + 20}{8 - 10x - 3x^2} \div \dfrac{6x^2 - 13x - 5}{9x^2 - 3x - 2}$

$= \dfrac{2x^2 + 13x + 20}{8 - 10x - 3x^2} \cdot \dfrac{9x^2 - 3x - 2}{6x^2 - 13x - 5}$

$= \dfrac{(2x+5)(x+4)}{(4+x)(2-3x)} \cdot \dfrac{(3x+1)(3x-2)}{(3x+1)(2x-5)}$

$= \dfrac{(2x+5)(x+4) \cdot (3x+1)(3x-2)}{(4+x)(2-3x) \cdot (3x+1)(2x-5)}$

$= -\dfrac{2x+5}{2x-5}$

89. $\dfrac{16x^2 - 9}{6 - 5x - 4x^2} \div \dfrac{16x^2 + 24x + 9}{4x^2 + 11x + 6}$

$= \dfrac{16x^2 - 9}{6 - 5x - 4x^2} \cdot \dfrac{4x^2 + 11x + 6}{16x^2 + 24x + 9}$

$= \dfrac{(4x+3)(4x-3)}{(2+x)(3-4x)} \cdot \dfrac{(x+2)(4x+3)}{(4x+3)(4x+3)}$

$= \dfrac{(4x+3)(4x-3) \cdot (x+2)(4x+3)}{(2+x)(3-4x) \cdot (4x+3)(4x+3)}$

$= -1$

91. $\dfrac{x^3 + y^3}{2x^3 + 2x^2 y} \div \dfrac{3x^3 - 3x^2 y + 3xy^2}{6x^2 - 6y^2}$

$= \dfrac{x^3 + y^3}{2x^3 + 2x^2 y} \cdot \dfrac{6x^2 - 6y^2}{3x^3 - 3x^2 y + 3xy^2}$

$= \dfrac{(x+y)(x^2 - xy + y^2)}{2x^2(x+y)} \cdot \dfrac{6(x+y)(x-y)}{3x(x^2 - xy + y^2)}$

$= \dfrac{(x+y)(x^2 - xy + y^2) \cdot 6(x+y)(x-y)}{2x^2(x+y) \cdot 3x(x^2 - xy + y^2)}$

$= \dfrac{(x+y)(x-y)}{x^3}$

93. $\dfrac{3x^2 + 10x - 8}{x^2 + 4x + 3} \cdot \dfrac{x^2 + 6x + 9}{3x^2 - 5x + 2} \div \dfrac{x^2 + x - 12}{x^2 - 1}$

$= \dfrac{3x^2 + 10x - 8}{x^2 + 4x + 3} \cdot \dfrac{x^2 + 6x + 9}{3x^2 - 5x + 2} \cdot \dfrac{x^2 - 1}{x^2 + x - 12}$

$= \dfrac{(3x-2)(x+4) \cdot (x+3)(x+3) \cdot (x-1)(x+1)}{(x+3)(x+1) \cdot (3x-2)(x-1) \cdot (x-3)(x+4)}$

$= \dfrac{x+3}{x-3}$

95. $\dfrac{x^{4n}-1}{x^{2n}+x^{n}-2} \div \dfrac{x^{2n}+1}{x^{2n}+3x^{n}+2}$

$= \dfrac{x^{4n}-1}{x^{2n}+x^{n}-2} \cdot \dfrac{x^{2n}+3x^{n}+2}{x^{2n}+1}$

$= \dfrac{(x^{2n}+1)(x^{2n}-1)}{(x^{n}+2)(x^{n}-1)} \cdot \dfrac{(x^{n}+1)(x^{n}+2)}{x^{2n}+1}$

$= \dfrac{(x^{2n}+1)(x^{n}-1)(x^{n}+1) \cdot (x^{n}+1)(x^{n}+2)}{(x^{n}+2)(x^{n}-1) \cdot (x^{2n}+1)}$

$= (x^{n}+1)(x^{n}+1) = (x^{n}+1)^{2}$

97. $p(x) = x^2 - 4$

Applying the Concepts

99. $\dfrac{5y^2-20}{3y^2-12y} \cdot \dfrac{9y^3+6y}{2y^2-4y} \div \dfrac{y^3+2y^2}{2y^2-8y}$

$= \dfrac{5y^2-20}{3y^2-12y} \cdot \dfrac{9y^3+6y}{2y^2-4y} \cdot \dfrac{2y^2-8y}{y^3+2y^2}$

$= \dfrac{5(y+2)(y-2) \cdot 3y(3y^2+2) \cdot 2y(y-4)}{3y(y-4) \cdot 2y(y-2) \cdot y^2(y+2)}$

$= \dfrac{5(3y^2+2)}{y^2}$

Section 6.2

Objective A Exercises

1. a) six
   b) four

3. The LCM is $12x^2y^4$.

$\dfrac{3}{4x^2y} = \dfrac{3}{4x^2y} \cdot \dfrac{3y^3}{3y^3} = \dfrac{9y^3}{12x^2y^4}$

$\dfrac{17}{12xy^4} = \dfrac{17}{12xy^4} \cdot \dfrac{x}{x} = \dfrac{17x}{12x^2y^4}$

5. The LCM is $6x^2(x-2)$.

$\dfrac{x-2}{3x(x-2)} = \dfrac{x-2}{3x(x-2)} \cdot \dfrac{2x}{2x} = \dfrac{2x^2-4x}{6x^2(x-2)}$

$\dfrac{3}{6x^2} = \dfrac{3}{6x^2} \cdot \dfrac{x-2}{x-2} = \dfrac{3x-6}{6x^2(x-2)}$

7. The LCM is $2x(x-5)$.

$\dfrac{3x-1}{2x(x-5)}$

$-3x = \dfrac{-3x}{1} = \dfrac{-3x}{1} \cdot \dfrac{2x(x-5)}{2x(x-5)} = -\dfrac{6x^3-30x^2}{2x(x-5)}$

9. The LCM is $(2x-3)(2x+3)$.

$\dfrac{3x}{2x-3} = \dfrac{3x}{2x-3} \cdot \dfrac{2x+3}{2x+3} = \dfrac{6x^2+9x}{(2x-3)(2x+3)}$

$\dfrac{5x}{2x+3} = \dfrac{5x}{2x+3} \cdot \dfrac{2x-3}{2x-3} = \dfrac{10x^2-15x}{(2x-3)(2x+3)}$

11. The LCM is $(x-3)(x+3)$.

$\dfrac{2x}{x^2-9} = \dfrac{2x}{(x-3)(x+3)}$

$\dfrac{x+1}{x-3} = \dfrac{x+1}{x-3} \cdot \dfrac{x+3}{x+3} = \dfrac{x^2+4x+3}{(x-3)(x+3)}$

13. $3x^2-12y^2 = 3(x+2y)(x-2y)$
$6x-12y = 6(x-2y)$

14. The LCM is $6(x+2y)(x-2y)$.

$$\frac{3}{3x^2-12y^2}=\frac{3}{3(x+2y)(x-2y)}\cdot\frac{2}{2}$$

$$=\frac{6}{6(x+2y)(x-2y)}$$

$$\frac{5}{6x-12y}=\frac{5}{6(x-2y)}\cdot\frac{x+2y}{x+2y}$$

$$=\frac{5x+10y}{6(x-2y)(x+2y)}$$

15. $x^2-1=(x+1)(x-1)$.

$x^2-2x+1=(x-1)(x-1)$

The LCM is

$(x+1)(x-1)(x-1)=(x+1)(x-1)^2$

$$\frac{3x}{x^2-1}=\frac{3x}{(x+1)(x-1)}\cdot\frac{x-1}{x-1}$$

$$=\frac{3x^2-3x}{(x+1)(x-1)^2}$$

$$\frac{5x}{x^2-2x+1}=\frac{5x}{(x-1)(x-1)}\cdot\frac{x+1}{x+1}$$

$$=\frac{5x^2+5x}{(x+1)(x-1)^2}$$

17. $8-x^3=-(x^3-8)=-(x-2)(x^2+2x+4)$

The LCM is $-(x-2)(x^2+2x+4)$.

$$\frac{x-3}{8-x^3}=-\frac{x-3}{(x-2)(x^2+2x+4)}$$

$$\frac{2}{4+2x+x^2}=\frac{2}{x^2+2x+4}\cdot\frac{x-2}{x-2}$$

$$=\frac{2x-4}{(x-2)(x^2+2x+4)}$$

19. $x^2+2x-3=(x+3)(x-1)$

$x^2+6x+9=(x+3)(x+3)$

The LCM is

$(x-1)(x+3)(x+3)=(x-1)(x+3)^2$.

$$\frac{2x}{x^2+2x-3}=\frac{2x}{(x+3)(x-1)}\cdot\frac{x+3}{x+3}$$

$$=\frac{2x^2+6x}{(x-1)(x+3)^2}$$

$$\frac{-x}{x^2+6x+9}=\frac{-x}{(x+3)(x+3)}\cdot\frac{x-1}{x-1}$$

$$=-\frac{x^2-x}{(x-1)(x+3)^3}$$

21. $4x^2-16x+15=(2x-3)(2x-5)$

$6x^2-19x+10=(2x-5)(3x-2)$

The LCM is $(2x-3)(2x-5)(3x-2)$.

$$\frac{-4x}{4x^2-16x+15}=\frac{-4x}{(2x-3)(2x-5)}\cdot\frac{3x-2}{3x-2}$$

$$=-\frac{12x^2-8x}{(2x-3)(2x-5)(3x-2)}$$

$$\frac{3x}{6x^2-19x+10}=\frac{3x}{(2x-5)(3x-2)}\cdot\frac{2x-3}{2x-3}$$

$$=\frac{6x^2-9x}{(2x-3)(2x-5)(3x-2)}$$

23. $6x^2-17x+12=(3x-4)(2x-3)$

$4-3x=-(3x-4)$

The LCM is $(3x-4)(2x-3)$.

$$\frac{5}{6x^2-17x+12}=\frac{5}{(3x-4)(2x-3)}$$

$$\frac{2x}{4-3x}=-\frac{2x}{3x-4}\cdot\frac{2x-3}{2x-3}$$

$$=-\frac{4x^2-6x}{(3x-4)(2x-3)}$$

$$\frac{x+1}{2x-3}=\frac{x+1}{2x-3}\cdot\frac{3x-4}{3x-4}=\frac{3x^2-x-4}{(3x-4)(2x-3)}$$

25. $15 - 2x - x^2 = -(x^2 + 2x - 15)$

$\qquad = -(x+5)(x-3)$

The LCM is $(x+5)(x-3)$.

$$\frac{2x}{x-3} = \frac{2x}{x-3} \cdot \frac{x+5}{x+5} = \frac{2x^2 + 10x}{(x-3)(x+5)}$$

$$\frac{-2}{x+5} = \frac{-2}{x+5} \cdot \frac{x-3}{x-3} = -\frac{2x-6}{(x-3)(x+5)}$$

$$\frac{x-1}{20 - x - x^2} = -\frac{x-1}{(x-3)(x+5)}$$

Objective B Exercises

27. True

29. The LCM is $4x^2$.

$$-\frac{3}{4x^2} + \frac{8}{4x^2} - \frac{3}{4x^2} = \frac{-3 + 8 - 3}{4x^2}$$

$$= \frac{2}{4x^2} = \frac{1}{2x^2}$$

31. The LCM is $3x^2 + x - 10$.

$$\frac{3x}{3x^2 + x - 10} - \frac{5}{3x^2 + x - 10} = \frac{3x - 5}{3x^2 + x - 10}$$

$$= \frac{3x-5}{(3x-5)(x+2)} = \frac{1}{x+2}$$

33. The LCM is $30a^2b^2$.

$$\frac{2}{5ab} - \frac{3}{10a^2b} + \frac{4}{15ab^2}$$

$$= \frac{2}{5ab} \cdot \frac{6ab}{6ab} - \frac{3}{10a^2b} \cdot \frac{3b}{3b} + \frac{4}{15ab^2} \cdot \frac{2a}{2a}$$

$$= \frac{12ab - 9b + 8a}{30a^2b^2}$$

35. The LCM is $40ab$.

$$\frac{3}{4ab} - \frac{2}{5a} + \frac{3}{10b} - \frac{5}{8ab}$$

$$= \frac{3}{4ab} \cdot \frac{10}{10} - \frac{2}{5a} \cdot \frac{8b}{8b} + \frac{3}{10b} \cdot \frac{4a}{4a} - \frac{5}{8ab} \cdot \frac{5}{5}$$

$$= \frac{30 - 16b + 12a - 25}{30ab}$$

$$= \frac{5 - 16b + 12a}{40ab}$$

37. The LCM is $12x$.

$$\frac{3x-4}{6x} - \frac{2x-5}{4x} = \frac{3x-4}{6x} \cdot \frac{2}{2} - \frac{2x-5}{4x} \cdot \frac{3}{3}$$

$$= \frac{2(3x-4) - 3(2x-5)}{12x} = \frac{6x - 8 - 6x + 15}{12x}$$

$$= \frac{7}{12x}$$

39. The LCM is $10x^2y^2$.

$$\frac{2y-4}{5xy^2} + \frac{3-2x}{10x^2y} = \frac{2y-4}{5xy^2} \cdot \frac{2x}{2x} + \frac{3-2x}{10x^2y} \cdot \frac{y}{y}$$

$$= \frac{2x(2y-4) + y(3-2x)}{10x^2y^2} = \frac{4xy - 8x + 3y - 2xy}{10x^2y^2}$$

$$= \frac{2xy - 8x + 3x}{10x^2y^2}$$

41. The LCM is $(a-2)(a+1)$.

$$\frac{3a}{a-2} - \frac{5a}{a+1} = \frac{3a}{a-2} \cdot \frac{a+1}{a+1} - \frac{5a}{a+1} \cdot \frac{a-2}{a-2}$$

$$= \frac{3a(a+1) - 5a(a-2)}{(a-2)(a+1)} = \frac{3a^2 + 3a - 5a^2 + 10a}{(a-2)(a+1)}$$

$$= \frac{-2a^2 + 13a}{(a-2)(a+1)} = -\frac{a(2a+13)}{(a-2)(a+1)}$$

43. The LCM is $(2x-5)(5x-2)$.

$$\frac{x}{2x-5}-\frac{2}{5x-2}=\frac{x}{2x-5}\cdot\frac{5x-2}{5x-2}-\frac{2}{5x-2}\cdot\frac{2x-5}{2x-5}$$

$$=\frac{x(5x-2)-2(2x-5)}{(2x-5)(5x-2)}=\frac{5x^2-2x-4x+10}{(2x-5)(5x-2)}$$

$$=\frac{5x^2-6x+10}{(2x-5)(5x-2)}$$

45. The LCM is $b(a-b)$.

$$\frac{1}{a-b}+\frac{1}{b}=\frac{1}{a-b}\cdot\frac{b}{b}+\frac{1}{b}\cdot\frac{a-b}{a-b}$$

$$=\frac{b+a-b}{b(a-b)}=\frac{a}{b(a-b)}$$

47. The LCM is $a(a-3)$.

$$\frac{6a}{a-3}-5+\frac{3}{a}=\frac{6a}{a-3}\cdot\frac{a}{a}-\frac{5}{1}\cdot\frac{a(a-3)}{a(a-3)}+\frac{3}{a}\cdot\frac{a-3}{a-3}$$

$$=\frac{6a^2-5a^2+15a+3a-9}{a(a-3)}$$

$$=\frac{a^2+18a-9}{a(a-3)}$$

49. The LCM is $x(6x-5)$.

$$\frac{5}{x}-\frac{5x}{5-6x}+2$$

$$=\frac{5}{x}\cdot\frac{6x-5}{6x-5}-\frac{-5x}{6x-5}\cdot\frac{x}{x}+\frac{2}{1}\cdot\frac{x(6x-5)}{x(6x-5)}$$

$$=\frac{5(6x-5)+5x(x)+2(x)(6x-5)}{x(6x-5)}$$

$$=\frac{30x-25+5x^2+12x^2-10x}{x(6x-5)}$$

$$=\frac{17x^2+20x-25}{x(6x-5)}$$

51. $x^2-6x+9=(x-3)(x-3)$

$x^2-9=(x+3)(x-3)$

The LCM is

$(x+3)(x-3)(x-3)=(x+3)(x-3)^2$.

$$\frac{1}{x^2-6x+9}-\frac{1}{x^2-9}$$

$$=\frac{1}{(x-3)(x-3)}\cdot\frac{x+3}{x+3}-\frac{1}{(x+3)(x-3)}\cdot\frac{x-3}{x-3}$$

$$=\frac{x+3-x+3}{(x+3)(x-3)^2}$$

$$=\frac{6}{(x+3)(x-3)^2}$$

53. $x^2+4x+4=(x+2)(x+2)$

The LCM is $(x+2)(x+2)=(x+2)^2$.

$$\frac{1}{x+2}-\frac{3x}{x^3+4x+4}$$

$$=\frac{1}{(x+2)}\cdot\frac{x+2}{x+2}-\frac{3x}{(x+2)(x+2)}$$

$$=\frac{x+2-3x}{(x+2)(x+2)}=\frac{-2x+2}{(x+2)^2}$$

$$=-\frac{2(x-1)}{(x+2)^2}$$

55. $x^2+2x-8=(x+4)(x-2)$

The LCM is $(x+4)(x-2)$.

$$\frac{-3x^2+8x+2}{x^2+2x-8}-\frac{2x-5}{x+4}$$

$$=\frac{-3x^2+8x+2}{(x+4)(x-2)}-\frac{2x-5}{x+4}\cdot\frac{x-2}{x-2}$$

$$=\frac{-3x^2+8x+2-2x^2+9x-10}{(x+4)(x-2)}$$

$$=\frac{-5x^2+17x-8}{(x+4)(x-2)}$$

$$=-\frac{5x^2-17x+8}{(x+4)(x-2)}$$

57. $x^{2n} - 1 = (x^n + 1)(x^n - 1)$

The LCM is $(x^n + 1)(x^n - 1)$.

$$\frac{x^n}{x^{2n} - 1} + \frac{2}{x^n + 1}$$

$$= \frac{2}{x^n - 1} \cdot \frac{x^n + 1}{x^n + 1} + \frac{x^n}{(x^n + 1)(x^n - 1)}$$

$$= \frac{2(x^n + 1) + x^n}{(x^n + 1)(x^n - 1)} = \frac{2x^n + 2 + x^n}{(x^n + 1)(x^n - 1)}$$

$$= \frac{3x^n + 2}{(x^n + 1)(x^n - 1)}$$

59. $x^{2n} - x^n - 6 = (x^n - 3)(x^n + 2)$

The LCM is
$(x^n - 3)(x^n + 2)$.

$$\frac{2x^n - 6}{x^{2n} - x^n - 6} + \frac{x^n}{x^n + 2}$$

$$= \frac{2x^n - 6}{(x^n - 3)(x^n + 2)} + \frac{x^n}{(x^n + 2)} \cdot \frac{x^n - 3}{x^n - 3}$$

$$= \frac{2x^n - 6 + x^{2n} - 3x^n}{(x^n - 3)(x^n + 2)}$$

$$= \frac{x^{2n} - x^n - 6}{(x^n - 3)(x^n + 2)}$$

$$= \frac{(x^n - 3)(x^n + 2)}{(x^n - 3)(x^n + 2)}$$

$$= 1$$

61. $4x^2 - 36 = 4(x^2 - 9) = 4(x - 3)(x + 3)$

The LCM is $4(x - 3)(x + 3)$.

$$\frac{x^2 + 4}{4x^2 - 36} - \frac{13}{x + 3}$$

$$= \frac{x^2 + 4}{4(x - 3)(x + 3)} - \frac{13}{x + 3} \cdot \frac{4(x - 3)}{4(x - 3)}$$

$$= \frac{x^2 + 4 - 13(4)(x - 3)}{4(x - 3)(x + 3)} = \frac{x^2 + 4 - 52x + 156}{4(x - 3)(x + 3)}$$

$$= \frac{x^2 - 52x + 160}{4(x - 3)(x + 3)}$$

63. $4x^2 + 9x + 2 = (4x + 1)(x + 2)$

The LCM is $(4x + 1)(x + 2)$.

$$\frac{3x - 4}{4x + 1} + \frac{3x + 6}{4x^2 + 9x + 2}$$

$$= \frac{3x - 4}{4x + 1} \cdot \frac{x + 2}{x + 2} + \frac{3x + 6}{(4x + 1)(x + 2)}$$

$$= \frac{3x^2 + 2x - 8 + 3x + 6}{(4x + 1)(x + 2)} = \frac{3x^2 + 5x - 2}{(4x + 1)(x + 2)}$$

$$= \frac{(3x - 1)(x + 2)}{(4x + 1)(x + 2)}$$

$$= \frac{3x - 1}{4x + 1}$$

65. $x^2 + x - 12 = (x + 4)(x - 3)$

$x^2 + 7x + 12 = (x + 4)(x + 3)$

The LCM is $(x + 4)(x - 3)(x + 3)$.

$$\frac{x + 1}{x^2 + x - 12} - \frac{x - 3}{x^2 + 7x + 12}$$

$$= \frac{x + 1}{(x + 4)(x - 3)} \cdot \frac{x + 3}{x + 3} - \frac{x - 3}{(x + 4)(x + 3)} \cdot \frac{x - 3}{x - 3}$$

$$= \frac{x^2 + 4x + 3 - x^2 + 6x - 9}{(x + 4)(x - 3)(x + 3)}$$

$$= \frac{10x - 6}{(x + 4)(x - 3)(x + 3)}$$

67. $x^2 - 2x - 15 = (x - 5)(x + 3)$

$5 - x = -(x - 5)$

The LCM is $(x - 5)(x + 3)$.

$$\frac{2x^2 - 2x}{x^2 - 2x - 15} - \frac{2}{x + 3} + \frac{x}{5 - x}$$

$$= \frac{2x^2 - 2x}{(x - 5)(x + 3)} - \frac{2}{x + 3} \cdot \frac{x - 5}{x - 5} + \frac{-(x)}{x - 5} \cdot \frac{x + 3}{x + 3}$$

$$= \frac{2x^2 - 2x - 2x + 10 - x^2 - 3x}{(x-5)(x+3)}$$

$$= \frac{x^2 - 7x + 10}{(x-5)(x+3)} = \frac{(x-5)(x-2)}{(x-5)(x+3)}$$

$$= \frac{x-2}{x+3}$$

69. $3x^2 - 11x - 20 = (3x+4)(x-5)$

The LCM is $(3x+4)(x-5)$

$$\frac{x}{3x+4} + \frac{3x+2}{x-5} - \frac{7x^2 + 24x + 28}{3x^2 - 11x - 20}$$

$$= \frac{x}{3x+4} \cdot \frac{x-5}{x-5} + \frac{3x+2}{x-5} \cdot \frac{3x+4}{3x+4} - \frac{7x^2 + 24x + 28}{(3x+4)(x-5)}$$

$$= \frac{x^2 - 5x + 9x^2 + 18x + 8 - 7x^2 - 24x - 28}{(3x+4)(x-5)}$$

$$= \frac{3x^2 - 11x - 20}{(2x-5)(x-2)} = \frac{(3x+4)(x-5)}{(3x+4)(x-5)}$$

$$= 1$$

71. $8x^2 - 10x + 3 = (4x-3)(2x-1)$

$1 - 2x = -(2x-1)$

The LCM is $(4x-3)(2x-1)$.

$$\frac{x+1}{1-2x} - \frac{x+3}{4x-3} + \frac{10x^2 + 7x - 9}{8x^2 - 10x + 3}$$

$$= \frac{-(x+1)}{2x-1} \cdot \frac{4x-3}{4x-3} - \frac{x+3}{4x-3} \cdot \frac{2x-1}{2x-1} + \frac{10x^2 + 7x - 9}{(4x-3)(2x-1)}$$

$$= \frac{-4x^2 - x + 3 - 2x^2 - 5x + 3 + 10x^2 + 7x - 9}{(4x-3)(2x-1)}$$

$$= \frac{4x^2 + x - 3}{(4x-3)(2x-1)} = \frac{(4x-3)(x+1)}{(4x-3)(2x-1)}$$

$$= \frac{x+1}{2x-1}$$

73. $8x^3 - 1 = (2x-1)(4x^2 + 2x + 1)$

The LCM is $(2x-1)(4x^2 + 2x + 1)$.

$$\frac{2x}{4x^2 + 2x + 1} + \frac{4x+1}{8x^3 - 1}$$

$$= \frac{2x}{4x^2 + 2x + 1} \cdot \frac{2x-1}{2x-1} + \frac{4x+1}{(2x-1)(4x^2 + 2x + 1)}$$

$$= \frac{4x^2 - 2x + 4x + 1}{(2x-1)(4x^2 + 2x + 1)} = \frac{4x^2 + 2x + 1}{(2x-1)(4x^2 + 2x + 1)}$$

$$= \frac{1}{2x-1}$$

75. $x^4 - 16 = (x^2 + 4)(x^2 - 4)$

The LCM is $(x^2 + 4)(x^2 - 4)$.

$$\frac{x^2 - 12}{x^4 - 16} + \frac{1}{x^2 - 4} - \frac{1}{x^2 + 4}$$

$$= \frac{x^2 - 12}{(x^2 + 4)(x^2 - 4)} + \frac{1}{x^2 - 4} \cdot \frac{x^2 + 4}{x^2 + 4} - \frac{1}{x^2 + 4} \cdot \frac{x^2 - 4}{x^2 - 4}$$

$$= \frac{x^2 - 12 + x^2 + 4 - x^2 + 4}{(x^2 + 4)(x^2 - 4)}$$

$$= \frac{x^2 - 4}{(x^2 + 4)(x^2 - 4)}$$

$$= \frac{1}{x^2 + 4}$$

77. The LCM is $9a^2$.

$$\left(\frac{a-3}{a^2}-\frac{a-3}{9}\right)\div\frac{a^2-9}{3a}$$

$$=\left(\frac{a-3}{a^2}\cdot\frac{9}{9}-\frac{a-3}{9}\cdot\frac{a^2}{a^2}\right)\div\frac{a^2-9}{3a}$$

$$=\left(\frac{9(a-3)-a^2(a-3)}{9a^2}\right)\div\frac{a^2-9}{3a}$$

$$=\frac{(9-a^2)(a-3)}{9a^2}\cdot\frac{3a}{a^2-9}$$

$$=\frac{(3-a)(3+a)(a-3)}{9a^2}\cdot\frac{3a}{(a+3)(a-3)}$$

$$=\frac{3-a}{3a}$$

79. $\dfrac{x^2-4x+4}{2x+1}\cdot\dfrac{2x^2+x}{x^3-4x}-\dfrac{3x-2}{x+1}$

$$=\frac{(x-2)(x-2)}{2x+1}\cdot\frac{x(2x+1)}{x(x+2)(x-2)}-\frac{3x-2}{x+1}$$

$$=\frac{x-2}{x+2}-\frac{3x-2}{x+1}$$

The LCM is $(x+2)(x+1)$.

$$\frac{x-2}{x+2}-\frac{3x-2}{x+1}$$

$$=\frac{x-2}{x+2}\cdot\frac{x+1}{x+1}-\frac{3x-2}{x+1}\cdot\frac{x+2}{x+2}$$

$$=\frac{x^2-x-2-3x^2-4x+4}{x(x-2)9x+2)}=\frac{-2x^2-5x+2}{(x+2)(x+1)}$$

$$=-\frac{2x^2+5x-2}{(x+2)(x+1)}$$

81. The LCM is ab.

$$\left(\frac{a-2b}{b}+\frac{b}{a}\right)\div\left(\frac{b+a}{a}-\frac{2a}{b}\right)$$

$$=\left(\frac{a-2b}{b}\cdot\frac{a}{a}+\frac{b}{a}\cdot\frac{b}{b}\right)\div\left(\frac{b+a}{a}\cdot\frac{b}{b}-\frac{2a}{b}\cdot\frac{a}{a}\right)$$

$$=\left(\frac{a^2-2ab+b^2}{ab}\right)\div\left(\frac{b^2+ab-2a^2}{ab}\right)$$

$$=\frac{(a-b)(a-b)}{ab}\cdot\frac{ab}{(b+2a)(b-a)}$$

$$=\frac{a-b}{-(b+2a)}$$

$$=-\frac{a-b}{b+2a}=\frac{b-a}{b+2a}$$

83. $\dfrac{2x}{x^2-x-6}-\dfrac{6x-6}{2x^2-9x+9}\div\dfrac{x^2+x-2}{2x-3}$

$$=\frac{2x}{x^2-x-6}-\frac{6(x-1)}{(2x-3)(x-3)}\div\frac{(x+2)(x-1)}{2x-3}$$

$$=\frac{2x}{(x-3)(x+2)}-\frac{6(x-1)}{(2x-3)(x-3)}\cdot\frac{2x-3}{(x+2)(x-1)}$$

$$=\frac{2x}{(x-3)(x+2)}-\frac{6}{(x-3)(x+2)}$$

$$=\frac{2x-6}{(x-3)(x+2)}=\frac{2(x-3)}{(x-3)(x+2)}$$

$$=\frac{2}{x+2}$$

Applying the Concepts

85. $f(x)=\dfrac{x}{x+2},\quad f(4)=\dfrac{2}{3}$

$$g(x)=\frac{4}{x-3},\quad g(4)=4$$

$$S(x)=\frac{x^2+x+8}{x^2-x-6},\quad S(4)=4\frac{2}{3}$$

$$f(4)+g(4)=\frac{2}{3}+4=4\frac{2}{3}=S(4)$$

Section 6.3

Objective A Exercises

1. A complex fraction is a fraction whose numerator or denominator contains one or more fractions.

3. The LCM is 3.

$$\frac{2-\dfrac{1}{3}}{4+\dfrac{11}{3}}=\frac{2-\dfrac{1}{3}}{4+\dfrac{11}{3}}\cdot\frac{3}{3}=\frac{2\cdot3-\dfrac{1}{3}\cdot3}{4\cdot3+\dfrac{11}{3}\cdot3}$$

$$=\frac{6-1}{12+11}=\frac{5}{23}$$

5. The LCM is 6.

$$\frac{3-\dfrac{2}{3}}{5+\dfrac{5}{6}}=\frac{3-\dfrac{2}{3}}{5+\dfrac{5}{6}}\cdot\frac{6}{6}=\frac{3\cdot6-\dfrac{2}{3}\cdot6}{5\cdot6+\dfrac{5}{6}\cdot6}$$

$$=\frac{18-4}{30+5}=\frac{14}{35}=\frac{2}{5}$$

7. The LCM is x^2.

$$\frac{1+\dfrac{1}{x}}{1-\dfrac{1}{x^2}}=\frac{1+\dfrac{1}{x}}{1-\dfrac{1}{x^2}}\cdot\frac{x^2}{x^2}=\frac{1\cdot x^2+\dfrac{1}{x}\cdot x^2}{1\cdot x^2-\dfrac{1}{x^2}\cdot x^2}$$

$$=\frac{x^2+x}{x^2-1}=\frac{x(x+1)}{(x-1)(x+1)}$$

$$=\frac{x}{x-1}$$

9. The LCM is a.

$$\frac{a-2}{\dfrac{4}{a}-a}=\frac{a-2}{\dfrac{4}{a}-a}\cdot\frac{a}{a}=\frac{a\cdot a-2\cdot a}{\dfrac{4}{a}\cdot a-a\cdot a}$$

$$=\frac{a^2-2a}{4-a^2}=\frac{a(a-2)}{(2-a)(2+a)}$$

$$=-\frac{a}{a+2}$$

11. The LCM is $x(x+h)$.

$$\frac{\dfrac{1}{x+h}-\dfrac{1}{x}}{h}=\frac{\dfrac{1}{x+h}-\dfrac{1}{x}}{h}\cdot\frac{x(x+h)}{x(x+h)}$$

$$=\frac{x-(x+h)}{hx(x+h)}=\frac{-h}{hx(x+h)}$$

$$=-\frac{1}{x(x+h)}$$

13. The LCM is x.

$$\frac{x-\dfrac{1}{x}}{x+\dfrac{1}{x}}=\frac{x-\dfrac{1}{x}}{x+\dfrac{1}{x}}\cdot\frac{x}{x}$$

$$=\frac{x^2-1}{x^2+1}$$

15. The LCM is a^2.

$$\frac{\dfrac{1}{a^2}-\dfrac{1}{a}}{\dfrac{1}{a^2}+\dfrac{1}{a}}=\frac{\dfrac{1}{a^2}-\dfrac{1}{a}}{\dfrac{1}{a^2}+\dfrac{1}{a}}\cdot\frac{a^2}{a^2}$$

$$=\frac{1-a}{1+a}$$

17. The LCM is $x+2$.

$$\frac{2-\dfrac{4}{x+2}}{5-\dfrac{10}{x+2}}=\frac{2-\dfrac{4}{x+2}}{5-\dfrac{10}{x+2}}\cdot\frac{x+2}{x+2}$$

$$=\frac{2(x+2)-4}{5(x+2)-10}=\frac{2x+4-4}{5x+10-10}$$

$$=\frac{2x}{5x}=\frac{2}{5}$$

19. The LCM is $2a - 3$.

$$\frac{\dfrac{3}{2a-3}+2}{\dfrac{-6}{2a-3}-4} = \frac{\dfrac{3}{2a-3}+2}{\dfrac{-6}{2a-3}-4} \cdot \frac{2a-3}{2a-3}$$

$$= \frac{3+2(2a-3)}{-6-4(2a-3)} = \frac{3+4a-6}{-6-8a+12}$$

$$= \frac{4a-3}{-8a+6} = \frac{4a-3}{-2(4a-3)}$$

$$= -\frac{1}{2}$$

21. The LCM is $(x-4)(x+1)$.

$$\frac{1-\dfrac{1}{x-4}}{1-\dfrac{6}{x+1}} = \frac{1-\dfrac{1}{x-4}}{1-\dfrac{6}{x+1}} \cdot \frac{(x-4)(x+1)}{(x-4)(x+1)}$$

$$= \frac{(x-4)(x+1)-(x+1)}{(x-4)(x+1)-6(x-4)}$$

$$= \frac{x^2-3x-4-x-1}{x^2-3x-4-6x+24}$$

$$= \frac{x^2-4x-5}{x^2-9x+20} = \frac{(x-5)(x+1)}{(x-5)(x-4)}$$

$$= \frac{x+1}{x-4}$$

23. The LCM is $(x-3)(2-x)$.

$$\frac{1-\dfrac{2}{x-3}}{1+\dfrac{3}{2-x}} = \frac{1-\dfrac{2}{x-3}}{1+\dfrac{3}{2-x}} \cdot \frac{(x-3)(2-x)}{(x-3)(2-x)}$$

$$= \frac{(x-3)(2-x)-2(2-x)}{(x-3)(2-x)+3(x-3)}$$

$$= \frac{2x-x^2-6+3x-4+2x}{2x-x^2-6+3x+3x-9}$$

$$= \frac{-x^2+7x-10}{-x^2+8x-15} = \frac{-(x^2-7x+10)}{-(x^2-8x+15)}$$

$$= \frac{(x-5)(x-2)}{(x-5)(x-3)}$$

$$= \frac{x-2}{x-3}$$

25. The LCM is $(2x+3)$.

$$\frac{x-4+\dfrac{9}{2x+3}}{x+3-\dfrac{5}{2x+3}} = \frac{x-4+\dfrac{9}{2x+3}}{x+3-\dfrac{5}{2x+3}} \cdot \frac{2x+3}{2x+3}$$

$$= \frac{x(2x+3)-4(2x+3)+9}{x(2x+3)+3(2x+3)-5}$$

$$= \frac{2x^2+3x-8x-12+9}{2x^2+3x+6x+9-5}$$

$$= \frac{2x^2-5x-3}{2x^2+9x+4} = \frac{(2x+1)(x-3)}{(2x+1)(x+4)}$$

$$= \frac{x-3}{x+4}$$

27. The LCM is $(x+4)(x-3)$.

$$\dfrac{x-3+\dfrac{10}{x+4}}{x+7+\dfrac{16}{x-3}} = \dfrac{x-3+\dfrac{10}{x+4}}{x+7+\dfrac{16}{x-3}} \cdot \dfrac{(x+4)(x-3)}{(x+4)(x-3)}$$

$$= \dfrac{x(x+4)(x-3)-3(x+4)(x-3)+10(x-3)}{x(x+4)(x-3)+7(x+4)(x-3)+16(x+4)}$$

$$= \dfrac{(x-3)(x^2+4x-3x-12+10)}{(x+4)(x^2-3x+7x-21+16)}$$

$$= \dfrac{(x-3)(x^2+x-2)}{(x+4)(x^2+4x-5)} = \dfrac{(x-3)(x+2)(x-1)}{(x+4)(x+5)(x-1)}$$

$$= \dfrac{(x-3)(x+2)}{(x+4)(x+5)}$$

29. The LCM is x^2.

$$\dfrac{1-\dfrac{1}{x}-\dfrac{6}{x^2}}{1-\dfrac{4}{x}+\dfrac{3}{x^2}} = \dfrac{1-\dfrac{1}{x}-\dfrac{6}{x^2}}{1-\dfrac{4}{x}+\dfrac{3}{x^2}} \cdot \dfrac{x^2}{x^2}$$

$$= \dfrac{x^2-x-6}{x^2-4x+3} = \dfrac{(x-3)(x+2)}{(x-3)(x-1)}$$

$$= \dfrac{x+2}{x-1}$$

31. The LCM is x^2.

$$\dfrac{1+\dfrac{1}{x}-\dfrac{12}{x^2}}{\dfrac{9}{x^2}+\dfrac{3}{x}-2} = \dfrac{1+\dfrac{1}{x}-\dfrac{12}{x^2}}{\dfrac{9}{x^2}+\dfrac{3}{x}-2} \cdot \dfrac{x^2}{x^2}$$

$$= \dfrac{x^2+x-12}{9+3x-2x^2} = \dfrac{(x-3)(x+4)}{(3-x)(3+2x)}$$

$$= -\dfrac{x+4}{2x+3}$$

33. The LCM is x^2y^2.

$$\dfrac{\dfrac{1}{y^2}-\dfrac{1}{xy}-\dfrac{2}{x^2}}{\dfrac{1}{y^2}-\dfrac{3}{xy}+\dfrac{2}{x^2}} = \dfrac{\dfrac{1}{y^2}-\dfrac{1}{xy}-\dfrac{2}{x^2}}{\dfrac{1}{y^2}-\dfrac{3}{xy}+\dfrac{2}{x^2}} \cdot \dfrac{x^2y^2}{x^2y^2}$$

$$= \dfrac{x^2-xy-2y^2}{x^2-3xy+2y^2} = \dfrac{(x+y)(x-2y)}{(x-y)(x-2y)}$$

$$= \dfrac{x+y}{x-y}$$

35. The LCM is $x(x+1)$.

$$\dfrac{\dfrac{x}{x+1}-\dfrac{1}{x}}{\dfrac{x}{x+1}+\dfrac{1}{x}} = \dfrac{\dfrac{x}{x+1}-\dfrac{1}{x}}{\dfrac{x}{x+1}+\dfrac{1}{x}} \cdot \dfrac{x(x+1)}{x(x+1)}$$

$$= \dfrac{x^2-(x+1)}{x^2+(x+1)} = \dfrac{x^2-x-1}{x^2+x+1}$$

37. The LCM is $a(a-2)$.

$$\dfrac{\dfrac{1}{a}-\dfrac{3}{a-2}}{\dfrac{2}{a}+\dfrac{5}{a-2}} = \dfrac{\dfrac{1}{a}-\dfrac{3}{a-2}}{\dfrac{2}{a}+\dfrac{5}{a-2}} \cdot \dfrac{a(a-2)}{a(a-2)}$$

$$= \dfrac{a-2-3a}{2a-4+5a} = \dfrac{-2a-2}{7a-4}$$

$$= -\dfrac{2a+2}{7a-4} = -\dfrac{2(a+1)}{7a-4}$$

39. The LCM is $(x-1)(x+1)$.

$$\dfrac{\dfrac{x-1}{x+1}-\dfrac{x+1}{x-1}}{\dfrac{x-1}{x+1}+\dfrac{x+1}{x-1}} = \dfrac{\dfrac{x-1}{x+1}-\dfrac{x+1}{x-1}}{\dfrac{x-1}{x+1}+\dfrac{x+1}{x-1}} \cdot \dfrac{(x-1)(x+1)}{(x-1)(x+1)}$$

$$= \dfrac{(x-1)(x-1)-(x+1)(x+1)}{(x-1)(x-1)+(x+1)(x+1)}$$

$$= \dfrac{x^2-2x+1-x^2-2x-1}{x^2-2x+1+x^2+2x+1} = \dfrac{-4x}{2x^2+2}$$

$$= \dfrac{-4x}{2(x^2+1)} = -\dfrac{2x}{x^2+1}$$

41. The LCM is a.

$$a + \frac{a}{a + \dfrac{1}{a}} = a + \frac{a}{a + \dfrac{1}{a}} \cdot \frac{a}{a}$$

$$= a + \frac{a^2}{a^2 + 1}$$

The LCM is $a^2 + 1$.

$$a + \frac{a^2}{a^2 + 1} = a \cdot \frac{a^2 + 1}{a^2 + 1} + \frac{a^2}{a^2 + 1}$$

$$= \frac{a^3 + a + a^2}{a^2 + 1}$$

$$= \frac{a^3 + a^2 + a}{a^2 + 1}$$

43. The LCM is $1 - a$.

$$a - \frac{a}{1 - \dfrac{a}{1-a}} = a - \frac{a}{1 - \dfrac{a}{1-a}} \cdot \frac{1-a}{1-a}$$

$$= a - \frac{a - a^2}{1 - a - a} = a - \frac{a - a^2}{1 - 2a}$$

The LCM is $1 - 2a$.

$$a - \frac{a - a^2}{1 - 2a} = a \cdot \frac{1 - 2a}{1 - 2a} - \frac{a - a^2}{1 - 2a}$$

$$= \frac{a - 2a^2 - a + a^2}{1 - 2a}$$

$$= \frac{-a^2}{1 - 2a} = -\frac{a^2}{1 - 2a}$$

45. The LCM is x.

$$3 - \frac{2}{1 - \dfrac{2}{3 - \dfrac{2}{x}}} = 3 - \frac{2}{1 - \dfrac{2}{3 - \dfrac{2}{x}} \cdot \dfrac{x}{x}}$$

$$= 3 - \frac{2}{1 - \dfrac{2x}{3x - 2}}$$

The LCM is $3x - 2$.

$$3 - \frac{2}{1 - \dfrac{2x}{3x - 2}} = 3 - \frac{2}{1 - \dfrac{2x}{3x - 2}} \cdot \frac{3x - 2}{3x - 2}$$

$$= 3 - \frac{6x - 4}{3x - 2 - 2x} = 3 - \frac{6x - 4}{x - 2}$$

The LCM is $x - 2$.

$$3 - \frac{6x - 4}{x - 2} = 3 \cdot \frac{x - 2}{x - 2} - \frac{6x - 4}{x - 2}$$

$$= \frac{3x - 6 - 6x + 4}{x - 2} = \frac{-3x - 2}{x - 2}$$

$$= -\frac{3x + 2}{x - 2}$$

47. $\dfrac{1}{1 - \dfrac{1}{a}} = \dfrac{1}{1 - \dfrac{1}{a}} \cdot \dfrac{a}{a} = \dfrac{a}{a - 1}$

The reciprocal is $\dfrac{a - 1}{a}$

Applying the Concepts

49. $(3 + 3^{-1})^{-1} = \left(3 + \dfrac{1}{3}\right)^{-1} = \left(\dfrac{10}{3}\right)^{-1} = \dfrac{3}{10}$

51. $\dfrac{x^{-1}}{x^{-1} + 2^{-1}} = \dfrac{\dfrac{1}{x}}{\dfrac{1}{x} + \dfrac{1}{2}} = \dfrac{\dfrac{1}{x}}{\dfrac{2 + x}{2x}} = \dfrac{1}{x} \cdot \dfrac{2x}{2 + x} = \dfrac{2}{2 + x}$

Section 6.4

Objective A Exercises

1. A ratio is the quotient of two quantities that have the same unit. A rate is the quotient of two quantities that have different units.

3. $\dfrac{x}{30} = \dfrac{3}{10}$.

$$\frac{x}{30} \cdot 30 = \frac{3}{10} \cdot 30$$

$$x = 9$$

The solution is 9.

5. $\dfrac{2}{x} = \dfrac{8}{30}$

$\dfrac{2}{x} \cdot 30x = \dfrac{8}{30} \cdot 30x$

$60 = 8x$

$\dfrac{15}{2} = x$

The solution is $\dfrac{15}{2}$.

7. $\dfrac{x+1}{10} = \dfrac{2}{5}$

$\dfrac{x+1}{10} \cdot 10 = \dfrac{2}{5} \cdot 10$

$x + 1 = 4$

$x = 3$

The solution is 3.

9. $\dfrac{4}{x+2} = \dfrac{3}{4}$

$\dfrac{4}{x+2} \cdot 4(x+2) = \dfrac{3}{4} \cdot 4(x+2)$

$16 = 3(x+2)$

$16 = 3x + 6$

$10 = 3x$

$x = \dfrac{10}{3}$

The solution is $\dfrac{10}{3}$.

11. $\dfrac{x}{4} = \dfrac{x-2}{8}$

$\dfrac{x}{4} \cdot 8 = \dfrac{x-2}{8} \cdot 8$

$2x = x - 2$

$x = -2$

The solution is -2.

13. $\dfrac{16}{2-x} = \dfrac{4}{x}$

$\dfrac{16}{2-x} \cdot x(2-x) = \dfrac{4}{x} \cdot x(2-x)$

$16x = 4(2-x)$

$16x = 8 - 4x$

$20x = 8$

$x = \dfrac{8}{20} = \dfrac{2}{5}$

The solution is $\dfrac{2}{5}$.

15. $\dfrac{8}{x-2} = \dfrac{4}{x+1}$

$\dfrac{8}{x-2} \cdot (x+1)(x+2) = \dfrac{4}{x+1} \cdot (x+1)(x-2)$

$8(x+1) = 4(x-2)$

$8x + 8 = 4x - 8$

$4x = -16$

$x = -4$

The solution is -4.

17. $\dfrac{x}{3} = \dfrac{x+1}{7}$

$\dfrac{x}{3} \cdot 21 = \dfrac{x+1}{7} \cdot 21$

$7x = 3(x+1)$

$7x = 3x + 3$

$4x = 3$

$x = \dfrac{3}{4}$

The solution is $\dfrac{3}{4}$.

19. $\dfrac{8}{3x-2} = \dfrac{2}{2x+1}$

$\dfrac{8}{3x-2} \cdot (3x-2)(2x+1) = \dfrac{2}{2x+1} \cdot (3x-2)(2x+1)$

$8(2x+1) = 2(3x-2)$

$16x+8 = 6x-4$

$10x = -12$

$x = \dfrac{-12}{10} = -\dfrac{6}{5}$

The solution is $-\dfrac{6}{5}$.

21. $\dfrac{3x+1}{3x-4} = \dfrac{x}{x-2}$

$\dfrac{3x+1}{3x-4} \cdot (3x-4)(x-2) = \dfrac{x}{x-2} \cdot (3x-4)(x-2)$

$(3x+1)(x-2) = x(3x-4)$

$3x^2 - 5x - 2 = 3x^2 - 4x$

$-2 = x$

The solution is -2.

23. True

Objective B Exercises

25. Strategy: Let x represent the number of grams of protein in a 454–gram box of pasta. Write and solve a proportion.

Solution: $\dfrac{56}{7} = \dfrac{454}{x}$

$\dfrac{56}{7} \cdot 7x = \dfrac{454}{x} \cdot 7x$

$56x = 3178$

$x = 56.75$

There are 56.75 g of protein in a 454-gram box of pasta.

27. Strategy: Let x represent the number of computers with defective USB ports. Write and solve a proportion.

Solution: $\dfrac{300}{15} = \dfrac{70{,}000}{x}$

$\dfrac{300}{15} \cdot 15x = \dfrac{70{,}000}{x} \cdot 15x$

$300x = 1{,}050{,}000$

$x = 3500$

There are 3500 computers with defective USB ports.

29. Strategy: Let x represent the cost in dollars. Write and solve a proportion.

Solution: $\dfrac{1}{0.346} = \dfrac{11}{x}$

$\dfrac{1}{0.346} \cdot 0.346x = \dfrac{11}{x} \cdot 0.346x$

$x = 3.81$

A gallon of milk would cost \$3.81.

31. Strategy: Let w represent the width of the room. Let l represent the length of the room. Write and solve two proportions to determine the dimensions of the room.

Solution: $\dfrac{\frac{1}{4}}{1} = \dfrac{\frac{9}{2}}{w}$, $\quad \dfrac{\frac{1}{4}}{1} = \dfrac{6}{l}$

$\dfrac{\frac{1}{4}}{1} \cdot w = \dfrac{\frac{9}{2}}{w} \cdot w \qquad \dfrac{\frac{1}{4}}{1} \cdot l = \dfrac{6}{l} \cdot l$

$\dfrac{1}{4}w = \dfrac{9}{2} \qquad\qquad \dfrac{1}{4}l = 6$

$w = 18 \qquad\qquad\qquad l = 24$

The room dimensions are 18 ft by 24 ft.

33. Strategy: Let x represent the number of miles needed to walk to lose one pound. Write and solve a proportion.

Solution:

$$\frac{4}{650} = \frac{x}{3500}$$

$$\frac{4}{650} \cdot (650)(3500) = \frac{x}{3500} \cdot (650)(3500)$$

$$14000 = 650x$$

$$21.54 = x$$

A person must walk 21.54 mi to lose one pound.

35. Strategy: Let x represent the number of minutes of talk time. Write and solve a proportion.

Solution: $\dfrac{30}{13} = \dfrac{x}{5}$

$$\frac{30}{13} \cdot (13)(5) = \frac{x}{5} \cdot (13)(5)$$

$$150 = 13x$$

$$11.5 = x$$

It would provide 12 minutes of talk time.

37. Strategy: Let x represent the value of 1 U.S. dollar in Euros. Write and solve a proportion.

Solution: $\dfrac{1}{0.59} = \dfrac{1.21}{x}$

$$\frac{1}{0.59} \cdot (0.59)(x) = \frac{1.21}{x} \cdot (0.59)(x)$$

$$x = 0.59(1.21) = 0.71$$

The value of one U.S. dollar in Euros is $.71.

Section 6.5

Objective A Exercises

1. -3 and 4

3. $\dfrac{x}{2} + \dfrac{5}{6} = \dfrac{x}{3}$

$$6\left(\frac{x}{2} + \frac{5}{6}\right) = 6\left(\frac{x}{3}\right)$$

$$3x + 5 = 2x$$

$$5 = -x$$

$$-5 = x$$

The solution is -5.

5. $\dfrac{8}{2x-1} = 2$

$$(2x-1)\left(\frac{8}{2x-1}\right) = (2x-1)(2)$$

$$8 = 4x - 2$$

$$10 = 4x$$

$$x = \frac{10}{4} = \frac{5}{2}$$

The solution is $\dfrac{5}{2}$.

7. $1 - \dfrac{3}{y} = 4$

$$y\left(1 - \frac{3}{y}\right) = 4(y)$$

$$y - 3 = 4y$$

$$-3 = 3y$$

$$-1 = y$$

The solution is -1.

9. $\dfrac{3}{x-2} = \dfrac{4}{x}$

$$x(x-2)\left(\frac{3}{x-2}\right) = x(x-2)\left(\frac{4}{x}\right)$$

$$3x = 4(x-2)$$

$$3x = 4x - 8$$

$$8 = x$$

The solution is 8.

11. $\dfrac{6}{2y+3} = \dfrac{6}{y}$

$y(2y+3)\left(\dfrac{6}{2y+3}\right) = y(2y+3)\left(\dfrac{6}{y}\right)$

$6y = 6(2y+3)$

$6y = 12y + 18$

$-6y = 18$

$y = -3$

The solution is −3.

13. $\dfrac{5}{y+3} - 2 = \dfrac{7}{y+3}$

$(y+3)\left(\dfrac{5}{y+3} - 2\right) = (y+3)\left(\dfrac{7}{y+3}\right)$

$5 - 2(y+3) = 7$

$5 - 2y - 6 = 7$

$-2y - 1 = 7$

$-2y = 8$

$y = -4$

The solution is −4.

15. $\dfrac{-4}{a-4} = 3 - \dfrac{a}{a-4}$

$(a-4)\left(\dfrac{-4}{a-4}\right) = (a-4)\left(3 - \dfrac{a}{a-4}\right)$

$-4 = 3a - 12 - a$

$-4 = 2a - 12$

$8 = 2a$

$4 = a$

$a = 4$ is not in the domain of the rational expressions. The equation has no solution.

17. $\dfrac{4}{x-5} = 9 - \dfrac{2x}{x-5}$

$(x-5)\left(\dfrac{4}{x-5}\right) = (x-5)\left(9 - \dfrac{2x}{x-5}\right)$

$4 = 9(x-5) - 2x$

$4 = 9x - 45 - 2x$

$4 = 7x - 45$

$49 = 7x$

$7 = x$

The solution is 7.

19. $\dfrac{4}{x-2} - 4 = \dfrac{-12}{x+2}$

$(x+2)(x-2)\left(\dfrac{4}{x-2} - 4\right) = (x+2)(x-2)\left(\dfrac{-12}{x+2}\right)$

$4(x+2) - 4(x+2)(x-2) = -12(x-2)$

$4x + 8 - 4x^2 + 16 = -12x + 24$

$-4x^2 + 4x + 24 = -12x + 24$

$0 = 4x^2 - 16x$

$0 = 4x(x-4)$

$4x = 0 \qquad x - 4 = 0$

$x = 0 \qquad x = 4$

The solutions are 0 and 4.

21. $\dfrac{4}{x-4} = x + \dfrac{x}{x-4}$

$(x-4)\left(\dfrac{4}{x-4}\right) = (x-4)\left(x + \dfrac{x}{x-4}\right)$

$4 = x(x-4) + x$

$4 = x^2 - 4x + x$

$0 = x^2 - 3x - 4$

$0 = (x+1)(x-4)$

$x + 1 = 0 \qquad x - 4 = 0$

$x = -1 \qquad x = 4$

$x = 4$ is not in the domain of the rational expressions. The solution is −1.

23. $\dfrac{2x}{x+2} + 3x = \dfrac{-5}{x+2}$

$(x+2)\left(\dfrac{2x}{x+2} + 3x\right) = (x+2)\left(\dfrac{-5}{x+2}\right)$

$2x + 3x(x+2) = -5$

$2x + 3x^2 + 6x = -5$

$3x^2 + 8x + 5 = 0$

$(3x+5)(x+1) = 0$

$3x + 5 = 0 \qquad x + 1 = 0$

$3x = -5 \qquad\quad x = -1$

$x = -\dfrac{5}{3}$

The solutions are $-\dfrac{5}{3}$ and -1.

25. $\dfrac{x}{2x-9} - 3x = \dfrac{10}{9-2x}$

$(2x-9)\left(\dfrac{x}{2x-9} - 3x\right) = (2x-9)\left(\dfrac{10}{9-2x}\right)$

$x - 3x(2x-9) = -10$

$x - 6x^2 + 27x = -10$

$-6x^2 + 28x + 10 = 0$

$-2(3x^2 - 14x - 5) = 0$

$-2(3x+1)(x-5) = 0$

$3x + 1 = 0 \qquad x - 5 = 0$

$3x = -1 \qquad\quad x = 5$

$x = -\dfrac{1}{3}$

The solutions are $-\dfrac{1}{3}$ and 5.

27. $\dfrac{5}{x-2} - \dfrac{2}{x+2} = \dfrac{3}{x^2-4}$

$\dfrac{5}{x-2} - \dfrac{2}{x+2} = \dfrac{3}{(x-2)(x+2)}$

$(x-2)(x+2)\left(\dfrac{5}{x-2} - \dfrac{2}{x+2}\right) = (x-2)(x+2)\left(\dfrac{3}{(x-2)(x+2)}\right)$

$5(x+2) - 2(x-2) = 3$

$5x + 10 - 2x + 4 = 3$

$3x + 14 = 3$

$3x = -11$

$x = -\dfrac{11}{3}$

The solution is $-\dfrac{11}{3}$.

29. $\dfrac{9}{x^2+7x+10}=\dfrac{5}{x+2}-\dfrac{3}{x+5}$

$\dfrac{9}{(x+2)(x+5)}=\dfrac{5}{x+2}-\dfrac{3}{x+5}$

$(x+2)(x+5)\left(\dfrac{9}{(x+2)(x+5)}\right)=(x+2)(x+5)\left(\dfrac{5}{x+2}-\dfrac{3}{x+5}\right)$

$9=5(x+5)-3(x+2)$

$9=5x+25-3x-6$

$9=2x+19$

$-10=2x$

$-5=x$

$x=-5$ is not in the domain of the rational expressions. The equation has no solution.

Objective B Exercises

31. Strategy: Let t represent the time required for an experienced bricklayer to do the job. time required for an inexperienced bricklayer is $2t$.

	Rate	Time	Part
Experienced bricklayer	$\dfrac{1}{t}$	6	$\dfrac{6}{t}$
Inexperienced bricklayer	$\dfrac{1}{2t}$	16	$\dfrac{16}{2t}$

The sum of the parts of the task completed by each bricklayer must equal 1.

Solution:

$\dfrac{6}{t}+\dfrac{16}{2t}=1$

$2t\left(\dfrac{6}{t}+\dfrac{16}{2t}\right)=1(2t)$

$12+16=2t$

$28=2t$

$14=t$

Working alone the experienced bricklayer can do the job in 14 h.

33. Strategy: Let t represent the time required by the second machine working alone.

	Rate	Time	Part
1st machine	$\dfrac{1}{40}$	30	$\dfrac{30}{40}$
2nd machine	$\dfrac{1}{t}$	15	$\dfrac{15}{t}$

The sum of the part of the task completed by the 1st machine and the part completed by the 2nd machine must equal 1.

Solution: $\dfrac{30}{40}+\dfrac{15}{t}=1$

$40t\left(\dfrac{30}{40}+\dfrac{15}{t}\right)=1(40t)$

$30t+600=40t$

$600=10t$

$60=t$

It would take the second machine 60 min.

35. Strategy: Let t represent the time required to fill the bottles working together.

	Rate	Time	Part
1st machine	$\dfrac{1}{10}$	t	$\dfrac{t}{10}$
2nd machine	$\dfrac{1}{12}$	t	$\dfrac{t}{12}$
3rd machine	$\dfrac{1}{15}$	t	$\dfrac{t}{15}$

The sum of the parts of the task completed by each machine must equal 1.

Solution: $\dfrac{t}{10}+\dfrac{t}{12}+\dfrac{t}{15}=1$

$60\left(\dfrac{t}{10}+\dfrac{t}{12}+\dfrac{t}{15}\right)=1(60)$

$6t+5t+4t=60$

$15t=60$

$t=4$

When all three machines are working together, it takes 4 h to fill the bottles.

37. Strategy: Let t represent the time required to empty the tank working together.

	Rate	Time	Part
Inlet pipe	$\dfrac{1}{45}$	t	$\dfrac{t}{45}$
Outlet pipe	$\dfrac{1}{30}$	t	$\dfrac{t}{30}$

The part completed by the outlet pipe minus the part complete by the inlet pipe equals 1.

Solution: $\dfrac{t}{30}-\dfrac{t}{45}=1$

$90\left(\dfrac{t}{30}-\dfrac{t}{45}\right)=1(90)$

$3t-2t=90$

$t=90$

It would take 90 min to empty the tank.

39. Strategy: Let t represent the time required for the clowns to blow up 76 balloons.

	Rate	Time	Part
1st clown	$\dfrac{1}{2}$	t	$\dfrac{t}{2}$
2nd clown	$\dfrac{1}{3}$	t	$\dfrac{t}{3}$
Balloon popping	$\dfrac{1}{5}$	t	$\dfrac{t}{5}$

The sum of the tasks completed by the two clowns minus the balloons popping is equal to 76. This gives us the number of balloons filled after t minutes.

Solution: $\dfrac{t}{2}+\dfrac{t}{3}-\dfrac{t}{5}=76$

$30\left(\dfrac{t}{2}+\dfrac{t}{3}-\dfrac{t}{5}\right)=76(30)$

$15t+10t-6t=2280$

$19t=2280$

$t=120$

It will take 120 min to have 76 balloons.

41. Strategy: Let t represent the time required to address 140 envelopes.

	Rate	Time	Part
1st clerk	$\dfrac{1}{70}$	t	$\dfrac{t}{70}$
2nd clerk	$\dfrac{3}{280}$	t	$\dfrac{3t}{280}$

The part of the task completed by the two clerks must equal 1.

Solution: $\dfrac{t}{70}+\dfrac{3t}{280}=1$

$280\left(\dfrac{t}{70}+\dfrac{3t}{280}\right)=1(280)$

$4t+3t=280$

$7t=280$

$t=40$

With both clerks working together, it will take 40 min (2400 s) to complete the task.

43. t is less than n.

Objective C Exercises

45. Strategy: Let r represent the rate of the runner.

The rate of the bicyclist is $r + 7$.

	Distance	Rate	Time
Runner	16	r	$\dfrac{16}{r}$
Bicyclist	30	$r+7$	$\dfrac{30}{r+7}$

The time traveled by the runner equals the time traveled by the bicyclist.

Solution: $\dfrac{16}{r} = \dfrac{30}{r+7}$

$r(r+7)\left(\dfrac{16}{r}\right) = r(r+7)\left(\dfrac{30}{r+7}\right)$

$16(r+7) = 30r$

$16r + 112 = 30r$

$112 = 14r$

$8 = r$

The rate of the runner is 8 mph.

47. Strategy: Let r represent the rate of the tortoise.

The rate of the hare is $180r$.

	Distance	Rate	Time
Tortoise	360	r	$\dfrac{360}{r}$
Hare	360	$180r$	$\dfrac{360}{180r}$

The time for the tortoise is 14 min 55 s (or 895 s) more than the hare.

Solution: $\dfrac{360}{180r} + 895 = \dfrac{360}{r}$

$\dfrac{2}{r} + 895 = \dfrac{360}{r}$

$r\left(\dfrac{2}{r} + 895\right) = r\left(\dfrac{360}{r}\right)$

$2 + 895r = 360$

$895r = 358$

$r = 0.4$

$180r = 180(0.4) = 72$

The rate of the tortoise is 0.4 ft/s.
The rate of the hare is 72 ft/s.

49. Strategy: Let r represent the rate of the jogger.

The rate of the cyclist is $2r$.

	Distance	Rate	Time
Runner	30	r	$\dfrac{30}{r}$
Cyclist	30	$2r$	$\dfrac{30}{2r}$

The time for the jogger is 3 h more than the time for the cyclist.

Solution: $\dfrac{30}{r} = \dfrac{30}{2r} + 3$

$2r\left(\dfrac{30}{r}\right) = 2r\left(\dfrac{30}{2r} + 3\right)$

$60 = 30 + 6r$

$30 = 6r$

$5 = r$

$2r = 2(5) = 10$

The rate of the cyclist is 10 mph.

51. Strategy: Let r represent the rate of the helicopter.

The rate of the commercial jet is $4r$.

	Distance	Rate	Time
Helicopter	105	r	$\dfrac{105}{r}$
Commercial jet	735	$4r$	$\dfrac{735}{4r}$

The total time was 2.2 h.

Solution: $\dfrac{105}{r} + \dfrac{735}{4r} = 2.2$

$4r\left(\dfrac{105}{r} + \dfrac{735}{4r}\right) = 4r(2.2)$

$420 + 735 = 8.8r$

$1155 = 8.8r$

$r = 131.25$

$4r = 4(131.25) = 525$

The rate of the jet was 525 mph.

53. Strategy: Let r represent the rate of the current.

	Distance	Rate	Time
With current	20	$7 + r$	$\dfrac{20}{7+r}$
Against current	8	$7 - r$	$\dfrac{8}{7-r}$

The time traveling with the current equals the time traveling against the current.

Solution: $\dfrac{20}{7+r} = \dfrac{8}{7-r}$

$(7+r)(7-r)\left(\dfrac{20}{7+r}\right) = (7+r)(7-r)\left(\dfrac{8}{7-r}\right)$

$20(7-r) = 8(7+r)$

$140 - 20r = 56 + 8r$

$84 = 28r$

$3 = r$

The rate of the current is 3 mph.

55. **Strategy:** Let r represent the rate of the wind.

	Distance	Rate	Time
With wind	3059	$550 + r$	$\dfrac{3059}{550+r}$
Against wind	2450	$550 - r$	$\dfrac{2450}{550-r}$

The time traveling with the wind equals the time traveling against the wind.

Solution: $\dfrac{3059}{550+r} = \dfrac{2450}{550-r}$

$(550+r)(550-r)\left(\dfrac{3059}{550+r}\right) = (550+r)(550-r)\left(\dfrac{2450}{550-r}\right)$

$3059(550 - r) = 2450(550 + r)$

$1{,}682{,}450 - 3059r = 1{,}347{,}500 + 2450r$

$1{,}682{,}450 = 1{,}347{,}500 + 5509r$

$334{,}950 = 5509r$

$r = 60.80$

The rate of the wind is 60.80 mph.

57. Strategy: Let r represent the rate of the current.

	Distance	Rate	Time
With current	16	$6 + r$	$\dfrac{16}{6+r}$
Against current	16	$6 - r$	$\dfrac{16}{6-r}$

The total time is 6 h.

Solution: $\dfrac{16}{6+r} + \dfrac{16}{6-r} = 6$

$(6+r)(6-r)\left(\dfrac{16}{6+r} + \dfrac{16}{6-r}\right) = (6+r)(6-r)(6)$

$16(6 - r) + 16(6 + r) = 6(36 - r^2)$

$96 - 16r + 96 + 16r = 216 - 6r^2$

$192 = 216 - 6r^2$

$0 = 24 - 6r^2$

$0 = 6(4 - r^2)$

$0 = 6(2 - r)(2 + r)$

$2 - r = 0 \quad 2 + r = 0$

$r = 2 \qquad r = -2$

The rate of the current cannot be a negative number. The rate of the current is 2 mph.

Applying the Concepts

59. a) $\dfrac{x-2}{y} = \dfrac{x+2}{5y}$

$5y\left(\dfrac{x-2}{y}\right) = 5y\left(\dfrac{x+2}{5y}\right)$

$5(x - 2) = x + 2$

$5x - 10 = x + 2$

$4x = 12$

$x = 3$

The solution is 3.

b) $\dfrac{x}{x+y} = \dfrac{2x}{4y}$

$$4y(x+y)\left(\dfrac{x}{x+y}\right) = 4y(x+y)\left(\dfrac{2x}{4y}\right)$$

$4xy = 2x(x+y)$

$4xy = 2x^2 + 2xy$

$2xy = 2x^2$

$2x^2 - 2xy = 0$

$2x(x-y) = 0$

$2x = 0 \qquad x - y = 0$

$x = 0 \qquad\quad x = y$

The solutions are 0 and y.

c) $\dfrac{x-y}{x} = \dfrac{2x}{9y}$

$$9xy\left(\dfrac{x-y}{x}\right) = 9xy\left(\dfrac{2x}{9y}\right)$$

$9y(x-y) = 2x^2$

$9xy - 9y^2 = 2x^2$

$2x^2 - 9xy + 9y^2 = 0$

$(2x-3y)(x-3y) = 0$

$2x - 3y = 0 \qquad x - 3y = 0$

$2x = 3y \qquad\qquad x = 3y$

$x = \dfrac{3}{2}y$

The solutions are $3y$ and $\dfrac{3}{2}y$.

61. $-\dfrac{a-b}{b}$

Section 6.6

Objective A Exercises

1. **Strategy:** Write the basic direct variation equation replacing the variables with the given values. Solve for k.
Write the direct variation equation replacing

k with its value. Substitute 5000 for s and solve for P.

Solution:
$P = ks$

$4000 = 250k$

$16 = k$

$P = 16s = 16(5000) = 80,000$

When the company sells 5000 products, its profit will be $80,000.

3. **Strategy:** Write the basic direct variation equation replacing the variables with the given values. Solve for k.
Write the direct variation equation replacing k with its value. Substitute 15 for d and solve for p.

Solution:
$p = kd$

$4.5 = 10k$

$0.45 = k$

$p = 0.45d = 0.45(15) = 6.75$
The pressure is 6.75 lb/in^2.

5. **Strategy:** Write the basic direct variation equation replacing the variables with the given values. Solve for k.
Write the direct variation equation replacing k with its value. Substitute 10 for t and solve for d.

Solution:
$d = kt^2$

$144 = k(3)^2$

$144 = 9k$

$16 = k$

$d = 16t^2 = 16(10^2) = 16(100) = 1600$
In 10 s the object will fall 1600 ft.

7. **Strategy:** Write the basic inverse variation equation replacing the variables with the given values. Solve for k.
Write the inverse variation equation replacing k with its value. Substitute 5 for n and solve for T.

Solution:

$$T = \frac{k}{n}$$

$$500 = \frac{k}{1}$$

$$500 = k$$

$$T = \frac{500}{n} = \frac{500}{5} = 100$$

It will take 5 computers 100 s to solve the same problem.

9. **Strategy:** Write the basic direct variation equation replacing the variables with the given values. Solve for k.
    Write the direct variation equation replacing k with its value. Substitute 15 for v and solve for L.

    Solution:

    $$L = kv^2$$

    $$640 = k(20)^2$$

    $$640 = 400k$$

    $$1.6 = k$$

    $$L = 1.6v^2 = 1.6(15)^2 = 360$$

    The load on the sail will be 360 lbs.

11. **Strategy:** Write the basic direct variation equation replacing the variables with the given values. Solve for k.
    Write the direct variation equation replacing k with its value. Substitute 230,000 for p and solve for c.

    Solution:

    $$c = kp$$

    $$15600 = 260,000k$$

    $$0.06 = k$$

    $$c = 0.06p = 0.06(230,000) = 13,800$$

    The commission on a $230,000 home is $13,800.

13. **Strategy:** Write the basic combined variation equation replacing the variables with the given values. Solve for k.

Write the combined variation equation replacing k with its value.
Substitute 24 for r, 180 for v and solve for I.

Solution:

$$I = \frac{kv}{r}$$

$$10 = \frac{110k}{11}$$

$$110 = 110k$$

$$1 = k$$

$$I = \frac{1v}{r} = \frac{180}{24} = 7.5$$

The current is 7.5 amps.

15. **Strategy:** Write the basic inverse variation equation replacing the variables with the given values. Solve for k.
    Write the inverse variation equation replacing k with its value. Substitute 5 for d and solve for I.

    Solution:

    $$I = \frac{k}{d^2}$$

    $$12 = \frac{k}{10^2}$$

    $$12 = \frac{k}{100}$$

    $$1200 = k$$

    $$I = \frac{1200}{d^2} = \frac{1200}{5^2} = \frac{1200}{25} = 48$$

    The intensity is 48 foot-candles when the distance is 5 ft.

17. If x is doubled then y is doubled.

Applying the Concepts

19. inversely

21. inversely

Chapter 6 Review Exercises

1. $\dfrac{a^6 b^5 + a^5 b^6}{a^5 b^4 - a^4 b^4} \cdot \dfrac{a-b}{a^2 - b^2}$

$= \dfrac{a^5 b^5 (a+b)}{a^4 b^4 (a-1)} \cdot \dfrac{a-b}{(a+b)(a-b)}$

$= \dfrac{a^5 b^5 (a+b)(a-b)}{a^4 b^4 (a-1)(a+b)(a-b)}$

$= \dfrac{ab}{a-1}$

2. The LCM is $(x-3)(x+2)$

$\dfrac{x}{x-3} - 4 - \dfrac{2x-5}{x+2}$

$= \dfrac{x}{x-3} \cdot \dfrac{x+2}{x+2} - 4 \cdot \dfrac{(x-3)(x+2)}{(x-3)(x+2)} - \dfrac{2x-5}{x+2} \cdot \dfrac{x-3}{x-3}$

$= \dfrac{x(x+2) - 4(x-3)(x+2) - (2x-5)(x-3)}{(x-3)(x+2)}$

$= \dfrac{x^2 + 2x - 4x^2 + 4x + 24 - 2x^2 + 11x - 15}{(x-3)(x+2)}$

$= \dfrac{x^2 + 2x - 4x^2 + 4x + 24 - 2x^2 + 11x - 15}{(x-3)(x+2)}$

$= \dfrac{-5x^2 + 17x + 9}{(x-3)(x+2)}$

$= -\dfrac{5x^2 - 17x - 9}{(x-3)(x+2)}$

3. $P(x) = \dfrac{x}{x-3}$

$P(4) = \dfrac{4}{4-3} = \dfrac{4}{1} = 4$

4. $\dfrac{3x-2}{x+6} = \dfrac{3x+1}{x+9}$

$(x+6)(x+9)\left(\dfrac{3x-2}{x+6}\right) = (x+6)(x+9)\left(\dfrac{3x+1}{x+9}\right)$

$(3x-2)(x+9) = (3x+1)(x+6)$

$3x^2 + 25x - 18 = 3x^2 + 19x + 6$

$25x - 18 = 19x + 6$

$6x = 24$

$x = 4$

The solution is 4.

5. $\dfrac{\dfrac{3x+4}{3x-4} + \dfrac{3x-4}{3x+4}}{\dfrac{3x-4}{3x+4} - \dfrac{3x+4}{3x-4}}$

The LCM is $(3x-4)(3x+4)$.

$\dfrac{\dfrac{3x+4}{3x-4} + \dfrac{3x-4}{3x+4}}{\dfrac{3x-4}{3x+4} - \dfrac{3x+4}{3x-4}} \cdot \dfrac{(3x-4)(3x+4)}{(3x-4)(3x+4)}$

$= \dfrac{(3x+4)^2 + (3x-4)^2}{(3x-4)^2 - (3x+4)^2}$

$= \dfrac{9x^2 + 24x + 16 + 9x^2 - 24x + 16}{9x^2 - 24x + 16 - 9x^2 - 24x - 16}$

$= \dfrac{18x^2 + 32}{-48x} = \dfrac{2(9x^2 + 16)}{-48x}$

$= -\dfrac{9x^2 + 16}{24x}$

6. The LCM is $(4x-1)(4x+1)$.

$\dfrac{4x}{4x-1} \cdot \dfrac{4x+1}{4x+1} = \dfrac{16x^2 + 4x}{(4x-1)(4x+1)}$

$\dfrac{3x-1}{4x+1} \cdot \dfrac{4x-1}{4x-1} = \dfrac{12x^2 - 7x + 1}{(4x-1)(4x+1)}$

7. $r = \dfrac{2}{3-r}$

$r(3-r) = \dfrac{2}{3-r} \cdot (3-r)$

$3r - r^2 = 2$

$0 = r^2 - 3r + 2$

$0 = (r-2)(r-1)$

$0 = r-2 \qquad 0 = r-1$

$2 = r \qquad\quad 1 = r$

The solutions are 1 and 2.

8. $P(x) = \dfrac{x^2 - 2}{3x^2 - 2x + 5}$

$P(-2) = \dfrac{(-2)^2 - 2}{3(-2)^2 - 2(-2) + 5}$

$= \dfrac{4-2}{12+4+5}$

$= \dfrac{2}{21}$

9. $\dfrac{10}{5x+3} = \dfrac{2}{10x-3}$

$(5x+3)(10x-3)\left(\dfrac{10}{5x+3}\right) = (5x+3)(10x-3)\left(\dfrac{2}{10x-3}\right)$

$10(10x-3) = 2(5x+3)$

$100x - 30 = 10x + 6$

$90x = 36$

$x = \dfrac{36}{90} = \dfrac{2}{5}$

10. $f(x) = \dfrac{2x-7}{3x^2 + 3x - 18}$

$3x^2 + 3x - 18 = 0$

$3(x^2 + x - 6) = 0$

$3(x+3)(x-2) = 0$

$x+3 = 0 \quad x - 2 = 0$

$x = -3 \qquad x = 2$

The domain is $\{x \mid x \neq -3, 2\}$.

11. $\dfrac{3x^4 + 11x^2 - 4}{3x^4 + 13x^2 + 4} = \dfrac{(3x^2 - 1)(x^2 + 4)}{(3x^2 + 1)(x^2 + 4)}$

$= \dfrac{3x^2 - 1}{3x^2 + 1}$

12. $g(x) = \dfrac{2x}{x-3}$

$x - 3 = 0$

$x = 3$

The domain is $\{x \mid x \neq 3\}$.

13. $\dfrac{x^3 - 8}{x^3 + 2x^2 + 4x} \cdot \dfrac{x^3 + 2x^2}{x^2 - 4}$

$= \dfrac{(x-2)(x^2 + 2x + 4)}{x(x^2 + 2x + 4)} \cdot \dfrac{x^2(x+2)}{(x-2)(x+2)}$

$= \dfrac{(x-2)(x^2 + 2x + 4)x^2(x+2)}{x(x^2 + 2x + 4)(x-2)(x+2)}$

$= x$

14. The LCM is $x^2 - 4$.

$\dfrac{3x^2 + 2}{x^2 - 4} - \dfrac{9 - x^2}{x^2 - 4} = \dfrac{3x^2 + 2 - 9x + x^2}{x^2 - 4}$

$= \dfrac{4x^2 - 9x + 2}{x^2 - 4} = \dfrac{(4x-1)(x-2)}{(x+2)(x-2)}$

$= \dfrac{4x-1}{x+2}$

15. $\dfrac{4}{x-3} = \dfrac{5}{x}$

$\dfrac{4}{x-3} \cdot x(x-3) = \dfrac{5}{x} \cdot x(x-3)$

$4x = 5(x-3)$

$4x = 5x - 15$

$-x = -15$

$x = 15$

The solution is 15.

16. $\dfrac{30}{x^2+5x+4}+\dfrac{10}{x+4}=\dfrac{4}{x+1}$

$\dfrac{30}{(x+4)(x+1)}+\dfrac{10}{x+4}=\dfrac{4}{x+1}$

$(x+4)(x+1)\left(\dfrac{30}{(x+4)(x+1)}+\dfrac{10}{x+4}\right)=(x+4)(x+1)\left(\dfrac{4}{x+1}\right)$

$30+10(x+1)=4(x+4)$

$30+10x+10=4x+16$

$10x+40=4x+16$

$6x=-24$

$x=-4$

$x=-4$ is not in the domain of the rational expressions. The equation has no solution.

17. $\dfrac{x+2+\dfrac{3}{x-4}}{x-3-\dfrac{2}{x-4}}$

$\dfrac{x+2+\dfrac{3}{x-4}}{x-3-\dfrac{2}{x-4}}\cdot\dfrac{x-4}{x-4}=\dfrac{(x+2)(x-4)+3}{(x-3)(x-4)-2}$

$=\dfrac{x^2-2x-8+3}{x^2-7x+12-2}$

$=\dfrac{x^2-2x-5}{x^2-7x+10}$

18. $x^2-9x+20=(x-4)(x-5)$

$4-x=-(x-4)$

The LCM is $(x-5)(x-4)$.

$\dfrac{x-3}{x-5}\cdot\dfrac{x-4}{x-4}=\dfrac{x^2-7x+12}{(x-5)(x-4)}$

$\dfrac{x}{x^2-9x+20}=\dfrac{x}{(x-5)(x-4)}$

$\dfrac{1}{4-x}=\dfrac{-1}{x-4}\cdot\dfrac{x-5}{x-5}=\dfrac{-x+5}{(x-5)(x-4)}$

$=-\dfrac{x-5}{(x-5)(x-4)}$

19. $\dfrac{6}{2x-3}=\dfrac{5}{x+5}+\dfrac{5}{2x^2+7x-15}$

$\dfrac{6}{2x-3}=\dfrac{5}{x+5}+\dfrac{5}{(2x-3)(x+5)}$

$(2x-3)(x+5)\left(\dfrac{6}{2x-3}\right)=(2x-3)(x+5)\left(\dfrac{5}{x+5}+\dfrac{5}{(2x-3)(x+5)}\right)$

$6(x+5)=5(2x-3)+5$

$6x+30=10x-15+5$

$6x+30=10x-10$

$40=4x$

$10=x$

The solution is 10.

20. $\dfrac{5}{x-4} - \dfrac{2x+1}{x^2-3x-4}$

$= \dfrac{5}{x-4} - \dfrac{2x+1}{(x-4)(x+1)}$

$= \dfrac{5}{x-4} \cdot \dfrac{x+1}{x+1} - \dfrac{2x+1}{(x-4)(x+1)}$

$= \dfrac{5x+5-2x-1}{(x-4)(x+1)} = \dfrac{3x+4}{(x-4)(x+1)}$

21. $\dfrac{27x^3-8}{9x^3+6x^2+4x} \div \dfrac{9x^2-12x+4}{9x^2-4}$

$= \dfrac{27x^3-8}{9x^3+6x^2+4x} \cdot \dfrac{9x^2-4}{9x^2-12x+4}$

$= \dfrac{(3x-2)(9x^2+6x+4)}{x(9x^2+6x+4)} \cdot \dfrac{(3x+2)(3x-2)}{(3x-2)(3x-2)}$

$= \dfrac{(3x-2)(9x^2+6x+4)(3x+2)(3x-2)}{x(9x^2+6x+4)(3x-2)(3x-2)}$

$= \dfrac{3x+2}{x}$

22. $3x^2-7x+2 = (3x-1)(x-2)$

The LCM is $(3x-1)(x-2)$.

$\dfrac{6x}{3x^2-7x+2} - \dfrac{2}{3x-1} + \dfrac{3x}{x-2}$

$= \dfrac{6x}{(3x-1)(x-2)} - \dfrac{2}{3x-1} \cdot \dfrac{x-2}{x-2} + \dfrac{3x}{x-2} \cdot \dfrac{3x-1}{3x-1}$

$= \dfrac{6x-2(x-2)+3x(3x-1)}{(3x-1)(x-2)}$

$= \dfrac{6x-2x+4+9x^2-3x}{(3x-1)(x-2)}$

$= \dfrac{9x^2+x+4}{(3x-1)(x-2)}$

23. $\dfrac{x^3-27}{x^2-9} = \dfrac{(x-3)(x^2+3x+9)}{(x-3)(x+3)} = \dfrac{x^2+3x+9}{x+3}$

24. The LCM is $(x-3)(x+2)$.

$\dfrac{2x}{x-3} - \dfrac{5}{x+2}$

$= \dfrac{2x}{x-3} \cdot \dfrac{x+2}{x+2} - \dfrac{5}{x+2} \cdot \dfrac{x-3}{x-3}$

$= \dfrac{2x^2+4x-5x+15}{(x-3)(x+2)}$

$= \dfrac{2x^2-x+15}{(x-3)(x+2)}$

25. Since $3x^2+4>0$ for all x, the domain is $\{x \mid x \in \text{real numbers}\}$.

26. The LCM is $(x+1)(x-2)$.

$\dfrac{x-4}{x+1} + \dfrac{x-3}{x-2}$

$= \dfrac{x-4}{x+1} \cdot \dfrac{x-2}{x-2} + \dfrac{x-3}{x-2} \cdot \dfrac{x+1}{x+1}$

$= \dfrac{x^2-6x+8+x^2-2x-3}{(x-2)(x+1)}$

$= \dfrac{2x^2-8x+5}{(x-2)(x+1)}$

27. $\dfrac{16-x^2}{x^3-2x^2-8x} = \dfrac{(4-x)(4+x)}{x(x^2-2x-8)}$

$= \dfrac{(4-x)(4+x)}{x(x-4)(x+2)}$

$= -\dfrac{x+4}{x(x+2)}$

28. $\dfrac{8x^3-27}{4x^2-9} = \dfrac{(2x-3)(4x^2+6x+9)}{(2x-3)(2x+3)}$

$= \dfrac{4x^2+6x+9}{2x+3}$

29. $\dfrac{16-x^2}{6x+12} \cdot \dfrac{x^2+5x+6}{x^2-8x+16}$

$= \dfrac{(4-x)(4+x)}{6(x+2)} \cdot \dfrac{(x+3)(x+2)}{(x-4)(x-4)}$

$= \dfrac{(4-x)(4+x)(x+3)(x+2)}{6(x+2)(x-4)(x-4)}$

$= -\dfrac{(x+4)(x+3)}{6(x-4)}$

30. $\dfrac{x^2-5x+4}{x^2-2x-8} \div \dfrac{x^2-4x+3}{x^2+8x+12}$

$= \dfrac{x^2-5x+4}{x^2-2x-8} \cdot \dfrac{x^2+8x+12}{x^2-4x+3}$

$= \dfrac{(x-4)(x-1)}{(x-4)(x+2)} \cdot \dfrac{(x+2)(x+6)}{(x-3)(x-1)}$

$= \dfrac{(x-4)(x-1)(x+2)(x+6)}{(x-4)(x+2)(x-3)(x-1)}$

$= \dfrac{x+6}{x-3}$

31. $\dfrac{8x^3-64}{4x^3+4x^2+x} \div \dfrac{x^2+2x+4}{4x^2-1}$

$= \dfrac{8x^3-64}{4x^3+4x^2+x} \cdot \dfrac{4x^2-1}{x^2+2x+4}$

$= \dfrac{8(x-2)(x^2+2x+4)}{x(2x+1)(2x+1)} \cdot \dfrac{(2x+1)(2x-1)}{x^2+2x+4}$

$= \dfrac{8(x-2)(x^2+2x+4)(2x+1)(2x-1)}{x(2x+1)(2x+1)(x^2+2x+4)}$

$= \dfrac{8(x-2)(2x-1)}{x(2x+1)}$

32. $\dfrac{3-x}{x^2+3x+9} \div \dfrac{x^2-9}{x^3-27}$

$= \dfrac{3-x}{x^2+3x+9} \cdot \dfrac{x^3-27}{x^2-9}$

$= \dfrac{3-x}{x^2+3x+9} \cdot \dfrac{(x-3)(x^2+3x+9)}{(x-3)(x+3)}$

$= \dfrac{(3-x)(x-3)(x^2+3x+9)}{(x^2+3x+9)(x-3)(x+3)}$

$= -\dfrac{x-3}{x+3}$

33. The LCM is $6a^3b^2$.

$\dfrac{3}{2a^3b^2} + \dfrac{5}{6a^2b}$

$\dfrac{3}{2a^3b^2} \cdot \dfrac{3}{3} + \dfrac{5}{6a^2b} \cdot \dfrac{ab}{ab} = \dfrac{9}{6a^3b^2} + \dfrac{5ab}{6a^3b^2}$

$= \dfrac{9+5ab}{6a^3b^2}$

34. $9x^2-4 = (3x-2)(3x+2)$

The LCM is $(3x-2)(3x+2)$.

$\dfrac{8}{9x^2-4} + \dfrac{5}{3x-2} - \dfrac{4}{3x+2}$

$= \dfrac{8}{(3x-3)(3x+2)} + \dfrac{5}{3x-2} \cdot \dfrac{3x+2}{3x+2} - \dfrac{4}{3x+2} \cdot \dfrac{3x-2}{3x-2}$

$= \dfrac{8+5(3x+2)-4(3x-2)}{(3x-2)(3x+2)}$

$= \dfrac{8+15x+10-12x+8}{(3x-2)(3x+2)}$

$= \dfrac{3x+26}{(3x-2)(3x+2)}$

35. $\dfrac{x-6+\dfrac{6}{x-1}}{x+3-\dfrac{12}{x-1}}$

$=\dfrac{x-6+\dfrac{6}{x-1}}{x+3-\dfrac{12}{x-1}}\cdot\dfrac{x-1}{x-1}=\dfrac{(x-6)(x-1)+6}{(x+3)(x-1)-12}$

$=\dfrac{x^2-7x+6+6}{x^2+2x-3-12}$

$=\dfrac{x^2-7x+12}{x^2+2x-15}=\dfrac{(x-3)(x-4)}{(x-3)(x+5)}$

$=\dfrac{x-4}{x+5}$

36. $\dfrac{x+\dfrac{3}{x-4}}{3+\dfrac{x}{x-4}}$

$=\dfrac{x+\dfrac{3}{x-4}}{3+\dfrac{x}{x-4}}\cdot\dfrac{x-4}{x-4}=\dfrac{x(x-4)+3}{3(x-4)+x}$

$=\dfrac{x^2-4x+3}{4x-12}=\dfrac{(x-3)(x-1)}{4(x-3)}$

$=\dfrac{x-1}{4}$

37. The LCM is $(x-3)(x+1)$.

$\dfrac{x+2}{x-3}=\dfrac{2x-5}{x+1}$

$(x-3)(x+1)\left(\dfrac{x+2}{x-3}\right)=(x-3)(x+1)\left(\dfrac{2x-5}{x+1}\right)$

$(x+1)(x+2)=(x-3)(2x-5)$

$x^2+3x+2=2x^2-11x+15$

$0=(x-13)(x-1)$

$x-13=0\quad x-1=0$

$x=13\qquad\quad x=1$

The solutions are 1 and 13.

38. $\dfrac{5x}{2x-3}+4=\dfrac{3}{2x-3}$

$(2x-3)\left(\dfrac{5x}{2x-3}+4\right)=(2x-3)\left(\dfrac{3}{2x-3}\right)$

$5x+4(2x-3)=3$

$5x+8x-12=3$

$13x=15$

$x=\dfrac{15}{13}$

The solution is $\dfrac{15}{13}$.

39. $I=\dfrac{V}{R}$

$I\cdot R=\dfrac{V}{R}\cdot R$

$IR=V$

$R=\dfrac{V}{I}$

40. $\dfrac{6}{x-3} - \dfrac{1}{x+3} = \dfrac{51}{x^2-9}$

$(x-3)(x+3)\left(\dfrac{6}{x-3} - \dfrac{1}{x+3}\right) = (x^2-9)\left(\dfrac{51}{x^2-9}\right)$

$6(x+3) - (x-3) = 51$

$6x + 18 - x + 3 = 51$

$5x + 21 = 51$

$5x = 30$

$x = 6$

The solution is 6.

41. Strategy: Let t represent the time required to empty the tub.

	Rate	Time	Part
Inlet pipe	$\dfrac{1}{24}$	t	$\dfrac{t}{24}$
Drain pipe	$\dfrac{1}{15}$	t	$\dfrac{t}{15}$

The tub is empty when the difference between the drain pipe part and the inlet pipe part equals 1.

Solution: $\dfrac{t}{15} - \dfrac{t}{24} = 1$

$120\left(\dfrac{t}{15} - \dfrac{t}{24}\right) = 1(120)$

$8t - 5t = 120$

$3t = 120$

$t = 40$

It would take 40 min to empty the tub with both pipes open.

42. Strategy: Let r represent the rate of the cyclist.

The rate of the bus is $3r$.

	Distance	Rate	Time
Cyclist	90	r	$\dfrac{90}{r}$
Bus	90	$3r$	$\dfrac{90}{3r}$

The difference in the time is 4 h.

Solution: $\dfrac{90}{r} - \dfrac{90}{3r} = 4$

$3r\left(\dfrac{90}{r}\right) - 3r\left(\dfrac{90}{3r}\right) = 4(3r)$

$270 - 90 = 12r$

$180 = 12r$

$r = 15$

$3r = 3(15) = 45$

The rate of the bus is 45 mph.

43. Strategy: Let r represent the rate of the helicopter.

The rate of the airplane is $r + 20$.

	Distance	Rate	Time
Helicopter	9	r	$\dfrac{9}{r}$
Airplane	10	$r + 20$	$\dfrac{10}{r+20}$

The time traveled by the helicopter equals the time traveled by the airplane.

Solution: $\dfrac{9}{r} = \dfrac{10}{r+20}$

$r(r+20)\left(\dfrac{9}{r}\right) = r(r+20)\left(\dfrac{10}{r+20}\right)$

$9(r+20) = 10r$

$9r + 180 = 10r$

$r = 180$

The rate of the helicopter is 180 mph.

44. Strategy: Let x represent the time. Write and solve a proportional equation.

Solution: $\dfrac{2}{5} = \dfrac{150}{x}$

$\dfrac{2}{5} \cdot 5x = \dfrac{150}{x} \cdot 5x$

$2x = 750$

$x = 375$

It will take 375 min to read 150 pages.

45. Strategy: Write the basic inverse variation equation replacing the variable with the given values. Solve for k.
Write the inverse variation equation replacing k with its value. Substitute 100 for R and solve for I.

Solution:

$$I = \frac{k}{R}$$

$$4 = \frac{k}{50}$$

$$200 = k$$

$$I = \frac{200}{R} = \frac{200}{100} = 2$$

The current is 2 amps.

46. Strategy: Let x represent the number of miles. Write and solve a proportion.

Solution: $\dfrac{2.5}{10} = \dfrac{12}{x}$

$$\frac{2.5}{10} \cdot 10x = \frac{12}{x} \cdot 10x$$

$$2.5x = 120$$

$$x = 48$$

48 mi would be represented.

47. Strategy: Write the basic direct variation equation replacing the variable with the given values. Solve for k.
Write the direct variation equation replacing k with its value. Substitute 65 for v and solve for 2.

Solution:

$$s = kv^2$$

$$170 = k(50)^2$$

$$170 = 2500k$$

$$0.068 = k$$

$$s = 0.068v^2 = 0.068(65^2) = 0.068(4225)$$

$$s = 287.3$$

The stopping distance for a car traveling at 65 mph is 287.3 ft.

48. Strategy: Let t represent the time required for an apprentice working alone to install a fan.

	Rate	Time	Part
Electrician	$\frac{1}{65}$	40	$\frac{40}{65}$
Apprentice	$\frac{1}{t}$	40	$\frac{40}{t}$

The sum of the part of the task completed by the electrician and the part completed by the apprentice equals 1.

Solution: $\dfrac{40}{65} + \dfrac{40}{t} = 1$

$$65t\left(\frac{40}{65} + \frac{40}{t}\right) = 1(65t)$$

$$40t + 2600 = 65t$$

$$2600 = 25t$$

$$t = 104$$

It would take the apprentice 104 min to compete the job alone.

Chapter 6 Test

1. $\dfrac{3}{x+1} = \dfrac{2}{x}$

$$x(x+1)\left(\frac{3}{x+1}\right) = x(x+1)\left(\frac{2}{x}\right)$$

$$3x = 2x + 2$$

$$x = 2$$

The solution is 2.

2. $\dfrac{x^2+x-6}{x^2+7x+12} \div \dfrac{x^2-3x+2}{x^2+6x+8}$

$= \dfrac{x^2+x-6}{x^2+7x+12} \cdot \dfrac{x^2+6x+8}{x^2-3x+2}$

$= \dfrac{(x+3)(x-2)}{(x+3)(x+4)} \cdot \dfrac{(x+2)(x+4)}{(x-2)(x-1)}$

$= \dfrac{(x+3)(x-2)(x+2)(x+4)}{(x+3)(x+4)(x-2)(x-1)}$

$= \dfrac{x+2}{x-1}$

3. The LCM is $(x+2)(x-3)$.

$\dfrac{2x-1}{x+2} - \dfrac{x}{x-3}$

$= \dfrac{2x-1}{x+2} \cdot \dfrac{x-3}{x-3} - \dfrac{x}{x-3} \cdot \dfrac{x+2}{x+2}$

$= \dfrac{2x^2-7x+3-x^2-2x}{(x+2)(x-3)}$

$= \dfrac{x^2-9x+3}{(x+2)(x-3)}$

4. $x^2+x-6 = (x+3)(x-2)$

$x^2-9 = (x+3)(x-3)$

The LCM is $(x-2)(x-3)(x+3)$.

$\dfrac{x+1}{x^2+x-6} = \dfrac{x+1}{(x-2)(x+3)} \cdot \dfrac{x-3}{x-3}$

$= \dfrac{x^2-2x-3}{(x-2)(x-3)(x+3)}$

$\dfrac{2x}{x^2-9} = \dfrac{2x}{(x-3)(x+3)} \cdot \dfrac{x-2}{x-2}$

$= \dfrac{2x^2-4x}{(x-2)(x-3)(x+3)}$

5. $\dfrac{4x}{2x-1} = 2 - \dfrac{1}{2x-1}$

$(2x-1)\left(\dfrac{4x}{2x-1}\right) = (2x-1)\left(2 - \dfrac{1}{2x-1}\right)$

$4x = 4x - 2 - 1$

$4x = 4x - 3$

$0 = -3$

There is no solution.

6. $x^2+3x-10 = (x+5)(x-2)$

The LCM is $(x+5)(x-2)$.

$\dfrac{3}{x+5} + \dfrac{2x}{x^2+3x-10}$

$= \dfrac{3}{x+5} \cdot \dfrac{x-2}{x-2} + \dfrac{2x}{(x+5)(x-2)}$

$= \dfrac{3x-6+2x}{(x+5)(x-2)}$

$= \dfrac{5x-6}{(x+5)(x-2)}$

7. $\dfrac{3x^2-12}{5x-15} \cdot \dfrac{2x^2-18}{x^2+5x+6}$

$= \dfrac{3(x+2)(x-2)}{5(x-3)} \cdot \dfrac{2(x+3)(x-3)}{(x+3)(x+2)}$

$= \dfrac{3(x+2)(x-2)2(x+3)(x-3)}{5(x-3)(x+3)(x+2)}$

$= \dfrac{6(x-2)}{5}$

8. $f(x) = \dfrac{3x^2-x+1}{x^2-9}$

$x^2-9 = 0$

$(x+3)(x-3) = 0$

$x+3 = 0 \quad x-3 = 0$

$x = -3 \qquad x = 3$

The domain is $\{x \mid x \neq -3, 3\}$.

9. $\dfrac{1-\dfrac{1}{x}-\dfrac{12}{x^2}}{1+\dfrac{6}{x}+\dfrac{9}{x^2}}$

$=\dfrac{1-\dfrac{1}{x}-\dfrac{12}{x^2}}{1+\dfrac{6}{x}+\dfrac{9}{x^2}}\cdot\dfrac{x^2}{x^2}=\dfrac{x^2-x-12}{x^2+6x+9}$

$=\dfrac{(x-4)(x+3)}{(x+3)(x+3)}$

$=\dfrac{x-4}{x+3}$

10. $\dfrac{1-\dfrac{1}{x+2}}{1-\dfrac{3}{x+4}}$

$=\dfrac{1-\dfrac{1}{x+2}}{1-\dfrac{3}{x+4}}\cdot\dfrac{(x+2)(x+4)}{(x+2)(x+4)}$

$=\dfrac{(x+2)(x+4)-(x+4)}{(x+2)(x+4)-3(x+2)}$

$=\dfrac{(x+4)(x+2-1)}{(x+2)(x+4-3)}$

$=\dfrac{(x+4)(x+1)}{(x+2)(x+1)}$

$=\dfrac{x+4}{x+2}$

11. $\dfrac{2x^2-x-3}{2x^2-5x+3}\div\dfrac{3x^2-x-4}{x^2-1}$

$=\dfrac{2x^2-x-3}{2x^2-5x+3}\cdot\dfrac{x^2-1}{3x^2-x-4}$

$=\dfrac{(2x-3)(x+1)}{(2x-3)(x-1)}\cdot\dfrac{(x+1)(x-1)}{(3x-4)(x+1)}$

$=\dfrac{(2x-3)(x+1)(x+1)(x-1)}{(2x-3)(x-1)(3x-4)(x+1)}$

$=\dfrac{x+1}{3x-4}$

12. The LCM is $(x+1)$.

$\dfrac{4x}{x+1}-x=\dfrac{2}{x+1}$

$(x+1)\left(\dfrac{4x}{x+1}-x\right)=(x+1)\left(\dfrac{2}{x+1}\right)$

$4x-x(x+1)=2$

$4x-x^2-x=2$

$x^2-3x+2=0$

$(x-2)(x-1)=0$

$x-2=0 \quad x-1=0$

$x=2 \qquad x=1$

The solutions are 1 and 2.

13. $\dfrac{2a^2-8a+8}{4+4a-3a^2}$

$=\dfrac{2(a^2-4a+4)}{(2-a)(2+3a)}=\dfrac{2(a-2)(a-2)}{(2-a)(2+3a)}$

$=-\dfrac{2(a-2)}{3a+2}$

14. $x - \dfrac{12}{x-3} = \dfrac{x}{x-3}$

$\left(x - \dfrac{12}{x-3} \right) \cdot (x-3) = \dfrac{x}{x-3} \cdot (x-3)$

$x(x-3) - 12 = x$

$x^2 - 3x - 12 = x$

$x^2 - 4x - 12 = 0$

$(x-6)(x+2) = 0$

$x - 6 = 0 \qquad x + 2 = 0$

$x = 6 \qquad\quad x = -2$

The solutions are -2 and 6.

15. $f(x) = \dfrac{3 - x^2}{x^3 - 2x^2 + 4}$

$f(-1) = \dfrac{3 - (-1)^2}{(-1)^3 - 2(-1)^2 + 4}$

$f(-1) = \dfrac{3 - 1}{-1 - 2 + 4}$

$f(-1) = 2$

16. $x^2 + 3x - 4 = (x+4)(x-1)$

$x^2 - 1 = (x+1)(x-1)$

The LCM is $(x+4)(x+1)(x-1)$.

$\dfrac{x+2}{x^2 + 3x - 4} - \dfrac{2x}{x^2 - 1}$

$= \dfrac{x+2}{(x+4)(x-1)} \cdot \dfrac{x+1}{x+1} - \dfrac{2x}{(x+1)(x-1)} \cdot \dfrac{x+4}{x+4}$

$= \dfrac{x^2 + 3x + 2 - 2x^2 - 8x}{(x+4)(x+1)(x-1)}$

$= \dfrac{-x^2 - 5x + 2}{(x+4)(x+1)(x-1)}$

17. $\dfrac{x+1}{2x+5} = \dfrac{x-3}{x}$

$x(2x+5)\left(\dfrac{x+1}{2x+5} \right) = x(2x+5)\left(\dfrac{x-3}{x} \right)$

$x(x+1) = (2x+5)(x-3)$

$x^2 + x = 2x^2 - x - 15$

$0 = x^2 - 2x - 15$

$0 = (x-5)(x+3)$

$x - 5 = 0 \quad x + 3 = 0$

$x = 5 \qquad\quad x = -3$

The solutions are -3 and 5.

18. Strategy: Let r represent the rate of the hiker.
The rate of the cyclist is $r + 7$.

	Distance	Rate	Time
Hiker	6	r	$\dfrac{6}{r}$
Cyclist	20	$r+7$	$\dfrac{20}{r+7}$

The time traveled by the hiker equals the time traveled by the cyclist.

Solution: $\dfrac{6}{r} = \dfrac{20}{r+7}$

$r(r+7)\left(\dfrac{6}{r} \right) = r(r+7)\left(\dfrac{20}{r+7} \right)$

$6(r+7) = 20r$

$6r + 42 = 20r$

$42 = 14r$

$r = 3$

$r + 7 = 3 + 7 = 10$

The rate of the cyclist is 10 mph.

19. Strategy: Write the basic combined variation equation replacing the variables with the given values. Solve for k. Write the combined variation equation replacing k with its value. Substitute 8000 for l, $\frac{1}{2}$ for d and solve for r.

Solution:

$$r = \frac{kl}{d^2}$$

$$3.2 = \frac{16{,}000k}{\left(\dfrac{1}{4}\right)^2}$$

$$0.0000125 = k$$

$$r = \frac{0.0000125l}{d^2} = \frac{0.0000125(8000)}{\left(\dfrac{1}{2}\right)^2} = 0.4$$

The resistance is 0.4 ohms.

20. Strategy: Let x represent the number of rolls. Write and solve a proportion.

Solution: $\dfrac{2}{45} = \dfrac{x}{315}$

$$\frac{2}{45} \cdot 315 = \frac{x}{315} \cdot 315$$

$$x = 14$$

The office requires 14 rolls of wallpaper.

21. Strategy: Let t represent the time required for both landscapers working together.

	Rate	Time	Part
1st landscaper	$\dfrac{1}{30}$	t	$\dfrac{t}{30}$
2nd landscaper	$\dfrac{1}{15}$	t	$\dfrac{t}{15}$

The sum of the parts of the task completed by each landscaper equals 1.

Solution: $\dfrac{t}{30} + \dfrac{t}{15} = 1$

$$30\left(\frac{t}{30} + \frac{t}{15}\right) = 1(30)$$

$$t + 2t = 30$$

$$3t = 30$$

$$t = 10$$

It would take 10 min to complete the task when both landscapers work together.

22. Strategy: Write the basic inverse variation equation replacing the variables with the given values. Solve for k. Write the inverse variation equation replacing k with its value. Substitute 5 for d and solve for I.

Solution:

$$I = \frac{k}{d^2}$$

$$50 = \frac{k}{8^2}$$

$$50 = \frac{k}{64}$$

$$3200 = k$$

$$I = \frac{3200}{d^2} = \frac{3200}{5^2} = \frac{3200}{25} = 128$$

The intensity is 128 decibels when the distance is 5 m.

Cumulative Review Exercises

1. $8 - 4[-3 - (-2)]^2 \div 5$

$$= 8 - 4[-3 + 2]^2 \div 5 = 8 - 4[-1]^2 \div 5$$

$$= 8 - 4(1) \div 5 = 8 - \frac{4}{5}$$

$$= \frac{36}{5}$$

2. $\dfrac{2x-3}{6} - \dfrac{x}{9} = \dfrac{x-4}{3}$

$18\left(\dfrac{2x-3}{6} - \dfrac{x}{9}\right) = 18\left(\dfrac{x-4}{3}\right)$

$3(2x-3) - 2x = 6(x-4)$

$6x - 9 - 2x = 6x - 24$

$4x - 9 = 6x - 24$

$2x = 15$

$x = \dfrac{15}{2}$

The solution is $\dfrac{15}{2}$.

3. $5 - |x-4| = 2$

$-|x-4| = -3$

$|x-4| = 3$

$x - 4 = 3 \qquad x - 4 = -3$

$x = 7 \qquad\quad x = 1$

The solutions are 1 and 7.

4. $f(x) = \dfrac{x}{x-3}$

$x - 3 = 0$

$x = 3$

The domain is $\{x \mid x \neq 3\}$.

5. $P(x) = \dfrac{x-1}{2x-3}$

$P(-2) = \dfrac{-2-1}{2(-2)-3} = \dfrac{-3}{-7}$

$P(-2) = \dfrac{3}{7}$

6. 3.5×10^{-8}

7. $\dfrac{(2a^{-2}b^3)^{-2}}{(4a)^{-1}} = \dfrac{2^{-2}a^4b^{-6}}{4^{-1}a^{-1}} = 2^{-2}4a^{4-(-1)}b^{-6}$

$= \dfrac{1}{4}4a^5b^{-6} = \dfrac{a^5}{b^6}$

8. $x - 3(1 - 2x) \geq 1 - 4(2 - 2x)$

$x - 3 + 6x \geq 1 - 8 + 8x$

$7x - 3 \geq -7 + 8x$

$-x - 3 \geq -7$

$-x \geq -4$

$x \leq 4$

$(-\infty, 4]$

9. $(2a^2 - 3a + 1)(-2a^2) = -4a^4 + 6a^3 - 2a^2$

10. $2x^2 + 3x - 2 = (2x - 1)(x + 2)$

11. $x^3y^3 - 27 = (xy)^3 - 3^3$

$= (xy - 3)(x^2y^2 + 3xy + 9)$

12. $\dfrac{x^4 + x^3y - 6x^2y^2}{x^3 - 2x^2y} = \dfrac{x^2(x^2 + xy - 6y^2)}{x^2(x - 2y)}$

$= \dfrac{x^2(x - 2y)(x + 3y)}{x^2(x - 2y)}$

$= x + 3y$

13. $3x - 2y = 6$

$-2y = -3x + 6$

$y = \dfrac{3}{2}x - 3$

$m = \dfrac{3}{2}$ and $(-2, -1)$

$y - y_1 = m(x - x_1)$

$y - (-1) = \dfrac{3}{2}[x - (-2)]$

$y + 1 = \dfrac{3}{2}(x + 2)$

$y + 1 = \dfrac{3}{2}x + 3$

$y = \dfrac{3}{2}x + 2$

14. $8x^2 - 6x - 9 = 0$

$(4x+3)(2x-3) = 0$

$4x+3 = 0 \quad 2x-3 = 0$

$4x = -3 \qquad 2x = 3$

$x = -\dfrac{3}{4} \qquad x = \dfrac{3}{2}$

The solutions are $-\dfrac{3}{4}$ and $\dfrac{3}{2}$.

15. $\dfrac{4x^3 + 2x^2 - 10x + 1}{x-2}$

$$
\begin{array}{r}
4x^2 + 10x + 10 \\
x-2 \overline{\smash{\big)}\, 4x^3 + 2x^2 - 10x + 1} \\
\underline{4x^3 - 8x^2} \\
10x^2 - 10x \\
\underline{10x^2 - 20x} \\
10x + 1 \\
\underline{10x - 20} \\
21
\end{array}
$$

$\dfrac{4x^3 + 2x^2 - 10x + 1}{x-2} = 4x^2 + 10x + 10 + \dfrac{21}{x-2}$

16. $\dfrac{16x^2 - 9y^2}{16x^2y - 12xy^2} \div \dfrac{4x^2 - xy - 3y^2}{12x^2y^2}$

$\dfrac{16x^2 - 9y^2}{16x^2y - 12xy^2} \cdot \dfrac{12x^2y^2}{4x^2 - xy - 3y^2}$

$= \dfrac{(4x-3y)(4x+3y)}{4xy(4x-3y)} \cdot \dfrac{12x^2y^2}{(4x+3y)(x-y)}$

$= \dfrac{(4x-3y)(4x+3y)12x^2y^2}{4xy(4x-3y)(4x+3y)(x-y)}$

$= \dfrac{3xy}{x-y}$

17. $2x^2 + 2x = 2x(x+1)$

$2x^4 - 2x^3 - 4x^2 = 2x^2(x^2 - x - 2)$

$\qquad\qquad\qquad = 2x^2(x-2)(x+1)$

The LCM is $2x^2(x-2)(x+1)$.

$\dfrac{xy}{2x^2 + 2x} = \dfrac{2xy}{2x(x+1)} \cdot \dfrac{x(x-2)}{x(x-2)}$

$= \dfrac{x^2y(x-2)}{2x^2(x-2)(x+1)} = \dfrac{x^3y - 2x^2y}{2x^2(x-2)(x+1)}$

$\dfrac{2}{2x^4 - 2x^3 - 4x^2} = \dfrac{2}{2x^2(x-2)(x+1)}$

18. $3x^2 - x - 2 = (3x+2)(x-1)$

$x^2 - 1 = (x+1)(x-1)$

The LCM is $(3x+2)(x+1)(x-1)$.

$\dfrac{5x}{3x^2 - x - 2} - \dfrac{2x}{x^2 - 1}$

$= \dfrac{5x}{(3x+2)(x-1)} \cdot \dfrac{x+1}{x+1} - \dfrac{2x}{(x+1)(x-1)} \cdot \dfrac{3x+2}{3x+2}$

$\dfrac{5x^2 + 5x - 6x^2 - 4x}{(3x+2)(x+1)(x-1)}$

$= \dfrac{-x^2 + x}{(3x+2)(x+1)(x-1)}$

$= \dfrac{-x(x-1)}{(3x+2)(x+1)(x-1)}$

$= -\dfrac{x}{(3x+2)(x+1)(x-1)}$

19. $-3x + 5y = -15$
x-intercept:
$(5, 0)$
y- intercept:
$(0, -3)$

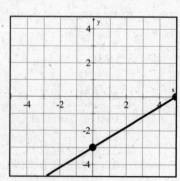

20. $x + y \le 3$ $\qquad$ $-2x + y > 4$

$\quad\;\; y \le 3 - x$ $\qquad$ $\;\; y > 2x + 4$

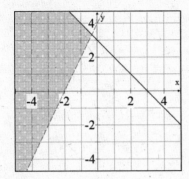

21. $\begin{vmatrix} 6 & 5 \\ 2 & -3 \end{vmatrix} = 6(-3) - 5(2) = -18 - 10 = -28$

22. $\dfrac{x - 4 + \dfrac{5}{x+2}}{x + 2 - \dfrac{1}{x+2}}$

$= \dfrac{x - 4 + \dfrac{5}{x+2}}{x + 2 - \dfrac{1}{x+2}} \cdot \dfrac{x+2}{x+2} = \dfrac{(x-4)(x+2) + 5}{(x+2)(x+2) - 1}$

$= \dfrac{x^2 - 2x - 8 + 5}{x^2 + 4x + 4 - 1}$

$= \dfrac{x^2 - 2x - 3}{x^2 + 4x + 3} = \dfrac{(x-3)(x+1)}{(x+3)(x+1)}$

$= \dfrac{x - 3}{x + 3}$

23. (1) $\;x + y + z = 3$

(2) $\;-2x + y + 3z = 2$

(3) $\;2x - 4y + z = -1$

Eliminate x. Multiply equation (1) by 2 and add to equation (2).

$\quad 2x + 2y + 2z = 6$

$\underline{-2x + y + 3z = 2}$

(4) $\;\;3y + 5z = 8$

Add equations (2) and (3).

$\quad -2x + y + 3z = 2$

$\underline{\;\;\; 2x - 4y + z = -1}$

(5) $\;-3y + 4z = 1$

Use equations (4) and (5) to solve for y and z.

$3y + 5z = 8$

$-3y + 4z = 1$

Eliminate y.

$\quad 3y + 5z = 8$

$\underline{-3y + 4z = 1}$

$\qquad\;\; 9z = 9$

$\qquad\quad\; z = 1$

Replace z with 1 in equation (4).

$3y + 5z = 8$

$3y + 5(1) = 8$

$\quad\;\; 3y = 3$

$\qquad y = 1$

Replace y with 1 and z with 1 in equation (1).

$x + y + z = 3$

$x + 1 + 1 = 3$

$x + 2 = 3$

$\quad\; x = 1$

The solution is (1, 1, 1).

24. $|3x - 2| > 4$

$3x - 2 < -4 \quad 3x - 2 > 4$

$3x < -2 \qquad\;\; 3x > 6$

$x < -\dfrac{2}{3} \qquad\quad\;\; x > 2$

$\left\{ x \,|\, x < -\dfrac{2}{3} \right\} \cup \left\{ x \,|\, x > 2 \right\}$

$= \left\{ x \,|\, x < -\dfrac{2}{3} \;or\; x > 2 \right\}$

25. $\dfrac{2}{x-3} = \dfrac{5}{2x-3}$

$(2x-3)(x-3)\left(\dfrac{2}{x-3}\right) = (2x-3)(x-3)\left(\dfrac{5}{2x-3}\right)$

$2(2x-3) = 5(x-3)$

$4x-6 = 5x-15$

$-6 = x-15$

$x = 9$

The solution is 9.

26. $x^2 - 36 = (x+6)(x-6)$

The LCM is $(x+6)(x-6)$.

$\dfrac{3}{x^2-36} = \dfrac{2}{x-6} - \dfrac{5}{x+6}$

$\dfrac{3}{(x+6)(x-6)} = \dfrac{2}{x-6} - \dfrac{5}{x+6}$

$(x+6)(x-6)\left(\dfrac{3}{(x+6)(x-6)}\right) = (x+6)(x-6)\left(\dfrac{2}{x-6} - \dfrac{5}{x+6}\right)$

$3 = 2(x+6) - 5(x-6)$

$3 = 2x+12-5x+30$

$3 = -3x+42$

$-39 = -3x$

$x = 13$

The solution is 13.

27. $I = \dfrac{E}{R+r}$

$I \cdot (R+r) = \dfrac{E}{R+r} \cdot (R+r)$

$IR + Ir = E$

$Ir = E - IR$

$r = \dfrac{E-IR}{I}$

28. $\left(1-x^{-1}\right)^{-1} = \left(1-\dfrac{1}{x}\right)^{-1} = \left(\dfrac{x-1}{x}\right)^{-1}$

$= \dfrac{x}{x-1}$

29. Strategy: Let x represent the smaller integer.
The larger integer is $15 - x$.

Solution: $5x = 5 + 2(15-x)$
$5x = 5 + 30 - 2x$
$7x = 35$
$x = 5$
$15 - x = 15 - 5 = 10$
The smaller integer is 5 and the larger integer is 10.

30. Strategy: Let x represent the number of pounds of almonds.

	Amount	Cost	Value
Almonds	x	5.40	$5.40x$
Peanuts	50	2.60	2.60(50)
Mixture	$50 + x$	4.00	$4(50 + x)$

The sum of the values before mixing equals the value after mixing.

Solution:
$$5.40x + 2.60(50) = 4(50 + x)$$
$$5.4x + 130 = 4x + 200$$
$$1.4x + 130 = 200$$
$$1.4x = 70$$
$$x = 50$$

50 lb of almonds must be mixed with peanuts.

31. Strategy: Let x represent the number of people expected to vote. Write and solve a proportion.

Solution: $\dfrac{3}{5} = \dfrac{x}{125,000}$

$$125,000 \cdot \dfrac{3}{5} = \dfrac{x}{125,000} \cdot 125,000$$

$$x = 75,000$$

75,000 people are expected to vote.

32. Strategy: Let t represent the time required by the new computer.
The time required by the old computer is $6t$.

	Rate	Time	Part
New computer	$\dfrac{1}{t}$	12	$\dfrac{12}{t}$
Older computer	$\dfrac{1}{6t}$	12	$\dfrac{12}{6t}$

The sum of the parts of the task completed by each computer equals 1.

Solution: $\dfrac{12}{t} + \dfrac{12}{6t} = 1$

$$6t\left(\dfrac{12}{t} + \dfrac{12}{6t} \right) = 1(6t)$$

$$72 + 12 = 6t$$

$$84 = 6t$$

$$t = 14$$

It would take the new computer 14 min to complete the task when working alone.

33. Strategy: Let r represent the rate of the wind.

	Distance	Rate	Time
With the wind	900	$300 + r$	$\dfrac{900}{300+r}$
Against the wind	600	$300 - r$	$\dfrac{600}{300-r}$

The time traveled with the wind equals the time against the wind.

Solution:
$$\frac{900}{300+r} = \frac{600}{300-r}$$
$$(300+r)(300-r)\left(\frac{900}{300+r}\right) = (300+r)(300-r)\left(\frac{600}{300-r}\right)$$
$$900(300-r) = 600(300+r)$$
$$270{,}000 - 900r = 180{,}000 + 600r$$
$$90{,}000 = 1500r$$
$$60 = r$$
The rate of the wind is 60 mph.

34. Strategy: To find the distance apart, calculate the number of times around the track each person has gone and find the difference in the distance.

Solution: The walker travels 3 mi in 1 h.
$$\frac{3 \, mi}{1 \, h} \cdot \frac{1 \, lap}{0.25 \, mi} = \frac{3}{0.25} = 12 \, laps$$
The jogger travels 5 miles in 1 hour.
$$\frac{5 \, mi}{1 \, h} \cdot \frac{1 \, lap}{0.25 \, mi} = \frac{5}{0.25} = 20 \, laps$$
Since the walker and jogger have completed integer multiples of laps, they are both at the starting point after one hour. Therefore they have no distance between them after 1 h.

Chapter 7: Exponents and Radicals

Prep Test

1. $48 = ? \cdot 3$
   What number multiplied by 3 equals 48?
   $? = 16$

2. $2^5 = 2 \cdot 2 \cdot 2 \cdot 2 \cdot 2 = 32$

3. $6\left(\dfrac{3}{2}\right) = \dfrac{6}{1}\left(\dfrac{3}{2}\right) = \dfrac{3 \cdot 2}{1}\left(\dfrac{3}{2}\right) = \dfrac{3 \cdot 3}{1 \cdot 1} = 9$

4. $\dfrac{1}{2} - \dfrac{2}{3} + \dfrac{1}{4} = \dfrac{6}{12} - \dfrac{8}{12} + \dfrac{3}{12}$
   $\qquad = \dfrac{6 - 8 + 3}{12}$
   $\qquad = \dfrac{1}{12}$

5. $(3 - 7x) - (4 - 2x)$
   $= 3 - 7x - 4 + 2x$
   $= -5x - 1$

6. $\dfrac{3x^5 y^6}{12x^4 y} = \dfrac{xy^5}{4}$

7. $(3x - 2)^2 = (3x - 2)(3x - 2)$
   $= 9x^2 - 6x - 6x + 4$
   $= 9x^2 - 12x + 4$

8. $(2 + 4x)(5 - 3x)$
   $= 10 - 6x + 20x - 12x^2$
   $= -12x^2 + 14x + 10$

9. $(6x - 1)(6x + 1)$
   $= 36x^2 + 6x - 6x - 1$
   $= 36x^2 - 1$

10. $x^2 - 14x - 5 = 10$
    $x^2 - 14x - 15 = 0$
    $(x - 15)(x + 1) = 0$
    $x - 15 = 0 \quad x + 1 = 0$
    $\qquad x = 15 \qquad x = -1$
    The solutions are -1 and 15.

Section 7.1

Objective A Exercises

1. (i) and (iii) are not real numbers.

3. $8^{1/3} = (2^3)^{1/3} = 2$

5. $9^{3/2} = (3^2)^{3/2} = 3^3 = 27$

7. $27^{-2/3} = (3^3)^{-2/3} = 3^{-2} = \dfrac{1}{3^2} = \dfrac{1}{9}$

9. $32^{2/5} = (2^5)^{2/5} = 2^2 = 4$

11. $(-25)^{5/2}$
    This is not a real number.
    The base of the exponent expression is a negative number and the denominator of the exponent is a positive even number.

13. $\left(\dfrac{25}{49}\right)^{-3/2}$
    $= \left(\dfrac{5^2}{7^2}\right)^{-3/2} = \left(\left(\dfrac{5}{7}\right)^2\right)^{-3/2}$
    $= \left(\dfrac{5}{7}\right)^{-3} = \dfrac{5^{-3}}{7^{-3}} = \dfrac{7^3}{5^3}$
    $= \dfrac{343}{125}$

15. $x^{1/2} x^{1/2} = x$

17. $y^{-1/4} y^{3/4} = y^{1/2}$

19. $x^{-2/3} \cdot x^{3/4} = x^{1/12}$

21. $a^{1/3} \cdot a^{3/4} \cdot a^{-1/2} = a^{7/12}$

23. $\dfrac{a^{1/2}}{a^{3/2}} = a^{-1} = \dfrac{1}{a}$

25. $\dfrac{y^{-3/4}}{y^{1/4}} = y^{-1} = \dfrac{1}{y}$

27. $\dfrac{y^{2/3}}{y^{-5/6}} = y^{9/6} = y^{3/2}$

29. $(x^2)^{-1/2} = x^{-1} = \dfrac{1}{x}$

31. $(x^{-2/3})^6 = x^{-4} = \dfrac{1}{x^4}$

33. $(a^{-1/2})^{-2} = a$

35. $(x^{-3/8})^{-4/5} = x^{3/10}$

37. $(a^{1/2} \cdot a)^2 = (a^{3/2})^2 = a^3$

39. $(x^{-1/2} \cdot x^{3/4})^{-2} = (x^{1/4})^{-2} = x^{-1/2} = \dfrac{1}{x^{1/2}}$

41. $(y^{-1/2} \cdot y^{2/3})^{2/3} = (y^{1/6})^{2/3} = y^{1/9}$

43. $(x^{-3}y^6)^{-1/3} = xy^{-2} = \dfrac{x}{y^2}$

45. $(x^{-2}y^{1/3})^{-3/4} = x^{3/2}y^{-1/4} = \dfrac{x^{3/2}}{y^{1/4}}$

47. $\left(\dfrac{x^{1/2}}{y^2}\right)^4 = \dfrac{x^2}{y^8}$

49. $\dfrac{x^{1/4} \cdot x^{-1/2}}{x^{2/3}} = \dfrac{x^{-1/4}}{x^{2/3}} = x^{-11/12} = \dfrac{1}{x^{11/12}}$

51. $\left(\dfrac{y^{2/3} \cdot y^{-5/6}}{y^{1/9}}\right)^9 = \left(\dfrac{y^{-1/6}}{y^{1/9}}\right)^9 = \left(y^{-5/18}\right)^9$
$$= y^{-5/2} = \dfrac{1}{y^{5/2}}$$

53. $\left(\dfrac{b^2 \cdot b^{-3/4}}{b^{-1/2}}\right)^{-1/2} = \left(\dfrac{b^{5/4}}{b^{-1/2}}\right)^{-1/2} = \left(b^{7/4}\right)^{-1/2}$
$$= b^{-7/8} = \dfrac{1}{b^{7/8}}$$

55. $(a^{2/3}b^2)^6(a^3b^3)^{1/3} = (a^4b^{12})(ab) = a^5b^{13}$

57. $(16x^{-2}y^4)^{-1/2}(xy^{1/2}) = (16)^{-1/2}(xy^{-2})(xy^{1/2})$
$$= (16)^{-1/2}x^2y^{-3/2}$$
$$= \dfrac{x^2}{16^{1/2}\,y^{3/2}}$$
$$= \dfrac{x^2}{4y^{3/2}}$$

59. $(x^{-2/3}y^{-3})^3(27x^{-3}y^6)^{-1/3} = (x^{-2}y^{-9})(27)^{-1/3}(xy^{-2})$
$$= (27)^{-1/3}x^{-1}y^{-11}$$
$$= \dfrac{1}{27^{1/3}xy^{11}}$$
$$= \dfrac{1}{3xy^{11}}$$

61. $\dfrac{\left(4a^{4/3}b^{-2}\right)^{-1/2}}{(a^{1/6}b^{-3/2})^2} = \dfrac{\left((4)^{-1/2}a^{-2/3}b\right)}{a^{1/3}b^{-3}} = \dfrac{b^4}{2a}$

63. $\left(\dfrac{x^{1/2}y^{-3/4}}{y^{2/3}}\right)^{-6} = \left(x^{1/2}y^{-17/12}\right)^{-6}$
$$= x^{-3}y^{17/2} = \dfrac{y^{17/2}}{x^3}$$

65. $\left(\dfrac{b^{-3}}{64a^{-1/2}}\right)^{-2/3} = \dfrac{b^2}{64^{-2/3}a^{1/3}}$
$$= \dfrac{64^{2/3}b^2}{a^{1/3}} = \dfrac{16b^2}{a^{1/3}}$$

67. $y^{3/2}(y^{1/2} - y^{1/2}) = y^{4/2} - y^{2/2} = y^2 - y$

69. $a^{-1/4}(a^{5/4} - a^{9/4}) = a^{4/4} - a^{8/4} = a - a^2$

71. $x^n \cdot x^{n/2} = x^{3n/2}$

73. $\dfrac{y^{n/2}}{y^{-n}} = y^{3n/2}$

Objective B Exercises

75. False

77. $3^{1/4} = \sqrt[4]{3}$

79. $a^{3/2} = (a^3)^{1/2} = \sqrt{a^3}$

81. $(2t)^{5/2} = \sqrt{(2t)^5} = \sqrt{32t^5}$

83. $-2x^{2/3} = -2(x)^{2/3} = -2\sqrt[3]{x^2}$

85. $(a^2b)^{2/3} = \sqrt[3]{(a^2b)^2} = \sqrt[3]{a^4b^2}$

87. $(a^2b^4)^{3/5} = \sqrt[5]{(a^2b^4)^3} = \sqrt[5]{a^6b^{12}}$

89. $(4x-3)^{3/4} = \sqrt[4]{(4x-3)^3}$

91. $x^{-2/3} = \dfrac{1}{x^{2/3}} = \dfrac{1}{\sqrt[3]{x^2}}$

93. $\sqrt{14} = 14^{1/2}$

95. $\sqrt[3]{x} = x^{1/3}$

97. $\sqrt[3]{x^4} = x^{4/3}$

99. $\sqrt[5]{b^3} = b^{3/5}$

101. $\sqrt[3]{2x^2} = (2x^2)^{1/3}$

103. $-\sqrt{3x^5} = -(3x^5)^{1/2}$

105. $3x\sqrt[3]{y^2} = 3xy^{2/3}$

107. $\sqrt{a^2 - 2} = (a^2 - 2)^{1/2}$

Objective C Exercises

109. Positive

111. Not a real number

113. $\sqrt{x^{16}} = x^8$

115. $-\sqrt{x^8} = -x^4$

117. $\sqrt[3]{x^3y^9} = xy^3$

119. $-\sqrt[3]{x^{15}y^3} = -x^5y$

121. $\sqrt{16a^4b^{12}} = 4a^2b^6$

123. The square root of a negative number is not a real number.

125. $\sqrt[3]{27x^9} = 3x^3$

127. $\sqrt[3]{-64x^9y^{12}} = -4x^3y^4$

129. $-\sqrt[4]{x^8y^{12}} = -x^2y^3$

131. $\sqrt[5]{x^{20}y^{10}} = x^4y^2$

133. $\sqrt[4]{81x^4y^{20}} = 3xy^5$

135. $\sqrt[5]{32a^5b^{10}} = 2ab^2$

Applying the Concepts

137.
a) False $\sqrt{(-2)^2} = \sqrt{4} = 2$
b) True
c) True
d) False $\sqrt[n]{a^n + b^n} = (a^n + b^n)^{1/n}$
e) False
   $(a^{1/2} + b^{1/2})^2 = a + 2a^{1/2}b^{1/2} + b$
f) False $\sqrt[m]{a^n} = a^{n/m}$

139. No. If $x \geq 0$, the statement is true.

However, if $x < 0$ then $\sqrt{x^2} = |x|$.

Section 7.2

Objective A Exercises

1. No

3. Yes

5. $\sqrt{x^4 y^3 z^5} = \sqrt{x^4 y^2 z^4 (yz)}$

$= \sqrt{x^4 y^2 z^4} \sqrt{yz}$

$= x^2 yz^2 \sqrt{yz}$

7. $\sqrt{8a^3 b^8} = \sqrt{4a^2 b^8 (2a)}$

$= \sqrt{4a^2 b^8} \sqrt{2a}$

$= 2ab^4 \sqrt{2a}$

9. $\sqrt{45x^2 y^3 z^5} = \sqrt{9x^2 y^2 z^4 (5yz)}$

$= \sqrt{9x^2 y^2 z^4} \sqrt{5yz}$

$= 3xyz^2 \sqrt{5yz}$

11. $\sqrt[4]{48x^4 y^5 z^6} = \sqrt[4]{16x^4 y^4 z^4 (3yz^2)}$

$= \sqrt[4]{16x^4 y^4 z^4} \sqrt[4]{3yz^2}$

$= 2xyz \sqrt[4]{3yz^2}$

13. $\sqrt[3]{a^{16} b^8} = \sqrt[3]{a^{15} b^6 (ab^2)}$

$= \sqrt[3]{a^{15} b^6} \sqrt[3]{ab^2}$

$= a^5 b^2 \sqrt[3]{ab^2}$

15. $\sqrt[3]{-125x^2 y^4} = \sqrt[3]{-125y^3 (x^2 y)}$

$= \sqrt[3]{-125y^3} \sqrt[3]{x^2 y}$

$= -5y \sqrt[3]{x^2 y}$

17. $\sqrt[3]{a^4 b^5 c^6} = \sqrt[3]{a^3 b^3 c^6 (ab^2)}$

$= \sqrt[3]{a^3 b^3 c^6} \sqrt[3]{ab^2}$

$= abc^2 \sqrt[3]{ab^2}$

19. $\sqrt[4]{16x^9 y^5} = \sqrt[4]{16x^8 y^4 (xy)}$

$= \sqrt[4]{16x^8 y^4} \sqrt[4]{xy}$

$= 2x^2 y \sqrt[4]{xy}$

Objective B Exercises

21. True

23. $2\sqrt{x} - 8\sqrt{x} = -6\sqrt{x}$

25. $\sqrt{8x} - \sqrt{32x} = \sqrt{4 \cdot 2x} - \sqrt{16 \cdot 2x}$

$= \sqrt{4} \sqrt{2x} - \sqrt{16} \sqrt{2x}$

$= 2\sqrt{2x} - 4\sqrt{2x}$

$= -2\sqrt{2x}$

27. $\sqrt{18b} + \sqrt{75b} = \sqrt{9 \cdot 2b} + \sqrt{25 \cdot 3b}$

$= \sqrt{9} \sqrt{2b} + \sqrt{25} \sqrt{3b}$

$= 3\sqrt{2b} + 5\sqrt{3b}$

29. $3\sqrt{8x^2 y^3} - 2x\sqrt{32y^3} = 3\sqrt{4 \cdot 2x^2 y^3} - 2x\sqrt{16 \cdot 2y^3}$

$= 3\sqrt{4x^2 y^2} \sqrt{2y} - 2x\sqrt{16y^2} \sqrt{2y}$

$= 3 \cdot 2xy\sqrt{2y} - 2x \cdot 4y\sqrt{2y}$

$= 6xy\sqrt{2y} - 8xy\sqrt{2y}$

$= -2xy\sqrt{2y}$

31. $2a\sqrt{27ab^5} + 3b\sqrt{3a^3 b} = 2a\sqrt{3^3 ab^5} + 3b\sqrt{3a^3 b}$

$= 2a\sqrt{3^2 b^4} \sqrt{3ab} + 3b\sqrt{a^2} \sqrt{3ab}$

$= 2a \cdot 3b^2 \sqrt{3ab} + 3ab\sqrt{3ab}$

$= 6ab^2 \sqrt{3ab} + 3ab\sqrt{3ab}$

33. $\sqrt[3]{16} - \sqrt[3]{54} = \sqrt[3]{8 \cdot 2} - \sqrt[3]{27 \cdot 2}$
$\quad = \sqrt[3]{8}\sqrt[3]{2} - \sqrt[3]{27}\sqrt[3]{2}$
$\quad = 2\sqrt[3]{2} - 3\sqrt[3]{2}$
$\quad = -\sqrt[3]{2}$

35. $2b\sqrt[3]{16b^2} + \sqrt[3]{128b^5} = 2b\sqrt[3]{8 \cdot 2b^2} + \sqrt[3]{64b^3 \cdot 2b^2}$
$\quad = 2b\sqrt[3]{8}\sqrt[3]{2b^2} + \sqrt[3]{64b^3}\sqrt[3]{2b^2}$
$\quad = 4b\sqrt[3]{2b^2} + 4b\sqrt[3]{2b^2}$
$\quad = 8b\sqrt[3]{2b^2}$

37. $3\sqrt[4]{32a^5} - a\sqrt[4]{162a} = 3\sqrt[4]{16a^4 \cdot 2a} - a\sqrt[4]{81 \cdot 2a}$
$\quad = 3\sqrt[4]{16a^4}\sqrt[4]{2a} - a\sqrt[4]{81}\sqrt[4]{2a}$
$\quad = 3 \cdot 2a\sqrt[4]{2a} - a \cdot 3\sqrt[4]{2a}$
$\quad = 6a\sqrt[4]{2a} - 3a\sqrt[4]{2a}$
$\quad = 3a\sqrt[4]{2a}$

39. $2\sqrt{50} - 3\sqrt{125} + \sqrt{98}$
$\quad = 2\sqrt{25 \cdot 2} - 3\sqrt{25 \cdot 5} + \sqrt{49 \cdot 2}$
$\quad = 2\sqrt{25}\sqrt{2} - 3\sqrt{25}\sqrt{5} + \sqrt{49}\sqrt{2}$
$\quad = 10\sqrt{2} - 15\sqrt{5} + 7\sqrt{2}$
$\quad = 17\sqrt{2} - 15\sqrt{5}$

41. $\sqrt{9b^3} - \sqrt{25b^3} + \sqrt{49b^3}$
$\quad = \sqrt{9b^2 \cdot b} - \sqrt{25b^2 \cdot b} + \sqrt{49b^2 \cdot b}$
$\quad = \sqrt{9b^2}\sqrt{b} - \sqrt{25b^2}\sqrt{b} + \sqrt{49b^2}\sqrt{b}$
$\quad = 3b\sqrt{b} - 5b\sqrt{b} + 7b\sqrt{b}$
$\quad = 5b\sqrt{b}$

43. $2x\sqrt{8xy^2} - 3y\sqrt{32x^3} + \sqrt{4x^3y^3}$
$\quad = 2x\sqrt{4y^2 \cdot 2x} - 3y\sqrt{16x^2 \cdot 2x} + \sqrt{4x^2y^2 \cdot xy}$
$\quad = 2x\sqrt{4y^2}\sqrt{2x} - 3y\sqrt{16x^2}\sqrt{2x} + \sqrt{4x^2y^2}\sqrt{xy}$
$\quad = 4xy\sqrt{2x} - 12xy\sqrt{2x} + 2xy\sqrt{xy}$
$\quad = -8xy\sqrt{2x} + 2xy\sqrt{xy}$

45. $\sqrt[3]{54xy^3} - 5\sqrt[3]{2xy^3} + y\sqrt[3]{128x}$
$\quad = \sqrt[3]{27y^3 \cdot 2x} - 5\sqrt[3]{y^3 \cdot 2x} + y\sqrt[3]{64 \cdot 2x}$
$\quad = \sqrt[3]{27y^3}\sqrt[3]{2x} - 5\sqrt[3]{y^3}\sqrt[3]{2x} + y\sqrt[3]{64}\sqrt[3]{2x}$
$\quad = 3y\sqrt[3]{2x} - 5y\sqrt[3]{2x} + 4y\sqrt[3]{2x}$
$\quad = 2y\sqrt[3]{2x}$

47. $2a\sqrt[4]{32b^5} - 3b\sqrt[4]{162a^4b} + \sqrt[4]{2a^4b^5}$
$\quad = 2a\sqrt[4]{16b^4 \cdot 2b} - 3b\sqrt[4]{81a^4 \cdot 2b} + \sqrt[4]{a^4b^4 \cdot 2b}$
$\quad = 2a\sqrt[4]{16b^4}\sqrt[4]{2b} - 3b\sqrt[4]{81a^4}\sqrt[4]{2b} + \sqrt[4]{a^4b^4}\sqrt[4]{2b}$
$\quad = 4ab\sqrt[4]{2b} - 9ab\sqrt[4]{2b} + ab\sqrt[4]{2b}$
$\quad = -4ab\sqrt[4]{2b}$

Objective C Exercises

49. $\sqrt{8}\sqrt{32} = \sqrt{256} = 16$

51. $\sqrt[3]{4}\sqrt[3]{8} = 2\sqrt[3]{4}$

53. $\sqrt{x^2y^5}\sqrt{xy} = \sqrt{x^3y^6}$
$\quad = \sqrt{x^2y^6 \cdot x} = xy^3\sqrt{x}$

55. $\sqrt{2x^2y}\sqrt{32xy} = \sqrt{64x^3y^2}$
$\quad = \sqrt{64x^2y^2 \cdot x} = 8xy\sqrt{x}$

57. $\sqrt[3]{x^2y}\sqrt[3]{16x^4y^2} = \sqrt[3]{16x^6y^3}$
$\quad = \sqrt[3]{8x^6y^3 \cdot 2} = 2x^2y\sqrt[3]{2}$

59. $\sqrt[4]{12ab^3}\sqrt[4]{4a^5b^2} = \sqrt[4]{48a^6b^5}$
$\quad = \sqrt[4]{16a^4b^4 \cdot 3a^2b} = 2ab\sqrt[4]{3a^2b}$

61. $\sqrt{3}(\sqrt{27} - \sqrt{3}) = \sqrt{81} - \sqrt{9} = 9 - 3 = 6$

63. $\sqrt{x}(\sqrt{x} - \sqrt{2}) = \sqrt{x^2} - \sqrt{2x} = x - \sqrt{2x}$

65. $\sqrt{2x}(\sqrt{8x}-\sqrt{32}) = \sqrt{16x^2}-\sqrt{64x}$

$\quad = \sqrt{16x^2}-\sqrt{64\cdot x} = 4x-8\sqrt{x}$

67. $(3-2\sqrt{5})(2+\sqrt{5}) = 6+3\sqrt{5}-4\sqrt{5}-2(\sqrt{5})^2$

$\quad = 6+3\sqrt{5}-4\sqrt{5}-10 = -4-\sqrt{5}$

69. $(-2+\sqrt{7})(3+5\sqrt{7}) = -6-10\sqrt{7}+3\sqrt{7}+5(\sqrt{7})^2$

$\quad = -6-10\sqrt{7}+3\sqrt{7}+35 = 29-7\sqrt{7}$

71. $(6+3\sqrt{2})(4-2\sqrt{2}) = 24-12\sqrt{2}+12\sqrt{2}-6(\sqrt{2})^2$

$\quad = 24-12\sqrt{2}+12\sqrt{2}-12 = 12$

73. $(5-2\sqrt{7})(5+2\sqrt{7}) = 25+10\sqrt{7}-10\sqrt{7}-4(\sqrt{7})^2$

$\quad = 25+10\sqrt{7}-10\sqrt{7}-28 = -3$

75. $(3-\sqrt{2x})(1+5\sqrt{2x})$

$\quad = 3+15\sqrt{2x}-\sqrt{2x}-5(\sqrt{2x})^2$

$\quad = 3+15\sqrt{2x}-\sqrt{2x}-10x$

$\quad = -10x+14\sqrt{2x}+3$

77. $(2+2\sqrt{x})(1+5\sqrt{x})$

$\quad = 2+10\sqrt{x}+2\sqrt{x}+10(\sqrt{x})^2$

$\quad = 2+10\sqrt{x}+2\sqrt{x}+10x$

$\quad = 10x+12\sqrt{x}+2$

79. $(2+\sqrt{x})^2 = (2+\sqrt{x})(2+\sqrt{x})$

$\quad = 4+2\sqrt{x}+2\sqrt{x}+(\sqrt{x})^2$

$\quad = x+4\sqrt{x}+4$

81. $(4-\sqrt{2x+1})^2 = (4-\sqrt{2x+1})(4-\sqrt{2x+1})$

$\quad = 16-4\sqrt{2x+1}-4\sqrt{2x+1}+(\sqrt{2x+1})^2$

$\quad = 2x+1-8\sqrt{2x+1}+16$

$\quad = 2x-8\sqrt{2x+1}+17$

83. $(\sqrt{6}-5\sqrt{3})(3\sqrt{6}+4\sqrt{3})$

$\quad = 3(\sqrt{6})^2+4\sqrt{18}-15\sqrt{18}-20(\sqrt{3})^2$

$\quad = 18+4\sqrt{9\cdot2}-15\sqrt{9\cdot2}-60$

$\quad = 18+12\sqrt{2}-45\sqrt{2}-60$

$\quad = -42-33\sqrt{2}$

85. True

Objective D Exercises

87. To rationalize the denominator of a radical expression means to rewrite the expression with no radicals in the denominator. To do this, multiply both the numerator and the denominator by the same expression, one that removes the radicals(s) from the denominator of the original expression.

89. $\sqrt[3]{4x}$

91. $\sqrt{3}+x$

93. $\dfrac{\sqrt{60y^4}}{\sqrt{12y}} = \sqrt{\dfrac{60y^4}{12y}}$

$\quad = \sqrt{5y^3} = \sqrt{y^2\cdot5y}$

$\quad = y\sqrt{5y}$

95. $\dfrac{\sqrt{65ab^4}}{\sqrt{5ab}} = \sqrt{\dfrac{65ab^4}{5ab}}$

$\quad = \sqrt{13b^3} = \sqrt{b^2\cdot13b}$

$\quad = b\sqrt{13b}$

97. $\dfrac{1}{\sqrt{2}} = \dfrac{1}{\sqrt{2}}\cdot\dfrac{\sqrt{2}}{\sqrt{2}} = \dfrac{\sqrt{2}}{\sqrt{2^2}} = \dfrac{\sqrt{2}}{2}$

99. $\dfrac{2}{\sqrt{3y}} = \dfrac{2}{\sqrt{3y}}\cdot\dfrac{\sqrt{3y}}{\sqrt{3y}} = \dfrac{2\sqrt{3y}}{\sqrt{9y^2}} = \dfrac{2\sqrt{3y}}{3y}$

101. $\dfrac{9}{\sqrt{3a}} = \dfrac{9}{\sqrt{3a}} \cdot \dfrac{\sqrt{3a}}{\sqrt{3a}} = \dfrac{9\sqrt{3a}}{\sqrt{9a^2}}$

$= \dfrac{9\sqrt{3a}}{3a} = \dfrac{3\sqrt{3a}}{a}$

103. $\sqrt{\dfrac{y}{2}} = \dfrac{\sqrt{y}}{\sqrt{2}} = \dfrac{\sqrt{y}}{\sqrt{2}} \cdot \dfrac{\sqrt{2}}{\sqrt{2}} = \dfrac{\sqrt{2y}}{\sqrt{4}} = \dfrac{\sqrt{2y}}{2}$

105. $\dfrac{5}{\sqrt[3]{9}} = \dfrac{5}{\sqrt[3]{9}} \cdot \dfrac{\sqrt[3]{3}}{\sqrt[3]{3}} = \dfrac{5\sqrt[3]{3}}{\sqrt[3]{27}} = \dfrac{5\sqrt[3]{3}}{3}$

107. $\dfrac{5}{\sqrt[3]{3y}} = \dfrac{5}{\sqrt[3]{3y}} \cdot \dfrac{\sqrt[3]{9y^2}}{\sqrt[3]{9y^2}} = \dfrac{5\sqrt[3]{9y^2}}{\sqrt[3]{27y^3}} = \dfrac{5\sqrt[3]{9y^2}}{3y}$

109. $\dfrac{\sqrt{15a^2b^5}}{\sqrt{30a^5b^3}} = \sqrt{\dfrac{15a^2b^5}{30a^5b^3}} = \sqrt{\dfrac{b^2}{2a^3}} = \dfrac{\sqrt{b^2}}{\sqrt{a^2 \cdot 2a}}$

$= \dfrac{b}{a\sqrt{2a}} = \dfrac{b}{a\sqrt{2a}} \cdot \dfrac{\sqrt{2a}}{\sqrt{2a}} = \dfrac{b\sqrt{2a}}{a\sqrt{4a^2}}$

$= \dfrac{b\sqrt{2a}}{2a^2}$

111. $\dfrac{\sqrt{12x^3y}}{\sqrt{20x^4y}} = \sqrt{\dfrac{12x^3y}{20x^4y}} = \sqrt{\dfrac{3}{5x}}$

$= \dfrac{\sqrt{3}}{\sqrt{5x}} = \dfrac{\sqrt{3}}{\sqrt{5x}} \cdot \dfrac{\sqrt{5x}}{\sqrt{5x}} = \dfrac{\sqrt{15x}}{\sqrt{25x^2}}$

$= \dfrac{\sqrt{15x}}{5x}$

113. $\dfrac{-2}{1-\sqrt{2}} = \dfrac{-2}{1-\sqrt{2}} \cdot \dfrac{1+\sqrt{2}}{1+\sqrt{2}}$

$= \dfrac{-2(1+\sqrt{2})}{1^2-(\sqrt{2})^2} = \dfrac{-2-2\sqrt{2}}{1-2} = \dfrac{-2-2\sqrt{2}}{-1}$

$= 2+2\sqrt{2}$

115. $\dfrac{-4}{3-\sqrt{2}} = \dfrac{-4}{3-\sqrt{2}} \cdot \dfrac{3+\sqrt{2}}{3+\sqrt{2}}$

$= \dfrac{-4(3+\sqrt{2})}{3^2-(\sqrt{2})^2} = \dfrac{-12-4\sqrt{2}}{9-2} = \dfrac{-12-4\sqrt{2}}{7}$

117. $\dfrac{5}{2-\sqrt{7}} = \dfrac{5}{2-\sqrt{7}} \cdot \dfrac{2+\sqrt{7}}{2+\sqrt{7}}$

$= \dfrac{5(2+\sqrt{7})}{2^2-(\sqrt{7})^2} = \dfrac{10+5\sqrt{7}}{4-7} = \dfrac{10+5\sqrt{7}}{-3}$

$= -\dfrac{10+5\sqrt{7}}{3}$

119. $\dfrac{-7}{\sqrt{x}-3} = \dfrac{-7}{\sqrt{x}-3} \cdot \dfrac{\sqrt{x}+3}{\sqrt{x}+3}$

$= \dfrac{-7(\sqrt{x}+3)}{(\sqrt{x})^2-3^2} = \dfrac{-(7\sqrt{x}+21)}{x-9}$

$= -\dfrac{7\sqrt{x}+21}{x-9}$

121. $\dfrac{\sqrt{3}+\sqrt{4}}{\sqrt{2}+\sqrt{3}} = \dfrac{\sqrt{3}+\sqrt{2^2}}{\sqrt{2}+\sqrt{3}} = \dfrac{\sqrt{3}+2}{\sqrt{2}+\sqrt{3}} \cdot \dfrac{\sqrt{2}-\sqrt{3}}{\sqrt{2}-\sqrt{3}}$

$= \dfrac{\sqrt{6}-(\sqrt{3})^2+2\sqrt{2}-2\sqrt{3}}{(\sqrt{2})^2-(\sqrt{3})^2}$

$= \dfrac{\sqrt{6}-3+2\sqrt{2}-2\sqrt{3}}{2-3}$

$= \dfrac{\sqrt{6}-3+2\sqrt{2}-2\sqrt{3}}{-1}$

$= -\sqrt{6}+3-2\sqrt{2}+2\sqrt{3}$

123. $\dfrac{2+3\sqrt{5}}{1-\sqrt{5}} = \dfrac{2+3\sqrt{5}}{1-\sqrt{5}} \cdot \dfrac{1+\sqrt{5}}{1+\sqrt{5}}$

$= \dfrac{2+2\sqrt{5}+3\sqrt{5}+3(\sqrt{5})^2}{1^2-(\sqrt{5})^2}$

$= \dfrac{2+5\sqrt{5}+15}{1-4}$

$= \dfrac{17+5\sqrt{4}}{-3}$

$= -\dfrac{17+5\sqrt{4}}{3}$

125. $\dfrac{2\sqrt{a}-\sqrt{b}}{4\sqrt{a}+3\sqrt{b}} = \dfrac{2\sqrt{a}-\sqrt{b}}{4\sqrt{a}+3\sqrt{b}} \cdot \dfrac{4\sqrt{a}-3\sqrt{b}}{4\sqrt{a}-3\sqrt{b}}$

$= \dfrac{8\sqrt{a^2}-6\sqrt{ab}-4\sqrt{ab}+3\sqrt{b^2}}{(4\sqrt{a})^2-(3\sqrt{b})^2}$

$= \dfrac{8a-10\sqrt{ab}+3b}{16a-9b}$

127. $\dfrac{3\sqrt{y}-y}{\sqrt{y}+2y} = \dfrac{3\sqrt{y}-y}{\sqrt{y}+2y} \cdot \dfrac{\sqrt{y}-2y}{\sqrt{y}-2y}$

$= \dfrac{3(\sqrt{y})^2-6y\sqrt{y}-y\sqrt{y}+2y^2}{(\sqrt{y})^2-(2y)^2}$

$= \dfrac{3y-7y\sqrt{y}+2y^2}{y-4y^2}$

$= \dfrac{3-7\sqrt{y}+2y}{1-4y}$

Applying the Concepts

129.
a) False $\sqrt[2]{3} \cdot \sqrt[3]{4} = \sqrt[6]{432}$
b) True
c) False $\sqrt[3]{x} \cdot \sqrt[3]{x} = x^{2/3}$
d) False $\sqrt{x} + \sqrt{y}$
e) False $\sqrt[2]{2} + \sqrt[3]{3}$
f) True

131. $\dfrac{\sqrt[4]{(a+b)^3}}{\sqrt{a+b}} = \dfrac{(a+b)^{3/4}}{(a+b)^{1/2}} = (a+b)^{1/4}$

$= \sqrt[4]{a+b}$

Section 7.3

Objective A Exercises

1. $\sqrt[3]{4x} = -2$
$(\sqrt[3]{4x})^3 = (-2)^3$
$4x = -8$
$x = -2$
The solution is -2.

3. $\sqrt{3x-2} = 5$
$(\sqrt{3x-2})^2 = (5)^2$
$3x-2 = 25$
$3x = 27$
$x = 9$
The solution is 9.

5. $\sqrt{4x-3}+9 = 4$
$\sqrt{4x-3} = -5$
No solution

7. $\sqrt[3]{2x-6} = 4$
$(\sqrt[3]{2x-6})^3 = (4)^3$
$2x-6 = 64$
$2x = 70$
$x = 35$
The solution is 35.

9. $\sqrt[4]{3x}+2 = 5$
$\sqrt[4]{3x} = 3$
$(\sqrt[4]{3x})^4 = (3)^4$
$3x = 81$
$x = 27$
The solution is 27.

11. $\sqrt[3]{2x-3}+5=2$

$\sqrt[3]{2x-3}=-3$

$\left(\sqrt[3]{2x-3}\right)^3=(-3)^3$

$2x-3=-27$

$2x=-24$

$x=-12$

The solution is −12.

13. $\sqrt{x}+2=x-4$

$\sqrt{x}=x-6$

$\left(\sqrt{x}\right)^2=(x-6)^2$

$x=x^2-12x+36$

$x^2-13x+36=0$

$(x-9)(x-4)=0$

$x-9=0 \quad x-4=0$

$x=9 \qquad x=4$

Check both solutions in the original equation

$\sqrt{x}+2=x-4$

$\sqrt{9}+2=9-4$

$5=5$

$\sqrt{4}+2=4-4$

$4=0$

The solution is 9.

15. $\sqrt{x}+\sqrt{x-5}=5$

$\sqrt{x}=5-\sqrt{x-5}$

$\left(\sqrt{x}\right)^2=\left(5-\sqrt{x-5}\right)^2$

$x=25-10\sqrt{x-5}+x-5$

$0=20-10\sqrt{x-5}$

$-20=-10\sqrt{x-5}$

$2=\sqrt{x-5}$

$(2)^2=\left(\sqrt{x-5}\right)^2$

$4=x-5$

$x=9$

The solution is 9.

17. $\sqrt{2x+5}-\sqrt{2x}=1$

$\sqrt{2x+5}=1+\sqrt{2x}$

$\left(\sqrt{2x+5}\right)^2=\left(1+\sqrt{2x}\right)^2$

$2x+5=1+2\sqrt{2x}+2x$

$5=1+2\sqrt{2x}$

$4=2\sqrt{2x}$

$2=\sqrt{2x}$

$(2)^2=\left(\sqrt{2x}\right)^2$

$4=2x$

$x=2$

The solution is 2.

19. $\sqrt{2x}-\sqrt{x-1}=1$

$\sqrt{2x}=1+\sqrt{x-1}$

$\left(\sqrt{2x}\right)^2=\left(1+\sqrt{x-1}\right)^2$

$2x=1+2\sqrt{x-1}+x-1$

$x=2\sqrt{x-1}$

$(x)^2=\left(2\sqrt{x-1}\right)^2$

$x^2=4(x-1)$

$x^2=4x-4$

$x^2-4x+4=0$

$(x-2)(x-2)=0$

$x-2=0 \quad x-2=0$

$x=2 \qquad x=2$

The solution is 2.

21. $\sqrt{2x+2} + \sqrt{x} = 3$

$\sqrt{2x+2} = 3 - \sqrt{x}$

$(\sqrt{2x+2})^2 = (3 - \sqrt{x})^2$

$2x + 2 = 9 - 6\sqrt{x} + x$

$x + 2 = 9 - 6\sqrt{x}$

$x - 7 = -6\sqrt{x}$

$(x-7)^2 = (-6\sqrt{x})^2$

$x^2 - 14x + 49 = 36x$

$x^2 - 50x + 49 = 0$

$(x-49)(x-1) = 0$

$x - 49 = 0 \quad x - 1 = 0$

$\quad x = 49 \qquad x = 1$

Check both solutions in the original equation

$\sqrt{2x+2} + \sqrt{x} \ne 3$

$\sqrt{2(49)+2} + \sqrt{49} \ne 3$

$\sqrt{100} + \sqrt{49} \ne 3$

$10 + 7 \ne 3$

$17 \ne 3$

$\sqrt{2(1)+2} + \sqrt{1} = 3$

$\sqrt{4} + \sqrt{1} = 3$

$2 + 1 = 3$

$3 = 3$

The solution is 1.

23. $\sqrt{x} < \sqrt{x+5}$.

Therefore $\sqrt{x} - \sqrt{x+5} < 0$ and cannot equal a positive number.

Objective B Exercises

25. Strategy: To find the distance the object will fall, substitute the given values for t and g in the equation and solve for d.

Solution: $t = \sqrt{\dfrac{2d}{g}}$

$3 = \sqrt{\dfrac{2d}{5.5}}$

$(3)^2 = \left(\sqrt{\dfrac{2d}{5.5}} \right)^2$

$9 = \dfrac{2d}{5.5}$

$49.5 = 2d$

$24.75 = d$

On the moon, the object will fall 24.75 ft in 3 s.

27. a) Strategy: To find the height of the water, evaluate the function for $t = 10$.

Solution: $h(t) = (88.18 - 3.18t)^{2/5}$

$h(10) = (88.18 - 3.18(10))^{2/5}$

$h(10) = (88.18 - 31.8)^{2/5}$

$h(10) = (56.38)^{2/5}$

$h(10) = 5.0$

The height of the water is 5.0 ft.

b) Strategy: To find how long it will take to empty the tank substitute the given value for h in the equation and solve for t.

Solution: $h(t) = (88.18 - 3.18t)^{2/5}$

$0 = (88.18 - 3.18t)^{2/5}$

$0 = ((88.18 - 3.18t)^{2/5})^{5/2}$

$0 = 88.18 - 3.18t$

$3.18t = 88.16$

$t = 27.7$

The tank will empty in 27.7 s.

29. Strategy: To find the length of the pendulum, substitute the given value for T in the equation and solve for L.

Solution: $T = 2\pi\sqrt{\dfrac{L}{32}}$

$$3 = 2\pi\sqrt{\dfrac{L}{32}}$$

$$\left(\dfrac{3}{2\pi}\right)^2 = \left(\sqrt{\dfrac{L}{32}}\right)^2$$

$$\left(\dfrac{3}{2\pi}\right)^2 = \dfrac{L}{32}$$

$$32\left(\dfrac{3}{2\pi}\right)^2 = L$$

$$7.3 = L$$

The length of the pendulum is 7.30 ft.

31. Strategy: Find the difference in the widths. Use the Pythagorean Theorem to find the width of the screen of a regular TV and then repeat the process to find the width of the screen for HDTV. Subtract the width of the regular TV from the width of the HDTV.

Solution: $c^2 = a^2 + b^2$
For the regular TV:
$27^2 = 16.2^2 + b^2$
$729 = 262.44 + b^2$
$466.56 = b^2$
$(466.56)^{1/2} = (b^2)^{1/2}$
$21.6 = b$

For the HDTV
$33^2 = 16.2^2 + b^2$
$1089 = 262.44 + b^2$
$826.56 = b^2$
$(826.56)^{1/2} = (b^2)^{1/2}$
$28.75 = b$

$28.75 - 21.6 = 7.15$
The HDTV is approximately 7.15 in. wider.

33. Strategy: Use the Pythagorean Theorem to determine the longest pole that can be placed in the box.

Solution:
Find the length diagonal of the bottom of the box.
$c^2 = a^2 + b^2$
$c^2 = 2^2 + 3^2$
$c^2 = 4 + 9$
$c^2 = 13$

This represents the value one of the legs of the right triangle needed to find the length of the pole.
$c^2 = a^2 + b^2$
$c^2 = 13 + 4^2$
$c^2 = 13 + 16$
$c^2 = 19$
$c = 5.4$

The longest pole can be 5.4 ft.

Applying the Concepts

35. $\sqrt{3x-2} = \sqrt{2x-3} + \sqrt{x-1}$

$\left(\sqrt{3x-2}\right)^2 = \left(\sqrt{2x-3} + \sqrt{x-1}\right)^2$

$3x - 2 = 2x - 3 + 2\sqrt{2x-3} \cdot \sqrt{x-1} + x - 1$

$3x - 2 = 3x - 4 + 2\sqrt{2x-3} \cdot \sqrt{x-1}$

$2 = 2\sqrt{2x-3} \cdot \sqrt{x-1}$

$1 = \sqrt{2x-3} \cdot \sqrt{x-1}$

$(1)^2 = \left(\sqrt{2x-3} \cdot \sqrt{x-1}\right)^2$

$1 = (2x-3)(x-1)$

$1 = 2x^2 - 5x + 3$

$0 = 2x^2 - 5x + 2$

$(2x-1)(x-2) = 0$

$x = \dfrac{1}{2} \qquad x = 2$

Check both solutions in the original equation.

$$\sqrt{3x-2} = \sqrt{2x-3} + \sqrt{x-1}$$

$$\sqrt{3\left(\frac{1}{2}\right)-2} = \sqrt{2\left(\frac{1}{2}\right)-3} + \sqrt{\left(\frac{1}{2}\right)-1}$$

$$\sqrt{-\frac{1}{2}} = \sqrt{-1} + \sqrt{-\frac{1}{2}}$$

Not real numbers.

$$\sqrt{3(2)-2} = \sqrt{2(2)-3} + \sqrt{(2)-1}$$

$$\sqrt{4} = \sqrt{1} + \sqrt{1}$$

$$2 = 1+1$$

$$2 = 2$$

The solution is 2.

Section 7.4

Objective A Exercises

1. An imaginary number is a number whose square is a negative number.
   A complex number is a number of the form $a + bi$, where a and b are real numbers and $i = \sqrt{-1}$.

3. $i\sqrt{a}$

5. $\sqrt{-25} = i\sqrt{25} = 5i$

7. $\sqrt{-98} = i\sqrt{98} = i\sqrt{49 \cdot 2} = 7i\sqrt{2}$

9. $3 + \sqrt{-45} = 3 + i\sqrt{45} = 3 + i\sqrt{9 \cdot 5} = 3 + 3i\sqrt{5}$

11. $-6 - \sqrt{-100} = -6 - i\sqrt{100} = -6 - 10i$

13. $-b + \sqrt{b^2 - 4ac}$

$$-4 + \sqrt{(4)^2 - 4(1)(5)} = -4 + \sqrt{16 - 20}$$

$$= -4 + \sqrt{-4} = -4 + i\sqrt{4}$$

$$= -4 + 2i$$

15. $-b + \sqrt{b^2 - 4ac}$

$$-(-4) + \sqrt{(-4)^2 - 4(2)(10)} = 4 + \sqrt{16 - 80}$$

$$= 4 + \sqrt{-64} = 4 + i\sqrt{64}$$

$$= 4 + 8i$$

17. $-b + \sqrt{b^2 - 4ac}$

$$-(-8) + \sqrt{(-8)^2 - 4(3)(6)} = 8 + \sqrt{64 - 72}$$

$$= 8 + \sqrt{-8} = 8 + i\sqrt{4 \cdot 2}$$

$$= 8 + 2i\sqrt{2}$$

19. $-b + \sqrt{b^2 - 4ac}$

$$-2 + \sqrt{(2)^2 - 4(4)(7)} = -2 + \sqrt{4 - 112}$$

$$= -2 + \sqrt{-108} = -2 + i\sqrt{36 \cdot 3}$$

$$= -2 + 6i\sqrt{3}$$

21. $-b + \sqrt{b^2 - 4ac}$

$$-5 + \sqrt{(5)^2 - 4(-2)(-6)} = -5 + \sqrt{25 - 48}$$

$$= -5 + \sqrt{-23}$$

$$= -5 + i\sqrt{23}$$

23. $-b + \sqrt{b^2 - 4ac}$

$$-4 + \sqrt{(4)^2 - 4(-3)(-6)} = -4 + \sqrt{16 - 72}$$

$$= -4 + \sqrt{-56} = -4 + i\sqrt{4 \cdot 14}$$

$$= -4 + 2i\sqrt{14}$$

Objective B Exercises

25. $(2 + 4i) + (6 - 5i) = 8 - i$

27. $(-2 - 4i) - (6 - 8i) = -8 + 4i$

29. $(8 - 2i) - (2 + 4i) = 6 - 6i$

31. $5 + (6 - 4i) = 11 - 4i$

33. $3i - (6 + 5i) = -6 - 2i$

35. The real parts of the complex numbers are additive inverses.

Objective C Exercises

37. $(7i)(-9i) = -63i^2 = -63(-1) = 63$

39. $\sqrt{-2}\sqrt{-8} = i\sqrt{2} \cdot i\sqrt{8} = i^2\sqrt{16} = -\sqrt{16} = -4$

41. $(5+2i)(5-2i) = 25 - 10i + 10i - 4i^2$
$\quad = 25 - 4i^2 = 25 - 4(-1) = 29$

43. $2i(6+2i) = 12i + 4i^2 = 12i + 4(-1)$
$\quad = -4 + 12i$

45. $-i(4-3i) = -4i + 3i^2 = -4i + 3(-1)$
$\quad = -3 - 4i$

47. $(5-2i)(3+i) = 15 + 5i - 6i - 2i^2$
$\quad = 15 - i - 2i^2$
$\quad = 15 - i - 2(-1)$
$\quad = 17 - i$

49. $(6+5i)(3+2i) = 18 + 12i + 15i + 10i^2$
$\quad = 18 + 27i + 10i^2$
$\quad = 18 + 27i + 10(-1)$
$\quad = 8 + 27i$

51. $(2+5i)^2 = 4 + 20i + 25i^2$
$\quad = 4 + 20i + 25(-1)$
$\quad = -21 + 20i$

53. $\left(\dfrac{6}{5} + \dfrac{3}{5}i\right)\left(\dfrac{2}{3} - \dfrac{1}{3}i\right) = \dfrac{4}{5} - \dfrac{2}{5}i + \dfrac{2}{5}i - \dfrac{1}{5}i^2$
$\quad = \dfrac{4}{5} - \dfrac{1}{5}i^2$
$\quad = \dfrac{4}{5} - \dfrac{1}{5}(-1)$
$\quad = \dfrac{4}{5} + \dfrac{1}{5} = 1$

55. True

Objective D Exercises

57. $\dfrac{3}{i} = \dfrac{3}{i} \cdot \dfrac{i}{i} = \dfrac{3i}{i^2} = \dfrac{3i}{-1} = -3i$

59. $\dfrac{2-3i}{-4i} = \dfrac{2-3i}{-4i} \cdot \dfrac{i}{i} = \dfrac{2i - 3i^2}{-4i^2}$
$\quad = \dfrac{2i - 3(-1)}{-4(-1)} = \dfrac{2i + 3}{4}$
$\quad = \dfrac{3}{4} + \dfrac{1}{2}i$

61. $\dfrac{4}{5+i} = \dfrac{4}{5+i} \cdot \dfrac{5-i}{5-i} = \dfrac{20 - 4i}{25 - i^2}$
$\quad = \dfrac{20 - 4i}{25 - (-1)} = \dfrac{20 - 4i}{26}$
$\quad = \dfrac{10}{13} - \dfrac{2}{13}i$

63. $\dfrac{2}{2-i} = \dfrac{2}{2-i} \cdot \dfrac{2+i}{2+i} = \dfrac{4 + 2i}{4 - i^2}$
$\quad = \dfrac{4 + 2i}{4 - (-1)} = \dfrac{4 + 2i}{5}$
$\quad = \dfrac{4}{5} + \dfrac{2}{5}i$

65. $\dfrac{1-3i}{3+i} = \dfrac{1-3i}{3+i} \cdot \dfrac{3-i}{3-i} = \dfrac{3 - i - 9i + 3i^2}{9 - i^2}$
$\quad = \dfrac{3 - 10i + 3(-1)}{9 - (-1)} = \dfrac{-10i}{10}$
$\quad = -i$

67. $\dfrac{3i}{1+4i} = \dfrac{3i}{1+4i} \cdot \dfrac{1-4i}{1-4i} = \dfrac{3i - 12i^2}{1 - 16i^2}$
$\quad = \dfrac{3i - 12(-1)}{1 - 16(-1)} = \dfrac{3i + 12}{17}$
$\quad = \dfrac{12}{17} + \dfrac{3}{17}i$

69. $\dfrac{2-3i}{3+i} = \dfrac{2-3i}{3+i} \cdot \dfrac{3-i}{3-i} = \dfrac{6-2i-9i+3i^2}{9-i^2}$

$= \dfrac{6-11i+3(-1)}{9-(-1)} = \dfrac{3-11i}{10}$

$= \dfrac{3}{10} - \dfrac{11}{10}i$

71. $\dfrac{5+3i}{3-i} = \dfrac{5+3i}{3-i} \cdot \dfrac{3+i}{3+i} = \dfrac{15+5i+9i+3i^2}{9-i^2}$

$= \dfrac{15+14i+3(-1)}{9-(-1)} = \dfrac{12+14i}{10}$

$= \dfrac{6+7i}{5}$

$= \dfrac{6}{5} + \dfrac{7}{5}i$

73. True

Applying the Concepts

75. a) $\quad 2x^2 + 18 = 0$

$2(3i)^2 + 18 = 0$

$2(9i^2) + 18 = 0$

$2(-9) + 18 = 0$

$\qquad\qquad 0 = 0$

Yes, $3i$ is a solution.

b) $\qquad\qquad x^2 - 6x + 10 = 0$

$(3+i)^2 - 6(3+i) + 10 = 0$

$9 + 6i + i^2 - 18 - 6i + 10 = 0$

$9 + (-1) - 8 = 0$

$\qquad\qquad\qquad 0 = 0$

Yes $3 + i$ is a solution.

Chapter 7 Review Exercises

1. $(16x^{-4}y^{12})^{1/4}(100x^6y^{-2})^{1/2}$

$= (16)^{1/4}x^{-1}y^3 \cdot 100^{1/2} x^3y^{-1}$

$= 20x^2y^2$

2. $\sqrt[4]{3x-5} = 2$

$(\sqrt[4]{3x-5})^4 = 2^4$

$3x - 5 = 16$

$3x = 21$

$x = 7$

The solution is 7.

3. $(6 - 5i)(4 + 3i) = 24 + 18i - 20i - 15i^2$

$= 24 - 2i - 15(-1)$

$= 39 - 2i$

4. $7y\sqrt[3]{x^2} = 7x^{2/3}y$

5. $(\sqrt{3} + 8)(\sqrt{3} - 2) = \sqrt{3^2} - 2\sqrt{3} + 8\sqrt{3} - 16$

$= 3 + 6\sqrt{3} - 16$

$= 6\sqrt{3} - 13$

6. $\sqrt{4x+9} + 10 = 11$

$\sqrt{4x+9} = 1$

$(\sqrt{4x+9})^2 = 1^2$

$4x + 9 = 1$

$4x = -8$

$x = -2$

The solution is -2.

7. $\dfrac{x^{-3/2}}{x^{7/2}} = x^{-10/2} = x^{-5} = \dfrac{1}{x^5}$

8. $\dfrac{8}{\sqrt{3y}} = \dfrac{8}{\sqrt{3y}} \cdot \dfrac{\sqrt{3y}}{\sqrt{3y}} = \dfrac{8\sqrt{3y}}{\sqrt{3^2 y^2}} = \dfrac{8\sqrt{3y}}{3y}$

9. $\sqrt[3]{-8a^6b^{12}} = -2a^2b^4$

10. $\sqrt{50a^4b^3} - ab\sqrt{18a^2b}$

$= \sqrt{25a^4b^2 \cdot 2b} - ab\sqrt{9a^2 \cdot 2b}$

$= 5a^2b\sqrt{2b} - 3a^2b\sqrt{2b}$

$= 2a^2b\sqrt{2b}$

11. $\dfrac{14}{4-\sqrt{2}} = \dfrac{14}{4-\sqrt{2}} \cdot \dfrac{4+\sqrt{2}}{4+\sqrt{2}} = \dfrac{56+14\sqrt{2}}{16-\sqrt{2}^2}$

$= \dfrac{56+14\sqrt{2}}{16-2} = \dfrac{56+14\sqrt{2}}{14}$

$= 4+\sqrt{2}$

12. $\dfrac{5+2i}{3i} = \dfrac{5+2i}{3i} \cdot \dfrac{-3i}{-3i} = \dfrac{-15i-6i^2}{-9i^2}$

$= \dfrac{-15i-6(-1)}{-9(-1)} = \dfrac{-15i+6}{9}$

$= \dfrac{2}{3} - \dfrac{5}{3}i$

13. $\sqrt{18a^3b^6} = \sqrt{9a^2b^6 \cdot 2a} = 3ab^3\sqrt{2a}$

14. $(17+8i) - (15-4i) = 2+12i$

15. $3x\sqrt[3]{54x^8y^{10}} - 2x^2y\sqrt[3]{16x^5y^7}$

$= 3x\sqrt[3]{27x^6y^9 \cdot 2x^2y} - 2x^2y\sqrt[3]{8x^3y^6 \cdot 2x^2y}$

$= 9x^3y^3\sqrt[3]{2x^2y} - 4x^3y^3\sqrt[3]{2x^2y}$

$= 5x^3y^3\sqrt[3]{2x^2y}$

16. $\sqrt[3]{16x^4y}\sqrt[3]{4xy^5} = \sqrt[3]{64x^5y^6}$

$= \sqrt[3]{64x^3y^6 \cdot x^2} = 4xy^2\sqrt[3]{x^2}$

17. $i(3-7i) = 3i-7i^2 = 3i-7(-1) = 7+3i$

18. $\dfrac{(4a^{-2/3}b^4)^{-1/2}}{(a^{-1/6}b^{3/2})^2} = \dfrac{(4)^{-1/2}\,a^{1/3}b^{-2}}{a^{-1/3}b^3}$

$= \dfrac{a^{2/3}}{2b^5}$

19. $\sqrt[5]{-64a^8b^{12}} = \sqrt[5]{-32a^5b^{10} \cdot 2a^3b^2}$

$= -2ab^2\sqrt[5]{2a^3b^2}$

20. $\dfrac{5+9i}{1-i} = \dfrac{5+9i}{1-i} \cdot \dfrac{1+i}{1+i} = \dfrac{5+5i+9i+9i^2}{1-i^2}$

$= \dfrac{5+14i+9(-1)}{1-(-1)} = \dfrac{-4+14i}{2}$

$= -2+7i$

21. $\sqrt{-12}\sqrt{-6} = i\sqrt{12} \cdot i\sqrt{6} = i^2\sqrt{72}$

$= -1\sqrt{36 \cdot 2} = -6\sqrt{2}$

22. $\sqrt{x-5} + \sqrt{x+6} = 11$

$\sqrt{x-5} = 11-\sqrt{x+6}$

$(\sqrt{x-5})^2 = (11-\sqrt{x+6})^2$

$x-5 = 121-22\sqrt{x+6}+x+6$

$-5 = 127-22\sqrt{x+6}$

$-132 = -22\sqrt{x+6}$

$6 = \sqrt{x+6}$

$(6)^2 = (\sqrt{x+6})^2$

$36 = x+6$

$30 = x$

The solution is 30.

23. $\sqrt[4]{81a^8b^{12}} = 3a^2b^3$

24. $\dfrac{9}{\sqrt[3]{3x}} = \dfrac{9}{\sqrt[3]{3x}} \cdot \dfrac{\sqrt[3]{9x^2}}{\sqrt[3]{9x^2}} = \dfrac{9\sqrt[3]{9x^2}}{\sqrt[3]{27x^3}}$

$= \dfrac{9\sqrt[3]{9x^2}}{3x} = \dfrac{3\sqrt[3]{9x^2}}{x}$

25. $(-8+3i) - (4-7i) = -12+10i$

26. $(2+\sqrt{2x-1})^2 = 4+4\sqrt{2x-1}+2x-1$

$= 2x+4\sqrt{2x-1}+3$

27. $4x\sqrt{12x^2y} + \sqrt{3x^4y} - x^2\sqrt{27y}$

$\quad = 4x\sqrt{4x^2 \cdot 3y} + \sqrt{x^4 \cdot 3y} - x^2\sqrt{9 \cdot 3y}$

$\quad = 8x^2\sqrt{3y} + x^2\sqrt{3y} - 3x^2\sqrt{3y}$

$\quad = 6x^2\sqrt{3y}$

28. $81^{-1/4} = (3^4)^{-1/4} = 3^{-1} = \dfrac{1}{3}$

29. $(a^{16})^{-5/8} = a^{-10} = \dfrac{1}{a^{10}}$

30. $-\sqrt{49x^6y^{16}} - 7x^3y^8$

31. $4a^{2/3} = 4\sqrt[3]{a^2}$

32. $(9x^2y^4)^{-1/2}(x^6y^6)^{1/3} = 9^{-1/2}x^{-1}y^{-2}x^2y^2$

$\quad = 9^{-1/2}x^1y^0 = 9^{-1/2}x$

$\quad = \dfrac{x}{9^{1/2}} = \dfrac{x}{3}$

33. $\sqrt[4]{x^6y^8z^{10}} = \sqrt[4]{x^4y^8z^8 \cdot x^2z^2}$

$\quad = xy^2z^2\sqrt[4]{x^2z^2}$

34. $\sqrt{54} + \sqrt{24} = \sqrt{9 \cdot 6} + \sqrt{4 \cdot 6}$

$\quad = 3\sqrt{6} + 2\sqrt{6} = 5\sqrt{6}$

35. $\sqrt{48x^5y} - x\sqrt{80x^2y} = \sqrt{16x^4 \cdot 3xy} - x\sqrt{16x^2 \cdot y}$

$\quad = 4x^2\sqrt{3xy} - 4x^2\sqrt{y}$

36. $\sqrt{32}\sqrt{50} = \sqrt{1600} = 40$

37. $\sqrt{3x}(3 + \sqrt{3x}) = 3\sqrt{3x} + \sqrt{(3x)^2}$

$\quad = 3\sqrt{3x} + 3x$

38. $\dfrac{\sqrt{125x^6}}{\sqrt{5x^3}} = \sqrt{\dfrac{125x^6}{5x^3}} = \sqrt{25x^3} = 5x\sqrt{x}$

39. $\dfrac{2 - 3\sqrt{7}}{6 - \sqrt{7}} = \dfrac{2 - 3\sqrt{7}}{6 - \sqrt{7}} \cdot \dfrac{6 + \sqrt{7}}{6 + \sqrt{7}}$

$\quad = \dfrac{12 + 2\sqrt{7} - 18\sqrt{7} - 3(\sqrt{7})^2}{6^2 - (\sqrt{7})^2}$

$\quad = \dfrac{12 - 16\sqrt{7} - 21}{36 - 7}$

$\quad = \dfrac{-9 - 16\sqrt{7}}{29}$

40. $\sqrt{-36} = i\sqrt{36} = 6i$

41. $-b + \sqrt{b^2 - 4ac}$

$\quad -(-8) + \sqrt{(-8)^2 - 4(1)(25)} = 8 + \sqrt{64 - 100}$

$\quad = 8 + \sqrt{-36} = 8 + i\sqrt{36}$

$\quad = 8 + 6i$

42. $-b + \sqrt{b^2 - 4ac}$

$\quad -2 + \sqrt{2^2 - 4(1)(9)} = -2 + \sqrt{4 - 36}$

$\quad = -2 + \sqrt{-32} = -2 + i\sqrt{16 \cdot 2}$

$\quad = -2 + 4i\sqrt{2}$

43. $(5 + 2i) + (4 - 3i) = 9 - i$

44. $(3 + 2\sqrt{5})(3 - 2\sqrt{5})$

$\quad = 9 - 6\sqrt{5} + 6\sqrt{5} - 4(\sqrt{5})^2$

$\quad = 9 - 20 = -11$

45. $(3 - 9i) - 7 = -4 - 9i$

46. $(4 - i)^2 = (4 - i)(4 - i)$

$\quad = 16 - 8i + i^2 = 16 - 8i + (-1)$

$\quad = 15 - 8i$

47. $\dfrac{-6}{i} = \dfrac{-6}{i} \cdot \dfrac{i}{i} = \dfrac{-6i}{i^2} = \dfrac{-6i}{-1} = 6i$

48. $\dfrac{7}{2-i} = \dfrac{7}{2-i} \cdot \dfrac{2+i}{2+i} = \dfrac{14+7i}{4-i^2}$

$= \dfrac{14+7i}{4-(-1)} = \dfrac{14+7i}{5}$

$= \dfrac{14}{5} + \dfrac{7}{5}i$

49. $\sqrt{2x-7} + 2 = 5$

$\sqrt{2x-7} = 3$

$\left(\sqrt{2x-7}\right)^2 = 3^2$

$2x - 7 = 9$

$2x = 16$

$x = 8$

The solution is 8.

50. $\sqrt[3]{9x} = -6$

$\left(\sqrt[3]{9x}\right)^3 = (-6)^3$

$9x = -216$

$x = -24$

The solution is −24.

51. Strategy: Use the Pythagorean Theorem to find the width of the rectangle.

Solution: $c^2 = a^2 + b^2$

$13^2 = 12^2 + b^2$

$169 = 144 + b^2$

$25 = b^2$

$(25)^{1/2} = (b^2)^{1/2}$

$5 = b$

The width of the rectangle is 5 in.

52. Strategy: To find the amount of power, substitute the given value for v in the equation and solve for P.

Solution: $v = 4.05\sqrt[3]{P}$

$20 = 4.05\sqrt[3]{P}$

$4.94 = \sqrt[3]{P}$

$(4.94)^3 = (\sqrt[3]{P})^3$

$120 \approx P$

The amount of power is 120 watts.

53. Strategy: To find the distance required, substitute the given values for v and a in the equation and solve for s.

Solution: $v = \sqrt{2as}$

$88 = \sqrt{2(16s)}$

$(88)^2 = (\sqrt{32s})^2$

$7744 = 32s$

$242 = s$

The distance required is 242 ft.

54. Strategy: To find the distance, use the Pythagorean Theorem.

Solution: $c^2 = a^2 + b^2$

$12^2 = 10^2 + b^2$

$144 = 100 + b^2$

$44 = b^2$

$(44)^{1/2} = (b^2)^{1/2}$

$6.63 = b$

The distance is 6.63 ft.

Chapter 7 Test

1. $\dfrac{1}{2}\sqrt[4]{x^3} = \dfrac{1}{2}x^{3/4}$

2. $\sqrt[3]{54x^7y^3} - x\sqrt[3]{128x^4y^3} - x^2\sqrt[3]{2xy^3}$

$= \sqrt[3]{27x^6y^3 \cdot 2x} - x\sqrt[3]{64x^3y^3 \cdot 2x} - x^2\sqrt[3]{y^3 \cdot 2x}$

$= 3x^2y\sqrt[3]{2x} - 4x^2y\sqrt[3]{2x} - x^2y\sqrt[3]{2x}$

$= -2x^2y\sqrt[3]{2x}$

3. $3y^{2/5} = 3\sqrt[5]{y^2}$

4. $(2+5i)(4-2i) = 8 - 4i + 20i - 10i^2$

$= 8 + 16i - 10(-1)$

$= 18 + 16i$

5. $(3 - 2\sqrt{x})^2 = (3 - 2\sqrt{x})(3 - 2\sqrt{x})$

$= 9 - 12\sqrt{x} + 4(\sqrt{x})^2$

$= 4x - 12\sqrt{x} + 9$

6. $\dfrac{r^{2/3}r^{-1}}{r^{-1/2}} = \dfrac{r^{-1/3}}{r^{-1/2}} = r^{1/6}$

7. $\sqrt{x + 12} - \sqrt{x} = 2$

$\sqrt{x + 12} = 2 + \sqrt{x}$

$(\sqrt{x + 12})^2 = (2 + \sqrt{x})^2$

$x + 12 = 4 + 4\sqrt{x} + x$

$12 = 4 + 4\sqrt{x}$

$8 = 4\sqrt{x}$

$2 = \sqrt{x}$

$(2)^2 = (\sqrt{x})^2$

$4 = x$

The solution is 4.

8. $\sqrt[4]{4a^5 b^3} \sqrt[4]{8a^3 b^7} = \sqrt[4]{32a^8 b^{10}}$

$= \sqrt[4]{16a^8 b^8 \cdot 2b^2} = 2a^2 b^2 \sqrt[4]{2b^2}$

9. $\sqrt{3x}(\sqrt{x} - \sqrt{25x}) = \sqrt{3x^2} - \sqrt{75x^2}$

$= \sqrt{x^2 \cdot 3} - \sqrt{25x^2 \cdot 3} = x\sqrt{3} - 5x\sqrt{3}$

$= -4x\sqrt{3}$

10. $(5 - 2i) - (8 - 4i) = -3 + 2i$

11. $\sqrt{32x^4 y^7} = \sqrt{16x^4 y^6 \cdot 2y} = 4x^2 y^3 \sqrt{2y}$

12. $(2\sqrt{3} + 4)(3\sqrt{3} - 1)$

$= 6\sqrt{3^2} - 2\sqrt{3} + 12\sqrt{3} - 4$

$= 18 + 10\sqrt{3} - 4$

$= 14 + 10\sqrt{3}$

13. $\sqrt{-5} \cdot \sqrt{-20} = i\sqrt{5} \cdot i\sqrt{20} = i^2 \sqrt{100}$

$= -1\sqrt{100} = -10$

14. $\dfrac{4 - 2\sqrt{5}}{2 - \sqrt{5}} = \dfrac{4 - 2\sqrt{5}}{2 - \sqrt{5}} \cdot \dfrac{2 + \sqrt{5}}{2 + \sqrt{5}}$

$= \dfrac{8 + 4\sqrt{5} - 4\sqrt{5} - 2(\sqrt{5})^2}{2^2 - (\sqrt{5})^2}$

$= \dfrac{8 - 10}{4 - 5} = \dfrac{-2}{-1} = 2$

15. $\sqrt{18a^3} + a\sqrt{50a} = \sqrt{9a^2 \cdot 2a} + a\sqrt{25 \cdot 2a}$

$= 3a\sqrt{2a} + 5a\sqrt{2a}$

$= 8a\sqrt{2a}$

16. $(\sqrt{a} - 3\sqrt{b})(2\sqrt{a} + 5\sqrt{b})$

$= 2\sqrt{a^2} + 5\sqrt{ab} - 6\sqrt{ab} - 15\sqrt{b^2}$

$= 2a - \sqrt{ab} - 15b$

17. $\dfrac{(2x^{1/3} y^{-2/3})^6}{(x^{-4} y^8)^{1/4}} = \dfrac{2^6 x^2 y^{-4}}{x^{-1} y^2} = \dfrac{64x^3}{y^6}$

18. $\dfrac{10x}{\sqrt[3]{5x^2}} = \dfrac{10x}{\sqrt[3]{5x^2}} \cdot \dfrac{\sqrt[3]{25x}}{\sqrt[3]{25x}} = \dfrac{10x\sqrt[3]{25x}}{\sqrt[3]{125x^3}}$

$= \dfrac{10x\sqrt[3]{25x}}{5x} = 2\sqrt[3]{25x}$

19. $\dfrac{2 + 3i}{1 - 2i} = \dfrac{2 + 3i}{1 - 2i} \cdot \dfrac{1 + 2i}{1 + 2i} = \dfrac{2 + 4i + 3i + 6i^2}{1 - 4i^2}$

$= \dfrac{2 + 7i + 6(-1)}{1 - 4(-1)} = \dfrac{-4 + 7i}{5}$

$= -\dfrac{4}{5} + \dfrac{7}{5}i$

20. $\sqrt[3]{2x - 2} + 4 = 2$

$\sqrt[3]{2x - 2} = -2$

$(\sqrt[3]{2x - 2})^3 = (-2)^3$

$2x - 2 = -8$

$2x = -6$

$x = -3$

The solution is -3.

21. $\left(\dfrac{4a^4}{b^2}\right)^{-3/2} = \dfrac{(4a^4)^{-3/2}}{(b^2)^{-3/2}} = \dfrac{4^{-3/2}\,a^{-6}}{b^{-3}} = \dfrac{b^3}{8a^6}$

22. $\sqrt[3]{27a^4b^3c^7} = \sqrt[3]{27a^3b^3c^6 \cdot ac}$

$= 3abc^2\sqrt[3]{ac}$

23. $\dfrac{\sqrt{32x^5y}}{\sqrt{2xy^3}} = \sqrt{\dfrac{32x^5y}{2xy^3}} = \sqrt{\dfrac{16x^4}{y^2}} = \dfrac{4x^2}{y}$

24. $\dfrac{5x}{\sqrt{5x}} = \dfrac{5x}{\sqrt{5x}} \cdot \dfrac{\sqrt{5x}}{\sqrt{5x}} = \dfrac{5x\sqrt{5x}}{\left(\sqrt{5x}\right)^2}$

$= \dfrac{5x\sqrt{5x}}{5x} = \sqrt{5x}$

25. Strategy: To find the distance the object has fallen substitute the value for v in the equation and solve for d.

Solution: $v = \sqrt{64d}$

$192 = \sqrt{64d}$

$(192)^2 = \left(\sqrt{64d}\right)^2$

$36{,}864 = 64d$

$576 = d$

The object has fallen 576 ft.

Cumulative Review Exercises

1. The Distributive Property

2. $f(x) = 3x^2 - 2x + 1$
$f(-3) = 3(-3)^2 - 2(-3) + 1$
$f(-3) = 27 + 6 + 1$
$f(-3) = 34$

3. $5 - \dfrac{2}{3}x = 4$

$5 - \dfrac{2}{3}x - 5 = 4 - 5$

$-\dfrac{2}{3}x = -1$

$\left(-\dfrac{3}{2}\right)\left(-\dfrac{2}{3}x\right) = -1\left(-\dfrac{3}{2}\right)$

$x = \dfrac{3}{2}$

The solution is $\dfrac{3}{2}$.

4. $2[4 - 2(3 - 2x)] = 4(1 - x)$
$2[4 - 6 + 4x] = 4 - 4x$
$2[-2 + 4x] = 4 - 4x$
$-4 + 8x = 4 - 4x$
$-4 + 12x = 4$
$12x = 8$

$x = \dfrac{2}{3}$

The solution is $\dfrac{2}{3}$.

5. $2 + |4 - 3x| = 5$
$|4 - 3x| = 3$

$\begin{array}{ll} 4 - 3x = -3 & \quad 4 - 3x = 3 \\ -3x = -7 & \quad -3x = -1 \\ x = \dfrac{7}{3} & \quad x = \dfrac{1}{3} \end{array}$

The solutions are $\dfrac{1}{3}$ and $\dfrac{7}{3}$.

6. $6x - 3(2x + 2) > 3 - 3(x + 2)$
$6x - 6x - 6 > 3 - 3x - 6$
$-6 > -3x - 3$
$-3 > -3x$
$1 < x$
$\{x \mid x > 1\}$

7. $|2x + 3| \le 9$
$-9 \le 2x + 3 \le 9$
$-9 - 3 \le 2x + 3 - 3 \le 9 - 3$
$-12 \le 2x \le 6$
$-6 \le x \le 3$
$\{x \mid -6 \le x \le 3\}$

8. $81x^2 - y^2 = (9x + y)(9x - y)$

9. $x^5 + 2x^3 - 3x = x(x^4 + 2x^2 - 3)$
$$= x(x^2 + 3)(x^2 - 1)$$
$$= x(x^2 + 3)(x + 1)(x - 1)$$

10. Find the slope of the line.
$$m = \frac{y_2 - y_1}{x_2 - x_1} = \frac{2 - 3}{-1 - 2} = \frac{-1}{-3} = \frac{1}{3}$$

Use the point-slope formula to find the equation of the line.
$$y - y_1 = m(x - x_1)$$
$$y - 3 = \frac{1}{3}(x - 2)$$
$$y - 3 = \frac{1}{3}x - \frac{2}{3}$$
$$y = \frac{1}{3}x + \frac{7}{3}$$

The equation of the line is $y = \frac{1}{3}x + \frac{7}{3}$.

11. $\begin{vmatrix} 1 & 2 & -3 \\ 0 & -1 & 2 \\ 3 & 1 & -2 \end{vmatrix} = 1\begin{vmatrix} -1 & 2 \\ 1 & -2 \end{vmatrix} - 2\begin{vmatrix} 0 & 2 \\ 3 & -2 \end{vmatrix} - 3\begin{vmatrix} 0 & -1 \\ 3 & 1 \end{vmatrix}$

$$= 1(0) - 2(-6) - 3(3)$$
$$= 3$$

12. $P = \dfrac{R - C}{n}$
$$P \cdot n = \frac{R - C}{n} \cdot n$$
$$nP = R - C$$
$$nP + C = R$$
$$C = R - nP$$

13. $(2^{-1}x^2y^{-6})(2^{-1}y^{-4})^{-2}$
$$= (2^{-1}x^2y^{-6})(2^2y^8)$$
$$= 2x^2y^2$$

14. $\dfrac{x^2y^3}{x^2 + 2x - 8} \cdot \dfrac{2x^2 - 7x + 6}{xy^4}$

$$= \frac{x^2y^3}{(x + 4)(x - 2)} \cdot \frac{(2x - 3)(x - 2)}{xy^4}$$

$$= \frac{x^2y^3(2x - 3)(x - 2)}{xy^4(x + 4)(x - 2)}$$

$$= \frac{x(2x - 3)}{y(x + 4)}$$

15. $\sqrt{40x^3} - x\sqrt{90x} = \sqrt{4x^2 \cdot 10x} - x\sqrt{9 \cdot 10x}$
$$= 2x\sqrt{10x} - 3x\sqrt{10x}$$
$$= -x\sqrt{10x}$$

16. $\dfrac{x}{x - 2} - 2x = \dfrac{-3}{x - 2}$

$$(x - 2)\left(\frac{x}{x - 2} - 2x\right) = (x - 2)\left(\frac{-3}{x - 2}\right)$$

$$x - 2x(x - 2) = -3$$
$$x - 2x^2 + 4x = -3$$
$$-2x^2 + 5x + 3 = 0$$
$$2x^2 - 5x - 3 = 0$$
$$(2x + 1)(x - 3) = 0$$

$2x + 1 = 0 \qquad x - 3 = 0$
$2x = -1 \qquad\qquad x = 3$
$x = -\dfrac{1}{2}$

The solutions are $-\dfrac{1}{2}$ and 3.

17. Solve $3x - 2y = -6$ for y.
$$3x - 2y = -6$$
$$-2y = -3x - 6$$
$$y = \frac{3}{2}x + 3$$

The y-intercept is $(0, 3)$.

The slope is $\dfrac{3}{2}$

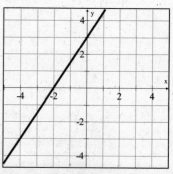

18. $3x + 2y \leq 4$
$2y \leq -3x + 4$
$y \leq -\dfrac{3}{2}x + 2$

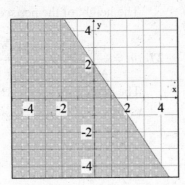

19. $\dfrac{2i}{3-i} = \dfrac{2i}{3-i} \cdot \dfrac{3+i}{3+i} = \dfrac{6i + 2i^2}{9 - i^2}$

$= \dfrac{6i + 2(-1)}{9 - (-1)} = \dfrac{6i - 2}{10}$

$= \dfrac{-1 + 3i}{5}$

$= -\dfrac{1}{5} + \dfrac{3}{5}i$

20. $\sqrt[3]{3x - 4} + 5 = 1$

$\sqrt[3]{3x - 4} = -4$

$(\sqrt[3]{3x - 4})^3 = (-4)^3$

$3x - 4 = -64$

$3x = -60$

$x = -20$

The solution is -20.

21. $\dfrac{x}{2x - 3} + \dfrac{4}{x + 4}$

$= \dfrac{x}{2x - 3} \cdot \dfrac{x + 4}{x + 4} + \dfrac{4}{x + 4} \cdot \dfrac{2x - 3}{2x - 3}$

$= \dfrac{x(x + 4) + 4(2x - 3)}{(x + 4)(2x - 3)}$

$= \dfrac{x^2 + 4x + 8x - 12}{(x + 4)(2x - 3)}$

$= \dfrac{x^2 + 12x - 12}{(x + 4)(2x - 3)}$

22. $2x - y = 4$
$-2x + 3y = 5$

$D = \begin{vmatrix} 2 & -1 \\ -2 & 3 \end{vmatrix} = 4$

$D_x = \begin{vmatrix} 4 & -1 \\ 5 & 3 \end{vmatrix} = 17$

$D_y = \begin{vmatrix} 2 & 4 \\ -2 & 5 \end{vmatrix} = 18$

$x = \dfrac{D_x}{D} = \dfrac{17}{4}$

$y = \dfrac{D_y}{D} = \dfrac{18}{4} = \dfrac{9}{2}$

The solution is $\left(\dfrac{17}{4}, \dfrac{9}{2}\right)$.

23. Strategy: Let x represent the number of 18¢ stamps.
The number of 13¢ stamps is $30 - x$.

Stamps	Number	Value	Total Value
18¢	x	18	$18x$
13¢	$30 - x$	13	$13(30 - x)$

The sum of the total values of each type of stamp equals the total value of all of the stamps (485¢).
$18x + 13(30 - x) = 485$

Solution: $18x + 13(30 - x) = 485$
$18x + 390 - 13x = 485$
$5x + 390 = 485$
$5x = 95$
$x = 19$

There are nineteen 18¢ stamps.

24. Strategy: Let x represent the unknown rate of the car.
The unknown rate of the plane is $5x$.

	Distance	Rate	Time
car	25	x	$\dfrac{25}{x}$
Plane	625	$5x$	$\dfrac{625}{5x}$

The total time of the trip was 3 h.

$$\frac{25}{x} + \frac{625}{5x} = 3$$

Solution: $\dfrac{25}{x} + \dfrac{625}{5x} = 3$

$$5x\left(\frac{25}{x} + \frac{625}{5x}\right) = 5x(3)$$

$$125 + 625 = 15x$$

$$750 = 15x$$

$$50 = x$$

$$250 = 5x$$

The rate of the plane is 250 mph.

25. Strategy: To find the time it takes light to travel from Earth to the moon, use the formula $D = RT$. Substitute in the given values for R and D and solve for T.

Solution: $D = RT$

$$1.86 \times 10^5 \cdot T = 232{,}500$$

$$T = 1.25 \times 10^0$$

$$T = 1.25$$

The time is 1.25 s.

26. Strategy: To find the height of the periscope, substitute in the given value for d and solve for h.

Solution: $d = \sqrt{1.5h}$

$$7 = \sqrt{1.5h}$$

$$(7)^2 = (\sqrt{1.5h})^2$$

$$49 = 1.5h$$

$$32.7 \approx h$$

The height of the periscope is 32.7 ft.

27. $m = \dfrac{y_2 - y_1}{x_2 - x_1} = \dfrac{400 - 0}{5000 - 0} = \dfrac{400}{5000} = 0.08$

The annual income is 8% of the investment.

Chapter 8: Quadratic Equations

Prep Test

1. $\sqrt{18} = \sqrt{9 \cdot 2} = 3\sqrt{2}$

2. $3i$

3. $\dfrac{3x-2}{x-1} - 1 = \dfrac{3x-2}{x-1} - \dfrac{x-1}{x-1}$

   $\quad = \dfrac{3x-2-x+1}{x-1}$

   $\quad = \dfrac{2x-1}{x-1}$

4. $b^2 - 4ac$

   $(-4)^2 - 4(2)(1) = 16 - 8$

   $\quad = 8$

5. $4x^2 + 28x + 49 = (2x+7)(2x+7)$

   Yes, it is a perfect square trinomial.

6. $4x^2 - 4x + 1 = (2x-1)(2x-1) = (2x-1)^2$

7. $9x^2 - 4 = (3x+2)(3x-2)$

8. $\{x\mid x < -1\} \cap \{x\mid x < 4\}$

9. $\quad x(x-1) = x + 15$

   $\quad\quad x^2 - x = x + 15$

   $\quad x^2 - 2x - 15 = 0$

   $\quad (x-5)(x+3) = 0$

   $x - 5 = 0 \quad x + 3 = 0$

   $\quad x = 5 \quad\quad x = -3$

   The solutions are -3 and 5.

10. $\quad\quad \dfrac{4}{x-3} = \dfrac{16}{x}$

    $x(x-3)\left(\dfrac{4}{x-3}\right) = x(x-3)\left(\dfrac{16}{x}\right)$

    $\quad\quad 4x = 16(x-3)$

    $\quad\quad 4x = 16x - 48$

    $\quad -12x = -48$

    $\quad\quad\quad x = 4$

    The solution is 4.

Section 8.1

Objective A Exercises

1. If $a = 0$ in a quadratic equation, there is no second-degree term. The equation is not be a quadratic equation.

3. $2x^2 - 4x - 5 = 0; a = 2, b = -4, c = -5$

5. $4x^2 - 5x + 6 = 0; a = 4, b = -5, c = 6$

7. $(x-3)(x-5) = 0$

   $x - 3 = 0 \quad x - 5 = 0$

   $\quad x = 3 \quad\quad x = 5$

   The solutions are 3 and 5.

9. $(x+7)(x+1) = 0$

   $x + 7 = 0 \quad x + 1 = 0$

   $\quad x = -7 \quad\quad x = -1$

   The solutions are -7 and -1.

11. $x^2 - 4x = 0$

    $x(x-4) = 0$

    $x = 0 \quad x - 4 = 0$

    $\quad\quad x = 4$

    The solutions are 0 and 4.

13. $t^2 - 25 = 0$

$(t+5)(t-5) = 0$

$t+5 = 0 \quad t-5 = 0$

$\quad t = -5 \qquad t = 5$

The solutions are -5 and 5.

15. $s^2 - s - 6 = 0$

$(s-3)(s+2) = 0$

$s-3 = 0 \quad s+2 = 0$

$\quad s = 3 \qquad s = -2$

The solutions are -2 and 3.

17. $y^2 - 6y + 9 = 0$

$(y-3)(y-3) = 0$

$y-3 = 0 \quad y-3 = 0$

$\quad y = 3 \qquad y = 3$

The solution is 3.

19. $9z^2 - 18z = 0$

$9z(z-2) = 0$

$9z = 0 \quad z-2 = 0$

$\quad z = 0 \qquad z = 2$

The solutions are 0 and 2.

21. $r^2 - 3r = 10$

$r^2 - 3r - 10 = 0$

$(r-5)(r+2) = 0$

$r-5 = 0 \quad r+2 = 0$

$\quad r = 5 \qquad r = -2$

The solutions are -2 and 5.

23. $v^2 + 10 = 7v$

$v^2 - 7v + 10 = 0$

$(v-5)(v-2) = 0$

$v-5 = 0 \quad v-2 = 0$

$\quad v = 5 \qquad v = 2$

The solutions are 2 and 5.

25. $2x^2 - 9x - 18 = 0$

$(2x+3)(x-6) = 0$

$2x+3 = 0 \quad x-6 = 0$

$2x = -3 \qquad x = 6$

$x = -\dfrac{3}{2}$

The solutions are $-\dfrac{3}{2}$ and 6.

27. $4z^2 - 9z + 2 = 0$

$(4z-1)(z-2) = 0$

$4z-1 = 0 \quad z-2 = 0$

$4z = 1 \qquad z = 2$

$z = \dfrac{1}{4}$

The solutions are $\dfrac{1}{4}$ and 2.

29. $3w^2 + 11w = 4$

$3w^2 + 11w - 4 = 0$

$(3w-1)(w+4) = 0$

$3w-1 = 0 \quad w+4 = 0$

$3w = 1 \qquad w = -4$

$w = \dfrac{1}{3}$

The solutions are -4 and $\dfrac{1}{3}$.

31. $6x^2 = 23x + 18$

$6x^2 - 23x - 18 = 0$

$(2x-9)(3x+2) = 0$

$2x-9 = 0 \quad 3x+2 = 0$

$2x = 9 \qquad 3x = -2$

$x = \dfrac{9}{2} \qquad x = -\dfrac{2}{3}$

The solutions are $-\dfrac{2}{3}$ and $\dfrac{9}{2}$.

33. $4 - 15u - 4u^2 = 0$

$(1 - 4u)(4 + u) = 0$

$1 - 4u = 0 \quad 4 + u = 0$

$-4u = -1 \qquad u = -4$

$u = \dfrac{1}{4}$

The solutions are $\dfrac{1}{4}$ and -4.

35. $x + 18 = x(x - 6)$

$x + 18 = x^2 - 6x$

$0 = x^2 - 7x - 18$

$0 = (x - 9)(x + 2)$

$x - 9 = 0 \quad x + 2 = 0$

$x = 9 \qquad x = -2$

The solutions are -2 and 9.

37. $4s(s + 3) = s - 6$

$4s^2 + 12s = s - 6$

$4s^2 + 11s + 6 = 0$

$(4s + 3)(s + 2) = 0$

$4s + 3 = 0 \quad s + 2 = 0$

$4s = -3 \qquad s = -2$

$s = -\dfrac{3}{4}$

The solutions are -2 and $-\dfrac{3}{4}$.

39. $u^2 - 2u + 4 = (2u - 3)(u + 2)$

$u^2 - 2u + 4 = 2u^2 + u - 6$

$0 = u^2 + 3u - 10$

$0 = (u + 5)(u - 2)$

$u + 5 = 0 \quad u - 2 = 0$

$u = -5 \quad u = 2$

The solutions are -5 and 2.

41. $(3x - 4)(x + 4) = x^2 - 3x - 28$

$3x^2 + 8x - 16 = x^2 - 3x - 28$

$2x^2 + 11x + 12 = 0$

$(2x + 3)(x + 4) = 0$

$2x + 3 = 0 \quad x + 4 = 0$

$2x = -3 \qquad x = -4$

$x = -\dfrac{3}{2}$

The solutions are -4 and $-\dfrac{3}{2}$.

43. $x^2 - 9bx + 14b^2 = 0$

$(x - 2b)(x - 7b) = 0$

$x - 2b = 0 \quad x - 7b = 0$

$x = 2b \qquad x = 7b$

The solutions are $2b$ and $7b$.

45. $x^2 - 6cx - 7c^2 = 0$

$(x + c)(x - 7c) = 0$

$x + c = 0 \quad x - 7c = 0$

$x = -c \qquad x = 7c$

The solutions are $-c$ and $7c$.

47. $2x^2 + 3bx + b^2 = 0$

$(2x + b)(x + b) = 0$

$2x + b = 0 \quad x + b = 0$

$2x = -b \qquad x = -b$

$x = -\dfrac{b}{2}$

The solutions are $-b$ and $-\dfrac{b}{2}$.

49. $3x^2 - 14ax + 8a^2 = 0$

$(3x - 2a)(x - 4a) = 0$

$3x - 2a = 0 \quad x - 4a = 0$

$3x = 2a \qquad x = 4a$

$x = \dfrac{2a}{3}$

The solutions are $4a$ and $\dfrac{2a}{3}$.

Objective B Exercises

51. $(x - r_1)(x - r_2) = 0$
$(x - 2)(x - 5) = 0$
$x^2 - 7x + 10 = 0$

53. $(x - r_1)(x - r_2) = 0$
$[x - (-2)][x - (-4)] = 0$
$(x + 2)(x + 4) = 0$
$x^2 + 6x + 8 = 0$

55. $(x - r_1)(x - r_2) = 0$
$(x - 6)[x - (-1)] = 0$
$(x - 6)(x + 1) = 0$
$x^2 - 5x - 6 = 0$

57. $(x - r_1)(x - r_2) = 0$
$(x - 3)[x - (-3)] = 0$
$(x - 3)(x + 3) = 0$
$x^2 - 9 = 0$

59. $(x - r_1)(x - r_2) = 0$
$(x - 4)(x - 4) = 0$
$x^2 - 8x + 16 = 0$

61. $(x - r_1)(x - r_2) = 0$
$(x - 0)(x - 5) = 0$
$x^2 - 5x = 0$

63. $(x - r_1)(x - r_2) = 0$
$(x - 0)(x - 3) = 0$
$x^2 - 3x = 0$

65. $(x - r_1)(x - r_2) = 0$
$(x - 3)\left(x - \dfrac{1}{2}\right) = 0$

$x^2 - \dfrac{7}{2}x + \dfrac{3}{2} = 0$

$2\left(x^2 - \dfrac{7}{2}x + \dfrac{3}{2}\right) = 0 \cdot 2$

$2x^2 - 7x + 3 = 0$

67. $(x - r_1)(x - r_2) = 0$
$\left(x - \left(-\dfrac{3}{4}\right)\right)(x - 2) = 0$

$\left(x + \dfrac{3}{4}\right)(x - 2) = 0$

$x^2 - \dfrac{5}{4}x - \dfrac{3}{2} = 0$

$4\left(x^2 - \dfrac{5}{4}x - \dfrac{3}{2}\right) = 0 \cdot 4$

$4x^2 - 5x - 6 = 0$

69. $(x - r_1)(x - r_2) = 0$
$\left(x - \left(-\dfrac{5}{3}\right)\right)[x - (-2)] = 0$

$\left(x + \dfrac{5}{3}\right)(x + 2) = 0$

$x^2 + \dfrac{11}{3}x + \dfrac{10}{3} = 0$

$3\left(x^2 + \dfrac{11}{3}x + \dfrac{10}{3}\right) = 0 \cdot 3$

$3x^2 + 11x + 10 = 0$

71. $(x - r_1)(x - r_2) = 0$
$\left(x - \left(-\dfrac{2}{3}\right)\right)\left(x - \dfrac{2}{3}\right) = 0$

$\left(x + \dfrac{2}{3}\right)\left(x - \dfrac{2}{3}\right) = 0$

$x^2 - \dfrac{4}{9} = 0$

$9\left(x^2 - \dfrac{4}{9}\right) = 0 \cdot 9$

$9x^2 - 4 = 0$

73. $(x - r_1)(x - r_2) = 0$

$$\left(x - \frac{1}{2}\right)\left(x - \frac{1}{3}\right) = 0$$

$$x^2 - \frac{5}{6}x + \frac{1}{6} = 0$$

$$6\left(x^2 - \frac{5}{6}x + \frac{1}{6}\right) = 0 \cdot 6$$

$$6x^2 - 5x + 1 = 0$$

75. $(x - r_1)(x - r_2) = 0$

$$\left(x - \frac{6}{5}\right)\left(x - \left(-\frac{1}{2}\right)\right) = 0$$

$$\left(x - \frac{6}{5}\right)\left(x + \frac{1}{2}\right) = 0$$

$$x^2 - \frac{7}{10}x - \frac{3}{5} = 0$$

$$10\left(x^2 - \frac{7}{10}x - \frac{3}{5}\right) = 0 \cdot 10$$

$$10x^2 - 7x - 6 = 0$$

77. $(x - r_1)(x - r_2) = 0$

$$\left(x - \left(-\frac{1}{4}\right)\right)\left(x - \left(-\frac{1}{2}\right)\right) = 0$$

$$\left(x + \frac{1}{4}\right)\left(x + \frac{1}{2}\right) = 0$$

$$x^2 + \frac{3}{4}x + \frac{1}{8} = 0$$

$$8\left(x^2 + \frac{3}{4}x + \frac{1}{8}\right) = 0 \cdot 8$$

$$8x^2 + 6x + 1 = 0$$

79. $(x - r_1)(x - r_2) = 0$

$$\left(x - \frac{3}{5}\right)\left(x - \left(-\frac{1}{10}\right)\right) = 0$$

$$\left(x - \frac{3}{5}\right)\left(x + \frac{1}{10}\right) = 0$$

$$x^2 - \frac{1}{2}x - \frac{3}{50} = 0$$

$$50\left(x^2 - \frac{1}{2}x - \frac{3}{50}\right) = 0 \cdot 50$$

$$50x^2 - 25x - 3 = 0$$

81. $c = 0$

Objective C Exercises

83. $y^2 = 49$

$$\sqrt{y^2} = \sqrt{49}$$

$$y = \pm\sqrt{49} = \pm 7$$

The solutions are -7 and 7.

85. $z^2 = -4$

$$\sqrt{z^2} = \sqrt{-4}$$

$$z = \pm\sqrt{-4} = \pm 2i$$

The solutions are $-2i$ and $2i$.

87. $s^2 - 4 = 0$

$$s^2 = 4$$

$$\sqrt{s^2} = \sqrt{4}$$

$$s = \pm\sqrt{4} = \pm 2$$

The solutions are -2 and 2.

89. $4x^2 - 81 = 0$

$4x^2 = 81$

$x^2 = \dfrac{81}{4}$

$\sqrt{x^2} = \sqrt{\dfrac{81}{4}}$

$x = \pm\sqrt{\dfrac{81}{4}} = \pm\dfrac{9}{2}$

The solutions are $-\dfrac{9}{2}$ and $\dfrac{9}{2}$.

91. $y^2 + 49 = 0$

$y^2 = -49$

$\sqrt{y^2} = \sqrt{-49}$

$x = \pm\sqrt{-49} = \pm 7i$

The solutions are $-7i$ and $7i$.

93. $v^2 - 48 = 0$

$v^2 = 48$

$\sqrt{v^2} = \sqrt{48}$

$v = \pm\sqrt{48} = \pm 4\sqrt{3}$

The solutions are $-4\sqrt{3}$ and $4\sqrt{3}$.

95. $r^2 - 75 = 0$

$r^2 = 75$

$\sqrt{r^2} = \sqrt{75}$

$r = \pm\sqrt{75} = \pm 5\sqrt{3}$

The solutions are $-5\sqrt{3}$ and $5\sqrt{3}$.

97. $z^2 + 18 = 0$

$z^2 = -18$

$\sqrt{z^2} = \sqrt{-18}$

$z = \pm\sqrt{-18} = \pm 3i\sqrt{2}$

The solutions are $-3i\sqrt{2}$ and $3i\sqrt{2}$.

99. $(x-1)^2 = 36$

$\sqrt{(x-1)^2} = \sqrt{36}$

$x - 1 = \pm\sqrt{36} = \pm 6$

$x - 1 = 6 \quad x - 1 = -6$

$x = 7 \qquad x = -5$

The solutions are -5 and 7.

101. $3(y+3)^2 = 27$

$(y+3)^2 = 9$

$\sqrt{(y+3)^2} = \sqrt{9}$

$y + 3 = \pm\sqrt{9} = \pm 3$

$y + 3 = 3 \quad y + 3 = -3$

$y = 0 \qquad y = -6$

The solutions are -6 and 0.

103. $5(z+2)^2 = 125$

$(z+2)^2 = 25$

$\sqrt{(z+2)^2} = \sqrt{25}$

$z + 2 = \pm\sqrt{25} = \pm 5$

$z + 2 = 5 \quad z + 2 = -5$

$z = 3 \qquad z = -7$

The solutions are -7 and 3.

105. $\left(v - \dfrac{1}{2}\right)^2 = \dfrac{1}{4}$

$\sqrt{\left(v - \dfrac{1}{2}\right)^2} = \sqrt{\dfrac{1}{4}}$

$v - \dfrac{1}{2} = \pm\sqrt{\dfrac{1}{4}} = \pm\dfrac{1}{2}$

$v - \dfrac{1}{2} = \dfrac{1}{2} \quad v - \dfrac{1}{2} = -\dfrac{1}{2}$

$v = 1 \qquad v = 0$

The solutions are 1 and 0.

107. $(x+5)^2 - 6 = 0$

$(x+5)^2 = 6$

$\sqrt{(x+5)^2} = \sqrt{6}$

$x+5 = \pm\sqrt{6}$

$x+5 = \sqrt{6} \qquad x+5 = -\sqrt{6}$

$x = -5+\sqrt{6} \qquad x = -5-\sqrt{6}$

The solutions are $-5-\sqrt{6}$ and $-5+\sqrt{6}$.

109. $(v-3)^2 + 45 = 0$

$(v-3)^2 = -45$

$\sqrt{(v-3)^2} = \sqrt{-45}$

$v-3 = \pm\sqrt{-45} = \pm 3i\sqrt{5}$

$v-3 = 3i\sqrt{5} \qquad v-3 = -3i\sqrt{5}$

$v = 3+3i\sqrt{5} \qquad v = 3-3i\sqrt{5}$

The solutions are $3-3i\sqrt{5}$ and $3+3i\sqrt{5}$.

111. $\left(u+\dfrac{2}{3}\right)^2 - 18 = 0$

$\left(u+\dfrac{2}{3}\right)^2 = 18$

$\sqrt{\left(u+\dfrac{2}{3}\right)^2} = \sqrt{18}$

$u+\dfrac{2}{3} = \pm\sqrt{18} = \pm 3\sqrt{2}$

$u+\dfrac{2}{3} = 3\sqrt{2} \qquad u+\dfrac{2}{3} = -3\sqrt{2}$

$u = -\dfrac{2}{3}+3\sqrt{2} \qquad u = -\dfrac{2}{3}-3\sqrt{2}$

$u = -\dfrac{2+9\sqrt{2}}{3} \qquad u = -\dfrac{2-9\sqrt{2}}{3}$

The solutions are $-\dfrac{2+9\sqrt{2}}{3}$ and

$-\dfrac{2-9\sqrt{2}}{3}$.

113. $(x-a)^2 = -b$

Two complex solutions

115. $(x-a)^2 = 0$

Two equal real solutions

Applying the Concepts

117. $(x-r_1)(x-r_2) = 0$

$(x-\sqrt{2})[x-(-\sqrt{2})] = 0$

$(x-\sqrt{2})(x+\sqrt{2}) = 0$

$x^2 - 2 = 0$

119. $(x-r_1)(x-r_2) = 0$

$(x-i)[x-(-i)] = 0$

$(x-i)(x+i) = 0$

$x^2 + 1 = 0$

121. $(x-r_1)(x-r_2) = 0$

$(x-2\sqrt{2})[x-(-2\sqrt{2})] = 0$

$(x-2\sqrt{2})(x+2\sqrt{2}) = 0$

$x^2 - 8 = 0$

123. $(x-r_1)(x-r_2) = 0$

$(x-i\sqrt{2})[x-(-i\sqrt{2})] = 0$

$(x-i\sqrt{2})(x+i\sqrt{2}) = 0$

$x^2 + 2 = 0$

125. $(2x-1)^2 = (2x+3)^2$

$\sqrt{(2x-1)^2} = \sqrt{(2x+3)^2}$

$2x-1 = \pm\sqrt{(2x+3)^2} = \pm(2x+3)$

$2x-1 = 2x+3 \qquad 2x-1 = -(2x+3)$

$-1 = 3 \qquad\qquad 4x = -2$

$\qquad\qquad\qquad\qquad x = -\dfrac{1}{2}$

The solution is $-\dfrac{1}{2}$.

127.
$$(2x+1)^2 = (x-7)^2$$
$$\sqrt{(2x+1)^2} = \sqrt{(x-7)^2}$$
$$2x+1 = \pm\sqrt{(x-7)^2} = \pm(x-7)$$
$$2x+1 = x-7 \qquad 2x+1 = -(x-7)$$
$$x = -8 \qquad\qquad 3x = 6$$
$$\qquad\qquad\qquad x = 2$$
The solutions are -8 and 2.

129. Because -2, 3 and 5 are solutions, by the Principle of Zero Products we have $(x+2)(x-3)(x-5) = 0$. Multiplying the left side gives a cubic polynomial.

Section 8.2

Objective A Exercises

1.
$$x^2 - 4x - 5 = 0$$
$$x^2 - 4x = 5$$
$$x^2 - 4x + 4 = 5 + 4$$
$$(x-2)^2 = 9$$
$$\sqrt{(x-2)^2} = \sqrt{9}$$
$$x - 2 = \pm\sqrt{9} = \pm3$$
$$x - 2 = 3 \quad x - 2 = -3$$
$$x = 5 \qquad x = -1$$
The solutions are -1 and 5.

3.
$$v^2 + 8v - 9 = 0$$
$$v^2 + 8v = 9$$
$$v^2 + 8v + 16 = 9 + 16$$
$$(v+4)^2 = 25$$
$$\sqrt{(v+4)^2} = \sqrt{25}$$
$$v + 4 = \pm\sqrt{25} = \pm5$$
$$v + 4 = 5 \quad v + 4 = -5$$
$$v = 1 \qquad v = -9$$
The solutions are -9 and 1.

5.
$$z^2 - 6z + 9 = 0$$
$$z^2 - 6z = -9$$
$$z^2 - 6z + 9 = -9 + 9$$
$$(z-3)^2 = 0$$
$$\sqrt{(z-3)^2} = \sqrt{0}$$
$$z - 3 = 0$$
$$z = 3$$
The solution is 3.

7.
$$r^2 + 4r - 7 = 0$$
$$r^2 + 4r = 7$$
$$r^2 + 4r + 4 = 7 + 4$$
$$(r+2)^2 = 11$$
$$\sqrt{(r+2)^2} = \sqrt{11}$$
$$r + 2 = \pm\sqrt{11}$$
$$r + 2 = \sqrt{11} \quad r + 2 = -\sqrt{11}$$
$$r = -2 + \sqrt{11} \qquad r = -2 - \sqrt{11}$$
The solutions are $-2 - \sqrt{11}$ and $-2 + \sqrt{11}$

9.
$$x^2 - 6x + 7 = 0$$
$$x^2 - 6x = -7$$
$$x^2 - 6x + 9 = -7 + 9$$
$$(x-3)^2 = 2$$
$$\sqrt{(x-3)^2} = \sqrt{2}$$
$$x - 3 = \pm\sqrt{2}$$
$$x - 3 = \sqrt{2} \quad x - 3 = -\sqrt{2}$$
$$x = 3 + \sqrt{2} \qquad x = 3 - \sqrt{2}$$
The solutions are $3 - \sqrt{2}$ and $3 + \sqrt{2}$.

11. $z^2 - 2z + 2 = 0$

$z^2 - 2z = -2$

$z^2 - 2z + 1 = -2 + 1$

$(z-1)^2 = -1$

$\sqrt{(z-1)^2} = \sqrt{-1}$

$z - 1 = \pm i$

$z - 1 = i \quad z - 1 = -i$

$z = 1 + i \quad z = 1 - i$

The solutions are $1 + i$ and $1 - i$.

13. $s^2 - 5s - 24 = 0$

$s^2 - 5s = 24$

$s^2 - 5s + \dfrac{25}{4} = 24 + \dfrac{25}{4}$

$\left(s - \dfrac{5}{2}\right)^2 = \dfrac{121}{4}$

$\sqrt{\left(s - \dfrac{5}{2}\right)^2} = \sqrt{\dfrac{121}{4}}$

$s - \dfrac{5}{2} = \pm\dfrac{11}{2}$

$s - \dfrac{5}{2} = \dfrac{11}{2} \quad s - \dfrac{5}{2} = -\dfrac{11}{2}$

$s = \dfrac{16}{2} = 8 \quad s = -\dfrac{6}{2} = -3$

The solutions are -3 and 8.

15. $x^2 + 5x - 36 = 0$

$x^2 + 5x = 36$

$x^2 + 5x + \dfrac{25}{4} = 36 + \dfrac{25}{4}$

$\left(x + \dfrac{5}{2}\right)^2 = \dfrac{169}{4}$

$\sqrt{\left(x + \dfrac{5}{2}\right)^2} = \sqrt{\dfrac{169}{4}}$

$x + \dfrac{5}{2} = \pm\dfrac{13}{2}$

$x + \dfrac{5}{2} = \dfrac{13}{2} \quad x + \dfrac{5}{2} = -\dfrac{13}{2}$

$x = \dfrac{8}{2} = 4 \quad x = -\dfrac{18}{2} = -9$

The solutions are -9 and 4.

17. $p^2 - 3p + 1 = 0$

$p^2 - 3p = -1$

$p^2 - 3p + \dfrac{9}{4} = -1 + \dfrac{9}{4}$

$\left(p - \dfrac{3}{2}\right)^2 = \dfrac{5}{4}$

$\sqrt{\left(p - \dfrac{3}{2}\right)^2} = \sqrt{\dfrac{5}{4}}$

$p - \dfrac{3}{2} = \pm\dfrac{\sqrt{5}}{2}$

$p - \dfrac{3}{2} = \dfrac{\sqrt{5}}{2} \quad p - \dfrac{3}{2} = -\dfrac{\sqrt{5}}{2}$

$p = \dfrac{3}{2} + \dfrac{\sqrt{5}}{2} \quad p = \dfrac{3}{2} - \dfrac{\sqrt{5}}{2}$

The solutions are $\dfrac{3 + \sqrt{5}}{2}$ and $\dfrac{3 - \sqrt{5}}{2}$.

19. $t^2 - t - 1 = 0$

$t^2 - t = 1$

$t^2 - t + \dfrac{1}{4} = 1 + \dfrac{1}{4}$

$\left(t - \dfrac{1}{2}\right)^2 = \dfrac{5}{4}$

$\sqrt{\left(t - \dfrac{1}{2}\right)^2} = \sqrt{\dfrac{5}{4}}$

$t - \dfrac{1}{2} = \pm \dfrac{\sqrt{5}}{2}$

$t - \dfrac{1}{2} = \dfrac{\sqrt{5}}{2} \qquad t - \dfrac{1}{2} = -\dfrac{\sqrt{5}}{2}$

$t = \dfrac{1}{2} + \dfrac{\sqrt{5}}{2} \qquad t = \dfrac{1}{2} - \dfrac{\sqrt{5}}{2}$

The solutions are $\dfrac{1 + \sqrt{5}}{2}$ and $\dfrac{1 - \sqrt{5}}{2}$.

21. $y^2 - 6y = 4$

$y^2 - 6y + 9 = 4 + 9$

$(y - 3)^2 = 13$

$\sqrt{(y - 3)^2} = \sqrt{13}$

$y - 3 = \pm\sqrt{13}$

$y - 3 = \sqrt{13} \qquad y - 3 = -\sqrt{13}$

$y = 3 + \sqrt{13} \qquad y = 3 - \sqrt{13}$

The solutions are $3 - \sqrt{13}$ and $3 + \sqrt{13}$.

23. $x^2 = 8x - 15$

$x^2 - 8x = -15$

$x^2 - 8x + 16 = -15 + 16$

$(x - 4)^2 = 1$

$\sqrt{(x - 4)^2} = \sqrt{1}$

$x - 4 = \pm 1$

$x - 4 = 1 \qquad x - 4 = -1$

$x = 5 \qquad x = 3$

The solutions are 3 and 5.

25. $v^2 = 4v - 13$

$v^2 - 4v = -13$

$v^2 - 4v + 4 = -13 + 4$

$(v - 2)^2 = -9$

$\sqrt{(v - 2)^2} = \sqrt{-9}$

$v - 2 = \pm 3i$

$v - 2 = 3i \qquad v - 2 = -3i$

$v = 2 + 3i \qquad v = 2 - 3i$

The solutions are $2 + 3i$ and $2 - 3i$.

27. $p^2 + 6p = -13$

$p^2 + 6p + 9 = -13 + 9$

$(p + 3)^2 = -4$

$\sqrt{(p + 3)^2} = \sqrt{-4}$

$p + 3 = \pm 2i$

$p + 3 = 2i \qquad p + 3 = -2i$

$p = -3 + 2i \qquad p = -3 - 2i$

The solutions are $-3 - 2i$ and $-3 + 2i$.

29. $y^2 - 2y = 17$

$y^2 - 2y + 1 = 17 + 1$

$(y - 1)^2 = 18$

$\sqrt{(y - 1)^2} = \sqrt{18}$

$y - 1 = \pm 3\sqrt{2}$

$y - 1 = 3\sqrt{2} \qquad y - 1 = -3\sqrt{2}$

$y = 1 + 3\sqrt{2} \qquad y = 1 - 3\sqrt{2}$

The solutions are $1 - 3\sqrt{2}$ and $1 + 3\sqrt{2}$.

31. $z^2 = z + 4$

$z^2 - z = 4$

$z^2 - z + \dfrac{1}{4} = 4 + \dfrac{1}{4}$

$\left(z - \dfrac{1}{2}\right)^2 = \dfrac{17}{4}$

$\sqrt{\left(z - \dfrac{1}{2}\right)^2} = \sqrt{\dfrac{17}{4}}$

$z - \dfrac{1}{2} = \pm\dfrac{\sqrt{17}}{2}$

$z - \dfrac{1}{2} = \dfrac{\sqrt{17}}{2} \qquad z - \dfrac{1}{2} = -\dfrac{\sqrt{17}}{2}$

$z = \dfrac{1}{2} + \dfrac{\sqrt{17}}{2} \qquad z = \dfrac{1}{2} - \dfrac{\sqrt{17}}{2}$

The solutions are $\dfrac{1 + \sqrt{17}}{2}$ and $\dfrac{1 - \sqrt{17}}{2}$.

33. $x^2 + 13 = 2x$

$x^2 - 2x = -13$

$x^2 - 2x + 1 = -13 + 1$

$(x - 1)^2 = -12$

$\sqrt{(x - 1)^2} = \sqrt{-12}$

$x - 1 = \pm 2i\sqrt{3}$

$x - 1 = 2i\sqrt{3} \qquad x - 1 = -2i\sqrt{3}$

$x = 1 + 2i\sqrt{3} \qquad x = 1 - 2i\sqrt{3}$

The solutions are $1 - 2i\sqrt{3}$ and $1 + 2i\sqrt{3}$.

35. $4x^2 - 4x + 5 = 0$

$4x^2 - 4x = -5$

$\dfrac{1}{4}\left(4x^2 - 4x\right) = \dfrac{1}{4}(-5)$

$x^2 - x = -\dfrac{5}{4}$

$x^2 - x + \dfrac{1}{4} = -\dfrac{5}{4} + \dfrac{1}{4}$

$\left(x - \dfrac{1}{2}\right)^2 = -1$

$\sqrt{\left(x - \dfrac{1}{2}\right)^2} = \sqrt{-1}$

$x - \dfrac{1}{2} = \pm i$

$x - \dfrac{1}{2} = i \qquad x - \dfrac{1}{2} = -i$

$x = \dfrac{1}{2} + i \qquad x = \dfrac{1}{2} - i$

The solutions are $\dfrac{1}{2} - i$ and $\dfrac{1}{2} + i$.

37. $9x^2 - 6x + 2 = 0$

$9x^2 - 6x = -2$

$\dfrac{1}{9}\left(9x^2 - 6x\right) = \dfrac{1}{9}(-2)$

$x^2 - \dfrac{2}{3}x = -\dfrac{2}{9}$

$x^2 - \dfrac{2}{3}x + \dfrac{1}{9} = -\dfrac{2}{9} + \dfrac{1}{9}$

$\left(x - \dfrac{1}{3}\right)^2 = -\dfrac{1}{9}$

$\sqrt{\left(x - \dfrac{1}{3}\right)^2} = \sqrt{-\dfrac{1}{9}}$

$x - \dfrac{1}{3} = \pm\dfrac{1}{3}i$

$$x - \frac{1}{3} = \frac{1}{3}i \qquad x - \frac{1}{3} = -\frac{1}{3}i$$

$$x = \frac{1}{3} + \frac{1}{3}i \qquad x = \frac{1}{3} - \frac{1}{3}i$$

The solutions are $\frac{1}{3} - \frac{1}{3}i$ and $\frac{1}{3} + \frac{1}{3}i$.

39. $2s^2 = 4s + 5$

$$2s^2 - 4s = 5$$

$$\frac{1}{2}\left(2s^2 - 4s\right) = \frac{1}{2}(5)$$

$$s^2 - 2s = \frac{5}{2}$$

$$s^2 - 2s + 1 = \frac{5}{2} + 1$$

$$(s - 1)^2 = \frac{7}{2}$$

$$\sqrt{(s - 1)^2} = \sqrt{\frac{7}{2}}$$

$$s - 1 = \pm\sqrt{\frac{7}{2}} = \pm\frac{\sqrt{14}}{2}$$

$$s - 1 = \frac{\sqrt{14}}{2} \qquad s - 1 = -\frac{\sqrt{14}}{2}$$

$$s = 1 + \frac{\sqrt{14}}{2} \qquad s = 1 - \frac{\sqrt{14}}{2}$$

The solutions are $\frac{2 - \sqrt{14}}{2}$ and $\frac{2 + \sqrt{14}}{2}$.

41. $2r^2 = 3 - r$

$$2r^2 + r = 3$$

$$\frac{1}{2}\left(2r^2 + r\right) = \frac{1}{2}(3)$$

$$r^2 + \frac{1}{2}r = \frac{3}{2}$$

$$r^2 + \frac{1}{2}r + \frac{1}{16} = \frac{3}{2} + \frac{1}{16}$$

$$\left(r + \frac{1}{4}\right)^2 = \frac{25}{16}$$

$$\sqrt{\left(r + \frac{1}{4}\right)^2} = \sqrt{\frac{25}{16}}$$

$$r + \frac{1}{4} = \pm\frac{5}{4}$$

$$r + \frac{1}{4} = \frac{5}{4} \qquad r + \frac{1}{4} = -\frac{5}{4}$$

$$r = \frac{4}{4} = 1 \qquad r = -\frac{6}{4} = -\frac{3}{2}$$

The solutions are $-\frac{3}{2}$ and 1.

43. $y - 2 = (y - 3)(y + 2)$

$$y - 2 = y^2 - y - 6$$

$$y^2 - 2y = 4$$

$$y^2 - 2y + 1 = 4 + 1$$

$$(y - 1)^2 = 5$$

$$\sqrt{(y - 1)^2} = \sqrt{5}$$

$$y - 1 = \pm\sqrt{5}$$

$$y - 1 = \sqrt{5} \qquad y - 1 = -\sqrt{5}$$

$$y = 1 + \sqrt{5} \qquad y = 1 - \sqrt{5}$$

The solutions are $1 - \sqrt{5}$ and $1 + \sqrt{5}$.

45. $6t - 2 = (2t - 3)(t - 1)$

$6t - 2 = 2t^2 - 5t + 3$

$2t^2 - 11t = -5$

$\frac{1}{2}(2t^2 - 11t) = \frac{1}{2}(-5)$

$t^2 - \frac{11}{2}t = -\frac{5}{2}$

$t^2 - \frac{11}{2}t + \frac{121}{16} = -\frac{5}{2} + \frac{121}{16}$

$\left(t - \frac{11}{4}\right)^2 = \frac{81}{16}$

$\sqrt{\left(t - \frac{11}{4}\right)^2} = \sqrt{\frac{81}{16}}$

$t - \frac{11}{4} = \pm\frac{9}{4}$

$t - \frac{11}{4} = \frac{9}{4} \qquad t - \frac{11}{4} = -\frac{9}{4}$

$t = \frac{20}{4} = 5 \qquad t = \frac{2}{4} = \frac{1}{2}$

The solutions are $\frac{1}{2}$ and 5.

47. $(x - 4)(x + 1) = x - 3$

$x^2 - 3x - 4 = x - 3$

$x^2 - 4x = 1$

$x^2 - 4x + 4 = 1 + 4$

$(x - 2)^2 = 5$

$\sqrt{(x - 2)^2} = \sqrt{5}$

$x - 2 = \pm\sqrt{5}$

$x - 2 = \sqrt{5} \qquad x - 2 = -\sqrt{5}$

$x = 2 + \sqrt{5} \qquad x = 2 - \sqrt{5}$

The solutions are $2 + \sqrt{5}$ and $2 - \sqrt{5}$.

49. $z^2 + 2z = 4$

$z^2 + 2z + 1 = 4 + 1$

$(z + 1)^2 = 5$

$\sqrt{(z + 1)^2} = \sqrt{5}$

$z + 1 = \pm\sqrt{5}$

$z + 1 = \sqrt{5} \qquad z + 1 = -\sqrt{5}$

$z = \sqrt{5} - 1 \qquad z = -\sqrt{5} - 1$

$z = 1.236 \qquad z = -3.236$

The solutions are -3.326 and 1.236.

51. $2x^2 = 4x - 1$

$2x^2 - 4x = -1$

$\frac{1}{2}(2x^2 - 4x) = \frac{1}{2}(-1)$

$x^2 - 2x = -\frac{1}{2}$

$x^2 - 2x + 1 = -\frac{1}{2} + 1$

$(x - 1)^2 = \frac{1}{2}$

$\sqrt{(x - 1)^2} = \sqrt{\frac{1}{2}}$

$x - 1 = \pm\sqrt{\frac{1}{2}}$

$x - 1 = \sqrt{\frac{1}{2}} \qquad x - 1 = -\sqrt{\frac{1}{2}}$

$x = \sqrt{\frac{1}{2}} + 1 \qquad x = -\sqrt{\frac{1}{2}} + 1$

$x = 1.707 \qquad x = -0.293$

The solutions are -0.293 and 1.707.

53. $4z^2 + 2z = 1$

$\dfrac{1}{4}\left(4z^2 + 2z\right) = \dfrac{1}{4}(1)$

$z^2 + \dfrac{1}{2}z = \dfrac{1}{4}$

$z^2 + \dfrac{1}{2}z + \dfrac{1}{16} = \dfrac{1}{4} + \dfrac{1}{16}$

$\left(z + \dfrac{1}{4}\right)^2 = \dfrac{5}{16}$

$\sqrt{\left(z + \dfrac{1}{4}\right)^2} = \sqrt{\dfrac{5}{16}}$

$z + \dfrac{1}{4} = \pm\dfrac{\sqrt{5}}{4}$

$z + \dfrac{1}{4} = \dfrac{\sqrt{5}}{4} \qquad z + \dfrac{1}{4} = -\dfrac{\sqrt{5}}{4}$

$z = \dfrac{\sqrt{5}}{4} - \dfrac{1}{4} \qquad z = -\dfrac{\sqrt{5}}{4} - \dfrac{1}{4}$

$z = 0.309 \qquad z = -0.809$

The solutions are 0.309 and -0.809.

55. $c \le 4$

Applying the Concepts

57. $x^2 - ax - 2a^2 = 0$

$x^2 - ax = 2a^2$

$x^2 - ax + \dfrac{1}{4}a^2 = 2a^2 + \dfrac{1}{4}a^2$

$\left(x - \dfrac{1}{2}a\right)^2 = \dfrac{9}{4}a^2$

$\sqrt{\left(x - \dfrac{1}{2}a\right)^2} = \sqrt{\dfrac{9}{4}a^2}$

$x - \dfrac{1}{2}a = \pm\dfrac{3}{2}a$

$x - \dfrac{1}{2}a = \dfrac{3}{2}a \qquad x - \dfrac{1}{2}a = -\dfrac{3}{2}a$

$x = \dfrac{4}{2}a = 2a \qquad x = -\dfrac{2}{2}a = -a$

The solutions are $-a$ and $2a$.

59. $x^2 + 3ax - 10a^2 = 0$

$x^2 + 3ax = 10a^2$

$x^2 + 3ax + \dfrac{9}{4}a^2 = 10a^2 + \dfrac{9}{4}a^2$

$\left(x + \dfrac{3}{2}a\right)^2 = \dfrac{49}{4}a^2$

$\sqrt{\left(x + \dfrac{3}{2}a\right)^2} = \sqrt{\dfrac{49}{4}a^2}$

$x + \dfrac{3}{2}a = \pm\dfrac{7}{2}a$

$x + \dfrac{3}{2}a = \dfrac{7}{2}a \qquad x + \dfrac{3}{2}a = -\dfrac{7}{2}a$

$x = \dfrac{4}{2}a = 2a \qquad x = -\dfrac{10}{2}a = -5a$

The solutions are $-5a$ and $2a$.

61. Strategy: Using the answer from Exercise 60, we know that it takes the ball 4.43 s to hit the ground. Use this time to find the horizontal distance the ball travels.

Solution: $t = 4.43$

$s = 44.5t = 44.5(4.43) = 197.2$ ft

No, the ball will have only gone 197.2 ft when it hits the ground.

Section 8.3

Objective A Exercises

1. The quadratic formula:

$$x = \dfrac{-b \pm \sqrt{b^2 - 4ac}}{2a}$$

a is the coefficient of x^2; b is the coefficient of x, and c is the constant term in the quadratic equation $ax^2 + bx + c, a \neq 0$.

3. $x^2 - 3x - 10 = 0$

$a = 1, b = -3, c = -10$

$$x = \frac{-b \pm \sqrt{b^2 - 4ac}}{2a}$$

$$x = \frac{-(-3) \pm \sqrt{(-3)^2 - 4(1)(-10)}}{2(1)}$$

$$x = \frac{3 \pm \sqrt{9 + 40}}{2} = \frac{3 \pm \sqrt{49}}{2}$$

$$x = \frac{3 \pm 7}{2}$$

$$x = \frac{3 + 7}{2} \qquad x = \frac{3 - 7}{2}$$

$$x = \frac{10}{2} = 5 \qquad x = -\frac{4}{2} = -2$$

The solutions are -2 and 5.

5. $y^2 + 5y - 36 = 0$

$a = 1, b = 5, c = -36$

$$y = \frac{-b \pm \sqrt{b^2 - 4ac}}{2a}$$

$$y = \frac{-5 \pm \sqrt{5^2 - 4(1)(-36)}}{2(1)}$$

$$y = \frac{-5 \pm \sqrt{25 + 144}}{2} = \frac{-5 \pm \sqrt{169}}{2}$$

$$y = \frac{-5 \pm 13}{2}$$

$$y = \frac{-5 + 13}{2} \qquad y = \frac{-5 - 13}{2}$$

$$y = \frac{8}{2} = 4 \qquad y = -\frac{18}{2} = -9$$

The solutions are -9 and 4.

7. $w^2 = 8w + 72$

$w^2 - 8w - 72 = 0$

$a = 1, b = -8, c = -72$

$$w = \frac{-b \pm \sqrt{b^2 - 4ac}}{2a}$$

$$w = \frac{-(-8) \pm \sqrt{(-8)^2 - 4(1)(-72)}}{2(1)}$$

$$w = \frac{8 \pm \sqrt{64 + 288}}{2} = \frac{8 \pm \sqrt{352}}{2}$$

$$w = \frac{8 \pm 4\sqrt{22}}{2}$$

$$w = \frac{8 + 4\sqrt{22}}{2} \qquad w = \frac{8 - 4\sqrt{22}}{2}$$

$$w = 4 + 2\sqrt{22} \qquad w = 4 - 2\sqrt{22}$$

The solutions are $4 - 2\sqrt{22}$ and $4 + 2\sqrt{22}$.

9. $v^2 = 24 - 5v$

$v^2 + 5v - 24 = 0$

$a = 1, b = 5, c = -24$

$$z = \frac{-b \pm \sqrt{b^2 - 4ac}}{2a}$$

$$z = \frac{-5 \pm \sqrt{5^2 - 4(1)(-24)}}{2(1)}$$

$$z = \frac{-5 \pm \sqrt{25 + 96}}{2} = \frac{-5 \pm \sqrt{121}}{2}$$

$$z = \frac{-5 \pm 11}{2}$$

$$z = \frac{-5 + 11}{2} \qquad z = \frac{-5 - 11}{2}$$

$$z = \frac{6}{2} = 3 \qquad z = -\frac{16}{2} = -8$$

The solutions are -8 and 3.

11. $2y^2 + 5y - 1 = 0$

$a = 2, b = 5, c = -1$

$$y = \frac{-b \pm \sqrt{b^2 - 4ac}}{2a}$$

$$y = \frac{-5 \pm \sqrt{5^2 - 4(2)(-1)}}{2(2)}$$

$$y = \frac{-5 \pm \sqrt{25 + 8}}{4} = \frac{-5 \pm \sqrt{33}}{4}$$

$$y = \frac{-5 + \sqrt{33}}{4} \qquad y = \frac{-5 - \sqrt{33}}{4}$$

The solutions are $\dfrac{-5 - \sqrt{33}}{4}$ and

$\dfrac{-5 + \sqrt{33}}{4}$.

13. $8s^2 = 10s + 3$

$8s^2 - 10s - 3 = 0$

$a = 8, b = -10, c = -3$

$$s = \frac{-b \pm \sqrt{b^2 - 4ac}}{2a}$$

$$s = \frac{-(-10) \pm \sqrt{(-10)^2 - 4(8)(-3)}}{2(8)}$$

$$s = \frac{10 \pm \sqrt{100 + 96}}{16} = \frac{10 \pm \sqrt{196}}{16}$$

$$s = \frac{10 \pm 14}{16}$$

$$s = \frac{10 + 14}{16} \qquad s = \frac{10 - 14}{16}$$

$$s = \frac{24}{16} = \frac{3}{2} \qquad s = -\frac{4}{16} = -\frac{1}{4}$$

The solutions are $-\dfrac{1}{4}$ and $\dfrac{3}{2}$.

15. $x^2 = 14x - 4$

$x^2 - 14x + 4 = 0$

$a = 1, b = -14, c = 4$

$$x = \frac{-b \pm \sqrt{b^2 - 4ac}}{2a}$$

$$x = \frac{-(-14) \pm \sqrt{(-14)^2 - 4(1)(4)}}{2(1)}$$

$$x = \frac{14 \pm \sqrt{196 - 16}}{2} = \frac{14 \pm \sqrt{180}}{2}$$

$$x = \frac{14 \pm 6\sqrt{5}}{2}$$

$$x = \frac{14 + 6\sqrt{5}}{2} \qquad x = \frac{14 - 6\sqrt{5}}{2}$$

$$x = 7 + 3\sqrt{5} \qquad x = 7 - 3\sqrt{5}$$

The solutions are $7 - 3\sqrt{5}$ and $7 + 3\sqrt{5}$.

17. $2z^2 - 2z - 1 = 0$

$a = 2, b = -2, c = -1$

$$z = \frac{-b \pm \sqrt{b^2 - 4ac}}{2a}$$

$$z = \frac{-(-2) \pm \sqrt{(-2)^2 - 4(2)(-1)}}{2(2)}$$

$$z = \frac{2 \pm \sqrt{4 + 8}}{4} = \frac{2 \pm \sqrt{12}}{4}$$

$$z = \frac{2 \pm 2\sqrt{3}}{4} = \frac{1 \pm \sqrt{3}}{2}$$

$$z = \frac{1 + \sqrt{3}}{2} \qquad z = \frac{1 - \sqrt{3}}{2}$$

The solutions are $\dfrac{1 - \sqrt{3}}{2}$ and $\dfrac{1 + \sqrt{3}}{2}$.

19. $z^2 + 2z + 2 = 0$
$a = 1, b = 2, c = 2$

$$z = \frac{-b \pm \sqrt{b^2 - 4ac}}{2a}$$

$$z = \frac{-2 \pm \sqrt{2^2 - 4(1)(2)}}{2(1)}$$

$$z = \frac{-2 \pm \sqrt{4 - 8}}{2} = \frac{-2 \pm \sqrt{-4}}{2}$$

$$z = \frac{-2 \pm 2i}{2} = -1 \pm i$$

$z = -1 + i \qquad z = -1 - i$

The solutions are $-1 - i$ and $-1 + i$.

21. $y^2 - 2y + 5 = 0$
$a = 1, b = -2, c = 5$

$$y = \frac{-b \pm \sqrt{b^2 - 4ac}}{2a}$$

$$y = \frac{-(-2) \pm \sqrt{(-2)^2 - 4(1)(5)}}{2(1)}$$

$$y = \frac{2 \pm \sqrt{4 - 20}}{2} = \frac{2 \pm \sqrt{-16}}{2}$$

$$y = \frac{2 \pm 4i}{2} = 1 \pm 2i$$

$y = 1 + 2i \qquad y = 1 - 2i$

The solutions are $1 - 2i$ and $1 + 2i$.

23. $s^2 - 4s + 13 = 0$
$a = 1, b = -4, c = 13$

$$s = \frac{-b \pm \sqrt{b^2 - 4ac}}{2a}$$

$$s = \frac{-(-4) \pm \sqrt{(-4)^2 - 4(1)(13)}}{2(1)}$$

$$s = \frac{4 \pm \sqrt{16 - 52}}{2} = \frac{4 \pm \sqrt{-36}}{2}$$

$$s = \frac{4 \pm 6i}{2} = 2 \pm 3i$$

$s = 2 + 3i \qquad s = 2 - 3i$

The solutions are $2 - 3i$ and $2 + 3i$..

25. $2w^2 - 2w - 5 = 0$
$a = 2, b = -2, c = -5$

$$w = \frac{-b \pm \sqrt{b^2 - 4ac}}{2a}$$

$$w = \frac{-(-2) \pm \sqrt{(-2)^2 - 4(2)(-5)}}{2(2)}$$

$$w = \frac{2 \pm \sqrt{4 + 40}}{4} = \frac{2 \pm \sqrt{44}}{4}$$

$$w = \frac{2 \pm 2\sqrt{11}}{4} = \frac{1 \pm \sqrt{11}}{2}$$

$$w = \frac{1 + \sqrt{11}}{2} \qquad w = \frac{1 - \sqrt{11}}{2}$$

The solutions are $\dfrac{1 - \sqrt{11}}{2}$ and $\dfrac{1 + \sqrt{11}}{2}$.

27. $2x^2 + 6x + 5 = 0$
$a = 2, b = 6, c = 5$

$$x = \frac{-b \pm \sqrt{b^2 - 4ac}}{2a}$$

$$x = \frac{-6 \pm \sqrt{6^2 - 4(2)(5)}}{2(2)}$$

$$x = \frac{-6 \pm \sqrt{36 - 40}}{4} = \frac{-6 \pm \sqrt{-4}}{4}$$

$$x = \frac{-6 \pm 2i}{4} = \frac{-3 \pm i}{2}$$

$$x = -\frac{3}{2} + \frac{1}{2}i \qquad x = -\frac{3}{2} - \frac{1}{2}i$$

The solutions are $-\dfrac{3}{2} - \dfrac{1}{2}i$ and $-\dfrac{3}{2} + \dfrac{1}{2}i$.

29. $4t^2 - 6t + 9 = 0$

$a = 4, b = -6, c = 9$

$$t = \frac{-b \pm \sqrt{b^2 - 4ac}}{2a}$$

$$t = \frac{-(-6) \pm \sqrt{(-6)^2 - 4(4)(9)}}{2(4)}$$

$$t = \frac{6 \pm \sqrt{36 - 144}}{8} = \frac{6 \pm \sqrt{-108}}{8}$$

$$t = \frac{6 \pm 6i\sqrt{3}}{8} = \frac{3 \pm 3i\sqrt{3}}{4}$$

$$t = \frac{3 + 3i\sqrt{3}}{4} \qquad t = \frac{3 - 3i\sqrt{3}}{4}$$

The solutions are $\dfrac{3 - 3i\sqrt{3}}{4}$ and $\dfrac{3 + 3i\sqrt{3}}{4}$.

31. $p^2 - 8p + 3 = 0$

$a = 1, b = -8, c = 3$

$$p = \frac{-b \pm \sqrt{b^2 - 4ac}}{2a}$$

$$p = \frac{-(-8) \pm \sqrt{(-8)^2 - 4(1)(3)}}{2(1)}$$

$$p = \frac{8 \pm \sqrt{64 - 12}}{2} = \frac{8 \pm \sqrt{52}}{2}$$

$$p = \frac{8 \pm 2\sqrt{13}}{2} = 4 \pm \sqrt{13}$$

$$p = 4 + \sqrt{13} \qquad p = 4 - \sqrt{13}$$

The solutions are 0.394 and 7.606.

33. $w^2 + 4w = 1$

$w^2 + 4w - 1 = 0$

$a = 1, b = 4, c = -1$

$$w = \frac{-b \pm \sqrt{b^2 - 4ac}}{2a}$$

$$w = \frac{-4 \pm \sqrt{4^2 - 4(1)(-1)}}{2(1)}$$

$$w = \frac{-4 \pm \sqrt{16 + 4}}{2} = \frac{-4 \pm \sqrt{20}}{2}$$

$$r = \frac{-4 \pm 2\sqrt{5}}{2} = -2 \pm \sqrt{5}$$

$$r = -2 + \sqrt{5} \qquad r = -2 - \sqrt{5}$$

The solutions are -4.236 and 0.236.

35. $2y^2 = y + 5$

$2y^2 - y - 5 = 0$

$a = 2, b = -1, c = -5$

$$y = \frac{-b \pm \sqrt{b^2 - 4ac}}{2a}$$

$$y = \frac{-(-1) \pm \sqrt{(-1)^2 - 4(2)(-5)}}{2(2)}$$

$$y = \frac{1 \pm \sqrt{1 + 40}}{4} = \frac{1 \pm \sqrt{41}}{4}$$

$$y = \frac{1 + \sqrt{41}}{4} \qquad y = \frac{1 - \sqrt{41}}{4}$$

The solutions are -1.351 and 1.851.

37. $3y^2 + y + 1 = 0$

$a = 3, b = 1, c = 1$

$b^2 - 4ac = 1^2 - 4(3)(1)$

$= 1 - 12 = -11$

$-11 < 0$

Since the discriminant is less than zero, the equation has two complex number solutions.

39. $4x^2 + 20x + 25 = 0$

$a = 4, b = 20, c = 25$

$b^2 - 4ac = (20)^2 - 4(4)(25)$

$= 400 - 400 = 0$

Since the discriminant is equal to zero, the equation has two equal real number solutions.

41. $3w^2 + 3w - 2 = 0$

$a = 3, b = 3, c = -2$

$b^2 - 4ac = 3^2 - 4(3)(-2)$

$= 9 + 24 = 33$

$33 > 0$

Since the discriminant is greater than zero, the equation has two unequal real number solutions.

43. $\sqrt{4ac}$

Applying the Concepts

45. Strategy: To determine if the arrow reaches a height of 275 ft, use the discriminant to determine if the equation has real number solutions.

Solution: $-16t^2 + 128t = 0$

$a = -16, b = 128, c = 0$

$b^2 - 4ac = (128)^2 - 4(-16)(0)$

$= 16{,}384 + 0 = 16{,}384$

$16{,}384 > 0$

Since the discriminant is greater than zero, the arrow reaches the height of 275 ft.

47. $x^2 - 6x + p = 0$

$x^2 - 6x = -p$

$x^2 - 6x + 9 = -p + 9$

$(x - 3)^2 = -p + 9$

$\sqrt{(x-3)^2} = \sqrt{9 - p}$

$x - 3 = \pm\sqrt{9 - p}$

$x = 3 \pm \sqrt{9 - p}$

x will have two real solutions if $9 - p > 0$.
Solving this inequality gives $p < 9$.
The values of p are $\{p \mid p < 9\}$.

49. $x^2 - 2x + p = 0$

$a = 1, b = -2, c = p$

$x = \dfrac{-b \pm \sqrt{b^2 - 4ac}}{2a}$

$x = \dfrac{-(-2) \pm \sqrt{(-2)^2 - 4(1)(p)}}{2(1)}$

$x = \dfrac{2 \pm \sqrt{4 - 4p}}{2}$

$x = 1 \pm \sqrt{1 - p}$

x will have two complex solutions when $1 - p < 0$.
Solving the inequality gives $p > 1$.
The values of p are $(1, \infty)$.

51. $x^2 + ix + 2 = 0$

$a = 1, b = i, c = 2$

$x = \dfrac{-b \pm \sqrt{b^2 - 4ac}}{2a}$

$x = \dfrac{-i \pm \sqrt{i^2 - 4(1)(2)}}{2(1)}$

$x = \dfrac{-i \pm \sqrt{-1 - 8}}{2} = \dfrac{i \pm \sqrt{-9}}{2}$

$x = \dfrac{-i \pm 3i}{2}$

$x = \dfrac{-i + 3i}{2} \qquad x = \dfrac{-i - 3i}{2}$

$x = i \qquad\qquad x = -2i$

The solutions are $-2i$ and i.

53. $2x^2 + bx - 2 = 0$

$a = 2, b = b, c = -2$

$$x = \frac{-b \pm \sqrt{b^2 - 4ac}}{2a}$$

$$x = \frac{-b \pm \sqrt{b^2 - 4(2)(-2)}}{2(2)}$$

$$x = \frac{-b \pm \sqrt{b^2 + 16}}{4}$$

If b is real, the quantity $\sqrt{b^2 + 16}$ can never be less than zero, and the quadratic equation will always have real number solutions.

Section 8.4

Objective A Exercises

1. Yes

3. No

5. $x^4 - 13x^2 + 36 = 0$

$(x^2)^2 - 13(x^2) + 36 = 0$

$u^2 - 13u + 36 = 0$

$(u - 4)(u - 9) = 0$

$u - 4 = 0 \quad u - 9 = 0$

$u = 4 \qquad u = 9$

Replace u with x^2.

$x^2 = 4 \qquad\quad x^2 = 9$

$\sqrt{x^2} = \sqrt{4} \quad \sqrt{x^2} = \sqrt{9}$

$x = \pm 2 \qquad\quad x = \pm 3$

The solutions are $-2, 2, -3$ and 3.

7. $z^4 - 6z^2 + 8 = 0$

$(z^2)^2 - 6(z^2) + 8 = 0$

$u^2 - 6u + 8 = 0$

$(u - 4)(u - 2) = 0$

$u - 4 = 0 \quad u - 2 = 0$

$u = 4 \qquad u = 2$

Replace u with z^2.

$z^2 = 4 \qquad\quad z^2 = 2$

$\sqrt{z^2} = \sqrt{4} \quad \sqrt{z^2} = \sqrt{2}$

$z = \pm 2 \qquad\quad z = \pm\sqrt{2}$

The solutions are $-2, 2, -\sqrt{2}$ and $\sqrt{2}$.

9. $p - 3p^{1/2} + 2 = 0$

$(p^{1/2})^2 - 3(p^{1/2}) + 2 = 0$

$u^2 - 3u + 2 = 0$

$(u - 2)(u - 1) = 0$

$u - 2 = 0 \quad u - 1 = 0$

$u = 2 \qquad u = 1$

Replace u with $p^{1/2}$.

$p^{1/2} = 2 \qquad\qquad p^{1/2} = 1$

$\left(p^{1/2}\right)^2 = 2^2 \quad \left(p^{1/2}\right)^2 = 1^2$

$p = 4 \qquad\qquad p = 1$

The solutions are 1 and 4.

11. $x - x^{1/2} - 12 = 0$

$(x^{1/2})^2 - (x^{1/2}) - 12 = 0$

$u^2 - u - 12 = 0$

$(u - 4)(u + 3) = 0$

$u - 4 = 0 \quad u + 3 = 0$

$u = 4 \qquad u = -3$

Replace u with $x^{1/2}$.

$x^{1/2} = 4 \qquad\qquad x^{1/2} = -3$

$\left(x^{1/2}\right)^2 = 4^2 \quad \left(x^{1/2}\right)^2 = (-3)^2$

$x = 16 \qquad\qquad x = 9$

9 does not check as a solution. The solution is 16.

13. $z^4 + 3z^2 - 4 = 0$

$(z^2)^2 + 3(z^2) - 4 = 0$

$u^2 + 3u - 4 = 0$

$(u + 4)(u - 1) = 0$

$u + 4 = 0 \quad u - 1 = 0$

$u = -4 \qquad u = 1$

Replace u with z^2.

$z^2 = -4 \qquad z^2 = 1$

$\sqrt{z^2} = \sqrt{-4} \quad \sqrt{z^2} = \sqrt{1}$

$z = \pm 2i \qquad z = \pm 1$

The solutions are -1, 1, $-2i$ and $2i$.

15. $x^4 + 12x^2 - 64 = 0$

$(x^2)^2 + 12(x^2) - 64 = 0$

$u^2 + 12u - 64 = 0$

$(u + 16)(u - 4) = 0$

$u + 16 = 0 \quad u - 4 = 0$

$u = -16 \qquad u = 4$

Replace u with x^2.

$x^2 = -16 \qquad x^2 = 4$

$\sqrt{x^2} = \sqrt{-16} \quad \sqrt{x^2} = \sqrt{4}$

$x = \pm 4i \qquad x = \pm 2$

The solutions are -2, 2, $-4i$ and $4i$.

17. $p + 2p^{1/2} - 24 = 0$

$(p^{1/2})^2 + 2(p^{1/2}) - 24 = 0$

$u^2 + 2u - 24 = 0$

$(u + 6)(u - 4) = 0$

$u + 6 = 0 \quad u - 4 = 0$

$u = -6 \qquad u = 4$

Replace u with $p^{1/2}$.

$p^{1/2} = -6 \qquad p^{1/2} = 4$

$\left(p^{1/2}\right)^2 = (-6)^2 \quad \left(p^{1/2}\right)^2 = 4^2$

$p = 36 \qquad p = 16$

36 does not check as a solution. The solution is 16.

19. $y^{2/3} - 9y^{1/3} + 8 = 0$

$(y^{1/3})^2 - 9(y^{1/3}) + 8 = 0$

$u^2 - 9u + 8 = 0$

$(u - 8)(u - 1) = 0$

$u - 8 = 0 \quad u - 1 = 0$

$u = 8 \qquad u = 1$

Replace u with $y^{1/3}$.

$y^{1/3} = 8 \qquad y^{1/3} = 1$

$(y^{1/3})^3 = 8^3 \quad (y^{1/3})^3 = 1^3$

$y = 512 \qquad y = 1$

The solutions are 1 and 512.

21. $9w^4 - 13w^2 + 4 = 0$

$9(w^2)^2 - 13(w^2) + 4 = 0$

$9u^2 - 13u + 4 = 0$

$(9u - 4)(u - 1) = 0$

$9u - 4 = 0 \quad u - 1 = 0$

$9u = 4 \qquad u = 1$

$u = \dfrac{4}{9}$

Replace u with w^2.

$w^2 = \dfrac{4}{9} \qquad w^2 = 1$

$\sqrt{w^2} = \sqrt{\dfrac{4}{9}} \quad \sqrt{w^2} = \sqrt{1}$

$w = \pm\dfrac{2}{3} \qquad w = \pm 1$

The solutions are -1, 1, $-\dfrac{2}{3}$ and $\dfrac{2}{3}$.

Objective B Exercises

23. Exercises 28, 29, 30, 34, 36, 38, 39, 42

25. $\sqrt{x+1} + x = 5$

$\sqrt{x+1} = 5 - x$

$\left(\sqrt{x+1}\right)^2 = (5 - x)^2$

$x + 1 = 25 - 10x + x^2$

$0 = x^2 - 11x + 24$

$(x - 3)(x - 8) = 0$

$x - 3 = 0 \quad x - 8 = 0$

$x = 3 \qquad x = 8$

8 does not check as a solution.
The solution is 3.

27. $x = \sqrt{x} + 6$

$x - 6 = \sqrt{x}$

$(x-6)^2 = (\sqrt{x})^2$

$x^2 - 12x + 36 = x$

$x^2 - 13x + 36 = 0$

$(x-9)(x-4) = 0$

$x - 9 = 0 \quad x - 4 = 0$

$x = 9 \qquad x = 4$

4 does not check as a solution.
The solution is 9.

29. $\sqrt{3w+3} = w + 1$

$(\sqrt{3w+3})^2 = (w+1)^2$

$3w + 3 = w^2 + 2w + 1$

$0 = w^2 - w - 2$

$(w-2)(w+1) = 0$

$w - 2 = 0 \quad w + 1 = 0$

$w = 2 \qquad w = -1$

The solutions are −1 and 2.

31. $\sqrt{4y+1} - y = 1$

$\sqrt{4y+1} = 1 + y$

$(\sqrt{4y+1})^2 = (1+y)^2$

$4y + 1 = 1 + 2y + y^2$

$0 = y^2 - 2y$

$y(y-2) = 0$

$y = 0 \quad y - 2 = 0$

$\qquad\quad y = 2$

The solutions are 0 and 2.

33. $\sqrt{10x+5} - 2x = 1$

$\sqrt{10x+5} = 1 + 2x$

$(\sqrt{10x+5})^2 = (1+2x)^2$

$10x + 5 = 1 + 4x + 4x^2$

$0 = 4x^2 - 6x - 4$

$2(2x+1)(x-2) = 0$

$2x + 1 = 0 \quad x - 2 = 0$

$2x = -1 \qquad x = 2$

$x = -\dfrac{1}{2}$

The solutions are $-\dfrac{1}{2}$ and 2.

35. $\sqrt{p+11} = 1 - p$

$(\sqrt{p+11})^2 = (1-p)^2$

$p + 11 = 1 - 2p + p^2$

$0 = p^2 - 3p - 10$

$(p-5)(p+2) = 0$

$p - 5 = 0 \quad p + 2 = 0$

$p = 5 \qquad p = -2$

5 does not check as a solution.
The solution is −2.

37. $\sqrt{x-1} - \sqrt{x} = -1$

$\sqrt{x-1} = \sqrt{x} - 1$

$(\sqrt{x-1})^2 = (\sqrt{x}-1)^2$

$x - 1 = x - 2\sqrt{x} + 1$

$2\sqrt{x} = 2$

$\sqrt{x} = 1$

$(\sqrt{x})^2 = 1^2$

$x = 1$

The solution is 1.

39. $\sqrt{2x-1} = 1 - \sqrt{x-1}$

$\left(\sqrt{2x-1}\right)^2 = \left(1 - \sqrt{x-1}\right)^2$

$2x - 1 = 1 - 2\sqrt{x-1} + x - 1$

$2\sqrt{x-1} = -x + 1$

$\left(2\sqrt{x-1}\right)^2 = (-x+1)^2$

$4(x-1) = x^2 - 2x + 1$

$4x - 4 = x^2 - 2x + 1$

$0 = x^2 - 6x + 5$

$(x-5)(x-1) = 0$

$x - 5 = 0 \quad x - 1 = 0$

$x = 5 \qquad x = 1$

5 does not check as a solution.
The solution is 1.

41. $\sqrt{t+3} + \sqrt{2t+7} = 1$

$\sqrt{t+3} = 1 - \sqrt{2t+7}$

$\left(\sqrt{t+3}\right)^2 = \left(1 - \sqrt{2t+7}\right)^2$

$t + 3 = 1 - 2\sqrt{2t+7} + 2t + 7$

$2\sqrt{2t+7} = t + 5$

$\left(2\sqrt{2t+7}\right)^2 = (t+5)^2$

$4(2t+7) = t^2 + 10t + 25$

$8t + 28 = t^2 + 10t + 25$

$0 = t^2 + 2t - 3$

$(t+3)(t-1) = 0$

$t + 3 = 0 \quad t - 1 = 0$

$t = -3 \qquad t = 1$

1 does not check as a solution.
The solution is −3.

Objective C exercises

43. $y + 2$

45. $x = \dfrac{10}{x-9}$

$(x-9)x = \left(\dfrac{10}{x-9}\right)(x-9)$

$x^2 - 9x = 10$

$x^2 - 9x - 10 = 0$

$(x-10)(x+1) = 0$

$x - 10 = 0 \quad x + 1 = 0$

$x = 10 \qquad x = -1$

The solutions are −1 and 10.

47. $\dfrac{y-1}{y+2} + y = 1$

$(y+2)\left(\dfrac{y-1}{y+2} + y\right) = 1(y+2)$

$y - 1 + y(y+2) = y + 2$

$y - 1 + y^2 + 2y = y + 2$

$y^2 + 2y - 3 = 0$

$(y+3)(y-1) = 0$

$y + 3 = 0 \quad y - 1 = 0$

$y = -3 \qquad y = 1$

The solutions are −3 and 1.

49. $\dfrac{3r+2}{r+2} - 2r = 1$

$(r+2)\left(\dfrac{3r+2}{r+2} - 2r\right) = 1(r+2)$

$3r + 2 - 2r(r+2) = r + 2$

$3r + 2 - 2r^2 - 4r = r + 2$

$-2r^2 - 2r = 0$

$-2r(r+1) = 0$

$-2r = 0 \quad r + 1 = 0$

$r = 0 \qquad r = -1$

The solutions are −1 and 0.

51. $\dfrac{2}{2x+1} + \dfrac{1}{x} = 3$

$x(2x+1)\left(\dfrac{2}{2x+1} + \dfrac{1}{x}\right) = 3x(2x+1)$

$2x + 2x + 1 = 6x^2 + 3x$

$0 = 6x^2 - x - 1$

$(2x-1)(3x+1) = 0$

$2x - 1 = 0 \quad 3x + 1 = 0$

$2x = 1 \quad\quad 3x = -1$

$x = \dfrac{1}{2} \quad\quad x = -\dfrac{1}{3}$

The solutions are $-\dfrac{1}{3}$ and $\dfrac{1}{2}$.

53. $\dfrac{16}{z-2} + \dfrac{16}{z+2} = 6$

$(z-2)(z+2)\left(\dfrac{16}{z-2} + \dfrac{16}{z+2}\right) = 6(z-2)(z+2)$

$16(z+2) + 16(z-2) = 6(z^2 - 4)$

$16z + 32 + 16z - 32 = 6z^2 - 24$

$0 = 6z^2 - 32z - 24$

$2(3z^2 - 16z - 12) = 0$

$2(3z+2)(z-6) = 0$

$3z + 2 = 0 \quad z - 6 = 0$

$3z = -2 \quad\quad z = 6$

$z = -\dfrac{2}{3}$

The solutions are $-\dfrac{2}{3}$ and 6.

55. $\dfrac{t}{t-2} + \dfrac{2}{t-1} = 4$

$(t-2)(t-1)\left(\dfrac{t}{t-2} + \dfrac{2}{t-1}\right) = 4(t-2)(t-1)$

$t(t-1) + 2(t-2) = 4(t^2 - 3t + 2)$

$t^2 - t + 2t - 4 = 4t^2 - 12t + 8$

$0 = 3t^2 - 13t + 12$

$(3t-4)(t-3) = 0$

$3t - 4 = 0 \quad t - 3 = 0$

$3t = 4 \quad\quad t = 3$

$t = \dfrac{4}{3}$

The solutions are $\dfrac{4}{3}$ and 3.

Applying the Concepts

57. $(x^2 - 7)^{1/2} = (x-1)^{1/2}$

$\left((x^2-7)^{1/2}\right)^2 = \left((x-1)^{1/2}\right)^2$

$x^2 - 7 = x - 1$

$x^2 - x - 6 = 0$

$(x-3)(x+2) = 0$

$x - 3 = 0 \quad x + 2 = 0$

$x = 3 \quad\quad x = -2$

The solutions are -2 and 3.

59. $\left(\sqrt{x} + 3\right)^2 - 4\sqrt{x} - 17 = 0$

Let $u = \sqrt{x} + 3$

$u^2 - 4(u-3) - 17 = 0$

$u^2 - 4u + 12 - 17 = 0$

$u^2 - 4u - 5 = 0$

$(u-5)(u+1) = 0$

$u - 5 = 0 \quad u + 1 = 0$

$u = 5 \quad\quad u = -1$

Replace u with $\sqrt{x} + 3$.

$$\sqrt{x}+3=5 \quad \sqrt{x}+3=-1$$

$$\sqrt{x}=2 \qquad \sqrt{x}=-4$$

$$\left(\sqrt{x}\right)^2=2^2 \quad \left(\sqrt{x}\right)^2=(-4)^2$$

$$x=4 \qquad \quad x=16$$

The solution is 4.

Section 8.5

Objective A Exercises

1. It must be true that $x-3>0$ and $x-5>0$ or the $x-3<0$ and $x-5<0$. In other words, either both factors are positive or both factors are negative.

3. a) $x=2$ True
   b) $x=-2$ False
   c) $x=-3$ False

5. a) $x<6$ False
   b) $-5\le x\le6$ True

7. $(x-4)(x+2)>0$

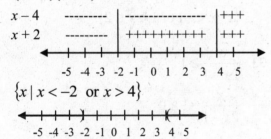

$\{x\,|\,x<-2 \text{ or } x>4\}$

9. $x^2-3x+2\ge0$
   $(x-2)(x-1)\ge0$

$\{x\,|\,x\le1 \text{ or } x\ge2\}$

11. $x^2-x-12<0$
   $(x-4)(x+3)<0$

$\{x\,|-3<x<4\}$

13. $(x-1)(x+2)(x-3)<0$

$\{x\,|\,x<-2 \text{ or } 1<x<3\}$

15. $(x+4)(x-2)(x-1)\ge0$

$\{x\,|-4\le x\le1 \text{ or } x\ge2\}$

17. $\dfrac{x-4}{x+2}>0$

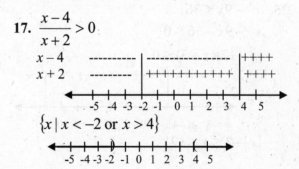

$\{x\,|\,x<-2 \text{ or } x>4\}$

19. $\dfrac{x-3}{x+1} \le 0$

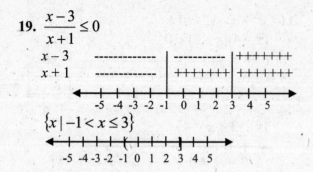

$\{x \mid -1 < x \le 3\}$

21. $\dfrac{(x-1)(x+2)}{x-3} \le 0$

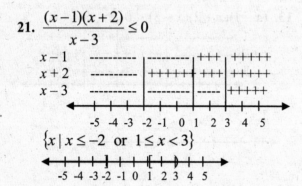

$\{x \mid x \le -2 \text{ or } 1 \le x < 3\}$

23. $x^2 - 16 > 0$

$(x+4)(x-4) > 0$

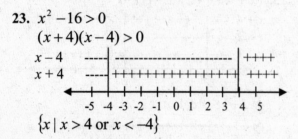

$\{x \mid x > 4 \text{ or } x < -4\}$

25. $x^2 - 9x \le 36$

$x^2 - 9x - 36 \le 0$

$(x-12)(x+3) \le 0$

$\{x \mid -3 \le x \le 12\}$

27. $4x^2 - 8x + 3 < 0$

$(2x-3)(2x-1) < 0$

$\left\{x \mid \dfrac{1}{2} < x < \dfrac{3}{2}\right\}$

29. $\dfrac{3}{x-1} < 2$

$\dfrac{3}{x-1} - 2 < 0$

$\dfrac{3}{x-1} - \dfrac{2x-2}{x-1} < 0$

$\dfrac{-2x+5}{x-1} < 0$

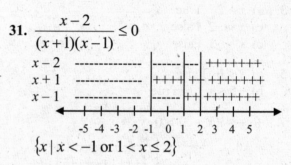

$\left\{x \mid x < 1 \text{ or } x > \dfrac{5}{2}\right\}$

31. $\dfrac{x-2}{(x+1)(x-1)} \le 0$

$\{x \mid x < -1 \text{ or } 1 < x \le 2\}$

33. $\dfrac{x}{2x-1} \ge 1$

$\dfrac{x}{2x-1} - 1 \ge 0$

$\dfrac{x}{2x-1} - \dfrac{2x-1}{2x-1} \ge 0$

$\dfrac{-x+1}{2x-1} \ge 0$

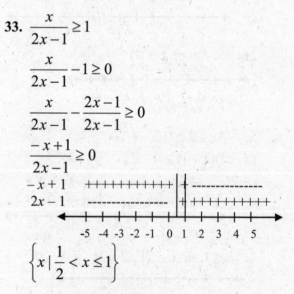

$\left\{x \mid \dfrac{1}{2} < x \le 1\right\}$

35. $\dfrac{x}{2-x} \le -3$

$\dfrac{x}{2-x} + 3 \le 0$

$\dfrac{x}{2-x} + \dfrac{6-3x}{2-x} \le 0$

$\dfrac{6-2x}{2-x} \le 0$

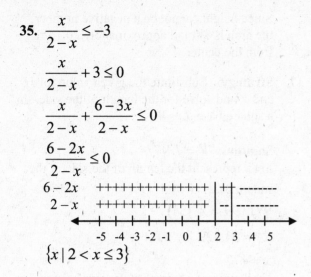

$\{x \mid 2 < x \le 3\}$

37. $\dfrac{3}{x-5} > \dfrac{1}{x+1}$

$\dfrac{3}{x-5} - \dfrac{1}{x+1} > 0$

$\dfrac{3(x+1)}{(x-5)(x+1)} - \dfrac{(x-5)}{(x-5)(x+1)} > 0$

$\dfrac{3x+3-x+5}{(x-5)(x+1)} > 0$

$\dfrac{2x+8}{(x-5)(x+1)} > 0$

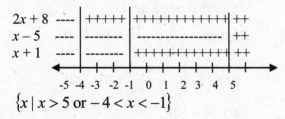

$\{x \mid x > 5 \text{ or } -4 < x < -1\}$

Applying the Concepts

39. $(x-1)(x+3)(x-2)(x-4) \ge 0$

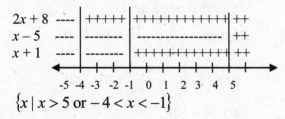

$\{x \mid x \le -3 \text{ or } 1 \le x \le 2 \text{ or } x \ge 4\}$

41. $(x^2 + 2x - 3)(x^2 + 3x + 2) \ge 0$

$(x-1)(x+3)(x+1)(x+2) \ge 0$

$\{x \mid x \le -3 \text{ or } -2 \le x \le -1 \text{ or } x \ge 1\}$

43. $\dfrac{x^2(3-x)(2x+1)}{(x+4)(x+2)} \ge 0$

$\left\{x \mid -4 < x < -2 \text{ or } -\dfrac{1}{2} \le x \le 3\right\}$

Section 8.6

Objective A Exercises

1. **Strategy:** To find the maximum safe speed, substitute for d and solve for v.

Solution: $d = 0.04v^2 + 0.5v$

$60 = 0.04v^2 + 0.5v$

$0 = 0.04v^2 + 0.5v - 60$

$v = \dfrac{-b \pm \sqrt{b^2 - 4ac}}{2a}$

$v = \dfrac{-0.5 \pm \sqrt{(0.5)^2 - 4(0.04)(-60)}}{2(0.04)}$

$v = \dfrac{-0.5 \pm \sqrt{9.85}}{0.08}$

$v \approx 33$

$v \approx -45$

Since the speed cannot be a negative number the maximum speed is 33 mph.

3. **Strategy:** To find the time it takes for the projectile to return to Earth, substitute for s and v_0 and solve for t.

Solution: $s = v_0 t - 16t^2$

$0 = 200t - 16t^2$

$0 = 8t(25 - 2t)$

$8t = 0 \quad 25 - 2t = 0$

$t = 0 \quad\quad t = 12.5$

The projectile will take 12.5 s to return to Earth.

5. a) **Strategy:** The maximum height occurs in the middle of the arch (when $x = 0$).

Solution: $h(x) = -\dfrac{3}{64}x^2 + 27$

$h(0) = -\dfrac{3}{64}(0)^2 + 27 = 27$

The maximum height of the arch is 27 ft.

b) **Strategy:** Let $x = 8$ in the function $h(x)$.

Solution: $h(x) = -\dfrac{3}{64}x^2 + 27$

$h(8) = -\dfrac{3}{64}(8)^2 + 27 = 24$

The height of the arch 8 ft to the right of the center is 24 ft.

c) **Strategy:** Let $h = 8$ in the height equation.

Solution: $h(x) = -\dfrac{3}{64}x^2 + 27$

$8 = -\dfrac{3}{64}x^2 + 27$

$-19 = -\dfrac{3}{64}x^2$

$x^2 = \dfrac{1216}{3}$

$x \approx \pm 20.13$

Since height cannot be a negative number, the arch is 8 ft tall approximately 20.13 ft from the center.

7. **Strategy:** Substitute the given value for H and V and solve for the length of the side. In a square base, $L = W$.

Solution: $V = LWH$

Let x represent the length of the side of the square base. $V = x^2 H$

$971{,}199 = x^2(31)$

$x^2 = 31{,}329$

$x = 177$

The length of a side of the square base is 177 m.

9. **Strategy:** Let t represent the time it takes the smaller pipe to fill the tank.

The time it takes the larger pipe to fill the tank is $t - 6$.

	Rate	Time	Part
Smaller pipe	$\dfrac{1}{t}$	4	$\dfrac{4}{t}$
Larger pipe	$\dfrac{1}{t-6}$	4	$\dfrac{4}{t-6}$

The sum of the parts of the task completed by each pipe equals 1.

Solution: $\dfrac{4}{t} + \dfrac{4}{t-6} = 1$

$t(t-6)\left(\dfrac{4}{t} + \dfrac{4}{t-6}\right) = 1t(t-6)$

$4(t-6) + 4t = t^2 - 6t$

$4t - 24 + 4t = t^2 - 6t$

$t^2 - 14t + 24 = 0$

$(t-12)(t-2) = 0$

$t - 12 = 0 \quad t - 2 = 0$

$t = 12 \quad\quad t = 2$

$t - 6 = 12 - 6 = 6$

$t - 6 = 2 - 6 = -4$

$t = 2$ is not possible since time cannot be a negative number. It will take the smaller pipe 12 min and the larger pipe 6 min.

11. Strategy: Let t represent the time it takes the faster computer working alone.

	Rate	Time	Part
Slower computer	$\dfrac{1}{t+4}$	3	$\dfrac{3}{t+4}$
Faster computer	$\dfrac{1}{t}$	1	$\dfrac{1}{t}$

The sum of the parts of the task completed by each computer equals 1.

Solution: $\dfrac{1}{t} + \dfrac{3}{t+4} = 1$

$$t(t+4)\left(\dfrac{1}{t} + \dfrac{3}{t+4}\right) = 1t(t+4)$$

$$t + 4 + 3t = t^2 + 4t$$

$$4t + 4 = t^2 + 4t$$

$$t^2 - 4 = 0$$

$$(t+2)(t-2) = 0$$

$$t + 2 = 0 \quad t - 2 = 0$$

$$t = -2 \qquad t = 2$$

Time cannot be a negative number. It will take the faster computer 2 h working alone.

13. Strategy: Let t represent the time it takes the experienced carpenter.
The time it takes the apprentice is $t+2$.

	Rate	Time	Part
Experienced carpenter	$\dfrac{1}{t}$	2	$\dfrac{2}{t}$
Apprentice carpenter	$\dfrac{1}{t+2}$	4	$\dfrac{4}{t+2}$

The sum of the parts of the task completed by each computer equals 1.

Solution: $\dfrac{2}{t} + \dfrac{4}{t+2} = 1$

$$t(t+2)\left(\dfrac{2}{t} + \dfrac{4}{t+2}\right) = 1t(t+2)$$

$$2(t+2) + 4t = t^2 + 2t$$

$$2t + 4 + 4t = t^2 + 2t$$

$$t^2 - 4t - 4 = 0$$

$$t = \dfrac{-b \pm \sqrt{b^2 - 4ac}}{2a}$$

$$t = \dfrac{-(-4) \pm \sqrt{(-4)^2 - 4(1)(-4)}}{2(1)}$$

$$t = \dfrac{4 \pm \sqrt{32}}{2} = 2 \pm 2\sqrt{2}$$

$$t \approx 4.8$$

$$t \approx -0.8$$

Time cannot be a negative number. It will take the apprentice carpenter $t + 2 = 6.8$ h working alone.

15. Strategy: Let r represent the rate of the wind.

	Distance	Rate	Time
With wind	4000	$1320 + r$	$\dfrac{4000}{1320 + r}$
Against wind	4000	$1320 - r$	$\dfrac{4000}{1320 - r}$

It took 0.5 h less time to make the return trip.

Solution: $\dfrac{4000}{1320 - r} - \dfrac{4000}{1320 + r} = 0.5$

$$(1320 - r)(1320 + r)\left(\dfrac{4000}{1320 - r} - \dfrac{4000}{1320 + r}\right) = 0.5(1320 - r)(1320 + r)$$

$$4000(1320 + r) - 4000(1320 - r) = 0.5(1{,}742{,}400 - r^2)$$

$$5{,}280{,}000 + 4000r - 5{,}280{,}000 + 4000r = 871{,}200 - 0.5r^2$$

$$8000r = 871{,}200 - 0.5r^2$$

$$0.5r^2 + 8000r - 871{,}200 = 0$$

$$r = \dfrac{-b \pm \sqrt{b^2 - 4ac}}{2a}$$

$$r = \dfrac{-(8000) \pm \sqrt{(8000)^2 - 4(0.5)(-871{,}200)}}{2(0.5)}$$

$$r = \dfrac{-8000 \pm \sqrt{65{,}742{,}400}}{1}$$

$$r \approx 108$$

$$r \cong -16{,}108$$

Since the rate cannot be a negative number. The rate of the wind was approximately 108 mph.

17. Strategy: Let r represent the rate of the jet stream.

	Distance	Rate	Time
With jet stream	3660	$630 + r$	$\dfrac{3660}{630 + r}$
Against jet stream	3660	$630 - r$	$\dfrac{3660}{630 - r}$

It took 1.75 h less time to make the trip flying with the jet stream.

Solution: $\dfrac{3660}{630 - r} - \dfrac{3660}{630 + r} = 1.75$

$$(630 + r)(630 - r)\left(\dfrac{3660}{630 - r} - \dfrac{3660}{630 + r}\right) = 1.75(630 + r)(630 - r)$$

$$3660(630 + r) - 3660(630 - r) = 1.75(396{,}900 - r^2)$$

$$2{,}305{,}800 + 3660r - 2{,}305{,}800 + 3660r = 694{,}575 - 1.75r^2$$

$$7320r = 694{,}575 - 1.75r^2$$

$$1.75r^2 = 7320r - 694{,}575 = 0$$

$$r = \frac{-b \pm \sqrt{b^2 - 4ac}}{2a}$$

$$r = \frac{-(7320) \pm \sqrt{(7320)^2 - 4(1.75)(-694{,}575)}}{2(1.75)}$$

$$r = \frac{-7320 \pm \sqrt{58{,}444{,}425}}{3.5}$$

$$r \approx 93$$

$$r \approx -4276$$

The rate cannot be a negative number. The rate of the jet stream is 93 mph.

19. Strategy: Let x represent the width of the rectangle.
The length of the rectangle is $x + 111$.
The area of the rectangle is 104,000 mi^2.

Solution: $A = LW$

$$104{,}000 = x(x+111)$$

$$104{,}000 = x^2 + 111x - 104{,}000$$

$$0 = x^2 + 111x - 104{,}000$$

$$x = \frac{-b \pm \sqrt{b^2 - 4ac}}{2a}$$

$$x = \frac{-111 \pm \sqrt{(111)^2 - 4(1)(-104{,}000)}}{2(1)}$$

$$x = \frac{-111 \pm \sqrt{428{,}321}}{3.5}$$

$$x \approx 272$$

$$x \approx -383$$

The width cannot be a negative number.
$x + 111 = 272 + 111 = 383$
The width is 272 mi.
The length is 383 mi.

21. Strategy: Let x represent the height of the triangle.
The base of the triangle is $5x - 1$.
The area of the triangle is 21 cm^2.

Solution: $A = \dfrac{1}{2}bh$

$$21 = \frac{1}{2}(5x - 1)(x)$$

$$42 = 5x^2 - x$$

$$0 = 5x^2 - x - 42$$

$$0 = (5x + 14)(x - 3)$$

$$5x + 14 = 0 \quad x - 3 = 0$$

$$5x = -14 \quad x = 3$$

$$x = -\frac{14}{5}$$

Since the height cannot be negative, $-\dfrac{14}{5}$ is

not a solution.
$5x - 1 = 5(3) - 1 = 14$
The height is 3 cm.
The base is 14 cm.

23. Strategy: Let x represent a side of the square base of the box.
The volume of the box is 49,000 cm^3.

Solution: $V = LWH$

$$49{,}000 = (x)(x)(10)$$

$$49{,}000 = 10x^2$$

$$x^2 = 4900$$

$$x = 70$$

The side of the original square is 20 cm more than side x.

$x + 20 = 70 + 20 = 90$

The dimensions of the original square base is 90 cm by 90 cm.

25. Strategy: Let x represent the width of the rectangle.

The length of the rectangle is $40 - x$.

The area of the rectangle is 300 ft^2.

Solution: $A = LW$

$$300 = x(40 - x)$$

$$300 = 40x - x^2$$

$$x^2 - 40x + 300 = 0$$

$$(x - 10(x - 30) = 0$$

$$x - 10 = 0 \quad x - 30 = 0$$

$$x = 10 \qquad x = 30$$

$$40 - x = 40 - 10 = 30$$

$$40 - x = 40 - 30 = 10$$

The dimensions of the rectangle are 10 ft by 30 ft.

Applying the Concepts

27. Strategy: To find the radius of the cone, substitute 11.25π for A and 6 for s in the equation $A = \pi r^2 + \pi rs$ and solve for r.

Solution: $A = \pi r^2 + \pi rs$

$$11.25\pi = \pi r^2 + \pi r(6)$$

$$0 = \pi r^2 + 6\pi r - 11.25\pi$$

$$a = \pi, b = 6\pi, c = -11.25\pi$$

$$r = \frac{-b \pm \sqrt{b^2 - 4ac}}{2a}$$

$$r = \frac{-6\pi \pm \sqrt{(6\pi)^2 - 4(\pi)(-11.25\pi)}}{2(\pi)}$$

$$r = \frac{-6\pi \pm \sqrt{36\pi^2 + 45\pi^2}}{2\pi}$$

$$r = \frac{-6\pi \pm \sqrt{81\pi^2}}{2\pi}$$

$$r = \frac{-6\pi \pm 9\pi}{2\pi}$$

$$r = 1.5$$

$$r = -7.5$$

The radius cannot be a negative number.

The radius of the cone is 1.5 in.

Chapter 8 Review Exercises

1. $2x^2 - 3x = 0$

$$x(2x - 3) = 0$$

$$2x - 3 = 0 \quad x = 0$$

$$2x = 3$$

$$x = \frac{3}{2}$$

The solutions are 0 and $\dfrac{3}{2}$.

2. $6x^2 + 9cx = 6c^2$

$$6x^2 + 9cx - 6c^2 = 0$$

$$3(2x^2 + 3cx - 2c^2) = 0$$

$$3(2x - c)(x + 2c) = 0$$

$$2x - c = 0 \quad x + 2c = 0$$

$$2x = c \qquad x = -2c$$

$$x = \frac{c}{2}$$

The solutions are $-2c$ and $\dfrac{c}{2}$.

3. $x^2 = 48$

$\sqrt{x^2} = \sqrt{48}$

$x = \pm\sqrt{48} = \pm 4\sqrt{3}$

The solutions are $4\sqrt{3}$ and $-4\sqrt{3}$.

4. $\left(x + \dfrac{1}{2}\right)^2 + 4 = 0$

$\left(x + \dfrac{1}{2}\right)^2 = -4$

$\sqrt{\left(x + \dfrac{1}{2}\right)^2} = \sqrt{-4}$

$x + \dfrac{1}{2} = \pm\sqrt{-4} = \pm 2i$

$x + \dfrac{1}{2} = 2i \quad x + \dfrac{1}{2} = -2i$

$x = 2i - \dfrac{1}{2} \quad x = -2i - \dfrac{1}{2}$

The solutions are $2i - \dfrac{1}{2}$ and $-2i - \dfrac{1}{2}$.

5. $x^2 + 4x + 3 = 0$

$x^2 + 4x = -3$

$x^2 + 4x + 4 = -3 + 4$

$(x + 2)^2 = 1$

$\sqrt{(x + 2)^2} = \sqrt{1}$

$x + 2 = \pm 1$

$x + 2 = 1 \quad x + 2 = -1$

$x = -1 \quad x = -3$

The solutions are -1 and -3.

6. $7x^2 - 14x + 3 = 0$

$7x^2 - 14x = -3$

$\dfrac{1}{7}(7x^2 - 14x) = \dfrac{1}{7}(-3)$

$x^2 - 2x = -\dfrac{3}{7}$

$x^2 - 2x + 1 = -\dfrac{3}{7} + 1$

$(x - 1)^2 = \dfrac{4}{7}$

$\sqrt{(x - 1)^2} = \sqrt{\dfrac{4}{7}}$

$x - 1 = \pm\sqrt{\dfrac{4}{7}} = \pm\dfrac{2\sqrt{7}}{7}$

$x - 1 = \dfrac{2\sqrt{7}}{7} \quad x - 1 = -\dfrac{2\sqrt{7}}{7}$

$x = 1 + \dfrac{2\sqrt{7}}{7} \quad x = 1 - \dfrac{2\sqrt{7}}{7}$

The solutions are $\dfrac{7 + 2\sqrt{7}}{7}$ and $\dfrac{7 - 2\sqrt{7}}{7}$.

7. $12x^2 - 25x + 12 = 0$

$a = 12, b = -25, c = 12$

$x = \dfrac{-b \pm \sqrt{b^2 - 4ac}}{2a}$

$x = \dfrac{-(-25) \pm \sqrt{(-25)^2 - 4(12)(12)}}{2(12)}$

$x = \dfrac{25 \pm \sqrt{625 - 576}}{24}$

$x = \dfrac{25 + 7}{24} = \dfrac{32}{24} = \dfrac{4}{3}$

$x = \dfrac{25 - 7}{24} = \dfrac{18}{24} = \dfrac{3}{4}$

$$x = \frac{25 \pm \sqrt{49}}{24}$$

$$x = \frac{25 \pm 7}{24}$$

The solutions are $\frac{4}{3}$ and $\frac{3}{4}$.

8. $x^2 - x + 8 = 0$

$a = 1, b = -1, c = 8$

$$x = \frac{-b \pm \sqrt{b^2 - 4ac}}{2a}$$

$$x = \frac{-(-1) \pm \sqrt{(-1)^2 - 4(1)(8)}}{2(1)}$$

$$x = \frac{1 \pm \sqrt{1 - 32}}{2}$$

$$x = \frac{1 \pm \sqrt{-31}}{2}$$

$$x = \frac{1 \pm i\sqrt{31}}{2}$$

The solutions are $\frac{1 + i\sqrt{31}}{2}$ and $\frac{1 - i\sqrt{31}}{2}$.

9. $(x - r_1)(x - r_2) = 0$

$(x - 0)(x - (-3)) = 0$

$x(x + 3) = 0$

$x^2 + 3x = 0$

10. $(x - r_1)(x - r_2) = 0$

$\left(x - \frac{3}{4}\right)\left(x - (-\frac{2}{3})\right) = 0$

$\left(x - \frac{3}{4}\right)\left(x + \frac{2}{3}\right) = 0$

$x^2 - \frac{1}{12}x - \frac{1}{2} = 0$

$12\left(x^2 - \frac{1}{12}x - \frac{1}{2}\right) = 0(12)$

$12x^2 - x - 6 = 0$

11. $x^2 - 2x + 8 = 0$

$x^2 - 2x = -8$

$x^2 - 2x + 1 = -8 + 1$

$(x - 1)^2 = -7$

$\sqrt{(x - 1)^2} = \sqrt{-7}$

$x - 1 = \pm\sqrt{-7} = \pm i\sqrt{7}$

$x = 1 \pm i\sqrt{7}$

The solutions are $1 + i\sqrt{7}$ and $1 - i\sqrt{7}$.

12. $(x - 2)(x + 3) = x - 10$

$x^2 + x - 6 = x - 10$

$x^2 = -4$

$\sqrt{x^2} = \sqrt{-4}$

$x = \pm\sqrt{-4}$

$x = \pm 2i$

The solutions are $2i$ and $-2i$.

13. $3x(x - 3) = 2x - 4$

$3x^2 - 9x = 2x - 4$

$3x^2 - 11x + 4 = 0$

$a = 3, b = -11, c = 4$

$$x = \frac{-b \pm \sqrt{b^2 - 4ac}}{2a}$$

$$x = \frac{-(-11) \pm \sqrt{(-11)^2 - 4(3)(4)}}{2(3)}$$

$$x = \frac{11 \pm \sqrt{121 - 48}}{6}$$

$$x = \frac{11 \pm \sqrt{73}}{6}$$

The solutions are $\frac{11 + \sqrt{73}}{6}$ and $\frac{11 - \sqrt{73}}{6}$.

14. $3x^2 - 5x + 3 = 0$

$a = 3, b = -5, c = 3$

$b^2 - 4ac$

$(-5)^2 - 4(3)(3) = 25 - 36 = -11$

$-11 > 0$

Since the discriminant is less than zero, the equation has two complex number solutions.

15. $(x+3)(2x-5) < 0$

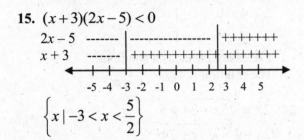

$\left\{ x \mid -3 < x < \dfrac{5}{2} \right\}$

16. $(x-2)(x+4)(2x+3) \le 0$

$x-2$	----- ⎮-------- ⎮--------------	+++++++
$x+4$	----- ⎮++++ +++++++	+++++++
$2x+3$	----- ⎮-------- ++++++	+++++++

-5 -4 -3 -2 -1 0 1 2 3 4 5

$\left\{ x \mid x \le -4 \; 1.\text{or} \; -\dfrac{3}{2} \le x \le 2 \right\}$

17. $x^{2/3} + x^{1/3} - 12 = 0$

$\left(x^{1/3} \right)^2 + x^{1/3} - 12 = 0$

$u^2 + u - 12 = 0$

$(u+4)(u-3) = 0$

$u+4 = 0 \quad u-3 = 0$

$u = -4 \qquad u = 3$

Replace u with $x^{1/3}$.

$x^{1/3} = -4 \qquad\quad x^{1/3} = 3$

$\left(x^{1/3} \right)^3 = (-4)^3 \quad \left(x^{1/3} \right)^3 = 3^3$

$x = -64 \qquad\qquad x = 27$

The solutions are -64 and 27.

18. $2(x-1) + 3\sqrt{x-1} - 2 = 0$

$2\left(\sqrt{x-1} \right)^2 + 3\sqrt{x-1} - 2 = 0$

$2u^2 + 3u - 2 = 0$

$(2u-1)(u+2) = 0$

$2u - 1 = 0 \qquad u + 2 = 0$

$2u = 1 \qquad\qquad u = -2$

$u = \dfrac{1}{2}$

Replace u with $\sqrt{x-1}$.

$\sqrt{x-1} = \dfrac{1}{2} \qquad\quad \sqrt{x-1} = 2$

$\left(\sqrt{x-1} \right)^2 = \left(\dfrac{1}{2} \right)^2 \quad \left(\sqrt{x-1} \right) = 2^2$

$x - 1 = \dfrac{1}{4} \qquad\qquad x - 1 = 4$

$x = \dfrac{5}{4} \qquad\qquad x = 5$

5 does not check as a solution.

The solution is $\dfrac{5}{4}$.

19. $3x = \dfrac{9}{x-2}$

$3x(x-2) = \dfrac{9}{x-2}(x-2)$

$3x^2 - 6x = 9$

$3x^2 - 6x - 9 = 0$

$3(x^2 - 2x - 3) = 0$

$3(x-3)(x+1) = 0$

$x - 3 = 0 \quad x + 1 = 0$

$x = 3 \qquad\quad x = -1$

The solutions are -1 and 3.

20. $\dfrac{3x+7}{x+2} + x = 3$

$\dfrac{3x+7}{x+2} = 3 - x$

$(x+2)\left(\dfrac{3x+7}{x+2}\right) = (3-x)(x+2)$

$3x + 7 = 3x + 6 - x^2 - 2x$

$x^2 + 2x + 1 = 0$

$(x+1)^2 = 0$

$\sqrt{(x+1)^2} = \sqrt{0}$

$x + 1 = 0$

$x = -1$

The solution is -1.

21. $\dfrac{x-2}{2x-3} \geq 0$

$\left\{ x \mid x < \dfrac{3}{2} \ \text{or} \ x \geq 2 \right\}$

22. $\dfrac{(2x-1)(x+3)}{x-4} \leq 0$

$\left\{ x \mid x \leq -3 \ \text{or} \ \dfrac{1}{2} \leq x < 4 \right\}$

23. $x = \sqrt{x} + 2$

$x - 2 = \sqrt{x}$

$(x-2)^2 = \left(\sqrt{x}\right)^2$

$x^2 - 4x + 4 = x$

$x^2 - 5x + 4 = 0$

$(x-4)(x-1) = 0$

$x - 4 = 0 \quad x - 1 = 0$

$x = 4 \qquad x = 1$

1 does not check as a solution.
The solution is 4.

24. $2x = \sqrt{5x+24} + 3$

$2x - 3 = \sqrt{5x+24}$

$(2x-3)^2 = \left(\sqrt{5x+24}\right)^2$

$4x^2 - 12x + 9 = 5x + 24$

$4x^2 - 17x - 15 = 0$

$(4x+3)(x-5) = 0$

$4x + 3 = 0 \quad x - 5 = 0$

$4x = -3 \qquad x = 5$

$x = -\dfrac{3}{4}$

$-\dfrac{3}{4}$ does not check as a solution.

The solution is 5.

25. $\dfrac{x-2}{2x+3} - \dfrac{x-4}{x} = 2$

$(2x+3)(x)\left(\dfrac{x-2}{2x+3} - \dfrac{x-4}{x}\right) = 2(2x+3)(x)$

$x(x-2) - (x-4)(2x+3) = 2x(2x+3)$

$x^2 - 2x - 2x^2 - 3x + 8x + 12 = 4x^2 + 6x$

$0 = 5x^2 + 3x - 12$

$a = 5, b = 3, c = -12$

$x = \dfrac{-b \pm \sqrt{b^2 - 4ac}}{2a}$

$x = \dfrac{-3 \pm \sqrt{3^2 - 4(5)(-12)}}{2(5)}$

$x = \dfrac{-3 \pm \sqrt{9 + 240}}{10}$

$x = \dfrac{-3 \pm \sqrt{249}}{10}$

The solutions are $\dfrac{-3 + \sqrt{249}}{10}$ and

$\dfrac{-3 - \sqrt{249}}{10}$.

26. $1 - \dfrac{x+4}{2-x} = \dfrac{x-3}{x+2}$

$(2-x)(x+2)\left(1 - \dfrac{x+4}{2-x}\right) = (2-x)(x+2)\dfrac{x-3}{x+2}$

$(2-x)(x+2) - (x+2)(x+4) = (2-x)(x-3)$

$4 - x^2 - x^2 - 6x - 8 = -x^2 + 5x - 6$

$x^2 + 11x - 2 = 0$

$a = 1, b = 11, c = -2$

$x = \dfrac{-b \pm \sqrt{b^2 - 4ac}}{2a}$

$x = \dfrac{-11 \pm \sqrt{11^2 - 4(1)(-2)}}{2(1)}$

$x = \dfrac{-11 \pm \sqrt{121 + 8}}{2}$

$x = \dfrac{-11 \pm \sqrt{129}}{2}$

The solutions are $\dfrac{-11 + \sqrt{129}}{2}$ and

$\dfrac{-11 - \sqrt{129}}{2}$.

27. $(x - r_1)(x - r_2) = 0$

$\left(x - \dfrac{1}{3}\right)(x - (-3)) = 0$

$\left(x - \dfrac{1}{3}\right)(x + 3) = 0$

$x^2 + \dfrac{8}{3}x - 1 = 0$

$3\left(x^2 + \dfrac{8}{3}x - 1\right) = 0(3)$

$3x^2 + 8x - 3 = 0$

28. $2x^2 + 9x = 5$

$2x^2 + 9x - 5 = 0$

$(2x - 1)(x + 5) = 0$

$2x - 1 = 0 \quad x + 5 = 0$

$2x = 1 \qquad x = -5$

$x = \dfrac{1}{2}$

The solutions are -5 and $\dfrac{1}{2}$.

29. $2(x+1)^2 - 36 = 0$

$2(x+1)^2 = 36$

$(x+1)^2 = 18$

$(x+1)^2 = \left(\sqrt{18}\right)^2$

$x + 1 = \pm\sqrt{18} = \pm 3\sqrt{2}$

$x = -1 \pm 3\sqrt{2}$

The solutions are $-1 + 3\sqrt{2}$ and $-1 - 3\sqrt{2}$.

30. $x^2 + 6x + 10 = 0$

$a = 1, b = 6, c = 10$

$x = \dfrac{-b \pm \sqrt{b^2 - 4ac}}{2a}$

$x = \dfrac{-6 \pm \sqrt{6^2 - 4(1)(10)}}{2(1)}$

$x = \dfrac{-6 \pm \sqrt{36 - 40}}{2} = \dfrac{-6 \pm \sqrt{-4}}{2}$

$x = \dfrac{-6 \pm 2i}{2}$

$x = -3 \pm i$

The solutions are $-3 + i$ and $-3 - i$.

31. $\dfrac{2}{x-4} + 3 = \dfrac{x}{2x-3}$

$(2x-3)(x-4)\left(\dfrac{2}{x-4} + 3\right) = (2x-3)(x-4)\dfrac{x}{2x-3}$

$2(2x-3) + 3(x-4)(2x-3) = x(x-4)$

$4x - 6 + 6x^2 - 33x + 36 = x^2 - 4x$

$5x^2 - 25x + 30 = 0$

$5(x^2 - 5x + 6) = 0$

$5(x-3)(x-2) = 0$

$x - 3 = 0 \quad x - 2 = 0$

$x = 3 \qquad x = 2$

The solutions are 2 and 3.

32. $x^4 - 28x^2 + 75 = 0$

$\left(x^2\right)^2 - 28x^2 + 75 = 0$

$u^2 - 28u + 75 = 0$

$(u-25)(u-3) = 0$

$u - 25 = 0 \quad u - 3 = 0$

$u = 25 \qquad u = 3$

Replace u with x^2.

$x^2 = 25 \qquad\qquad x^2 = 3$

$\sqrt{x^2} = \sqrt{25} \quad \sqrt{x^2} = \sqrt{3}$

$x = \pm 5 \qquad\qquad x = \pm\sqrt{3}$

The solutions are $-5, 5, \sqrt{3}$ and $-\sqrt{3}$.

33. $\sqrt{2x-1} + \sqrt{2x} = 3$

$\sqrt{2x-1} = 3 - \sqrt{2x}$

$\left(\sqrt{2x-1}\right)^2 = \left(3 - \sqrt{2x}\right)^2$

$2x - 1 = 9 - 6\sqrt{2x} - 2x$

$-10 = -6\sqrt{2x}$

$5 = 3\sqrt{2x}$

$5^2 = \left(3\sqrt{2x}\right)^2$

$25 = 18x$

$x = \dfrac{25}{18}$

The solution is $\dfrac{25}{18}$.

34. $2x^{2/3} + 3x^{1/3} - 2 = 0$

$2\left(x^{1/3}\right)^2 + 3x^{1/3} - 2 = 0$

$2u^2 + 3u - 2 = 0$

$(2u-1)(u+2) = 0$

$2u - 1 = 0 \quad u + 2 = 0$

$2u = 1 \qquad u = -2$

$u = \dfrac{1}{2}$

Replace u with $x^{1/3}$.

$$x^{1/3} = \frac{1}{2} \qquad x^{1/3} = -2$$

$$\left(x^{1/3}\right)^3 = \left(\frac{1}{2}\right)^3 \quad \left(x^{1/3}\right)^3 = (-2)^3$$

$$x = \frac{1}{8} \qquad x = -8$$

The solutions are -8 and $\frac{1}{8}$.

35. $\sqrt{3x-2} + 4 = 3x$

$\sqrt{3x-2} = 3x - 4$

$\left(\sqrt{3x-2}\right)^2 = (3x-4)^2$

$3x - 2 = 9x^2 - 24x + 16$

$9x^2 - 27x + 18 = 0$

$9(x^2 - 3x + 2) = 0$

$(x-2)(x-1) = 0$

$x - 2 = 0 \quad x - 1 = 0$

$x = 2 \qquad x = 1$

1 does not check as a solution.
The solution is 2.

36. $x^2 - 10x + 7 = 0$

$x^2 - 10x = -7$

$x^2 - 10x + 25 = -7 + 25$

$(x-5)^2 = 18$

$\sqrt{(x-5)^2} = \sqrt{18}$

$x - 5 = \pm 3\sqrt{2}$

$x = 5 \pm 3\sqrt{2}$

The solutions are $5 + 3\sqrt{2}$ and $5 - 3\sqrt{2}$.

37. $\dfrac{2x}{x-4} + \dfrac{6}{x+1} = 11$

$(x-4)(x+1)\left(\dfrac{2x}{x-4} + \dfrac{6}{x+1}\right) = 11(x-4)(x+1)$

$2x(x+1) + 6(x-4) = 11(x-4)(x+1)$

$2x^2 + 2x + 6x - 24 = 11x^2 - 33x - 44$

$0 = 9x^2 - 41x - 20$

$0 = (9x+4)(x-5)$

$9x + 4 = 0 \quad x - 5 = 0$

$9x = -4 \qquad x = 5$

$x = -\dfrac{4}{9}$

The solutions are $-\dfrac{4}{9}$ and 5.

38. $9x^2 - 3x = 1$

$9x^2 - 3x - 1 = 0$

$a = 9, b = -3, c = -1$

$x = \dfrac{-b \pm \sqrt{b^2 - 4ac}}{2a}$

$x = \dfrac{-(-3) \pm \sqrt{(-3)^2 - 4(9)(-1)}}{2(9)}$

$x = \dfrac{3 \pm \sqrt{9+36}}{18} = \dfrac{3 \pm \sqrt{45}}{18}$

$x = \dfrac{3 \pm 3\sqrt{5}}{18}$

$x = \dfrac{1 \pm \sqrt{5}}{6}$

The solutions are $\dfrac{1+\sqrt{5}}{6}$ and $\dfrac{1-\sqrt{5}}{6}$.

39. $2x = 4 - 3\sqrt{x-1}$

$2x - 4 = 3\sqrt{x-1}$

$(2x-4)^2 = \left(3\sqrt{x-1}\right)^2$

$4x^2 - 16x + 16 = 9x - 9$

$4x^2 - 25x + 25 = 0$

$(4x-5)(x-5) = 0$

$4x - 5 = 0 \quad x - 5 = 0$

$4x = 5 \qquad x = 5$

$x = \dfrac{5}{4}$

$\dfrac{5}{4}$ does not check as a solution.

The solution is 5.

40. $1 - \dfrac{x+3}{3-x} = \dfrac{x-4}{x+3}$

$(3-x)(x+3)\left(1 - \dfrac{x+3}{3-x}\right) = (3-x)(x+3)\dfrac{x-4}{x+3}$

$(3-x)(x+3) - (x+3)(x+3) = (3-x)(x-4)$

$3x + 9 - x^2 - 3x - x^2 - 6x - 9 = -x^2 + 7x - 12$

$x^2 + 13x - 12 = 0$

$a = 1, b = 13, c = -12$

$x = \dfrac{-b \pm \sqrt{b^2 - 4ac}}{2a}$

$x = \dfrac{-13 \pm \sqrt{13^2 - 4(1)(-12)}}{2(1)}$

$x = \dfrac{-13 \pm \sqrt{169 + 48}}{2}$

$x = \dfrac{-13 \pm \sqrt{217}}{2}$

The solutions are $\dfrac{-13 + \sqrt{217}}{2}$ and

$-\dfrac{13 - \sqrt{217}}{2}$.

41. $2x^2 - 5x = 6$

$2x^2 - 5x - 6 = 0$

$a = 2, b = -5, c = -6$

$b^2 - 4ac = (-5)^2 - 4(2)(-6) = 73$

$73 > 0$

Since the discriminant is greater than zero the equation has two unequal real number solutions.

42. $x^2 - 3x \le 10$

$x^2 - 3x - 10 \le 0$

$(x-5)(x+2) \le 0$

The zeros are −2 and 5. The factors have opposite signs between the zeros. The solution set is $\{x \mid -2 \le x \le 5\}$.

43. Strategy: Let r represent the rate of the rowing in calm water.

	Distance	Rate	Time
With current	16	$r+2$	$\dfrac{16}{r+2}$
Against current	16	$r-2$	$\dfrac{16}{r-2}$

The total time traveled was 6 h.

Solution: $\dfrac{16}{r+2} + \dfrac{16}{r-2} = 6$

$(r-2)(r+2)\left(\dfrac{16}{r+2} + \dfrac{16}{r-2}\right) = 6(r-2)(r+2)$

$16(r-2) + 16(r+2) = 6r^2 - 24$

$16r - 32 + 16r + 32 = 6r^2 - 24$

$0 = 6r^2 - 32r - 24$

$0 = 2(3r^2 - 16r - 12)$

$0 = (3r+2)(r-6)$

$3r + 2 = 0 \quad r - 6 = 0$

$3r = -2 \qquad r = 6$

$r = -\dfrac{2}{3}$

The rate cannot be a negative number. The rowing rate in calm water is 6 mph.

44. Strategy: Let x represent the width of the rectangle.
The length of the rectangle is $2x + 2$.
The area of the rectangle is 60 cm^2.

Solution: $A = LW$

$60 = x(2x + 2)$

$60 = 2x^2 + 2x$

$0 = 2x^2 + 2x - 60$

$0 = 2(x^2 + x - 30)$

$0 = 2(x + 6)(x - 5)$

$x + 6 = 0 \quad x - 5 = 0$

$x = -6 \quad\quad x = 5$

The width cannot be a negative number.
$2x + 2 = 2(5) + 2 = 12$
The width is 5 cm.
The length is 12 cm.

45. Strategy: Let x represent the first integer.
The second consecutive even integer: $x + 2$.
The third consecutive even integer: $x + 4$.
The sum of the squares of the three consecutive even integers is 56.

Solution: $x^2 + (x + 2)^2 + (x + 4)^2 = 56$

$x^2 + x^2 + 4x + 4 + x^2 + 8x + 16 = 56$

$3x^2 + 12x + 20 = 56$

$3x^2 + 12x - 36 = 0$

$3(x^2 + 4x - 12) = 0$

$3(x + 6)(x - 2) = 0$

$x + 6 = 0 \quad x - 2 = 0$

$x = -6 \quad\quad x = 2$

$x = -6,\ x + 2 = -4,\ x + 4 = -2$

$x = 2,\ x + 2 = 4,\ x + 4 = 6$

The integers are $-6, -4$ and -2 or 2, 4 and 6.

46. Strategy: Let t represent the time it takes the new computer to print the payroll.
The time it takes the older computer to print the payroll is $t + 12$.

	Rate	Time	Part
New computer	$\dfrac{1}{t}$	8	$\dfrac{8}{t}$
Older computer	$\dfrac{1}{t+12}$	8	$\dfrac{8}{t+12}$

The sum of the parts of the task completed equals 1.

Solution: $\dfrac{8}{t} + \dfrac{8}{t+12} = 1$

$t(t+12)\left(\dfrac{8}{t} + \dfrac{8}{t+12}\right) = 1t(t+12)$

$8(t + 12) + 8t = t^2 + 12t$

$8t + 96 + 8t = t^2 + 12t$

$t^2 - 4t - 96 = 0$

$(t - 12)(t + 8) = 0$

$t - 12 = 0 \quad t + 8 = 0$

$t = 12 \quad\quad t = -8$

$t = -8$ is not possible since time cannot be a negative number. Working alone, the new computer can print the payroll in 12 min.

47. Strategy: Let r represent the rate of the first car.
The rate of the second car is $r + 10$.

	Distance	Rate	Time
1st car	200	r	$\dfrac{200}{r}$
2nd car	200	$r + 10$	$\dfrac{200}{r+10}$

The second car's time is one hour less than the first car's time.

Solution: $\dfrac{200}{r+10} = \dfrac{200}{r} - 1$

$(r)(r+10)\left(\dfrac{200}{r+10}\right) = (r)(r+10)\left(\dfrac{200}{r} - 1\right)$

$200r = 200(r+10) - r(r+10)$

$200r = 200r + 2000 - r^2 - 10r$

$0 = r^2 - 10r - 2000$

$0 = (r-40)(r+50)$

$r - 40 = 0 \quad r + 50 = 0$

$r = 40 \qquad r = -50$

The rate cannot be a negative number.
The rate of the first car is 40 mph.
The rate of the second car is 50 mph.

Chapter 8 Test

1. $3x^2 + 10x = 8$

$3x^2 + 10x - 8 = 0$

$(3x-2)(x+4) = 0$

$(3x-2)(x+4) = 0$

$3x - 2 = 0 \quad x + 4 = 0$

$3x = 2 \qquad x = -4$

$x = \dfrac{2}{3}$

The solutions are -4 and $\dfrac{2}{3}$.

2. $6x^2 - 5x - 6 = 0$

$2x - 3 = 0 \quad 3x + 2 = 0$

$2x = 3 \qquad 3x = -2$

$x = \dfrac{3}{2} \qquad x = -\dfrac{2}{3}$

The solutions are $\dfrac{3}{2}$ and $-\dfrac{2}{3}$.

3. $(x - r_1)(x - r_2) = 0$

$(x-3)(x-(-3)) = 0$

$(x-3)(x+3) = 0$

$x^2 - 9 = 0$

4. $(x - r_1)(x - r_2) = 0$

$\left(x - \dfrac{1}{2}\right)(x - (-4)) = 0$

$\left(x - \dfrac{1}{2}\right)(x + 4) = 0$

$x^2 + \dfrac{7}{2}x - 2 = 0$

$2\left(x^2 + \dfrac{7}{2}x - 2\right) = 0(2)$

$2x^2 + 7x - 4 = 0$

5. $3(x-2)^2 - 24 = 0$

$3(x-2)^2 = 24$

$(x-2)^2 = 8$

$\sqrt{(x-2)^2} = \sqrt{8}$

$x - 2 = \pm 2\sqrt{2}$

$x = 2 \pm 2\sqrt{2}$

The solutions are $2 + 2\sqrt{2}$ and $2 - 2\sqrt{2}$.

6. $x^2 - 6x - 2 = 0$

$x^2 - 6x = 2$

$x^2 - 6x + 9 = 2 + 9$

$(x-3)^2 = 11$

$\sqrt{(x-3)^2} = \sqrt{11}$

$x - 3 = \pm\sqrt{11}$

$x = 3 \pm \sqrt{11}$

The solutions are $3 + \sqrt{11}$ and $3 - \sqrt{11}$.

7. $3x^2 - 6x = 2$

$\dfrac{1}{3}\left(3x^2 - 6x\right) = 2\left(\dfrac{1}{3}\right)$

$x^2 - 2x = \dfrac{2}{3}$

$x^2 - 2x + 1 = \dfrac{2}{3} + 1$

$(x - 1)^2 = \dfrac{5}{3}$

$\sqrt{(x-1)^2} = \sqrt{\dfrac{5}{3}}$

$x - 1 = \pm\dfrac{\sqrt{15}}{3}$

$x = 1 \pm \dfrac{\sqrt{15}}{3}$

$x = \dfrac{3 \pm \sqrt{15}}{3}$

The solutions are $\dfrac{3 + \sqrt{15}}{3}$ and $\dfrac{3 - \sqrt{15}}{3}$.

8. $2x^2 - 2x = 1$

$2x^2 - 2x - 1 = 0$

$a = 2,\, b = -2,\, c = -1$

$x = \dfrac{-b \pm \sqrt{b^2 - 4ac}}{2a}$

$x = \dfrac{-(-2) \pm \sqrt{(-2)^2 - 4(2)(-1)}}{2(2)}$

$x = \dfrac{2 \pm \sqrt{4 + 8}}{4} = \dfrac{2 \pm \sqrt{12}}{4}$

$x = \dfrac{2 \pm 2\sqrt{3}}{4}$

$x = \dfrac{1 \pm \sqrt{3}}{2}$

The solutions are $\dfrac{1 + \sqrt{3}}{2}$ and $\dfrac{1 - \sqrt{3}}{2}$.

9. $x^2 + 4x + 12 = 0$

$a = 1,\, b = 4,\, c = 12$

$x = \dfrac{-b \pm \sqrt{b^2 - 4ac}}{2a}$

$x = \dfrac{-4 \pm \sqrt{4^2 - 4(1)(12)}}{2(1)}$

$x = \dfrac{-4 \pm \sqrt{16 - 48}}{2} = \dfrac{-4 \pm \sqrt{-32}}{2}$

$x = \dfrac{-4 \pm 4i\sqrt{2}}{2}$

$x = -2 \pm 2i\sqrt{2}$

The solutions are $-2 + 2i\sqrt{2}$ and $-2 - 2i\sqrt{2}$.

10. $2x + 7x^{1/2} - 4 = 0$

$2\left(x^{1/2}\right)^2 + 7x^{1/2} - 4 = 0$

$2u^2 + 7u - 4 = 0$

$(2u - 1)(u + 4) = 0$

$2u - 1 = 0 \quad u + 4 = 0$

$2u = 1 \qquad u = -4$

$u = \dfrac{1}{2}$

Replace u with $x^{1/2}$.

$x^{1/2} = \dfrac{1}{2} \qquad\qquad x^{1/2} = -4$

$\left(x^{1/2}\right)^2 = \left(\dfrac{1}{2}\right)^2 \quad \left(x^{1/2}\right)^2 = (-4)^2$

$x = \dfrac{1}{4} \qquad\qquad x = 16$

16 does not check as a solution.

The solution is $\dfrac{1}{4}$.

11. $x^4 - 4x^2 + 3 = 0$

$(x^2)^2 - 4x^2 + 3 = 0$

$u^2 - 4u + 3 = 0$

$(u-1)(u-3) = 0$

$u - 1 = 0 \quad u - 3 = 0$

$u = 1 \qquad u = 3$

Replace u with x^2.

$x^2 = 1 \qquad x^2 = 3$

$\sqrt{x^2} = \sqrt{1} \quad \sqrt{x^2} = \sqrt{3}$

$x = \pm 1 \qquad x = \pm\sqrt{3}$

The solutions are -1, 1, $\sqrt{3}$ and $-\sqrt{3}$.

12. $\sqrt{2x+1} + 5 = 2x$

$\sqrt{2x+1} = 2x - 5$

$\left(\sqrt{2x+1}\right)^2 = (2x-5)^2$

$2x + 1 = 4x^2 - 20x + 25$

$4x^2 - 22x + 24 = 0$

$2(2x^2 - 11x + 12) = 0$

$2(2x-3)(x-4) = 0$

$2x - 3 = 0 \quad x - 4 = 0$

$2x = 3 \qquad x = 4$

$x = \dfrac{3}{2}$

$\dfrac{3}{2}$ does not check as a solution.

The solution is 4.

13. $\sqrt{x-2} = \sqrt{x} - 2$

$(\sqrt{x-2})^2 = \left(\sqrt{x} - 2\right)^2$

$x - 2 = x - 4\sqrt{x} + 4$

$-6 = -4\sqrt{x}$

$3 = 2\sqrt{x}$

$3^2 = \left(2\sqrt{x}\right)^2$

$9 = 4x$

$x = \dfrac{9}{4}$

$\dfrac{9}{4}$ does not check as a solution.

There is no solution.

14. $\dfrac{2x}{x-3} + \dfrac{5}{x-1} = 1$

$(x-3)(x-1)\left(\dfrac{2x}{x-3} + \dfrac{5}{x-1}\right) = 1(x-3)(x-1)$

$2x(x-1) + 5(x-3) = 1(x-3)(x-1)$

$2x^2 - 2x + 5x - 15 = x^2 - 4x + 3$

$x^2 + 7x - 18 = 0$

$(x+9)(x-2) = 0$

$x + 9 = 0 \quad x - 2 = 0$

$x = -9 \qquad x = 2$

The solutions are -9 and 2.

15. $(x-2)(x+4)(x-4) < 0$

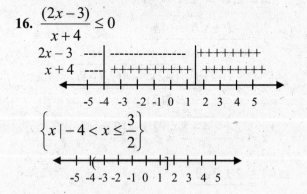

$\{x \mid x < -4 \text{ or } 2 < x < 4\}$

16. $\dfrac{(2x-3)}{x+4} \le 0$

$\left\{x \mid -4 < x \le \dfrac{3}{2}\right\}$

17. $9x^2 + 24x = -16$

$9x^2 + 24x + 16 = 0$

$a = 9, b = 24, c = 16$

$b^2 - 4ac = (24)^2 - 4(9)(16) = 0$

Since the discriminant is equal to zero the equation has two equal real number solutions.

18. Strategy: To find the time when the ball hits the basket substitute 10 ft for h in the equation and solve for t.

Solution: $h = -16t^2 + 32t + 6.5$

$10 = -16t^2 + 32t + 6.5$

$0 = -16t^2 + 32t - 3.5$

$0 = 16t^2 - 32t + 3.5$

$a = 16, b = -32, c = 3.5$

$t = \dfrac{-b \pm \sqrt{b^2 - 4ac}}{2a}$

$t = \dfrac{-(-32) \pm \sqrt{(-32)^2 - 4(16)(3.5)}}{2(16)}$

$t = \dfrac{32 \pm \sqrt{800}}{32}$

$t = 1.88$

$t = 0.12$

We need to find the time it takes to reach the basket after the ball has reached its peak. This occurs 1.88 s after the ball has been released.

19. Strategy: Let t represent the time it takes Cora to stain a bookcase.
The time it takes Clive to stain the bookcase is $t + 6$.

	Rate	Time	Part
Cora	$\dfrac{1}{t}$	4	$\dfrac{4}{t}$
Clive	$\dfrac{1}{t+6}$	4	$\dfrac{4}{t+6}$

The sum of the parts of the task completed equals 1.

Solution: $\dfrac{4}{t} + \dfrac{4}{t+6} = 1$

$t(t+6)\left(\dfrac{4}{t} + \dfrac{4}{t+6}\right) = 1t(t+6)$

$4(t+6) + 4t = t^2 + 6t$

$4t + 24 + 4t = t^2 + 6t$

$t^2 - 2t - 24 = 0$

$(t-6)(t+4) = 0$

$t - 6 = 0 \quad t + 4 = 0$

$t = 6 \qquad t = -4$

$t = -4$ is not possible since time cannot be a negative number. Working alone, it will take Cora 6 h to stain the bookcase.

20. Strategy: Let r represent the rate of the canoe in calm water.

	Distance	Rate	Time
With current	6	$r+2$	$\dfrac{6}{r+2}$
Against current	6	$r-2$	$\dfrac{6}{r-2}$

The total time traveled was 4 h.

Solution: $\dfrac{6}{r+2} + \dfrac{6}{r-2} = 4$

$(r-2)(r+2)\left(\dfrac{6}{r+2} + \dfrac{6}{r-2}\right) = 4(r-2)(r+2)$

$6(r-2) + 6(r+2) = 4r^2 - 16$

$6r - 12 + 6r + 12 = 4r^2 - 16$

$0 = 4r^2 - 12r - 16$

$0 = 4(r^2 - 3r - 4)$

$0 = 4(r-4)(r+1)$

$r - 4 = 0 \quad r + 1 = 0$

$r = 4 \qquad r = -1$

The rate cannot be a negative number. The rate of the canoe in calm water is 4 mph.

Cumulative Review Exercises

1. $2a^2 - b^2 \div c^2$

$2(3)^2 - (-4)^2 \div (-2)^2 = 2(9) - 16 \div 4$

$= 18 - 16 \div 4 = 18 - 4$

$= 14$

2. $\dfrac{2x-3}{4} - \dfrac{x+4}{6} = \dfrac{3x-2}{8}$

$24\left(\dfrac{2x-3}{4} - \dfrac{x+4}{6}\right) = 24\left(\dfrac{3x-2}{8}\right)$

$6(2x-3) - 4(x+4) = 3(3x-2)$

$12x - 18 - 4x - 16 = 9x - 6$

$8x - 34 = 9x - 6$

$-x = 28$

$x = -28$

The solution is -28.

3. $(3, -4)$ and $(-1, 2)$

$m = \dfrac{y_2 - y_1}{x_2 - x_1} = \dfrac{2 - (-4)}{-1 - 3} = \dfrac{2+4}{-4} = \dfrac{6}{-4}$

$m = -\dfrac{3}{2}$

4. $x - y = 1$

$y = x - 1$

$m = 1$ and $(1, 2)$

$y - y_1 = m(x - x_1)$

$y - 2 = 1(x - 1)$

$y - 2 = x - 1$

$y = x + 1$

5. $-3x^3 y + 6x^2 y^2 - 9xy^3 = -3xy(x^2 - 2xy + 3y^2)$

6. $6x^2 - 7x - 20 = (2x - 5)(3x + 4)$

7. $a^n x + a^n y - 2x - 2y = a^n(x + y) - 2(x + y)$

$= (a^n - 2)(x + y)$

8.

$$\begin{array}{r} x^2 - 3x - 4 \\ 3x - 4 \overline{) 3x^3 - 13x^2 + 0x + 10} \\ \underline{3x^3 - 4x^2} \\ -9x^2 + 0x \\ \underline{-9x^2 + 12x} \\ -12x + 10 \\ \underline{-12x + 16} \\ -6 \end{array}$$

$(3x^3 - 13x^2 + 10) \div (3x - 4) = x^2 - 3x - 4 + \dfrac{-6}{3x - 4}$

9. $\dfrac{x^2 + 2x + 1}{8x^2 + 8x} \cdot \dfrac{4x^3 - 4x^2}{x^2 - 1}$

$= \dfrac{(x+1)(x+1)}{8x(x+1)} \cdot \dfrac{4x^2(x-1)}{(x+1)(x-1)}$

$= \dfrac{(x+1)(x+1) \cdot 4x^2(x-1)}{8x(x+1) \cdot (x+1)(x-1)}$

$= \dfrac{x}{2}$

10. $(-2, 3)$ and $2, 5)$

$d = \sqrt{(x_1 - x_2)^2 + (y_1 - y_2)^2}$

$d = \sqrt{(-2 - 2)^2 + (3 - 5)^2}$

$d = \sqrt{(-4)^2 + (-2)^2} = \sqrt{16 + 4} = \sqrt{20}$

$d = 2\sqrt{5}$

The distance between the points is $2\sqrt{5}$.

11. $S = \dfrac{n}{2}(a + b)$

$2S = n(a + b)$

$2S = na + nb$

$2S - na = nb$

$\dfrac{2S - na}{n} = b$

12. $-2i(7 - 4i) = -14i + 8i^2 = -8 - 14i$

13. $a^{-1/2}(a^{1/2} - a^{3/2}) = a^0 - a^1$
$= 1 - a$

14. $\dfrac{\sqrt[3]{8x^4y^5}}{\sqrt[3]{16xy^6}} = \sqrt[3]{\dfrac{8x^4y^5}{16xy^6}} = \sqrt[3]{\dfrac{x^3}{2y}}$

$= \dfrac{\sqrt[3]{x^3}}{\sqrt[3]{2y}}$

$= \dfrac{x}{\sqrt[3]{2y}} \cdot \dfrac{\sqrt[3]{4y^2}}{\sqrt[3]{4y^2}} = \dfrac{x\sqrt[3]{4y^2}}{\sqrt[3]{8y^3}}$

$= \dfrac{x\sqrt[3]{4y^2}}{2y}$

15. $\dfrac{x}{x+2} - \dfrac{4x}{x+3} = 1$

$(x+2)(x+3)\left(\dfrac{x}{x+2} - \dfrac{4x}{x+3}\right) = 1(x+2)(x+3)$

$x(x+3) - 4x(x+2) = 1(x+2)(x+3)$

$x^2 + 3x - 4x^2 - 8x = x^2 + 5x + 6$

$0 = 4x^2 + 10x + 6$

$0 = 2(2x^2 + 5x + 3)$

$(2x+3)(x+1) = 0$

$2x + 3 = 0 \quad x + 1 = 0$

$2x = -3 \qquad x = -1$

$x = -\dfrac{3}{2}$

The solutions are -1 and $-\dfrac{3}{2}$.

16. $\dfrac{x}{2x+3} - \dfrac{3}{4x^2-9} = \dfrac{x}{2x-3}$

$(2x+3)(2x-3)\left(\dfrac{x}{2x+3} - \dfrac{3}{(2x+3)(2x-3)}\right) = (2x+3)(2x-3)\dfrac{x}{2x-3}$

$x(2x-3) - 3 = x(2x+3)$

$2x^2 - 3x - 3 = 2x^2 + 3x$

$-3 = 6x$

$x = -\dfrac{1}{2}$

The solution is $-\dfrac{1}{2}$.

17. $x^4 - 6x^2 + 8 = 0$

$(x^2)^2 - 6x^2 + 8 = 0$

$u^2 - 6u + 8 = 0$

$(u-4)(u-2) = 0$

$u - 4 = 0 \quad u - 2 = 0$

$u = 4 \qquad u = 2$

Replace u with x^2.

$x^2 = 4 \qquad\qquad x^2 = 2$

$\sqrt{x^2} = \sqrt{4} \quad \sqrt{x^2} = \sqrt{2}$

$x = \pm 2 \qquad\qquad x = \pm\sqrt{2}$

The solutions are $-2, 2, \sqrt{2}$ and $-\sqrt{2}$.

18. $\sqrt{3x+1}-1=x$

$\sqrt{3x+1}=x+1$

$\left(\sqrt{3x+1}\right)^2=\left(x+1\right)^2$

$3x+1=x^2+2x+1$

$0=x^2-x$

$0=x(x-1)$

$x-1=0 \quad x=0$

$x=1$

The solutions are 1 and 0.

19. $|3x-2|<8$

$-8<3x-2<8$

$-8+2<3x-2+2<8+2$

$-6<3x<10$

$-2<x<\dfrac{10}{3}$

$\left\{x \mid -2<x<\dfrac{10}{3}\right\}$

20. $6x-5y=15$

$6x-5(0)=15$

$6x=15$

$x=\dfrac{15}{6}=\dfrac{5}{2}$

The x-intercept is $\left(\dfrac{5}{2},0\right)$.

$6(0)-5y=15$

$-5y=15$

$y=-3$

The y-intercept is $(0,-3)$.

21. Solve each inequality for y.

$x+y\le 3 \qquad 2x-y<4$

$y\le 3-x \qquad -y<4-2x$

$\qquad\qquad\qquad y>2x-4$

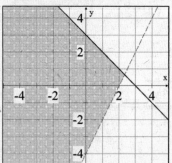

22. $x+y+z=2$

$-x+2y-3z=-9$

$x-2y-2z=-1$

$$D=\begin{vmatrix} 1 & 1 & 1 \\ -1 & 2 & -3 \\ 1 & -2 & -2 \end{vmatrix}=-15$$

$$D_x=\begin{vmatrix} 2 & 1 & 1 \\ 4 & 2 & -2 \\ -1 & -2 & -2 \end{vmatrix}=-15$$

$$D_y=\begin{vmatrix} 1 & 2 & 1 \\ -1 & -9 & -3 \\ 1 & -1 & -2 \end{vmatrix}=15$$

$$D_z=\begin{vmatrix} 1 & 1 & 2 \\ -1 & 2 & -9 \\ 1 & -2 & -1 \end{vmatrix}=-30$$

$x=\dfrac{D_x}{D}=\dfrac{-15}{-15}=1$

$y=\dfrac{D_y}{D}=\dfrac{15}{-15}=-1$

$z=\dfrac{D_z}{D}=\dfrac{-30}{-15}=2$

The solution is $(1,-1,2)$.

23. $f(x) = \dfrac{2x-3}{x^2-1}$

$f(-2) = \dfrac{2(-2)-3}{(-2)^2-1} = \dfrac{-4-3}{4-1} = -\dfrac{7}{3}$

24. $f(x) = \dfrac{x-2}{x^2-2x-15}$

$f(x) = \dfrac{x-2}{(x-5)(x+3)}$

$x - 5 = 0 \quad x + 3 = 0$

$x = 5 \qquad x = -3$

$\{x \mid x \neq -3, 5\}$

25. $x^3 + x^2 - 6x < 0$

$x(x^2 + x - 6) < 0$

$x(x+3)(x-2) < 0$

$\{x \mid x < -3 \text{ or } 0 < x < 2\}$

26. $\dfrac{(x-1)(x-5)}{x+3} \geq 0$

$\{x \mid -3 < x \leq 1 \text{ or } x \geq 5\}$

27. Strategy: Let P represent the length of the piston rod, T the tolerance and m the given length. Solve the absolute value inequality $|m - p| \leq T$ for m.

Solution: $|m - p| \leq T$

$\left| m - 9\dfrac{3}{8} \right| \leq \dfrac{1}{64}$

$-\dfrac{1}{64} \leq m - 9\dfrac{3}{8} \leq \dfrac{1}{64}$

$-\dfrac{1}{64} + 9\dfrac{3}{8} \leq m - 9\dfrac{3}{8} + 9\dfrac{3}{8} \leq \dfrac{1}{64} + 9\dfrac{3}{8}$

$9\dfrac{23}{64} \leq m \leq 9\dfrac{25}{64}$

The lower limit is $9\dfrac{23}{64}$ in.

The upper limit is $9\dfrac{25}{64}$ in.

28. Strategy: The base of the triangle is $x + 8$. The height of the triangle is $2x - 4$.

Solution: $A = \dfrac{1}{2}bh$

$A = \dfrac{1}{2}(x+8)(2x-4)$

$A = \dfrac{1}{2}(2x^2 + 12x - 32)$

$A = (x^2 + 6x - 16) \text{ ft}^2$

29. $2x^2 + 4x + 3 = 0$

$a = 2, b = 4, c = 3$

$b^2 - 4ac = 4^2 - 4(2)(3) = 16 - 24 = -8$

$-8 < 0$

Since the discriminant is less than zero, the equation has two complex number solutions.

30. $(0, 250)$ and $(30, 0)$

$m = \dfrac{y_2 - y_1}{x_2 - x_1} = \dfrac{250 - 0}{0 - 30} = \dfrac{250}{-30} = -\dfrac{25}{3}$

$m = -\dfrac{25,000}{3}$

The building depreciates $\dfrac{\$25,000}{3}$, or about $8333, each year.

Chapter 9: Functions and Relations

Prep Test

1. $-\dfrac{b}{2a}$

$-\dfrac{(-4)}{2(2)} = -\dfrac{-4}{4} = -(-1) = 1$

2. $y = -x^2 + 2x + 1$

$y = -(-2)^2 + 2(-2) + 1 = -4 - 4 + 1 = -7$

3. $f(x) = x^2 - 3x + 2$

$f(-4) = (-4)^2 - 3(-4) + 2$

$f(-4) = 16 + 12 + 2$

$f(-4) = 30$

4. $p(r) = r^2 - 5$

$p(2 + h) = (2 + h)^2 - 5$

$= 4 + 4h + h^2 - 5$

$= h^2 + 4h - 1$

5. $0 = 3x^2 - 7x - 6$

$0 = (3x + 2)(x - 3)$

$0 = 3x + 2 \quad 0 = x - 3$

$-2 = 3x \qquad 3 = x$

$-\dfrac{2}{3} = x$

The solutions are $-\dfrac{2}{3}$ and 3.

6. $0 = x^2 - 4x + 1$

$a = 1 \quad b = -4 \quad c = 1$

$x = \dfrac{-b \pm \sqrt{b^2 - 4ac}}{2a}$

$x = \dfrac{-(-4) \pm \sqrt{(-4)^2 - 4(1)(1)}}{2(1)}$

$x = \dfrac{4 \pm \sqrt{16 - 4}}{2} = \dfrac{4 \pm \sqrt{12}}{2} = \dfrac{4 \pm 2\sqrt{3}}{2}$

$x = \dfrac{4}{2} \pm \dfrac{2\sqrt{3}}{2} = 2 \pm \sqrt{3}$

The solutions are $2 + \sqrt{3}$ and $2 - \sqrt{3}$.

7. $x = 2y + 4$

$2y + 4 = x$

$2y = x - 4$

$\left(\dfrac{1}{2}\right)2y = \left(\dfrac{1}{2}\right)(x - 4)$

$y = \dfrac{1}{2}x - 2$

8. Domain: $\{-2, 3, 4, 6\}$

Range: $\{4, 5, 6\}$

Yes, the relation is a function.

9. Values for which $x - 8 = 0$ are excluded from the domain.

$x - 8 = 0$

$x = 8$

10.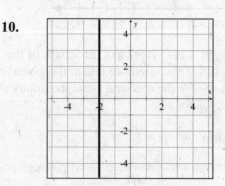

Section 9.1

Objective A Exercises

1. A quadratic function is a function of the
form $f(x) = ax^2 + bx + c$, $a \neq 0$.

3. The vertex of a parabola is the point with the
smallest y-coordinate or the largest
y-coordinate. When $a > 0$, the parabola
opens up and the vertex of the parabola is
the point with the smallest y-coordinate.
When $a < 0$, the parabola opens down and
the vertex of the parabola is the point with
the largest y-coordinate.

5. -5

7. $x = 7$

9. $y = x^2 - 2x - 4$

$$-\frac{b}{2a} = -\frac{-2}{2(1)} = -\frac{-2}{2} = -(-1) = 1$$

$$y = (1)^2 - 2(1) - 4 = -5$$

Vertex:
$(1, -5)$
Axis of
symmetry:
$x = 1$

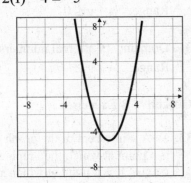

11. $y = -x^2 + 2x - 3$

$$-\frac{b}{2a} = -\frac{2}{2(-1)} = -\frac{2}{-2} = -(-1) = 1$$

$$y = -(1)^2 + 2(1) - 3 = -2$$

Vertex: $(1, -2)$
Axis of
symmetry: $x = 1$

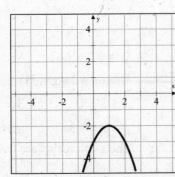

13. $f(x) = x^2 - x - 6$

$$-\frac{b}{2a} = -\frac{(-1)}{2(1)} = -\frac{-1}{2} = \frac{1}{2}$$

$$y = \left(\frac{1}{2}\right)^2 - \frac{1}{2} - 6 = -\frac{25}{4}$$

Vertex:
$\left(\frac{1}{2}, -\frac{25}{4}\right)$

Axis of
symmetry:

$x = \dfrac{1}{2}$

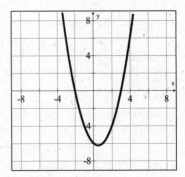

15. $F(x) = x^2 - 3x + 2$

$$-\frac{b}{2a} = -\frac{(-3)}{2(1)} = -\frac{-3}{2} = \frac{3}{2}$$

$$y = \left(\frac{3}{2}\right)^2 - 3\left(\frac{3}{2}\right) + 2 = -\frac{1}{4}$$

Vertex:
$\left(\frac{3}{2}, -\frac{1}{4}\right)$

Axis of
symmetry:

$x = \dfrac{3}{2}$

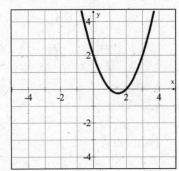

17. $y = -2x^2 + 6x$

$$-\frac{b}{2a} = -\frac{6}{2(-2)} = -\frac{6}{-4} = \frac{3}{2}$$

$$y = -2\left(\frac{3}{2}\right)^2 + 6\left(\frac{3}{2}\right) = \frac{9}{2}$$

Vertex: $\left(\frac{3}{2}, \frac{9}{2}\right)$

Axis of
symmetry:

$x = \dfrac{3}{2}$

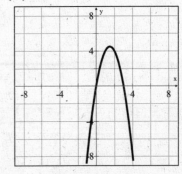

19. $y = -\dfrac{1}{4}x^2 - 1$

$$-\dfrac{b}{2a} = -\dfrac{(0)}{2\left(-\dfrac{1}{4}\right)} = -\dfrac{0}{-\dfrac{1}{2}} = 0$$

$$y = -\left(\dfrac{1}{4}\right)0^2 - 1 = -1$$

Vertex:
$(0,-1)$
Axis of
symmetry:
$x = 0$

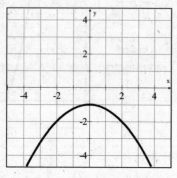

21. $P(x) = -\dfrac{1}{2}x^2 + 2x - 3$

$$-\dfrac{b}{2a} = -\dfrac{2}{2\left(-\dfrac{1}{2}\right)} = -\dfrac{2}{-1} = -(-2) = 2$$

$$y = \left(-\dfrac{1}{2}\right)2^2 + 2(2) - 3 = -1$$

Vertex:
$(2,-1)$
Axis of
symmetry:
$x = 2$

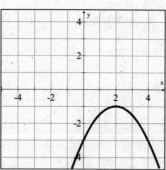

23. $y = -\dfrac{1}{2}x^2 + x - 3$

$$-\dfrac{b}{2a} = -\dfrac{1}{2\left(-\dfrac{1}{2}\right)} = -\dfrac{1}{-1} = -(-1) = 1$$

$$y = \left(-\dfrac{1}{2}\right)1^2 + 1 - 3 = -\dfrac{5}{2}$$

Vertex: $\left(1, -\dfrac{5}{2}\right)$

Axis of symmetry:
$x = 1$

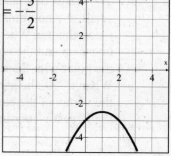

25. Domain: $\{x \mid x \in \text{real numbers}\}$
Range: $\{y \mid y \geq 2\}$

27. Domain: $\{x \mid x \in \text{real numbers}\}$
Range: $\{y \mid y \geq -5\}$

29. Domain: $\{x \mid x \in \text{real numbers}\}$
Range: $\{y \mid y \leq 0\}$

31. Domain: $\{x \mid x \in \text{real numbers}\}$
Range: $\{y \mid y \geq -7\}$

Objective B Exercises

33. a) A y-intercept of the graph of a parabola
is a point at which the graph crosses the
y-axis. It is a point at which $x = 0$.

b) The graph of a parabola has one
y-intercept.

35. a) The zeros are -4 and 1.
b) The solutions are -4 and 1.

37. $y = x^2 - 9$
$0 = x^2 - 9$
$0 = (x + 3)(x - 3)$
$x + 3 = 0 \qquad x - 3 = 0$
$\quad x = -3 \qquad\quad x = 3$
The x-intercepts are $(-3, 0)$ and $(3, 0)$.

39. $y = 3x^2 + 6x$

$0 = 3x^2 + 6x$

$0 = 3x(x+2)$

$3x = 0 \qquad x + 2 = 0$

$x = 0 \qquad x = -2$

The x-intercepts are $(0, 0)$ and $(-2, 0)$.

41. $y = x^2 - 2x - 8$

$0 = x^2 - 2x - 8$

$0 = (x-4)(x+2)$

$x - 4 = 0 \qquad x + 2 = 0$

$x = 4 \qquad x = -2$

The x-intercepts are $(4, 0)$ and $(-2, 0)$.

43. $y = 2x^2 - 5x - 3$

$0 = 2x^2 - 5x - 3$

$0 = (x-3)(2x+1)$

$x - 3 = 0 \qquad 2x + 1 = 0$

$x = 3 \qquad 2x = -1$

$$x = -\frac{1}{2}$$

The x-intercepts are $(3, 0)$ and $\left(-\frac{1}{2}, 0\right)$.

45. $y = x^2 + 4x - 3$

$0 = x^2 + 4x - 3$

$a = 1 \quad b = 4 \quad c = -3$

$$x = \frac{-b \pm \sqrt{b^2 - 4ac}}{2a}$$

$$x = \frac{-4 \pm \sqrt{(4)^2 - 4(1)(-3)}}{2(1)}$$

$$x = \frac{-4 \pm \sqrt{16+12}}{2} = \frac{-4 \pm \sqrt{28}}{2} = \frac{-4 \pm 2\sqrt{7}}{2}$$

$x = -2 \pm \sqrt{7}$

The x-intercepts are $\left(-2 + \sqrt{7}, 0\right)$ and $\left(-2 - \sqrt{7}, 0\right)$.

47. $y = -x^2 - 4x - 5$

$0 = -x^2 - 4x - 5$

$a = -1 \quad b = -4 \quad c = -5$

$$x = \frac{-b \pm \sqrt{b^2 - 4ac}}{2a}$$

$$x = \frac{-(-4) \pm \sqrt{(-4)^2 - 4(-1)(-5)}}{2(-1)}$$

$$x = \frac{4 \pm \sqrt{16-20}}{-2} = \frac{4 \pm \sqrt{-4}}{-2} = \frac{4 \pm 2i}{-2}$$

$x = -2 \pm i$

There are no real solutions. The parabola has no x-intercepts.

49. $f(x) = x^2 - 4x - 5$

$x^2 - 4x - 5 = 0$

$(x-5)(x+1) = 0$

$x - 5 = 0 \qquad x + 1 = 0$

$x = 5 \qquad x = -1$

The zeros are 5 and -1.

51. $f(x) = 3x^2 - 2x - 8$

$3x^2 - 2x - 8 = 0$

$(3x+4)(x-2) = 0$

$3x + 4 = 0 \qquad x - 2 = 0$

$3x = -4 \qquad x = 2$

$$x = -\frac{4}{3}$$

The zeros are $-\frac{4}{3}$ and 2.

53. $h(x) = 4x^2 - 4x + 1$

$4x^2 - 4x + 1 = 0$

$(2x-1)(2x-1) = 0$

$2x - 1 = 0 \qquad 2x - 1 = 0$

$2x = 1 \qquad 2x = 1$

$$x = \frac{1}{2} \qquad x = \frac{1}{2}$$

The zero is $\frac{1}{2}$.

55. $f(x) = -3x^2 + 4x$

$-3x^2 + 4x = 0$

$x(-3x + 4) = 0$

$-3x + 4 = 0 \qquad x = 0$

$-3x = -4$

$x = \dfrac{4}{3}$

The zeros are $\dfrac{4}{3}$ and 0.

57. $f(x) = -3x^2 + 12$

$-3x^2 + 12 = 0$

$-3(x^2 - 4) = 0$

$-3(x + 2)(x - 2) = 0$

$x + 2 = 0 \qquad x - 2 = 0$

$x = -2 \qquad x = 2$

The zeros are -2 and 2.

59. $f(x) = 2x^2 - 54$

$2x^2 - 54 = 0$

$2(x^2 - 27) = 0$

$x^2 - 27 = 0$

$x^2 = 27$

$x = \pm\sqrt{27} = \pm 3\sqrt{3}$

The zeros are $3\sqrt{3}$ and $-3\sqrt{3}$.

61. $f(x) = x^2 - 2x - 17$

$x^2 - 2x - 17 = 0$

$a = 1 \quad b = -2 \quad c = -17$

$x = \dfrac{-b \pm \sqrt{b^2 - 4ac}}{2a}$

$x = \dfrac{-(-2) \pm \sqrt{(-2)^2 - 4(1)(-17)}}{2(1)}$

$x = \dfrac{2 \pm \sqrt{4 + 68}}{2} = \dfrac{2 \pm \sqrt{72}}{2} = \dfrac{2 \pm 6\sqrt{2}}{2}$

$x = 1 \pm 3\sqrt{2}$

The zeros are $1 + 3\sqrt{2}$ and $1 - 3\sqrt{2}$.

63. $f(x) = x^2 + 4x + 5$

$x^2 + 4x + 5 = 0$

$a = 1 \quad b = 4 \quad c = 5$

$x = \dfrac{-b \pm \sqrt{b^2 - 4ac}}{2a}$

$x = \dfrac{-4 \pm \sqrt{(4)^2 - 4(1)(5)}}{2(1)}$

$x = \dfrac{-4 \pm \sqrt{16 - 20}}{2} = \dfrac{-4 \pm \sqrt{-4}}{2} = \dfrac{-4 \pm 2i}{2}$

$x = -2 \pm i$

The zeros are $-2 + i$ and $-2 - i$.

65. $f(x) = x^2 + 4x + 13$

$x^2 + 4x + 13 = 0$

$a = 1 \quad b = 4 \quad c = 13$

$x = \dfrac{-b \pm \sqrt{b^2 - 4ac}}{2a}$

$x = \dfrac{-4 \pm \sqrt{(4)^2 - 4(1)(13)}}{2(1)}$

$x = \dfrac{-4 \pm \sqrt{16 - 52}}{2} = \dfrac{-4 \pm \sqrt{-36}}{2} = \dfrac{-4 \pm 6i}{2}$

$x = -2 \pm 3i$

The zeros are $-2 + 3i$ and $-2 - 3i$.

67. $y = -x^2 - x + 3$

$a = -1 \quad b = -1 \quad c = 3$

$b^2 - 4ac$

$(-1)^2 - 4(-1)(3) = 1 + 12 = 13$

$13 > 0$

Since the discriminant is greater than zero, the parabola has two x-intercepts.

69. $y = x^2 - 10x + 25$

$a = 1 \quad b = -10 \quad c = 25$

$b^2 - 4ac$

$(-10)^2 - 4(1)(25) = 100 - 100 = 0$

Since the discriminant is equal to zero, the parabola has one x-intercept.

71. $y = -2x^2 + x - 1$

$a = -2 \quad b = 1 \quad c = -1$

$b^2 - 4ac$

$(1)^2 - 4(-2)(-1) = 1 - 8 = -7$

$-7 < 0$

Since the discriminant is less than zero, the parabola has no x-intercepts.

73. $y = 4x^2 - x - 2$

$a = 4 \quad b = -1 \quad c = -2$

$b^2 - 4ac$

$(-1)^2 - 4(4)(-2) = 1 + 32 = 33$

$33 > 0$

Since the discriminant is greater than zero, the parabola has two x-intercepts.

75. $y = 2x^2 + x + 4$

$a = 2 \quad b = 1 \quad c = 4$

$b^2 - 4ac$

$(1)^2 - 4(2)(4) = 1 - 32 = -31$

$-31 < 0$

Since the discriminant is less than zero, the parabola has no x-intercepts.

77. $y = 4x^2 + 2x - 5$

$a = 4 \quad b = 2 \quad c = -5$

$b^2 - 4ac$

$(2)^2 - 4(4)(-5) = 4 + 80 = 84$

$84 > 0$

Since the discriminant is greater than zero, the parabola has two x-intercepts.

79. a) $a > 0$

b) $a = 0$

c) $a < 0$

Objective C Exercises

81. To find the minimum or maximum value of a quadratic function, find the x-coordinate of the vertex. Then evaluate the function at the value of this x-coordinate.

83. a) Since $a > 0$, the parabola opens up. The function has a minimum value.

b) Since $a < 0$, the parabola opens down. The function has a maximum value.

c) Since $a > 0$, the parabola opens up. The function has a minimum value.

85. $f(x) = 2x^2 + 4x$

$x = -\dfrac{b}{2a} = -\dfrac{4}{2(2)} = -1$

$f(x) = 2x^2 + 4x$

$f(-1) = 2(-1)^2 + 4(-1) = 2 - 4 = -2$

Since $a > 0$, the function has a minimum value. The minimum value of the function is -2.

87. $f(x) = -2x^2 + 4x - 5$

$x = -\dfrac{b}{2a} = -\dfrac{4}{2(-2)} = 1$

$f(x) = -2x^2 + 4x - 5$

$f(1) = -2(1)^2 + 4(1) - 5 = -2 + 4 - 5 = -3$

Since $a < 0$, the function has a maximum value. The maximum value of the function is -3.

89. $f(x) = -2x^2 - 3x$

$$x = -\frac{b}{2a} = -\frac{-3}{2(-2)} = -\frac{3}{4}$$

$$f(x) = -2x^2 - 3x$$

$$f\left(-\frac{3}{4}\right) = -2\left(-\frac{3}{4}\right)^2 - 3\left(-\frac{3}{4}\right) = -\frac{9}{8} + \frac{9}{4}$$

$$= \frac{9}{8}$$

Since $a < 0$, the function has a maximum value. The maximum value of the function is $\frac{9}{8}$.

91. $f(x) = 3x^2 + 3x - 2$

$$x = -\frac{b}{2a} = -\frac{3}{2(3)} = -\frac{1}{2}$$

$$f(x) = 3x^2 + 3x - 2$$

$$f\left(-\frac{1}{2}\right) = 3\left(-\frac{1}{2}\right)^2 + 3\left(-\frac{1}{2}\right) - 2 = \frac{3}{4} - \frac{3}{2} - 2$$

$$= -\frac{11}{4}$$

Since $a > 0$, the function has a minimum value. The minimum value of the function is $-\frac{11}{4}$.

93. $f(x) = -x^2 - x + 2$

$$x = -\frac{b}{2a} = -\frac{-1}{2(-1)} = -\frac{1}{2}$$

$$f(x) = -x^2 - x + 2$$

$$f\left(-\frac{1}{2}\right) = -\left(-\frac{1}{2}\right)^2 - \left(-\frac{1}{2}\right) + 2 = -\frac{1}{4} + \frac{1}{2} + 2$$

$$= \frac{9}{4}$$

Since $a < 0$, the function has a maximum value. The maximum value of the function is $\frac{9}{4}$.

95. $f(x) = 3x^2 + 5x + 2$

$$x = -\frac{b}{2a} = -\frac{5}{2(3)} = -\frac{5}{6}$$

$$f(x) = 3x^2 + 5x + 2$$

$$f\left(-\frac{5}{6}\right) = 3\left(-\frac{5}{6}\right)^2 + 5\left(-\frac{5}{6}\right) + 2 = \frac{25}{12} - \frac{25}{6} + 2$$

$$= -\frac{1}{12}$$

Since $a > 0$, the function has a minimum value. The minimum value of the function is $-\frac{1}{12}$.

Objective D Exercises

97. Strategy: To find the price that will give the maximum revenue, find the P-coordinate of the vertex.

Solution:

$$P = -\frac{b}{2a} = -\frac{125}{2\left(-\frac{1}{4}\right)} = 250$$

A price of $250 will give the maximum revenue.

99. Strategy: To find the time it takes the plane to reach its maximum height, find the t-coordinate of the vertex. To find the maximum height, evaluate the function at the t-coordinate of the vertex.

Solution:

$$t = -\frac{b}{2a} = -\frac{119}{2(-1.42)} \approx 42$$

$$h(t) = -1.42t^2 + 119t + 6000$$

$$h(42) = -1.42(42)^2 + 119(42) + 6000$$

$$= -2504.88 + 4998 + 6000 = 8493.12$$

The maximum height of the plane is about 8500 m.

101.   Strategy: To find the distance from one end of the bridge where the cable is at its minimum height, find the x-coordinate of the vertex. To find the minimum height, evaluate the function at the x-coordinate of the vertex.

Solution:

$$x = -\frac{b}{2a} = -\frac{-0.8}{2(0.25)} = 1.6$$

$$h(x) = 0.25x^2 - 0.8x + 25$$

$$h(1.6) = 0.25(1.6)^2 - 0.8(1.6) + 25$$

$$= 0.64 - 1.28 + 25 = 24.36$$

The cable is at its minimum height 1.6 ft from the end of the bridge.
The minimum height is 24.36 ft.

103.   Strategy: Let x represent one number. The other number is $20 - x$. Their product is $x(20 - x)$. To find one number, find the x-coordinate of the vertex. To find the second number, evaluate $20 - x$ at the x-coordinate of the vertex.

Solution:

$$x(20 - x) = 20x - x^2$$

$$x = -\frac{b}{2a} = -\frac{20}{2(-1)} = 10$$

$$20 - x = 20 - 10 = 10$$

The two numbers are 10 and 10.

105.   Strategy: Let x represent one number. The other number is $x - 14$. Their product is $x(x - 14)$. To find one number, find the x-coordinate of the vertex. To find the second number, evaluate $x - 14$ at the x-coordinate of the vertex.

Solution:

$$x(x - 14) = x^2 - 14x$$

$$x = -\frac{b}{2a} = -\frac{-14}{2(1)} = 7$$

$$x - 14 = 7 - 14 = -7$$

The two numbers are 7 and -7.

107.   Strategy: Let x represent width of the rectangular corral. The length is $200 - 2x$. The area is $x(200 - 2x)$. To find the width, find the x-coordinate of the vertex. To find the length, evaluate $200 - 2x$ at the x-coordinate of the vertex.

Solution:

$$x(200 - 2x) = 200x - 2x^2$$

$$x = -\frac{b}{2a} = -\frac{200}{2(-2)} = 50$$

$$200 - 2x = 200 - 2(50) = 100$$

The width is 50 ft and the length is 100 ft.

109.   Strategy: Let x represent the width of the ball fields. The length is $\dfrac{2100 - 3x}{2}$.

The area is $x\left(\dfrac{2100 - 3x}{2}\right)$.

To find the width, find the x-coordinate of the vertex. To find the length, evaluate $\dfrac{2100 - 3x}{2}$ at the x-coordinate of the vertex.

Solution:

$$x\left(\frac{2100 - 3x}{2}\right) = x(1050 - 1.5x) = 1050x - 1.5x^2$$

$$x = -\frac{b}{2a} = -\frac{1050}{2(-1.5)} = 350$$

$$\frac{2100 - 3x}{2} = \frac{2100 - 1050}{2} = \frac{1050}{2} = 525$$

The dimensions are 350 ft by 525 ft.

Applying the Concepts

111.   Strategy: To find the value of k:
Find the x-coordinate of the vertex.
The y-coordinate will be 0 since the vertex is on the x-axis. Substitute the value for x and y in the original equation and solve for k.

Solution:

$y = x^2 - 8x + k$

$x = -\dfrac{b}{2a} = -\dfrac{-8}{2(1)} = 4$

$x = 4 \quad y = 0$

$0 = (4)^2 - 8(4) + k$

$0 = 16 - 32 + k$

$0 = -16 + k$

$16 = k$

Section 9.2

Objective A Exercises

1. A vertical line intersects the graph no more than once. Yes, the graph is a function.

3. A vertical line intersects the graph more than once. No, the graph is not a function.

5. A vertical line intersects the graph no more than once. Yes, the graph is a function.

7. No

9. $f(x) = 1 - x^3$

Domain: $\{x \mid x \in \text{real numbers}\}$
Range: $\{y \mid y \in \text{real numbers}\}$

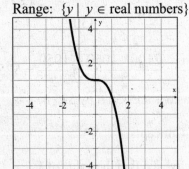

11. $f(x) = 3 \mid 2 - x \mid$

Domain: $\{x \mid x \in \text{real numbers}\}$
Range: $\{y \mid y \geq 0\}$

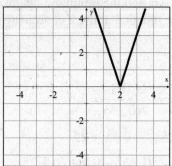

13. $f(x) = \sqrt{4 - x}$

Domain: $\{x \mid x \leq 4\}$
Range: $\{y \mid y \geq 0\}$

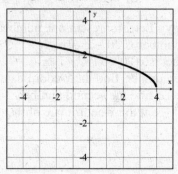

15. $f(x) = x^3 + 4x^2 + 4x$

Domain: $\{x \mid x \in \text{real numbers}\}$
Range: $\{y \mid y \in \text{real numbers}\}$

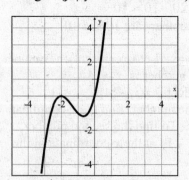

17. $f(x) = |2x + 2|$

Domain: $\{x \mid x \in \text{real numbers}\}$
Range: $\{y \mid y \geq 0\}$

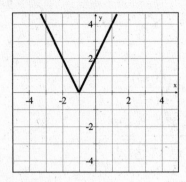

19. $f(x) = -\sqrt{x + 2}$

Domain: $\{x \mid x \geq -2\}$
Range: $\{y \mid y \leq 0\}$

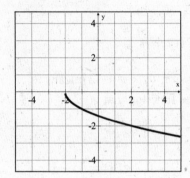

Applying the Concepts

21. $f(x) = \sqrt{x - 2}$

$f(a) = 4 = \sqrt{a - 2}$

$4^2 = \left(\sqrt{a - 2}\right)^2$

$16 = a - 2$

$18 = a$

23. $f(a,b) = a + b$

$g(a,b) = a \cdot b$

$f(2,5) = 2 + 5 = 7$

$g(2,5) = 2 \cdot 5 = 10$

$f(2,5) + g(2,5) = 7 + 10 = 17$

25. $f(14) = 8$

27. $f(x) = (x + 2)(x - 2)$

$\{x \mid -2 < x < 2\}$

29. $f(x) = |2x - 2|$

$f(x)$ is smallest when $2x - 2 = 0$.

$x = 1$

Section 9.3

Objective A Exercises

1. $f(x) = x^2 + 4 \qquad g(x) = \sqrt{x + 4}$

a) Yes
b) Yes
c) Yes
d) No

3. $f(2) - g(2) = (2 \cdot 2^2 - 3) - (-2 \cdot 2 + 4)$

$= (2 \cdot 4 - 3) - (-4 + 4)$

$= (8 - 3) - (0)$

$= 5$

5. $f(0) + g(0) = (2 \cdot 0^2 - 3) + (-2 \cdot 0 + 4)$

$= (0 - 3) + (0 + 4)$

$= -3 + 4$

$= 1$

7. $(f \cdot g)(2) = f(2) \cdot g(2)$

$= (2 \cdot 2^2 - 3) \cdot (-2 \cdot 2 + 4)$

$= (2 \cdot 4 - 3) \cdot (-4 + 4)$

$= (8 - 3) \cdot (0)$

$= 0$

9. $\left(\dfrac{f}{g}\right)(4) = \dfrac{f(4)}{g(4)}$

$= \dfrac{2\cdot(4)^2 - 3}{-2\cdot(4)+4}$

$= \dfrac{2\cdot 16 - 3}{-8+4}$

$= \dfrac{29}{-4} = -\dfrac{29}{4}$

11. $f(1) + g(1) = (2\cdot 1^2 + 3\cdot 1 - 1) + (2\cdot 1 - 4)$

$= (2\cdot 1 + 3 - 1) + (2\cdot 1 - 4)$

$= (2+3-1) + (2-4)$

$= 4-2$

$= 2$

13. $f(4) - g(4)$

$= (2\cdot(4)^2 + 3\cdot(4) - 1) - (2\cdot(4) - 4)$

$= (2\cdot 16 + 12 - 1) - (2\cdot(4) - 4)$

$= (32 + 12 - 1) - (8 - 4)$

$= 43 - 4$

$= 39$

15. $(f\cdot g)(1) = f(1)\cdot g(1)$

$= (2\cdot(1)^2 + 3\cdot(1) - 1)\cdot(2\cdot(1) - 4)$

$= (2\cdot 1 + 3 - 1)\cdot(2\cdot(1) - 4)$

$= (2 + 3 - 1)\cdot(2 - 4)$

$= 4\cdot(-2)$

$= -8$

17. $\left(\dfrac{f}{g}\right)(2) = \dfrac{f(2)}{g(2)}$

$= \dfrac{2\cdot(2)^2 + 3\cdot(2) - 1}{2\cdot(2) - 4}$

$= \dfrac{2\cdot 4 + 6 - 1}{4 - 4}$

$= \dfrac{13}{0}$

Undefined

19. $f(2) - g(2)$

$= (2^2 + 3\cdot(2) - 5) - (2^3 - 2\cdot(2) + 3)$

$= (4 + 6 - 5) - (8 - 4 + 3)$

$= 5 - 7$

$= -2$

21. $\left(\dfrac{f}{g}\right)(-2) = \dfrac{f(-2)}{g(-2)}$

$= \dfrac{(-2)^2 + 3\cdot(-2) - 5}{(-2)^3 - 2\cdot(-2) + 3}$

$= \dfrac{4 - 6 - 5}{-8 + 4 + 3}$

$= \dfrac{-7}{-1}$

$= 7$

Objective B Exercises

23. The expression $(f\circ g)(-2)$ means to evaluate the function f at $g(-2)$.

25. $f(-3) = 5$

27. $f(x) = 2x - 3 \qquad g(x) = 4x - 1$

$f(0) = 2(0) - 3 = 0 - 3 = -3$

$g(-3) = 4(-3) - 1 = -12 - 1 = -13$

$g[f(0)] = -13$

29. $f(x) = 2x - 3 \qquad g(x) = 4x - 1$

$f(-2) = 2(-2) - 3 = -4 - 3 = -7$

$g(-7) = 4(-7) - 1 = -28 - 1 = -29$

$g[f(-2)] = -29$

31. $f(x) = 2x - 3 \qquad g(x) = 4x - 1$

$g(2x - 3) = 4(2x - 3) - 1$

$= 8x - 12 - 1$

$= 8x - 13$

$g[f(x)] = 8x - 13$

33. $h(x) = 2x + 4 \qquad f(x) = \frac{1}{2}x + 2$

$h(0) = 2(0) + 4 = 0 + 4 = 4$

$f(4) = \frac{1}{2}(4) + 2 = 2 + 2 = 4$

$f[h(0)] = 4$

35. $h(x) = 2x + 4 \qquad f(x) = \frac{1}{2}x + 2$

$h(-1) = 2(-1) + 4 = -2 + 4 = 2$

$f(2) = \frac{1}{2}(2) + 2 = 1 + 2 = 3$

$f[h(-1)] = 3$

37. $h(x) = 2x + 4 \qquad f(x) = \frac{1}{2}x + 2$

$f(2x + 4) = \frac{1}{2}(2x + 4) + 2$

$= x + 2 + 2$

$= x + 4$

$f[h(x)] = x + 4$

39. $g(x) = x^2 + 3 \qquad h(x) = x - 2$

$g(0) = 0^2 + 3 = 3$

$h(3) = 3 - 2 = 1$

$h[g(0)] = 1$

41. $g(x) = x^2 + 3 \qquad h(x) = x - 2$

$g(-2) = (-2)^2 + 3 = 4 + 3 = 7$

$h(7) = 7 - 2 = 5$

$h[g(-2)] = 5$

43. $g(x) = x^2 + 3 \qquad h(x) = x - 2$

$h(x^2 + 3) = x^2 + 3 - 2$

$= x^2 + 1$

$h[g(x)] = x^2 + 1$

45. $f(x) = x^2 + x + 1 \qquad h(x) = 3x + 2$

$f(0) = (0)^2 + 0 + 1 == 1$

$h(1) = 3(1) + 2 = 3 + 2 = 5$

$h[f(0)] = 5$

47. $f(x) = x^2 + x + 1 \qquad h(x) = 3x + 2$

$f(-2) = (-2)^2 + (-2) + 1 = 4 - 2 + 1 = 3$

$h(3) = 3(3) + 2 = 9 + 2 = 11$

$h[f(-2)] = 11$

49. $f(x) = x^2 + x + 1 \qquad h(x) = 3x + 2$

$h(x^2 + x + 1) = 3(x^2 + x + 1) + 2$

$= 3x^2 + 3x + 3 + 2$

$= 3x^2 + 3x + 5$

$h[f(x)] = 3x^2 + 3x + 5$

51. $f(x) = x - 2 \qquad g(x) = x^3$

$g(-1) = (-1)^3 = -1$

$f(-1) = -1 - 2 = -3$

$f[g(-1)] = -3$

53. $f(x) = x - 2 \qquad g(x) = x^3$

$f(-1) = -1 - 2 = -3$

$g(-3) = (-3)^3 = -27$

$g[f(-1)] = -27$

55. $f(x) = x - 2 \qquad g(x) = x^3$

$g(x - 2) = (x - 2)^3$

$= (x - 2)(x - 2)(x - 2)$

$= x^3 - 6x^2 + 12x - 8$

$g[f(x)] = x^3 - 6x^2 + 12x - 8$

57. a) Strategy: Selling price equals the cost plus the mark up. If the cost is x and the mark up is 60%, then $S = x + 0.60x$.

If $M(x) = \dfrac{50x + 10,000}{x}$ is the cost per camera, then $(S \circ M)(x)$ is the selling price per camera. Find $(S \circ M)(x)$.

Solution:
$$(S \circ M)(x) = S(M(x))$$
$$= M(x) + 0.60(M(x)) = 1.60(M(x))$$
$$= 1.60\left(\frac{50x + 10,000}{x}\right)$$
$$= \frac{80x + 16,000}{x}$$
$$S(M(x)) = 80 + \frac{16,000}{x}$$

b)
$$(S \circ M)(5000) = 80 + \frac{16000}{5000}$$
$$= 80 + 3.2$$
$$= \$83.20$$

c) When 5000 digital cameras are manufactured, the camera store sells each camera for $83.20.

59. a) $I(n) = 12,500n \quad n(m) = 4m$
$$(I \circ n)(m) = I(n(m))$$
$$= 12,500(4m)$$
$$= 50,000m$$

b) $(I \circ n)(3) = 50,000(3) = \$150,000$

c) The garage's income from conversions done during a 3 month period is $150,000.

61. rebate: $r(p) = p - 1500$
discounted price: $d(p) = 0.90p$

a) If the dealer takes the rebate first and then the discount, we are finding
$d(r(p)) = 0.90(p - 1500) = 0.90p - 1350.$

b) If the dealer takes the discount first and then the rebate, we are finding
$r(d(p)) = 0.90p - 1500.$

c) As a buyer, you would prefer the dealer to use $r(d(p))$ since the cost would be less.

Applying the Concepts

63. $f(1) = 2$ and $g(2) = 0$
$g[f(1)] = g(2) = 0$

65. $(f \circ g)(3) = f(g(3))$
$g(3) = 5$ and $f(5) = -2$
$(f \circ g)(3) = f(g(3)) = -2$

67. $g(0) = -4$ and $f(-4) = 7$
$f[g(0)] = f(-4) = 7$

69. $g(x) = x^2 - 1$
$$g(3 + h) - g(3) = (3 + h)^2 - 1 - (3^2 - 1)$$
$$= 9 + 6h + h^2 - 1 - 8$$
$$g(3 + h) - g(3) = h^2 + 6h$$

71. $g(x) = x^2 - 1$
$$\frac{g(1 + h) - g(1)}{h} = \frac{(1 + h)^2 - 1 - (1^2 - 1)}{h}$$
$$= \frac{1 + 2h + h^2 - 1 - 0}{h}$$
$$= \frac{2h + h^2}{h} = 2 + h$$
$$\frac{g(1 + h) - g(1)}{h} = 2 + h$$

73. $g(x) = x^2 - 1$
$$\frac{g(a + h) - g(a)}{h} = \frac{(a + h)^2 - 1 - (a^2 - 1)}{h}$$
$$= \frac{a^2 + 2ah + h^2 - 1 - a^2 + 1}{h}$$
$$= \frac{2ah + h^2}{h} = 2a + h$$
$$\frac{g(a + h) - g(a)}{h} = 2a + h$$

75. $f(x) = 2x \quad g(x) = 3x - 1 \quad h(x) = x - 2$

$f(1) = 2(1) = 2$

$h(2) = 2 - 2 = 0$

$g(0) = 3(0) - 1 = 0 - 1 = -1$

$g(h[f(1)]) = -1$

77. $f(x) = 2x \quad g(x) = 3x - 1 \quad h(x) = x - 2$

$g(0) = 3(0) - 1 = 0 - 1 = -1$

$h(-1) = -1 - 2 = -3$

$f(-3) = 2(-3) = -6$

$f(h[g(0)]) = -6$

79. $f(x) = 2x \quad g(x) = 3x - 1 \quad h(x) = x - 2$

$h(x) = x - 2$

$f(x - 2) = 2(x - 2) = 2x - 4$

$g(2x - 4) = 3(2x - 4) - 1 = 6x - 13$

$g(f[h(x)]) = 6x - 13$

Section 9.4

Objective A Exercises

1. A function is a 1-1 function if, for any a and b in the domain of f, $f(a) = f(b)$ implies that $a = b$.

3. a) Yes
b) No

5. Yes, the graph represents a 1-1 function.

7. No, the graph is not a 1-1 function.

9. Yes, the graph represents a 1-1 function.

11. No, the graph is not a 1-1 function.

13. No, the graph is not a 1-1 function.

15. No, the graph is not a 1-1 function.

Objective B Exercises

17. The coordinates of each ordered pair of the inverse of a function are in the reverse order of the coordinates of the ordered pairs of the original function.

19. No

21. a) $f^{-1}(5) = 4$
b) $f^{-1}(4) = 3$
c) $f^{-1}(7) = 6$

23. inverse function: {(0, 1), (3, 2), (8, 3), (15, 4)}

25. There is no inverse because the numbers 5 and −5 would each be paired with two different values in the range.

27. inverse function:
{(−2, 0), (5, −1), (3, 3), (6, −4)}

29. There is no inverse because the number 3 would be paired with three different values of the range.

31. $f(x) = 4x - 8$

$y = 4x - 8$

$x = 4y - 8$

$x + 8 = 4y$

$\dfrac{1}{4}x + 2 = y$

$f^{-1}(x) = \dfrac{1}{4}x + 2$

33. $f(x) = 2x + 4$

$y = 2x + 4$

$x = 2y + 4$

$x - 4 = 2y$

$\dfrac{1}{2}x - 2 = y$

$f^{-1}(x) = \dfrac{1}{2}x - 2$

35. $f(x) = \dfrac{1}{2}x - 1$

$y = \dfrac{1}{2}x - 1$

$x = \dfrac{1}{2}y - 1$

$x + 1 = \dfrac{1}{2}y$

$2x + 2 = y$

$f^{-1}(x) = 2x + 2$

37. $f(x) = -2x + 2$

$y = -2x + 2$

$x = -2y + 2$

$x - 2 = -2y$

$-\dfrac{1}{2}x + 1 = y$

$f^{-1}(x) = -\dfrac{1}{2}x + 1$

39. $f(x) = \dfrac{2}{3}x + 4$

$y = \dfrac{2}{3}x + 4$

$x = \dfrac{2}{3}y + 4$

$x - 4 = \dfrac{2}{3}y$

$\dfrac{3}{2}x - 6 = y$

$f^{-1}(x) = \dfrac{3}{2}x - 6$

41. $f(x) = -\dfrac{1}{3}x + 1$

$y = -\dfrac{1}{3}x + 1$

$x = -\dfrac{1}{3}y + 1$

$x - 1 = -\dfrac{1}{3}y$

$-3x + 3 = y$

$f^{-1}(x) = -3x + 3$

43. $f(x) = 2x - 5$

$y = 2x - 5$

$x = 2y - 5$

$x + 5 = 2y$

$\dfrac{1}{2}x + \dfrac{5}{2} = y$

$f^{-1}(x) = \dfrac{1}{2}x + \dfrac{5}{2}$

45. $f(x) = 5x - 2$

$y = 5x - 2$

$x = 5y - 2$

$x + 2 = 5y$

$\dfrac{1}{5}x + \dfrac{2}{5} = y$

$f^{-1}(x) = \dfrac{1}{5}x + \dfrac{2}{5}$

47. $f(x) = 6x - 3$

$y = 6x - 3$

$x = 6y - 3$

$x + 3 = 6y$

$\dfrac{1}{6}x + \dfrac{1}{2} = y$

$f^{-1}(x) = \dfrac{1}{6}x + \dfrac{1}{2}$

49. $f(x) = 3x - 5$

$y = 3x - 5$

$x = 3y - 5$

$x + 5 = 3y$

$\dfrac{1}{3}x + \dfrac{5}{3} = y$

$f^{-1}(x) = \dfrac{1}{3}x + \dfrac{5}{3}$

$f^{-1}(0) = \dfrac{1}{3}(0) + \dfrac{5}{3}$

$f^{-1}(0) = \dfrac{5}{3}$

51. $f(x) = 3x - 5$

$y = 3x - 5$

$x = 3y - 5$

$x + 5 = 3y$

$\dfrac{1}{3}x + \dfrac{5}{3} = y$

$f^{-1}(x) = \dfrac{1}{3}x + \dfrac{5}{3}$

$f^{-1}(4) = \dfrac{1}{3}(4) + \dfrac{5}{3}$

$f^{-1}(4) = \dfrac{9}{3} = 3$

53. Using the vertical-line test the graph is a function. Using the horizontal-line test, the graph is 1-1 and therefore does have an inverse.

55. $f(g(x)) = f\left(\dfrac{x}{4}\right) = 4\left(\dfrac{x}{4}\right) = x$

$g(f(x)) = g(4x) = \dfrac{4x}{4} = x$

Yes, the functions are inverses of each other.

57. $f(h(x)) = f\left(\dfrac{1}{3x}\right) = 3\left(\dfrac{1}{3x}\right) = \dfrac{1}{x}$

$h(f(x)) = h(3x) = \dfrac{1}{3(3x)} = \dfrac{1}{9x}$

No, the functions are not inverses of each other.

59. $f(g(x)) = f(3x + 2)$

$\quad = \dfrac{1}{3}(3x + 2) - \dfrac{2}{3} = x + \dfrac{2}{3} - \dfrac{2}{3} = x$

$g(f(x)) = g\left(\dfrac{1}{3}x - \dfrac{2}{3}\right)$

$\quad = 3\left(\dfrac{1}{3}x - \dfrac{2}{3}\right) + 2 = x - 2 + 2 = x$

Yes, the functions are inverses of each other.

61. $f(g(x)) = f(2x + 3)$

$\quad = \dfrac{1}{2}(2x + 3) - \dfrac{3}{2} = x + \dfrac{3}{2} - \dfrac{3}{2} = x$

$g(f(x)) = g\left(\dfrac{1}{2}x - \dfrac{3}{2}\right)$

$\quad = 2\left(\dfrac{1}{2}x - \dfrac{3}{2}\right) + 3 = x - 3 + 3 = x$

Yes, the functions are inverses of each other.

63. $f(x) = \dfrac{x}{16}$

$y = \dfrac{x}{16}$

$x = \dfrac{y}{16}$

$16x = y$

$f^{-1}(x) = 16x$

The inverse function converts pounds to ounces.

65. $f(x) = x + 30$

$y = x + 30$

$x = y + 30$

$x - 30 = y$

$f^{-1}(x) = x - 30$

The inverse function converts a dress size in France to a dress size in the United States.

67. $f(x) = 90x + 65$

$y = 90x + 65$

$x = 90y + 65$

$x - 65 = 90y$

$\dfrac{1}{90}x - \dfrac{13}{18} = y$

$f^{-1}(x) = \dfrac{1}{90}x - \dfrac{13}{18}$

The inverse function gives the training intensity percent for a given target heart rate.

Applying the Concepts

69.

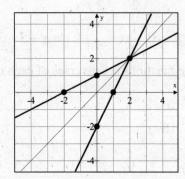

71.

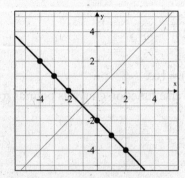

The inverse is the same graph.

73.

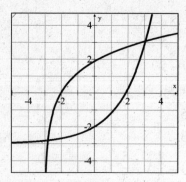

75. Inverse of the function:

Grade	Score
A	90-100
B	80-89
C	70-79
D	60-69
F	0-59

No, the inverse of the grading scale is not a function because each grade is paired with more than one score.

77. A constant function is defined as $y = b$, where b is a constant value. The inverse of this function would be $x = a$, where a is a constant. This is not a function.

Chapter 9 Review Exercises

1. Yes, the graph is a function. It passes the vertical-line test.

2. Yes, the graph is a 1-1 function. It passes both the vertical-line test and the horizontal-line test.

3. $f(x) = 3x^3 - 2$

Domain: $\{x | \, x \in \text{real numbers}\}$

Range: $\{y | \, y \in \text{real numbers}\}$

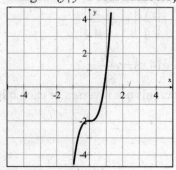

4. $f(x) = \sqrt{x+4}$

Domain: $\{x \mid x \geq -4\}$
Range: $\{y \mid y \geq 0\}$

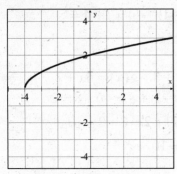

5. $f(x) = \mid x \mid -3$

Domain: $\{x \mid x \in \text{real numbers}\}$
Range: $\{y \mid y \geq -3\}$

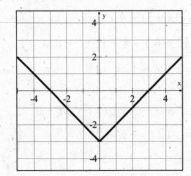

6. $f(x) = x^2 - 2x + 3$

$$-\frac{b}{2a} = -\frac{-2}{2(1)} = -\frac{-2}{2} = -(-1) = 1$$

$$y = 1^2 - 2(1) + 3 = 2$$

Vertex:
$(1, 2)$
Axis of
symmetry:
$x = 1$

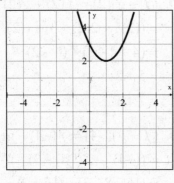

7. $y = -3x^2 + 4x + 6$

$a = -3 \quad b = 4 \quad c = 6$

$b^2 - 4ac$

$(4)^2 - 4(-3)(6) = 16 + 72 = 88$

$88 > 0$

Since the discriminant is greater than zero, the parabola has two x-intercepts.

8. $y = 2x^2 + x + 5$

$a = 2 \quad b = 1 \quad c = 5$

$b^2 - 4ac$

$(1)^2 - 4(2)(5) = 1 - 40 = -39$

$-39 < 0$

Since the discriminant is less than zero, the parabola has no x-intercepts.

9. $y = 3x^2 + 9x$

$0 = 3x^2 + 9x$

$0 = 3x(x + 3)$

$3x = 0 \quad x + 3 = 0$

$x = 0 \qquad x = -3$

The x-intercepts are $(0, 0)$ and $(-3, 0)$.

10. $f(x) = x^2 - 6x + 7$

$0 = x^2 - 6x + 7$

$a = 1 \quad b = -6 \quad c = 7$

$$x = \frac{-b \pm \sqrt{b^2 - 4ac}}{2a}$$

$$x = \frac{-(-6) \pm \sqrt{(-6)^2 - 4(1)(7)}}{2(1)}$$

$$x = \frac{6 \pm \sqrt{36 - 28}}{2} = \frac{6 \pm \sqrt{8}}{2} = \frac{6 \pm 2\sqrt{2}}{2}$$

$$x = 3 \pm \sqrt{2}$$

The x-intercepts are $(3 + \sqrt{2}, 0)$ and $(3 - \sqrt{2}, 0)$.

11. $f(x) = 2x^2 - 7x - 15$

$2x^2 - 7x - 15 = 0$

$(2x + 3)(x - 5) = 0$

$2x + 3 = 0 \qquad x - 5 = 0$

$2x = -3 \qquad x = 5$

$x = -\dfrac{3}{2}$

The zeros are $-\dfrac{3}{2}$ and 5.

12. $f(x) = x^2 - 2x + 10$

$x^2 - 2x + 10 = 0$

$a = 1 \quad b = -2 \quad c = 10$

$x = \dfrac{-b \pm \sqrt{b^2 - 4ac}}{2a}$

$x = \dfrac{-(-2) \pm \sqrt{(-2)^2 - 4(1)(10)}}{2(1)}$

$x = \dfrac{2 \pm \sqrt{4 - 40}}{2} = \dfrac{2 \pm \sqrt{-36}}{2} = \dfrac{2 \pm 6i}{2}$

$x = 1 \pm 3i$

The zeros are $1 + 3i$ and $1 - 3i$.

13. $f(x) = -2x^2 + 4x + 1$

$x = -\dfrac{b}{2a} = -\dfrac{4}{2(-2)} = 1$

$f(x) = -2x^2 + 4x + 1$

$f(1) = -2(1)^2 + 4(1) + 1 = -2 + 4 + 1 = 3$

The maximum value of the function is 3.

14. $f(x) = x^2 - 7x + 8$

$x = -\dfrac{b}{2a} = -\dfrac{-7}{2(1)} = \dfrac{7}{2}$

$f(x) = x^2 - 7x + 8$

$f\left(\dfrac{7}{2}\right) = \left(\dfrac{7}{2}\right)^2 - 7\left(\dfrac{7}{2}\right) + 8 = \dfrac{49}{4} - \dfrac{49}{2} + 8 = \dfrac{17}{4}$

The minimum value of the function is $\dfrac{17}{4}$.

15. $f(x) = x^2 + 4 \quad g(x) = 4x - 1$

$g(0) = 4(0) - 1 = -1$

$f(-1) = (-1)^2 + 4 = 1 + 4 = 5$

$f[g(0)] = 5$

16. $f(x) = 6x + 8 \quad g(x) = 4x + 2$

$f(-1) = 6(-1) + 8 = -6 + 8 = 2$

$g(2) = 4(2) + 2 = 8 + 2 = 10$

$g[f(-1)] = 10$

17. $f(x) = 3x^2 - 4 \quad g(x) = 2x + 1$

$f(g(x)) = f(2x + 1)$

$= 3(2x + 1)^2 - 4 = 3(2x + 1)(2x + 1) - 4$

$= 3(4x^2 + 4x + 1) - 4$

$= 12x^2 + 12x + 3 - 4 = 12x^2 + 12x - 1$

$f[g(x)] = 12x^2 + 12x - 1$

18. $f(x) = 2x^2 + x - 5 \quad g(x) = 3x - 1$

$g(f(x)) = g(2x^2 + x - 5)$

$= 3(2x^2 + x - 5) - 1$

$= 6x^2 + 3x - 15 - 1$

$= 6x^2 + 3x - 16$

$g[f(x)] = 6x^2 + 3x - 16$

19. $(f + g)(2) = f(2) + g(2)$

$= ((2)^2 + 2(2) - 3) + ((2)^2 - 2)$

$= (4 + 4 - 3) + (4 - 2)$

$= 5 + 2$

$= 7$

20. $(f - g)(-4) = f(-4) - g(-4)$

$= ((-4)^2 + 2(-4) - 3) - ((-4)^2 - 2)$

$= (16 - 8 - 3) - (16 - 2)$

$= 5 - 14$

$= -9$

21. $(f \cdot g)(-4) = f(-4) \cdot g(-4)$

$\quad = ((-4)^2 + 2(-4) - 3) \cdot ((-4)^2 \div 2)$

$\quad = (16 - 8 - 3) \cdot (16 - 2)$

$\quad = 5 \cdot 14$

$\quad = 70$

22. $\left(\dfrac{f}{g}\right)(3) = \dfrac{f(3)}{g(3)}$

$\quad = \dfrac{(3)^2 + 2(3) - 3}{(3)^2 - 2}$

$\quad = \dfrac{9 + 6 - 3}{9 - 2}$

$\quad = \dfrac{12}{7}$

23. $f(x) = -6x + 4$

$\quad y = -6x + 4$

$\quad x = -6y + 4$

$\quad x - 4 = -6y$

$\quad -\dfrac{1}{6}x + \dfrac{2}{3} = y$

$\quad f^{-1}(x) = -\dfrac{1}{6}x + \dfrac{2}{3}$

24. $f(x) = \dfrac{2}{3}x - 12$

$\quad y = \dfrac{2}{3}x - 12$

$\quad x = \dfrac{2}{3}y - 12$

$\quad x + 12 = \dfrac{2}{3}y$

$\quad \dfrac{3}{2}x + 18 = y$

$\quad f^{-1}(x) = \dfrac{3}{2}x + 18$

25. $f(g(x)) = f(-4x + 5)$

$\quad = -\dfrac{1}{4}(-4x + 5) + \dfrac{5}{4}$

$\quad = x - \dfrac{5}{4} + \dfrac{5}{4} = x$

$g(f(x)) = g\left(-\dfrac{1}{4}x + \dfrac{5}{4}\right)$

$\quad = -4\left(-\dfrac{1}{4}x + \dfrac{5}{4}\right) + 5$

$\quad = x - 5 + 5 = x$

Yes, the functions are inverses of each other.

26. $f(g(x)) = f(2x + 1)$

$\quad = \dfrac{1}{2}(2x + 1) = x + \dfrac{1}{2}$

$g(f(x)) = g\left(\dfrac{1}{2}x\right)$

$\quad = 2\left(\dfrac{1}{2}x\right) + 1 = x + 1$

No, the functions are not inverses of each other.

27. $p(x) = 0.4x + 15$

$\quad p = 0.4x + 15$

$\quad x = 0.4p + 15$

$\quad x - 15 = 0.4p$

$\quad 2.5x - 37.5 = p$

$\quad p^{-1}(x) = 2.5x - 37.5$

The inverse function gives the diver's depth below the surface of the water for a given pressure on the diver.

28. Strategy: To find the number of gloves to make for a maximum profit, find the x-coordinate of the vertex. To find the maximum profit, evaluate the function at the x-coordinate of the vertex.

Solution:

$$x = -\frac{b}{2a} = -\frac{100}{2(-1)} = -\frac{100}{-2} = -(-50) = 50$$

$$P(x) = -x^2 + 100x + 2500$$

$$P(50) = -(50)^2 + 100(50) + 2500$$

$$= -2500 + 5000 + 2500 = 5000$$

The company should make 50 baseball gloves each month to maximize profit. The maximum profit is $5000.

29. Strategy: Let x represent width of the rectangle.
The length is $14 - x$.
The area is $x(14 - x)$.
To find the width, find the x-coordinate of the vertex. To find the length, evaluate $14 - x$ at the x-coordinate of the vertex.

Solution:

$$x(14 - x) = 14x - x^2$$

$$x = -\frac{b}{2a} = -\frac{14}{2(-1)} = -\frac{14}{-2} = -(-7) = 7$$

$$14 - x = 14 - 7 = 7$$

The dimensions are 7 ft by 7 ft.

Chapter 9 Test

1. $f(x) = x^2 - 6x + 4$

$$-\frac{b}{2a} = -\frac{-6}{2(1)} = -\frac{-6}{2} = -(-3) = 3$$

$$y = 3^2 - 6(3) + 4 = -5$$

Vertex:
$(3, -5)$
Axis of symmetry:
$x = 3$

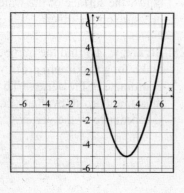

2. $f(x) = \left|\frac{1}{2}x\right| - 2$

Domain: $\{x \mid x \in \text{real numbers}\}$
Range: $\{y \mid y \geq -2\}$

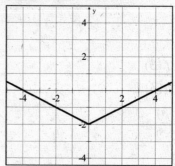

3. $f(x) = -\sqrt{3 - x}$

Domain: $\{x \mid x \leq 3\}$
Range: $\{y \mid y \leq 0\}$

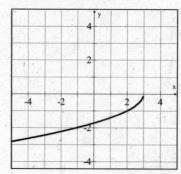

4. $f(x) = x^3 - 3x + 2$

Domain: $\{x \mid x \in \text{real numbers}\}$
Range: $\{y \mid y \in \text{real numbers}\}$

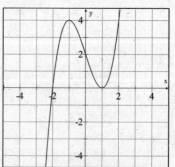

5. $y = 3x^2 - 4x + 6$

$a = 3 \quad b = -4 \quad c = 6$

$b^2 - 4ac$

$(-4)^2 - 4(3)(6) = 16 - 72 = -56$

$-56 < 0$

Since the discriminant is less than zero the parabola has no x-intercepts.

6. $y = 3x^2 - 7x - 6$

$0 = 3x^2 - 7x - 6$

$0 = (3x + 2)(x - 3)$

$3x + 2 = 0 \quad x - 3 = 0$

$3x = -2 \qquad x = 3$

$x = -\dfrac{2}{3}$

The x-intercepts are $(-\dfrac{2}{3}, 0)$ and $(3, 0)$.

7. $f(x) = -x^2 + 8x - 7$

$x = -\dfrac{b}{2a} = -\dfrac{8}{2(-1)} = -(-4) = 4$

$f(x) = -x^2 + 8x - 7$

$f(4) = -(4)^2 + 8(4) - 7 = -16 + 32 - 7 = 9$

The maximum value of the function is 9.

8. $f(x) = 2x^2 + 4x - 2$

Domain: $\{x \mid x \in \text{real numbers}\}$

Range: $\{y \mid y \geq -7\}$

9. $f(x) = x^2 + 2x - 3 \quad g(x) = x^3 - 1$

$(f - g)(2) = f(2) - g(2)$

$= (2^2 + 2(2) - 3) - (2^3 - 1)$

$= (4 + 4 - 3) - (8 - 1)$

$= 5 - 7$

$= -2$

10. $f(x) = x^3 + 1 \quad g(x) = 2x - 3$

$(f \cdot g)(-3) = f(-3) \cdot g(-3)$

$= ((-3)^3 + 1) \cdot (2(-3) - 3)$

$= (-27 + 1) \cdot (-6 - 3)$

$= (-26) \cdot (-9)$

$= 234$

11. $f(x) = 4x - 5 \quad g(x) = x^2 + 3x + 4$

$\left(\dfrac{f}{g}\right)(-2) = \dfrac{f(-2)}{g(-2)}$

$= \dfrac{4(-2) - 5}{(-2)^2 + 3(-2) + 4}$

$= \dfrac{-8 - 5}{4 - 6 + 4} = \dfrac{-13}{2}$

$= -\dfrac{13}{2}$

12. $f(x) = x^2 + 4 \quad g(x) = 2x^2 + 2x + 1$

$(f - g)(-4) = f(-4) - g(-4)$

$= ((-4)^2 + 4) - (2(-4)^2 + 2(-4) + 1)$

$= (16 + 4) - (32 - 8 + 1)$

$= 20 - 25$

$= -5$

13. $f(x) = 2x - 7 \quad g(x) = x^2 - 2x - 5$

$g(2) = 2^2 - 2(2) - 5 = 4 - 4 - 5 = -5$

$f(-5) = 2(-5) - 7 = -10 - 7 = -17$

$f[g(2)] = -17$

14. $f(x) = x^2 + 1 \quad g(x) = x^2 + x + 1$

$f(-2) = (-2)^2 + 1 = 4 + 1 = 5$

$g(5) = (5)^2 + 5 + 1 = 25 + 5 + 1 = 31$

$g[f(-2)] = 31$

15. $f(x) = x^2 - 1 \quad g(x) = 3x + 2$

$g(f(x)) = g(x^2 - 1)$

$= 3(x^2 - 1) + 2 = 3x^2 - 3 + 2$

$= 3x^2 - 1$

$g[f(x)] = 3x^2 - 1$

16. $f(x) = 2x^2 - 7 \quad g(x) = x - 1$

$f(g(x)) = f(x - 1)$

$= 2(x - 1)^2 - 7 = 2(x - 1)(x - 1) - 7$

$= 2(x^2 - 2x + 1) - 7$

$= 2x^2 - 4x + 2 - 7$

$= 2x^2 - 4x - 5$

$f[g(x)] = 2x^2 - 4x - 5$

17. There is no inverse because the numbers 4 and 5 would be paired with two different values of the range.

18. Inverse function: $\{(6, 2), (5, 3), (4, 4), (3, 5)\}$

19. $f(x) = 4x - 2$

$y = 4x - 2$

$x = 4y - 2$

$x + 2 = 4y$

$\dfrac{1}{4}x + \dfrac{1}{2} = y$

$f^{-1}(x) = \dfrac{1}{4}x + \dfrac{1}{2}$

20. $f(x) = \dfrac{1}{4}x - 4$

$y = \dfrac{1}{4}x - 4$

$x = \dfrac{1}{4}y - 4$

$x + 4 = \dfrac{1}{4}y$

$4x + 16 = y$

$f^{-1}(x) = 4x + 16$

21. $f(g(x)) = f(2x - 4)$

$= \dfrac{1}{2}(2x - 4) + 2 = x - 2 + 2 = x$

$g(f(x)) = g\left(\dfrac{1}{2}x + 2\right)$

$= 2\left(\dfrac{1}{2}x + 2\right) - 4 = x + 4 - 4 = x$

Yes, the functions are inverses of each other.

22. $f(g(x)) = f\left(\dfrac{3}{2}x + 3\right)$

$= \dfrac{2}{3}\left(\dfrac{3}{2}x + 3\right) - 2 = x + 2 - 2 = x$

$g(f(x)) = g\left(\dfrac{2}{3}x - 2\right)$

$= \dfrac{3}{2}\left(\dfrac{2}{3}x - 2\right) + 3 = x - 3 + 3 = x$

Yes, the functions are inverses of each other.

23. No, the graph is not a 1-1 function. It does not pass the horizontal-line test.

24. $C(x) = 1.25x + 5$

$C = 1.25x + 5$

$x = 1.25C + 5$

$x - 5 = 1.25C$

$0.8x - 4 = C$

$C^{-1}(x) = 0.8x - 4$

The inverse function gives the number of miles to a certain location for a given cost.

25. Strategy: To find the number of speakers for a minimum production cost, find the x-coordinate of the vertex. To find the minimum cost, evaluate the function at the x-coordinate of the vertex.

Solution:

$x = -\dfrac{b}{2a} = -\dfrac{-50}{2} = -\dfrac{-50}{2} = -(-25) = 25$

$C(x) = x^2 - 50x + 675$

$C(25) = (25)^2 - 50(25) + 675$

$= 625 - 1250 + 675 = 50$

The company should make 25 speakers each day to minimize production costs.
The minimum daily production cost is $50.

26. Strategy: Let x represent one number. The other number is $28 - x$.
Their product is $x(28 - x)$.
To find one number, find the x-coordinate of the vertex. To find the second number, evaluate $28 - x$ at the x-coordinate of the vertex.

Solution:

$$x(28 - x) = 28x - x^2$$

$$x = -\frac{b}{2a} = -\frac{28}{2(-1)} = -\frac{28}{-2} = -(-14) = 14$$

$$28 - x = 28 - 14 = 14$$

The two numbers are 14 and 14.

27. Strategy: Let x represent width of the rectangle. The length is $100 - x$.
The area is $x(100 - x)$.
To find the width, find the x-coordinate of the vertex. To find the length, evaluate $100 - x$ at the x-coordinate of the vertex.

Solution:

$$x(100 - x) = 100x - x^2$$

$$x = -\frac{b}{2a} = -\frac{100}{2(-1)} = -\frac{100}{-2} = -(-50) = 50$$

$$100 - x = 100 - 50 = 50$$

$$A = l \cdot w = 50 \cdot 50 = 2500$$

The dimensions of the rectangle are 50 cm by 50 cm. The area is 2500 cm^2.

Cumulative Review Exercises

1. $-3a + \left| \dfrac{3b - ab}{3b - c} \right|$

$$-3(2) + \left| \frac{3(2) - (2)(2)}{3(2) - (-2)} \right|$$

$$= -6 + \left| \frac{6 - 4}{6 + 2} \right| = -6 + \left| \frac{2}{8} \right| = -6 + \left| \frac{1}{4} \right|$$

$$= -6 + \frac{1}{4} = -\frac{23}{4}$$

2.

$$\begin{array}{ccccccccccc} & & (&) & & & & & & & \\ \hline -5 & -4 & -3 & -2 & -1 & 0 & 1 & 2 & 3 & 4 & 5 \end{array}$$

3. $\dfrac{3x - 1}{6} - \dfrac{5 - x}{4} = \dfrac{5}{6}$

$$12\left(\frac{3x - 1}{6} - \frac{5 - x}{4} \right) = 12\left(\frac{5}{6} \right)$$

$$2(3x - 1) - 3(5 - x) = 10$$

$$6x - 2 - 15 + 3x = 10$$

$$9x - 17 = 10$$

$$9x = 27$$

$$x = 3$$

The solution is 3.

4. $4x - 2 < -10$ or $3x - 1 > 8$

$$\begin{array}{ll} 4x - 2 < -10 & 3x - 1 > 8 \\ 4x < -8 & 3x > 9 \\ x < -2 & x > 3 \end{array}$$

$\{x \mid x < -2\}$ or $\{x \mid x > 3\}$
$\{x \mid x < -2\} \cup \{x \mid x > 3\}$
$\quad = \{x \mid x < -2 \text{ or } x > 3\}$

5. $|8 - 2x| \geq 0$

$$\begin{array}{ll} 8 - 2x \leq 0 & 8 - 2x \geq 0 \\ -2x \leq -8 & -2x \geq -8 \\ x \geq 4 & x \leq 4 \end{array}$$

$\{x \mid x \geq 4\}$ or $\{x \mid x \leq 4\}$
$\{x \mid x \geq 4\} \cup \{x \mid x \leq 4\}$
$\quad = \{x \mid x \in \text{real numbers}\}$

6. $\left(\dfrac{3a^3b}{2a}\right)^2\left(\dfrac{a^2}{-3b^2}\right)^3 = \left(\dfrac{3a^2b}{2}\right)^2\left(\dfrac{a^2}{-3b^2}\right)^3$

$\quad = \left(\dfrac{3^2a^4b^2}{2^2}\right)\left(\dfrac{a^6}{(-3)^3b^6}\right) = \dfrac{9a^{10}b^2}{4(-27)b^6}$

$\quad = \dfrac{9a^{10}}{-108b^4} = -\dfrac{a^{10}}{12b^4}$

7. $(x-4)(2x^2+4x-1)$
$\quad = x(2x^2+4x-1)-4(2x^2+4x-1)$
$\quad = 2x^3+4x^2-x-8x^2-16x+4$
$\quad = 2x^3-4x^2-17x+4$

8. $6x-2y=-3$
$\quad 4x+\ y=5$

$\quad 6x-2y=-3$
$\quad \underline{8x+2y=\ 10}$
$\quad 14x\quad\ =7$
$\qquad\qquad x=\dfrac{1}{2}$

$\quad 6\left(\dfrac{1}{2}\right)-2y=-3$
$\qquad 3-2y=-3$
$\qquad\ -2y=-6$
$\qquad\qquad y=3$

The solution is $\left(\dfrac{1}{2},3\right)$.

9. $x^3y+x^2y^2-6xy^3 = xy(x^2+xy-6y^2)$
$\qquad\qquad\qquad\qquad = xy(x+3y)(x-2y)$

10. $(b+2)(b-5)=2b+14$
$\quad b^2-3b-10=2b+14$
$\quad b^2-5b-24=0$
$\quad (b-8)(b+3)=0$
$\quad b-8=0\qquad b+3=0$
$\qquad b=8\qquad\quad b=-3$
The solutions are -3 and 8.

11. $x^2-2x>15$
$\quad x^2-2x-15>0$
$\quad (x-5)(x+3)>0$
$\quad \{x\,|\,x<-3 \text{ or } x>5\}$

12. $\dfrac{x^2+4x-5}{2x^2-3x+1}-\dfrac{x}{2x-1}$

$\quad = \dfrac{(x+5)(x-1)}{(2x-1)(x-1)}-\dfrac{x}{2x-1}$

$\quad = \dfrac{x+5}{2x-1}-\dfrac{x}{2x-1} = \dfrac{x+5-x}{2x-1}$

$\quad = \dfrac{5}{2x-1}$

13. $\dfrac{5}{x^2+7x+12} = \dfrac{9}{x+4}-\dfrac{2}{x+3}$

$\quad \dfrac{5}{(x+4)(x+3)} = \dfrac{9}{x+4}\cdot\dfrac{x+3}{x+3}-\dfrac{2}{x+3}\cdot\dfrac{x+4}{x+4}$

$\quad \dfrac{5}{(x+4)(x+3)} = \dfrac{9x+27}{(x+4)(x+3)}-\dfrac{2x+8}{(x+4)(x+3)}$

$\quad 5=(9x+27)-(2x+8)$

$\quad 5=7x+19$

$\quad -14=7x$

$\quad -2=x$

The solution is -2.

14. $\dfrac{4-6i}{2i} = \dfrac{4-6i}{2i}\cdot\dfrac{i}{i} = \dfrac{4i-6i^2}{2i^2}$

$\quad = \dfrac{4i+6}{-2} = -3-2i$

15. $f(x)=\dfrac{1}{4}x^2$

$\quad -\dfrac{b}{2a} = -\dfrac{0}{2\left(\dfrac{1}{4}\right)} = -\dfrac{0}{\dfrac{1}{4}} = 0$

$\quad y=\left(\dfrac{1}{4}\right)0^2=0$

Vertex:
$(0,0)$
Axis of
symmetry:
$x=0$

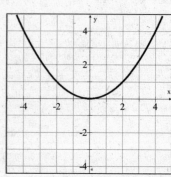

16. $3x - 4y \geq 8$

$-4y \geq -3x + 8$

$y \leq \dfrac{3}{4}x - 2$

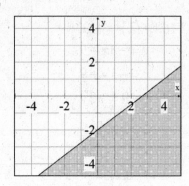

17. $m = \dfrac{y_2 - y_1}{x_2 - x_1} = \dfrac{-6 - 4}{2 - (-3)} = \dfrac{-10}{5} = -2$

$y - y_1 = m(x - x_1)$

$y - 4 = -2(x - (-3))$

$y - 4 = -2(x + 3)$

$y - 4 = -2x - 6$

$y = -2x - 2$

18. The product of the slopes of perpendicular lines is -1.

$2x - 3y = 6$

$-3y = -2x + 6$

$y = \dfrac{2}{3}x - 2$

$m_1 \cdot m_2 = -1$

$\dfrac{2}{3} \cdot m_2 = -1$

$m_2 = -\dfrac{3}{2}$

$y - y_1 = m(x - x_1)$

$y - 1 = -\dfrac{3}{2}(x - (-3))$

$y - 1 = -\dfrac{3}{2}(x + 3)$

$y - 1 = -\dfrac{3}{2}x - \dfrac{9}{2}$

$y = -\dfrac{3}{2}x - \dfrac{7}{2}$

19. $3x^2 = 3x - 1$

$3x^2 - 3x + 1 = 0$

$a = 3 \quad b = -3 \quad c = 1$

$x = \dfrac{-b \pm \sqrt{b^2 - 4ac}}{2a}$

$x = \dfrac{-(-3) \pm \sqrt{(-3)^2 - 4(3)(1)}}{2(3)}$

$x = \dfrac{3 \pm \sqrt{9 - 12}}{6} = \dfrac{3 \pm \sqrt{-3}}{6} = \dfrac{3 \pm i\sqrt{3}}{6}$

$x = \dfrac{1}{2} \pm \dfrac{i\sqrt{3}}{6}$

The zeros are $\dfrac{1}{2} + \dfrac{i\sqrt{3}}{6}$ and $\dfrac{1}{2} - \dfrac{i\sqrt{3}}{6}$.

20. $\sqrt{8x + 1} = 2x - 1$

$\left(\sqrt{8x + 1}\right)^2 = (2x - 1)^2$

$8x + 1 = 4x^2 - 4x + 1$

$0 = 4x^2 - 12x$

$0 = 4x(x - 3)$

$4x = 0 \qquad x - 3 = 0$

$x = 0 \qquad\quad x = 3$

Check both solutions in the original equation:

$\sqrt{8(0) + 1} = 2(0) - 1$

$\sqrt{1} = -1$

$1 \neq -1$

$\sqrt{8(3) + 1} = 2(3) - 1$

$\sqrt{25} = 5$

$5 = 5$

The solution is 3.

21. $f(x) = 2x^2 - 3$

$$x = -\frac{b}{2a} = -\frac{0}{2(2)} = -\frac{0}{4} = 0$$

$$f(x) = 2x^2 - 3$$

$$f(0) = 2(0)^2 - 3 = 0 - 3$$

$$= -3$$

Since $a > 0$, the function has a minimum value. The minimum value of the function is -3.

22. $f(x) = |3x - 4|$

$$f(0) = |3(0) - 4| = |-4| = 4$$

$$f(1) = |3(1) - 4| = |-1| = 1$$

$$f(2) = |3(2) - 4| = |2| = 2$$

$$f(3) = |3(3) - 4| = |5| = 5$$

The range is $\{1, 2, 4, 5\}$.

23. Yes

24. $\sqrt[3]{5x - 2} = 2$

$$\left(\sqrt[3]{5x - 2}\right)^3 = 2^3$$

$$5x - 2 = 8$$

$$5x = 10$$

$$x = 2$$

The solution is 2.

25. $g(x) = 3x - 5 \quad h(x) = \frac{1}{2}x + 4$

$$h(2) = \frac{1}{2}(2) + 4 = 1 + 4 = 5$$

$$g(5) = 3(5) - 5 = 15 - 5 = 10$$

$$g[h(2)] = 10$$

26. $f(x) = -3x + 9$

$$y = -3x + 9$$

$$x = -3y + 9$$

$$x - 9 = -3y$$

$$-\frac{1}{3}x + 3 = y$$

$$f^{-1}(x) = -\frac{1}{3}x + 3$$

27. Strategy: Let x represent the cost per pound of the mixture.

	Amount	Cost	Value
$4.50 tea	30	4.50	30(4.50)
$3.60 tea	45	3.60	45(3.60)
Mixture	75	x	75x

The sum of the values before mixing is equal to the value after mixing.

Solution:
$$30(4.50) + 45(3.60) = 75x$$
$$135 + 162 = 75x$$
$$297 = 75x$$
$$x = 3.96$$
The cost per pound of the mixture is $3.96.

28. Strategy: Let x represent the number of pounds of 80% copper alloy.

	Amount	Percent	Quantity
80%	x	0.80	0.80x
20%	50	0.20	0.20(50)
40%	$50 + x$	0.40	0.40(50 + x)

The sum of the quantities before mixing is equal to the quantity after mixing.

Solution:
$$0.80x + 0.20(50) = 0.40(50 + x)$$
$$0.80x + 10 = 20 + 0.40x$$
$$0.40x + 10 = 20$$
$$0.40x = 10$$
$$x = 25$$
25 lb of the 80% copper alloy must be used.

29. **Strategy:** Let x represent the additional amount of insecticide.
The total amount of insecticide is $x + 6$.
To find the additional amount of insecticide write and solve a proportion.

Solution: $\dfrac{6}{16} = \dfrac{x+6}{28}$.

$\dfrac{3}{8} = \dfrac{x+6}{28}$

$\dfrac{3}{8} \cdot 56 = \dfrac{x+6}{28} \cdot 56$

$21 = 2x + 12$

$9 = 2x$

$4.5 = x$

An additional 4.5 oz of insecticide are required.

30. **Strategy:** Let x represent the time it takes for the smaller pipe to fill the take.
The time it takes the larger pipe to fill the tank is $x - 8$.

	Rate	Time	Part
Smaller pipe	$\dfrac{1}{t}$	3	$\dfrac{3}{t}$
Larger pipe	$\dfrac{1}{t-8}$	3	$\dfrac{3}{t-8}$

The sum of the parts of the task completed must equal 1.

Solution:

$\dfrac{3}{t} + \dfrac{3}{t-8} = 1$

$t(t-8)\left(\dfrac{3}{t} + \dfrac{3}{t-8}\right) = 1(t(t-8))$

$3(t-8) + 3t = t^2 - 8t$

$3t - 24 + 3t = t^2 - 8t$

$6t - 24 = t^2 - 8t$

$0 = t^2 - 14t + 24$

$0 = (t-2)(t-12)$

$t - 2 = 0 \quad t - 12 = 0$

$\quad t = 2 \qquad t = 12$

The solution $t = 2$ is not possible since the time for the larger pipe would then be a negative number.
$t - 8 = 2 - 8 = -6$
It takes the larger pipe $t - 8 = 12 - 8 = 4$ min to fill the tank.

31. **Strategy:** Write the basic direct variation equation, replacing the variable with the given values. Solve for k.
Write the direct variation equation, replacing k with its value. Substitute 40 for f and solve for d.

Solution:

$d = kf \qquad\qquad d = \dfrac{3}{5}f$

$30 = k(50) \qquad\quad = \dfrac{3}{5}(40)$

$\dfrac{3}{5} = k \qquad\qquad = 24$

A force of 40 lb will stretch the spring 24 in.

32. **Strategy:** Write the basic inverse variation equation, replacing the variable with the given values. Solve for k.
Write the inverse variation equation, replacing k with its value. Substitute 1.5 for L and solve for f.

Solution:

$f = \dfrac{k}{L} \qquad\qquad f = \dfrac{120}{L}$

$60 = \dfrac{k}{2} \qquad\quad = \dfrac{120}{1.5}$

$120 = k \qquad\qquad = 80$

The frequency is 80 vibrations/min.

Chapter 10: Exponential and Logarithmic Functions

Prep Test

1. $3^{-2} = \dfrac{1}{3^2} = \dfrac{1}{9}$

2. $\left(\dfrac{1}{2}\right)^{-4} = \left(\dfrac{2}{1}\right)^4 = 2^4 = 16$

3. $\dfrac{1}{8} = \dfrac{1}{2^3} = 2^{-3}$

4. $f(x) = x^4 + x^3$
$f(-1) = (-1)^4 + (-1)^3 = 1 + (-1) = 0$
$f(3) = (3)^4 + (3)^3 = 81 + 27 = 108$

5. $3x + 7 = x - 5$
$2x + 7 = -5$
$2x = -12$
$x = -6$
The solution is -6.

6. $16 = x^2 - 6x$
$0 = x^2 - 6x - 16$
$0 = (x - 8)(x + 2)$
$x - 8 = 0 \qquad x + 2 = 0$
$x = 8 \qquad\quad x = -2$
The solutions are -2 and 8.

7. $A(1 + r)^n$
$5000(1 + 0.04)^6 = 5000(1.04)^6$
$\qquad\qquad\qquad\qquad = 6326.60$

8. $f(x) = x^2 - 1$
$x = -\dfrac{b}{2a} = \dfrac{0}{2(1)} = \dfrac{0}{2} = 0$
$f(0) = (0)^2 - 1 = -1$

Vertex: $(0, -1)$.
Axis of symmetry: $x = 0$.

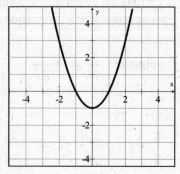

Section 10.1

Objective A Exercises

1. An exponential function with base b is defined by $f(x) = b^x$, $b > 0$, $b \neq 1$, and x is any real number.

3. $f(x) = b^x$, $b > 0$, $b \neq 1$
(iii) cannot be the base.

5. $f(x) = 3^x$
a) $f(2) = 3^2 = 9$
b) $f(0) = 3^0 = 1$
c) $f(-2) = 3^{-2} = \dfrac{1}{3^2} = \dfrac{1}{9}$

7. $g(x) = 2^{x+1}$
a) $g(3) = 2^{3+1} = 2^4 = 16$
b) $g(1) = 2^{1+1} = 2^2 = 4$
c) $g(-3) = 2^{-3+1} = 2^{-2} = \dfrac{1}{2^2} = \dfrac{1}{4}$

9. a) $P(0) = \left(\dfrac{1}{2}\right)^{2(0)} = \left(\dfrac{1}{2}\right)^0 = 1$

b) $P\left(\dfrac{3}{2}\right) = \left(\dfrac{1}{2}\right)^{2(3/2)} = \left(\dfrac{1}{2}\right)^3 = \dfrac{1}{8}$

c) $P(-2) = \left(\dfrac{1}{2}\right)^{2(-2)} = \left(\dfrac{1}{2}\right)^{-4} = 2^4 = 16$

11. $G(x) = e^{x/2}$

   a) $G(4) = e^{4/2} = e^2 = 7.3891$

   b) $G(-2) = e^{-2/2} = e^{-1} = \dfrac{1}{e^1} = 0.3679$

   c) $G\left(\dfrac{1}{2}\right) = e^{(1/2)/2} = e^{1/4} = 1.2840$

13. $H(r) = e^{-r+3}$

   a) $H(-1) = e^{-(-1)+3} = e^4 = 54.5982$

   b) $H(3) = e^{-3+3} = e^0 = 1$

   c) $H(5) = e^{-5+3} = e^{-2} = \dfrac{1}{e^2} = 0.1353$

15. $F(x) = 2^{x^2}$

   a) $F(2) = 2^{2^2} = 2^4 = 16$

   b) $F(-2) = 2^{(-2)^2} = 2^4 = 16$

   c) $F\left(\dfrac{3}{4}\right) = 2^{(3/4)^2} = 2^{9/16} = 1.4768$

17. $f(x) = e^{-x^2/2}$

   a) $f(-2) = e^{-2^2/2} = e^{-2} = \dfrac{1}{e^2} = 0.1353$

   b) $f(2) = e^{-2^2/2} = e^{-2} = \dfrac{1}{e^2} = 0.1353$

   c) $f(-3) = e^{-3^2/2} = e^{-9/2} = \dfrac{1}{e^{9/2}} = 0.0111$

19. $f(a) > f(b)$

Objective B Exercises

21. $f(x) = 3^x$

x	y
0	1
-1	⅓
1	3

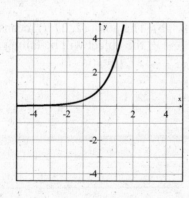

23. $f(x) = 2^{x+1}$

x	y
0	2
-1	1
1	4

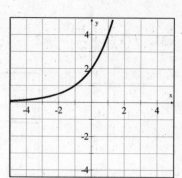

25. $f(x) = \left(\dfrac{1}{3}\right)^x$

x	y
0	1
-1	3
1	⅓

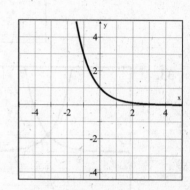

27. $f(x) = 2^{-x} + 1$

x	y
0	2
-1	3
1	³⁄₂

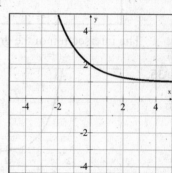

29. $f(x) = \left(\dfrac{1}{3}\right)^{-x}$

x	y
0	1
-1	⅓
1	3

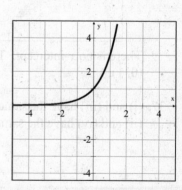

31. $f(x) = \left(\dfrac{1}{2}\right)^{-x} + 2$

x	y
0	3
−1	$\frac{5}{2}$
1	4

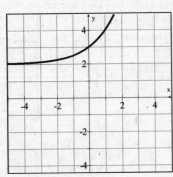

33.

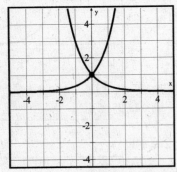

The intersection of the graphs is (0, 1).

35.

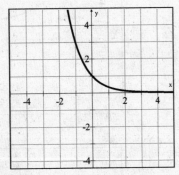

There is no x-intercept. The y-intercept is (0, 1).

37. (i) and (iii) have the same graphs.
(ii) and (iv) have the same graphs.

Applying the Concepts

39. $P(x) = \left(\sqrt{3}\right)^{x}$

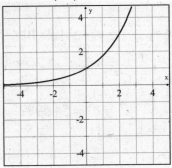

41. $f(x) = \pi^{x}$

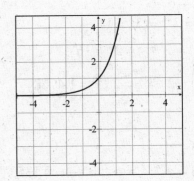

43. a)

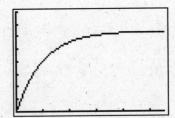

Xmin = 0, Xmax = 5.5, Xscl = 1
Ymin = 0, Ymax = 40, Yscl = 5

b) The point (2, 27.7) means that after 2 s the object will be falling at a speed of 27.7 ft/s.

Section 10.2

Objective A Exercises

1. a) A common logarithm is a logarithm with a base of 10.

   b) $\log 4z$

3. $\log_5 25 = 2$

5. $\log_4 \dfrac{1}{16} = -2$

7. $\log_{10} x = y$

9. $\log_a w = x$

11. $3^2 = 9$

13. $10^{-2} = 0.01$

15. $e^y = x$

17. $b^v = u$

19. $\log_3 81 = x$
    $3^x = 81$
    $x = 4$
    $\log_3 81 = 4$

21. $\log_2 128 = x$
    $2^x = 128$
    $x = 7$
    $\log_2 128 = 7$

23. $\log 100 = x$
    $10^x = 100$
    $x = 2$
    $\log 100 = 2$

25. $\ln e^3 = x$
    $3 \ln e = x$
    $3(1) = x$
    $x = 3$
    $\ln e^3 = 3$

27. $\log_8 1 = x$
    $8^x = 1$
    $x = 0$
    $\log_8 1 = 0$

29. $\log_5 625 = x$
    $5^x = 625$
    $x = 4$
    $\log_5 625 = 4$

31. $\log_3 x = 2$
    $3^2 = x$
    $x = 9$

33. $\log_4 x = 3$
    $4^3 = x$
    $x = 64$

35. $\log_7 x = -1$
    $7^{-1} = x$
    $x = \dfrac{1}{7}$

37. $\log_6 x = 0$
    $6^0 = x$
    $x = 1$

39. $\log x = 2.5$
    $10^{2.5} = x$
    $x = 316.23$

41. $\log x = -1.75$
    $10^{-1.75} = x$
    $x = 0.02$

43. $\ln x = 2$

$e^2 = x$

$x = 7.39$

45. $\ln x = -\dfrac{1}{2}$

$e^{-1/2} = x$

$x = 0.61$

47. $x > 1$

Objective B Exercises

49. $\log_b(xy) = \log_b(x) + \log_b(y)$

51. False

53. Logarithm of One Property

$\log_{12} 1 = 0$

55. Logarithm of the Base Property

$\ln e = 1$

57. Inverse Property of Logarithms

$\log_3 3^x = x$

59. Inverse Property of Logarithms

$e^{\ln v} = v$

61. Inverse Property of Logarithms

$2^{\log_2(x^2+1)} = x^2 + 1$

63. Inverse Property of Logarithms

$\log_5 5^{x^2-x-1} = x^2 - x - 1$

65. $\log_8(xz) = \log_8 x + \log_8 z$

67. $\log_3 x^5 = 5\log_3 x$

69. $\log_b\left(\dfrac{r}{s}\right) = \log_b r - \log_b s$

71. $\log_3(x^2 y^6) = \log_3 x^2 + \log_3 y^6$

$= 2\log_3 x + 6\log_3 y$

73. $\log_7\left(\dfrac{u^3}{v^4}\right) = \log_7 u^3 - \log_7 v^4$

$= 3\log_7 u - 4\log_7 v$

75. $\log_2(rs)^2 = 2\log_2(rs) = 2[\log_2 r + \log_2 s]$

77. $\ln(x^2 yz) = \ln x^2 + \ln y + \ln z$

$= 2\ln x + \ln y + \ln z$

79. $\log_5\left(\dfrac{xy^2}{z^4}\right) = \log_5 xy^2 - \log_5 z^4$

$= \log_5 x + \log_5 y^2 - \log_5 z^4$

$= \log_5 x + 2\log_5 y - 4\log_5 z$

81. $\log_8\left(\dfrac{x^2}{yz^2}\right) = \log_8 x^2 - \log_8 yz^2$

$= \log_8 x^2 - (\log_8 y + \log_8 z^2)$

$= \log_8 x^2 - \log_8 y - \log_8 z^2$

$= 2\log_8 x - \log_8 y - 2\log_8 z$

83. $\log_4 \sqrt{x^3 y} = \log_4 (x^3 y)^{1/2} = \dfrac{1}{2}\log_4 (x^3 y)$

$= \dfrac{1}{2}[\log_4 x^3 + \log_4 y]$

$= \dfrac{1}{2}[3\log_4 x + \log_4 y]$

$= \dfrac{3}{2}\log_4 x + \dfrac{1}{2}\log_4 y$

85. $\log_7 \sqrt{\dfrac{x^3}{y}} = \log_7 \left(\dfrac{x^3}{y}\right)^{1/2} = \dfrac{1}{2}\log_7 \dfrac{x^3}{y}$

$= \dfrac{1}{2}[\log_7 x^3 - \log_7 y]$

$= \dfrac{1}{2}[3\log_7 x - \log_7 y]$

$= \dfrac{3}{2}\log_7 x - \dfrac{1}{2}\log_7 y$

87. $\log_3\left(\dfrac{t}{\sqrt{x}}\right) = \log_3\left(\dfrac{t}{x^{1/2}}\right)$

$= \log_3 t - \log_3 x^{1/2}$

$= \log_3 t - \dfrac{1}{2}\log_3 x$

89. $\log_3 x^3 + \log_3 y^2 = \log_3(x^3 y^2)$

91. $\ln x^4 - \ln y^2 = \ln\left(\dfrac{x^4}{y^2}\right)$

93. $3\log_7 x = \log_7 x^3$

95. $3\ln x + 4\ln y = \ln x^3 + \ln y^4 = \ln(x^3 y^4)$

97. $2(\log_4 x + \log_4 y) = 2\log_4(xy)$

$= \log_4(xy)^2$

$= \log_4(x^2 y^2)$

99. $2\log_3 x - \log_3 y + 2\log_3 z$

$= \log_3 x^2 - \log_3 y + \log_3 z^2$

$= \log_3\left(\dfrac{x^2}{y}\right) + \log_3 z^2$

$= \log_3\left(\dfrac{x^2 z^2}{y}\right)$

101. $\ln x - (2\ln y + \ln z) = \ln x - (\ln y^2 + \ln z)$

$= \ln x - \ln(y^2 z)$

$= \ln\left(\dfrac{x}{y^2 z}\right)$

103. $\dfrac{1}{2}(\log_6 x - \log_6 y) = \dfrac{1}{2}\log_6\left(\dfrac{x}{y}\right)$

$= \log_6\left(\dfrac{x}{y}\right)^{1/2}$

$= \log_6\sqrt{\dfrac{x}{y}}$

105. $2(\log_4 s - 2\log_4 t + \log_4 r)$

$= 2(\log_4 s - \log_4 t^2 + \log_4 r)$

$= 2\left(\log_4\dfrac{s}{t^2} + \log_4 r\right)$

$= 2\log_4\left(\dfrac{sr}{t^2}\right)$

$= \log_4\left(\dfrac{sr}{t^2}\right)^2$

$= \log_4\dfrac{s^2 r^2}{t^4}$

107. $\ln x - 2(\ln y + \ln z)$

$= \ln x - 2\ln(yz)$

$= \ln x - \ln(yz)^2$

$= \ln\left(\dfrac{x}{(yz)^2}\right)$

$= \ln\dfrac{x}{y^2 z^2}$

109. $\dfrac{1}{2}(3\log_4 x - 2\log_4 y + \log_4 z)$

$= \dfrac{1}{2}(\log_4 x^3 - \log_4 y^2 + \log_4 z)$

$= \dfrac{1}{2}\left(\log_4\left(\dfrac{x^3}{y^2}\right) + \log_4 z\right)$

$= \log_4\left(\dfrac{x^3 z}{y^2}\right)^{1/2}$

$= \log_4\sqrt{\dfrac{x^3 z}{y^2}}$

111. $\dfrac{1}{2}\log_2 x - \dfrac{2}{3}\log_2 y + \dfrac{1}{2}\log_2 z$

$= \log_2 x^{1/2} - \log_2 y^{2/3} + \log_2 z^{1/2}$

$= \log_2\left(\dfrac{x^{1/2}}{y^{2/3}}\right) + \log_5 z^{1/2}$

$= \log_2\left(\dfrac{x^{1/2}z^{1/2}}{y^{2/3}}\right)$

$= \log_2 \dfrac{\sqrt{xy}}{\sqrt[3]{y^2}}$

Objective C Exercises

113. $\log_8 6 = \dfrac{\log_{10} 6}{\log_{10} 8} = 0.8617$

115. $\log_5 30 = \dfrac{\log_{10} 30}{\log_{10} 5} = 2.1133$

117. $\log_3 0.5 = \dfrac{\log_{10} 0.5}{\log_{10} 3} = -0.6309$

119. $\log_7 1.7 = \dfrac{\log_{10} 1.7}{\log_{10} 7} = 0.2727$

121. $\log_5 15 = \dfrac{\log_{10} 15}{\log_{10} 5} = 1.6826$

123. $\log_{12} 120 = \dfrac{\log_{10} 120}{\log_{10} 12} = 1.9266$

125. $\log_4 2.55 = \dfrac{\log_{10} 2.55}{\log_{10} 4} = 0.6752$

127. $\log_5 67 = \dfrac{\log_{10} 67}{\log_{10} 5} = 2.6125$

129. $\log_5 x = \dfrac{\log_{10} x}{\log_{10} 5}$

Applying the Concepts

131.   a) False

$3^{-2} = \dfrac{1}{3^2} = \dfrac{1}{9} \neq -9$

b) True

c) False

$\log x^{-1} = -\log x \neq \dfrac{1}{\log x}$

d) True

e) False

$\log(x \cdot y) = \log x + \log y \neq \log x \cdot \log y$

f) True

Section 10.3

Objective A Exercises

1. They have the same graph.

3. $f(x) = \log_4 x$

$y = \log_4 x$ is equivalent to $x = 4^y$.

x	y
$\frac{1}{16}$	-2
$\frac{1}{4}$	-1
1	0
4	1

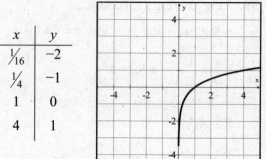

5. $f(x) = \log_3(2x-1)$

$y = \log_3(2x-1)$ is equivalent to

$2x - 1 = 3^y$ or $x = \dfrac{1}{2}(3^y + 1)$.

x	y
$\frac{2}{3}$	-1
1	0
2	1
5	2

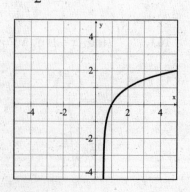

7. $f(x) = 3\log_2 x$

$y = 3\log_2 x$ is equivalent to $x = 2^{y/3}$.

x	y
$\frac{1}{2}$	-3
1	0
2	3
4	6

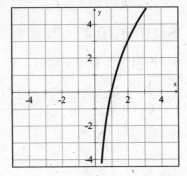

9. $f(x) = -\log_2 x$

$y = -\log_2 x$ is equivalent to $x = 2^{-y}$.

x	y
2	-1
1	0
$\frac{1}{2}$	1
$\frac{1}{4}$	2

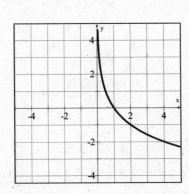

11. $f(x) = \log_2(x-1)$

$y = \log_2(x-1)$ is equivalent to $x - 1 = 2^y$

or $x = 2^y + 1$.

x	y
$\frac{3}{2}$	-1
2	0
3	1
5	2

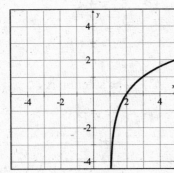

13. $f(x) = -\log_2(x-1)$

$y = -\log_2(x-1)$ is equivalent to

$x - 1 = 2^{-y}$ or $x = 2^{-y} + 1$.

x	y
3	-1
2	0
$\frac{3}{2}$	1
$\frac{5}{4}$	2

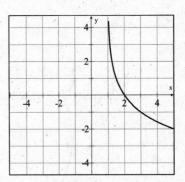

15. They intersect at the point (1, 0).

Applying the Concepts

17. $f(x) = x - \log_2(1-x)$

$y = x - \log_2(1-x)$

$y = x - \dfrac{\log(1-x)}{\log 2}$

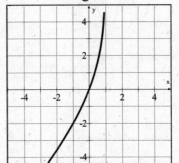

19. $f(x) = \dfrac{x}{2} - 2\log_2(x+1)$

$y = \dfrac{x}{2} - 2\log_2(x+1)$

$y = \dfrac{x}{2} - \log_2(x+1)^2$

$y = \dfrac{x}{2} - \dfrac{\log(x+1)^2}{\log 2}$

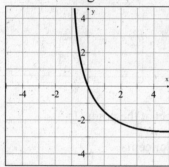

21. $f(x) = x^2 - 10\ln(x-1)$

$y = x^2 - 10\ln(x-1)$

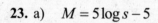

23. a) $M = 5\log s - 5$

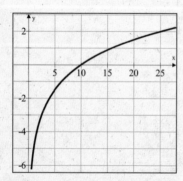

b) The point (25.1, 2) means that a star that is 25.1 parsecs from Earth has a distance modulus of 2.

Section 10.4

Objective A Exercises

1. An exponential equation is one in which a variable occurs in an exponent.

3. $x < 0$

5. $5^{4x-1} = 5^{x-2}$

$4x - 1 = x - 2$

$3x - 1 = -2$

$3x = -1$

$x = -\dfrac{1}{3}$

The solution is $-\dfrac{1}{3}$.

7. $8^{x-4} = 8^{5x+8}$

$x - 4 = 5x + 8$

$-4 = 4x + 8$

$4x = -12$

$x = -3$

The solution is -3.

9. $9^x = 3^{x+1}$

$3^{2x} = 3^{x+1}$

$2x = x + 1$

$x = 1$

The solution is 1.

11. $8^{x+2} = 16^x$

$(2^3)^{x+2} = 2^{4x}$

$2^{3x+6} = 2^{4x}$

$3x + 6 = 4x$

$x = 6$

The solution is 6.

13. $16^{2-x} = 32^{2x}$

$(2^4)^{2-x} = (2^5)^{2x}$

$2^{8-4x} = 2^{10x}$

$8 - 4x = 10x$

$8 = 14x$

$x = \dfrac{8}{14} = \dfrac{4}{7}$

The solution is $\dfrac{4}{7}$.

15. $25^{3-x} = 125^{2x-1}$

$(5^2)^{3-x} = (5^3)^{2x-1}$

$5^{6-2x} = 5^{6x-3}$

$6 - 2x = 6x - 3$

$6 = 8x - 3$

$9 = 8x$

$x = \dfrac{9}{8}$

The solution is $\dfrac{9}{8}$.

17. $5^x = 6$

$\log 5^x = \log 6$

$x \log 5 = \log 6$

$x = \dfrac{\log 6}{\log 5}$

$x = 1.1133$

The solution is 1.1133.

19. $8^{x/4} = 0.4$

$\log 8^{x/4} = \log 0.4$

$\dfrac{x}{4} \log 8 = \log 0.4$

$\dfrac{x}{4} = \dfrac{\log 0.4}{\log 8}$

$x = 4 \cdot \dfrac{\log 0.4}{\log 8}$

$x = -1.7626$

The solution is -1.7626.

21. $2^{3x} = 5$

$\log 2^{3x} = \log 5$

$3x \log 2 = \log 5$

$3x = \dfrac{\log 5}{\log 2}$

$3x = 2.3219$

$x = 0.7740$

The solution is 0.7740.

23. $2^{-x} = 7$

$\log 2^{-x} = \log 7$

$-x \log 2 = \log 7$

$-x = \dfrac{\log 7}{\log 2}$

$-x = 2.8074$

$x = -2.8074$

The solution is -2.8074.

25. $2^{x-1} = 6$

$\log 2^{x-1} = \log 6$

$(x-1) \log 2 = \log 6$

$x - 1 = \dfrac{\log 6}{\log 2}$

$x = \dfrac{\log 6}{\log 2} + 1$

$x = 3.5850$

The solution is 3.5850.

27. $3^{2x-1} = 4$

$\log 3^{2x-1} = \log 4$

$(2x-1) \log 3 = \log 4$

$2x - 1 = \dfrac{\log 4}{\log 3}$

$2x - 1 = 1.2619$

$2x = 2.2619$

$x = 1.1309$

The solution is 1.1309.

29. $\left(\dfrac{1}{2}\right)^{x+1} = 3$

$\log\left(\dfrac{1}{2}\right)^{x+1} = \log 3$

$(x+1)\log\left(\dfrac{1}{2}\right) = \log 3$

$x + 1 = \dfrac{\log 3}{\log \dfrac{1}{2}}$

$x = \dfrac{\log 3}{\log \dfrac{1}{2}} - 1$

$x = -2.5850$

The solution is -2.5850.

31. $3 \cdot 2^x = 7$

$\log(3 \cdot 2^x) = \log 7$

$\log 3 + \log 2^x = \log 7$

$\log 3 + x \log 2 = \log 7$

$x \log 2 = \log 7 - \log 3$

$x = \dfrac{\log 7 - \log 3}{\log 2}$

$x = 1.2224$

The solution is 1.2224.

33. $7 = 10\left(\dfrac{1}{2}\right)^{x/8}$

$\log 7 = \log 10 \left(\dfrac{1}{2}\right)^{x/8}$

$\log 7 = \log 10 + \log\left(\dfrac{1}{2}\right)^{x/8}$

$\log 7 = \log 10 + \dfrac{x}{8}\log\dfrac{1}{2}$

$\log 7 - \log 10 = \dfrac{x}{8}\log\dfrac{1}{2}$

$\dfrac{\log 7 - \log 10}{\log \dfrac{1}{2}} = \dfrac{x}{8}$

$0.5146 = \dfrac{x}{8}$

$4.1166 = x$

The solution is 4.1166.

35. $15 = 12(e)^{0.05x}$

$\ln 15 = \ln 12(e)^{0.05x}$

$\ln 15 = \ln 12 + \ln(e)^{0.05x}$

$\ln 15 = \ln 12 + 0.05x$

$\ln 15 - \ln 12 = 0.05x$

$0.2231 = 0.05x$

$4.4629 = x$

The solution is 4.4629.

Objective B Exercises

37. $\log x = \log(1 - x)$

$x = 1 - x$

$2x = 1$

$x = \dfrac{1}{2}$

The solution is $\dfrac{1}{2}$.

39. $\ln(3x + 2) = \ln(5x + 4)$

$3x + 2 = 5x + 4$

$-2x = 2$

$x = -1$

When we substitute $x = -1$ in either side of the equation, we get a logarithm of a negative number.

Because the logarithm of a negative number is not a real number, there is no solution.

41. $\log_2(8x) - \log_2(x^2 - 1) = \log_2 3$

$\log_2 \dfrac{8x}{x^x - 1} = \log_2 3$

$\dfrac{8x}{x^2 - 1} = 3$

$(x^2 - 1)\dfrac{8x}{x^2 - 1} = 3(x^2 - 1)$

$8x = 3x^2 - 3$

$0 = 3x^2 - 8x - 3$

$0 = (3x + 1)(x - 3)$

$3x + 1 = 0 \qquad x - 3 = 0$

$\quad 3x = -1 \qquad x = 3$

$\qquad x = -\dfrac{1}{3}$

$-\dfrac{1}{3}$ does not check as a solution.

The solution is 3.

43. $\log_9 x + \log_9(2x - 3) = \log_9 2$

$\log_9(x(2x - 3)) = \log_9 2$

$x(2x - 3) = 2$

$2x^2 - 3x = 2$

$2x^2 - 3x - 2 = 0$

$(2x + 1)(x - 2) = 0$

$2x + 1 = 0 \qquad x - 2 = 0$

$\quad 2x = -1 \qquad x = 2$

$\qquad x = -\dfrac{1}{2}$

$-\dfrac{1}{2}$ does not check as a solution.

The solution is 2.

45. $\log_2(2x - 3) = 3$

$2x - 3 = 2^3$

$2x - 3 = 8$

$2x = 11$

$x = \dfrac{11}{2}$

The solution is $\dfrac{11}{2}$.

47. $\ln(3x + 2) = 4$

$3x + 2 = e^4$

$3x + 2 = 54.5982$

$3x = 52.5982$

$x = 17.5327$

The solution is 17.5327.

49. $\log_2(x + 1) + \log_2(x + 3) = 3$

$\log_2((x + 1)(x + 3)) = 3$

$\log_2(x^2 + 4x + 3) = 3$

$x^2 + 4x + 3 = 2^3$

$x^2 + 4x + 3 = 8$

$x^2 + 4x - 5 = 0$

$(x + 5)(x - 1) = 0$

$x + 5 = 0 \qquad x - 1 = 0$

$\quad x = -5 \qquad x = 1$

-5 does not check as a solution.

The solution is 1.

51. $\log_5(2x) - \log_5(x-1) = 1$

$$\log_5 \frac{2x}{x-1} = 1$$

$$\frac{2x}{x-1} = 5^1$$

$$(x-1)\frac{2x}{x-1} = 5(x-1)$$

$$2x = 5x - 5$$

$$-3x = -5$$

$$x = \frac{-5}{-3} = \frac{5}{3}$$

The solution is $\frac{5}{3}$.

53. $\log_8(6x) = \log_8 2 + \log_8(x-4)$

$$\log_8(6x) = \log_8(2(x-4))$$

$$6x = 2x - 8$$

$$4x = -8$$

$$x = -2$$

-2 does not check as a solution. The equation has no solution.

55. $x - 2 < x$ and therefore $\log(x - 2) < \log x$. This means that $\log(x-2) - \log x < 0$ and could not equal the positive number 3.

Applying the Concepts

57. $3^{x+1} = 2^{x-2}$

$$\log 3^{x+1} = \log 2^{x-2}$$

$$(x+1)\log 3 = (x-2)\log 2$$

$$x+1 = \frac{\log 2}{\log 3}(x-2)$$

$$x+1 = 0.6309297536(x-2)$$

$$x+1 = 0.6309297536x - 1.261859507$$

$$0.3690702464x = -2.261859507$$

$$x = -6.1285$$

The solution is -6.1285.

59. $7^{2x-1} = 3^{2x+3}$

$$\log 7^{2x-1} = \log 3^{2x+3}$$

$$(2x-1)\log 7 = (2x+3)\log 3$$

$$2x-1 = \frac{\log 3}{\log 7}(2x+3)$$

$$2x-1 = 0.5645750341(2x+3)$$

$$2x-1 = 1.129150068x + 1.693725102$$

$$0.870849932x = 2.693725102$$

$$x = 3.0932$$

The solution is 3.0935

61. a) $s = 312.5\ln\dfrac{e^{0.32t} + e^{-0.32t}}{2}$

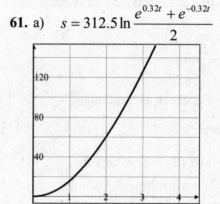

b) Use a graphing calculator to find t when $s = 100$.

$t \approx 2.64$

It will take 2.64s for the object to fall 100 ft.

Section 10.5

Objective A Exercises

1. Strategy: To find the value of the investment, use the compound interest formula.

$$P = 1000, \; n = 8, \; i = \frac{8\%}{4} = \frac{0.08}{4} = 0.02$$

Solution: $A = P(1+i)^n$

$$A = 1000(1+0.02)^8$$

$$A = 1000(1.02)^8$$

$$A \approx 1171.66$$

The value of the investment after 2 years is $1172.

3. **Strategy**: To find how many years it will take for the investment to be worth $15,000, solve the compound interest formula for n.

$$A = 15{,}000, \ P = 5000, \ i = \frac{6\%}{12} = \frac{0.06}{12} = 0.005$$

Solution: $A = P(1+i)^n$

$$15000 = 5000(1+0.005)^n$$

$$3 = (1.005)^n$$

$$\log 3 = \log(1.005)^n$$

$$\log 3 = n \log 1.005$$

$$\frac{\log 3}{\log 1.005} = n$$

$$n \approx 220$$

$$\frac{n}{12} = \frac{220}{12} \approx 18$$

The investment will be worth $15,000 in approximately 18 years.

5. a) **Strategy**: To find the technetium level, use the exponential decay formula.

$$A_0 = 30, \ k = 6, \ t = 3$$

Solution: $A = A_0(0.5)^{t/k}$

$$A = 30(0.5)^{3/6}$$

$$A \approx 21.2$$

The technetium level is 21.2 mg after 3 h.

b) **Strategy**: To find out how long it will take the technetium level to reach 20 mg, use the exponential decay formula.

$$A_0 = 30, \ A = 20, \ k = 6$$

Solution: $A = A_0(0.5)^{t/k}$

$$20 = 30(0.5)^{t/6}$$

$$\frac{2}{3} = 0.5^{t/6}$$

$$\log \frac{2}{3} = \log 0.5^{t/6}$$

$$\log \frac{2}{3} = \frac{t}{6} \log 0.5$$

$$\frac{6 \log \frac{2}{3}}{\log 0.5} = t$$

$$t \approx 3.5$$

The technetium level is 20 mg after 3.5 h.

7. **Strategy**: To find the half life, use the exponential decay formula.

$$A_0 = 25, \ A = 18.95, \ t = 1$$

Solution: $A = A_0(0.5)^{t/k}$

$$18.95 = 25(0.5)^{1/k}$$

$$0.758 = 0.5^{1/k}$$

$$\log 0.758 = \log 0.5^{1/k}$$

$$\log 0.758 = \frac{1}{k} \log 0.5$$

$$k \log 0.758 = \log 0.5$$

$$k = \frac{\log 0.5}{\log 0.758}$$

$$k \approx 2.5$$

The half life is 2.5 years.

9. **Strategy**: To determine the intensity of the earthquake, use the Richter scale equation.

$$M = 8.9$$

Solution: $M = \log \dfrac{I}{I_0}$

$$8.9 = \log \frac{I}{I_0}$$

$$10^{8.9} = \frac{I}{I_0}$$

$$I = 10^{8.9} I_0$$

$$I \approx 794{,}328{,}235 I_0$$

The intensity of the earthquake was $794{,}328{,}235 I_0$.

11. Strategy: To determine the how many times stronger the Honshu earthquake was, use the Richter scale equation.
$M_1 = 6.9$ $M_2 = 6.4$

Solution: $M = \log \dfrac{I}{I_0} = \log I - \log I_0$

$6.9 = \log I_1 - \log I_0$

$6.4 = \log I_2 - \log I_0$

Subtract the equations.

$0.5 = \log I_1 - \log I_2$

$0.5 = \log \dfrac{I_1}{I_2}$

$10^{0.5} = \dfrac{I_1}{I_2}$

$I_1 = 10^{0.5} I_2$

$I_1 \approx 3.2 I_2$

The Honshu earthquake was 3.2 times stronger than the Quetta earthquake.

13. Strategy: To determine the magnitude of the earthquake for the seismogram given, use the given equation.
$A = 23$ $t = 24$

Solution: $M = \log A + 3\log 8t - 2.92$

$M = \log 23 + 3\log 8(24) - 2.92$

$M = \log 23 + 3\log 192 - 2.92$

$M \approx 5.3$

The magnitude of the earthquake was 5.3.

15. Strategy: To determine the magnitude of the earthquake for the seismogram given, use the given equation.
$A = 28$ $t = 28$

Solution: $M = \log A + 3\log 8t - 2.92$

$M = \log 28 + 3\log 8(28) - 2.92$

$M = \log 28 + 3\log 224 - 2.92$

$M \approx 5.6$

The magnitude of the earthquake was 5.6.

17. Strategy: To find the pH, replace H^+ with its given value and solve for pH.

Solution: $pH = -\log(H^+)$

$pH = -\log(3.98 \times 10^{-9})$

$pH \approx 8.4$

The pH of baking soda is 8.4.

19. Strategy: To find the hydrogen ion concentration, replace pH with its given value and solve for H^+.

Solution: $pH = -\log(H^+)$

$5.3 < -\log(H^+) < 6.6$

$-5.3 > \log(H^+) > -6.6$

$10^{-5.3} > H^+ > 10^{-6.6}$

$5 \times 10^{-6} > H^+ > 2.5 \times 10^{-7}$

The range of hydrogen ion concentration for peanuts is 2.5×10^{-7} to 5.0×10^{-6}

21. Strategy: To find the number of decibels, replace I with its given value in the equation and solve for D.

Solution: $D = 10(\log I + 16)$

$D = 10(\log(630) + 16)$

$D \approx 10(18.7993)$

$D = 187.993$

The blue whale sounds emit 188 decibels.

23. Strategy: To find the intensity, replace D with its given value in the equation and solve for I.

Solution: $D = 10(\log I + 16)$

$25 = 10(\log(I) + 16)$

$25 = 10\log I + 160$

$-135 = 10\log I$

$-13.5 = \log I$

$10^{-13.5} = I$

$I \approx 3.16 \times 10^{-14}$

The intensity is 3.16×10^{-14} watts/cm².

25. Strategy: To find the percent solve the equation for P.

$d = 0.005 \quad k = 20$

Solution: $\log P = -kd$

$\log P = -20(0.005)$

$\log P = -0.1$

$P = 10^{-0.1}$

$P \approx 0.7943$

79.4% of the light will pass through the glass.

27. Strategy: To find the thickness of copper needed, replace I and I_o with the given values then solve for x. $I = 0.25 \quad I_0 = 1$

Solution: $I = I_0 e^{-3.2x}$

$0.25 = e^{-3.2x}$

$\ln 0.25 = \ln e^{-3.2x}$

$-1.39 \approx -3.2x$

$x \approx 0.43$

The thickness of the copper is 0.4 cm.

29. a) Strategy: Evaluate the given function at $x = 375$ ft.

$$f(x) = \left(\frac{0.5774v + 155.3}{v}\right)x + 565.3\ln\left(\frac{v - 0.2747x}{v}\right) + 3.5$$

Solution:

$$f(375) = \left(\frac{0.5774(160) + 155.3}{160}\right)375 + 565.3\ln\left(\frac{160 - 0.2747(375)}{160}\right) + 3.5$$

$$f(375) = \left(\frac{247.684}{160}\right)375 + 565.3\ln\left(\frac{56.9875}{160}\right) + 3.5$$

$$f(375) \approx 580.5094 - 583.5829 + 3.5$$

$$f(375) \approx 0.43$$

The ball will hit 0.43 ft from the bottom of the fence.

b) Strategy: Increase the speed by 4% so that $v = 166.4$ ft/s.

$$f(375) = \left(\frac{0.5774(166.4) + 155.3}{166.4}\right)375 + 565.3\ln\left(\frac{166.4 - 0.2747(375)}{166.4}\right) + 3.5$$

$$f(375) = 566.5100 - 545.5868 + 3.5$$

$$f(375) = 24.4$$

The height of the ball is 24.4 feet so it will clear the 15-foot fence by approximately 9 ft.

31. Strategy: Determine the value for x for which $h(x) = 0$ (or when it hits the ground).

$$h(x) = \left(\frac{-21.33}{v^2}\right)x^2 + 0.5774x + 3.5$$

Solution:

$$0 = \left(\frac{-21.33}{v^2}\right)x^2 + 0.5774x + 3.5$$

$$x \approx 699$$

Use a graphing calculator to determine the ball hits the ground at $x = 699$ ft if air resistance is ignored.
This is 324 ft further than the ball would travel if we did not ignore air resistance.

Applying the Concepts

33. $A = A_0(0.5)^{t/713,000,000}$

$$\frac{A}{A_0} = (0.5)^{4,280,000,000/713,000,000}$$

$$\frac{A}{A_0} = (0.5)^{4280/713}$$

$$\frac{A}{A_0} \approx 0.016$$

35. a) Strategy: To find the value of the investments after 3 years, use the given equation.

Solution: $A = A_0 e^{rt}$

$$A = 5000e^{(0.06)(3)}$$

$$A = 5000e^{0.18}$$

$$A \approx 5986.09$$

The value of the investment will be worth $5986.09.

b) Strategy: To find the interest rate needed to grow an investment from $1000 to $1250 in 2 years, use the given equation.

Solution: $A = A_0 e^{rt}$

$$1250 = 1000e^{2r}$$

$$1.25 = e^{2r}$$

$$\ln 1.25 = \ln e^{2r}$$

$$0.2231 = 2r$$

$$r \approx 0.112$$

The rate must be 11.2%.

Chapter 10 Review Exercises

1. $f(2) = e^{2-2} = e^0 = 1$

2. $5^2 = 25$

3. $f(x) = 3^{-x} + 2$

x	y
0	3
−1	5
1	$\frac{7}{3}$
2	$\frac{19}{9}$

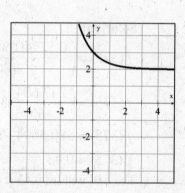

4. $f(x) = \log_3(x - 1)$

$y = \log_3(x - 1)$ is equivalent to

$x - 1 = 3^y$ or $x = 3^y + 1$.

x	y
$\frac{4}{3}$	−1
2	0
4	1

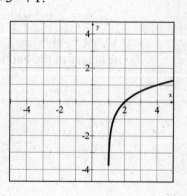

5. $\log_3 \sqrt[5]{x^2 y^4} = \log_3 (x^2 y^4)^{1/5} = \dfrac{1}{5} \log_3 (x^2 y^4)$

$\quad = \dfrac{1}{5}[\log_3 x^2 + \log_3 y^4]$

$\quad = \dfrac{1}{5}[2\log_3 x + 4\log_3 y]$

$\quad = \dfrac{2}{5}\log_3 x + \dfrac{4}{5}\log_3 y$

6. $2\log_3 x - 5\log_3 y = \log_3 x^2 - \log_3 y^5$

$\quad = \log_3\left(\dfrac{x^2}{y^5}\right)$

7. $\quad 27^{2x+4} = 81^{x-3}$

$\quad (3^3)^{2x+4} = (3^4)^{x-3}$

$\quad 6x + 12 = 4x - 12$

$\quad 2x = -24$

$\quad x = -12$

The solution is -12.

8. $\log_5 \dfrac{7x+2}{3x} = 1$

Rewrite in exponential form.

$\quad 5^1 = \dfrac{7x+2}{3x}$

$\quad 15x = 7x + 2$

$\quad 8x = 2$

$\quad x = \dfrac{1}{4}$

The solution is $\dfrac{1}{4}$.

9. $\log_6 22 = \dfrac{\log_{10} 22}{\log_{10} 6} \approx 1.7251$

10. $\log_2 x = 5$

Rewrite in exponential form.

$\quad 2^5 = x$

$\quad x = 32$

The solution is 32.

11. $\log_3 (x+2) = 4$

Rewrite in exponential form.

$\quad 3^4 = x + 2$

$\quad 81 = x + 2$

$\quad x = 79$

The solution is 79.

12. $\log_{10} x = 3$

Rewrite in exponential form.

$\quad 10^3 = x$

$\quad 1x = 1000$

The solution is 1000.

13. $\dfrac{1}{3}(\log_7 x + 4\log_7 y) = \dfrac{1}{3}(\log_7 x + \log_7 y^4)$

$\quad = \dfrac{1}{3}(\log_7 (xy^4)) = \log_7 (xy^4)^{1/3}$

$\quad = \log_7 \sqrt[3]{xy^4}$

14. $\log_8 \sqrt{\dfrac{x^5}{y^3}} = \log_8 \left(\dfrac{x^5}{y^3}\right)^{1/2}$

$\quad = \dfrac{1}{2}(\log_8 x^5 - \log_8 y^3)$

$\quad = \dfrac{5}{2}\log_8 x - \dfrac{3}{2}\log_8 y$

15. $\log_2 32 = 5$

16. $\log_3 1.6 = \dfrac{\log_{10} 1.6}{\log_{10} 3} \approx 0.4278$

17. $\quad 3^{x+2} = 5$

$\quad \log 3^{x+2} = \log 5$

$\quad (x+2)\log 3 = \log 5$

$\quad x + 2 = \dfrac{\log 5}{\log 3}$

$\quad x = \dfrac{\log 5}{\log 3} - 2$

$\quad x = -0.535$

The solution is -0.535.

18. $f(-3) = \left(\dfrac{2}{3}\right)^{-3+2} = \left(\dfrac{2}{3}\right)^{-1} = \dfrac{3}{2}$

19. $\log_2(x+3) - \log_2(x-1) = 3$

$\log_2 \dfrac{x+3}{x-1} = 3$

$\dfrac{x+3}{x-1} = 2^3$

$\dfrac{x+3}{x-1} = 8$

$(x-1)\dfrac{x+3}{x-1} = 8(x-1)$

$x+3 = 8x - 8$

$3 = 7x - 8$

$11 = 7x$

$x = \dfrac{11}{7}$

The solution is $\dfrac{11}{7}$.

20. $\log_3(2x+3) + \log_3(x-2) = 2$

$\log_3((2x+3)(x-2)) = 2$

$2x^2 - x - 6 = 3^2$

$2x^2 - x - 6 = 9$

$2x^2 - x - 15 = 0$

$(2x+5)(x-3) = 0$

$2x + 5 = 0 \quad x - 3 = 0$

$2x = -5 \qquad x = 3$

$x = -\dfrac{5}{2}$

$-\dfrac{5}{2}$ does not check as a solution.

The solution is 3.

21. $f(x) = \left(\dfrac{2}{3}\right)^{x+1}$

x	y
-1	1
-2	$\frac{3}{2}$
0	$\frac{2}{3}$
1	$\frac{4}{9}$

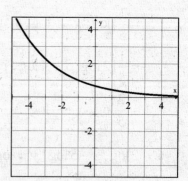

22. $f(x) = \log_2(2x-1)$

$y = \log_2(2x-1)$ is equivalent to

$2x - 1 = 2^y$ or $x = \dfrac{2^y + 1}{2}$

x	y
$\frac{3}{4}$	-1
1	0
$\frac{3}{2}$	1

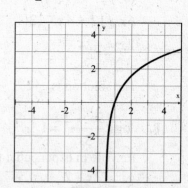

23. $\log_6 36 = x$

$6^x = 36$

$x = 2$

24. $\dfrac{1}{3}(\log_2 x - \log_2 y) = \dfrac{1}{3}\log_2\left(\dfrac{x}{y}\right)$

$= \log_2\left(\dfrac{x}{y}\right)^{1/3} = \log_2 \sqrt[3]{\dfrac{x}{y}}$

25. $9^{2x} = 3^{x+3}$

$(3^2)^{2x} = 3^{x+3}$

$3^{4x} = 3^{x+3}$

$4x = x + 3$

$3x = 3$

$x = 1$

The solution is 1.

26. $5 \cdot 3^{x/2} = 12$

$3^{x/2} = 2.4$

$\log 3^{x/2} = \log 2.4$

$\dfrac{x}{2} \log 3 = \log 2.4$

$\dfrac{x}{2} = \dfrac{\log 2.4}{\log 3}$

$\dfrac{x}{2} \approx 0.7969$

$x = 1.5938$

The solution is 1.5938.

27. $\log_5 x = -1$

Rewrite in exponential form.

$5^{-1} = x$

$\dfrac{1}{5} = x$

The solution is $\dfrac{1}{5}$.

28. $\log_3 81 = 4$

29. $\log x + \log(x - 2) = \log 15$

$\log(x(x - 2)) = \log 15$

$x(x - 2) = 15$

$x^2 - 2x - 15 = 0$

$(x - 5)(x + 3) = 0$

$x - 5 = 0 \quad x + 3 = 0$

$x = 5 \qquad x = -3$

-3 does not check as a solution.

The solution is 5.

30. $\log_5 \sqrt[3]{x^2 y} = \log_3 (x^2 y)^{1/3} = \dfrac{1}{3} \log_3 (x^2 y)$

$\quad = \dfrac{1}{3}[\log_3 x^2 + \log_3 y]$

$\quad = \dfrac{1}{3}[2 \log_3 x + \log_3 y]$

$\quad = \dfrac{2}{3} \log_3 x + \dfrac{1}{3} \log_3 y$

31. $\quad 6e^{-2x} = 17$

$\ln 6e^{-2x} = \ln 17$

$\ln 6 + \ln e^{-2x} = \ln 17$

$-2x = \ln 17 - \ln 6$

$-2x = 1.0415$

$x = -0.5207$

The solution is -0.5207

32. $f(-3) = 7^{-3+2} = 7^{-1} = \dfrac{1}{7}$

33. $\log_2 16 = x$

$2^x = 16$

$x = 4$

34. $\log_6 x = \log_6 2 + \log_6(2x - 3)$

$\log_6 x = \log_6(2(2x - 3)$

$x = 2(2x - 3)$

$x = 4x - 6$

$-3x = -6$

$x = 2$

The solution is 2.

35. $\log_2 5 = x$

$x = \dfrac{\log 5}{\log 2} \approx 2.3219$

36. $4^x = 8^{x-1}$

$(2^2)^x = (2^3)^{x-1}$

$2^{2x} = 2^{3x-3}$

$2x = 3x - 3$

$-x = -3$

$x = 3$

The solution is 3.

37. $\log_5 x = 4$

Rewrite in exponential form.

$x = 5^4 = 625$

38. $3\log_b x - 7\log_b y = \log_b x^3 - \log_b y^7$

$= \log_b\left(\dfrac{x^3}{y^7}\right)$

39. $f(x) = 5^{-x-1}$

$f(-2) = 5^{-(-2)-1} = 5^{2-1} = 5^1 = 5$

40. $5^{x-2} = 7$

$\log 5^{x-2} = \log 7$

$(x-2)\log 5 = \log 7$

$x - 2 = \dfrac{\log 7}{\log 5}$

$x = \dfrac{\log 7}{\log 5} + 2$

$x \approx 3.2091$

The solution is 3.2091.

41. Strategy: To find the value of the investment, use the compound interest formula.

$P = 4000, n = 24, i = \dfrac{8\%}{12} = \dfrac{0.08}{12} = 0.00\overline{6}$

Solution: $A = P(1+i)^n$

$A = 4000(1 + 0.00\overline{6})^{24}$

$A = 4000(1.00\overline{6})^{24}$

$A = 4691.55$

The value of the investment after 2 years is $4692.

42. Strategy: To determine the magnitude of the earthquake, use the Richter scale equation.

Solution: $M = \log\dfrac{I}{I_0}$

$M = \log\dfrac{199{,}526{,}232 I_0}{I_0}$

$M = \log 199{,}526{,}232$

$M \cong 8.3$

The magnitude of the earthquake was 8.3.

43. Strategy: To find the half life use the exponential decay formula.

$A_0 = 25, A = 15, t = 20$

Solution: $A = A_0(0.5)^{t/k}$

$15 = 25(0.5)^{20/k}$

$0.6 = 0.5^{20/k}$

$\log 0.6 = \log 0.5^{20/k}$

$\log 0.6 = \dfrac{20}{k}\log 0.5$

$k\log 0.6 = 20\log 0.5$

$k = \dfrac{20\log 0.5}{\log 0.6}$

$k \cong 27$

The half life is 27 days.

44. Strategy: To find the number of decibels replace I with its given value in the equation and solve for D.

Solution: $D = 10(\log I + 16)$

$D = 10(\log(5 \times 10^{-6}) + 16)$

$D \approx 10(10.6990)$

$D = 106.99$

The sound emitted from a busy street corner is 107 decibels.

Chapter 10 Test

1. $f(0) = \left(\dfrac{2}{3}\right)^0 = 1$

2. $f(-2) = 3^{-2+1} = 3^{-1} = \dfrac{1}{3}$

3. $f(x) = 2^x - 3$

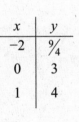

x	y
-2	$-11/4$
0	-2
2	1

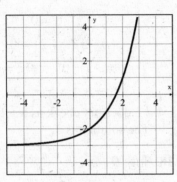

4. $f(x) = 2^x + 2$

x	y
-2	$9/4$
0	3
1	4

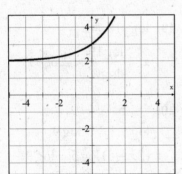

5. $\log_4 16 = x$

$4^x = 16$

$x = 2$

6. $\log_3 x = -2$

$x = 3^{-2} = \dfrac{1}{3^2} = \dfrac{1}{9}$

7. $f(x) = \log_2(2x)$

$y = \log_2(2x)$ is equivalent to

$2x = 2^y$ or $x = \dfrac{2^y}{2}$.

X	y
$1/4$	-1
$1/2$	0
1	1
2	2

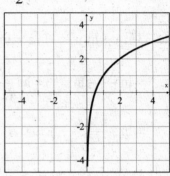

8. $f(x) = \log_3(x+1)$

$y = \log_3(x+1)$ is equivalent to

$x+1 = 3^y$ or $x = 3^y - 1$.

x	y
$-2/3$	-1
0	0
2	1

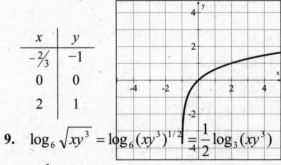

9. $\log_6 \sqrt{xy^3} = \log_6 (xy^3)^{1/2} = \dfrac{1}{2}\log_3(xy^3)$

$= \dfrac{1}{2}[\log_6 x + \log_6 y^3]$

$= \dfrac{1}{2}[\log_6 x + 3\log_6 y]$

$= \dfrac{1}{2}\log_6 x + \dfrac{3}{2}\log_6 y$

10. $\dfrac{1}{2}(\log_3 x - \log_3 y) = \dfrac{1}{2}\log_3\left(\dfrac{x}{y}\right)$

$= \log_3 \sqrt{\dfrac{x}{y}}$

11. $\ln \dfrac{x}{\sqrt{z}} = \ln x - \ln \sqrt{z} = \ln x - \ln z^{1/2}$

$= \ln x - \dfrac{1}{2}\ln z$

12. $3\ln x - \ln y - \dfrac{1}{2}\ln z = \ln x^3 - \ln y - \ln z^{1/2}$

$= \ln \dfrac{x^3}{y} - \ln z^{1/2} = \ln \dfrac{x^3}{yz^{1/2}}$

$= \ln \dfrac{x^3}{y\sqrt{z}}$

13. $3^{7x+1} = 3^{4x-5}$

$7x + 1 = 4x - 5$

$3x = -6$

$x = -2$

The solution is -2.

14. $8^x = 2^{x-6}$

$(2^3)^x = 2^{x-6}$

$3x = x - 6$

$2x = -6$

$x = -3$

The solution is -3.

15. $3^x = 17$

$\log 3^x = \log 17$

$x \log 3 = \log 17$

$x = \dfrac{\log 17}{\log 3}$

$x \approx 2.5789$

The solution is 2.5789.

16. $\log x + \log(x - 4) = \log 12$

$\log(x(x - 4)) = \log 12$

$x(x - 4) = 12$

$x^2 - 4x - 12 = 0$

$(x - 6)(x + 2) = 0$

$x - 6 = 0 \quad x + 2 = 0$

$x = 6 \qquad x = -2$

-2 does not check as a solution.
The solution is 6.

17. $\log_6 x + \log_6(x - 1) = 1$

$\log_6(x(x - 1)) = 1$

$x(x - 1) = 6^1$

$x^2 - x - 6 = 0$

$(x - 3)(x + 2) = 0$

$x - 3 = 0 \quad x + 2 = 0$

$x = 3 \qquad x = -2$

-2 does not check as a solution.
The solution is 3.

18. $\log_5 9 = x$

$x = \dfrac{\log 9}{\log 5} \approx 1.3652$

19. $\log_3 19 = x$

$x = \dfrac{\log 19}{\log 3} \approx 2.6801$

20. $5^{2x-5} = 9$

$\log 5^{2x-5} = \log 9$

$(2x - 5)\log 5 = \log 9$

$2x - 5 = \dfrac{\log 9}{\log 5}$

$2x - 5 \approx 1.3652$

$2x = 6.3652$

$x = 3.1826$

The solution is 3.1826.

21. $2e^{x/4} = 9$

$\ln 2e^{x/4} = \ln 9$

$\ln 2 + \ln e^{x/4} = \ln 9$

$\dfrac{x}{4} = \ln 9 - \ln 2$

$\dfrac{x}{4} \approx 1.5041$

$x = 6.0163$

The solution is 6.0163.

22. $\log_5(30x) - \log_5(x+1) = 2$

$$\log_5 \dfrac{30x}{x+1} = 2$$

$$\dfrac{30x}{x+1} = 5^2$$

$$\dfrac{30x}{x+1} = 25$$

$$(x+1)\dfrac{30x}{x+1} = 25(x+1)$$

$$30x = 25x + 25$$

$$5x = 25$$

$$x = 5$$

The solution is 5.

23. Strategy: To find the approximate age of the shard, use the given equation.

$A_0 = 250,\ A = 170$

Solution: $A = A_0(0.5)^{t/5570}$

$170 = 250(0.5)^{t/5570}$

$0.68 = 0.5^{t/5570}$

$\log 0.68 = \log 0.5^{t/5570}$

$\log 0.68 = \dfrac{t}{5570} \log 0.5$

$\dfrac{5570 \log 0.68}{\log 0.5} = t$

$t \approx 3099$

The shard is approximately 3099 years old.

24. Strategy: To find the intensity, replace D with its given value in the equation and solve for I.

Solution: $D = 10(\log I + 16)$

$75 = 10(\log(I) + 16)$

$75 = 10 \log I + 160$

$-85 = 10 \log I$

$-8.5 = \log I$

$10^{-8.5} = I$

$I \approx 3.16 \times 10^{-9}$

The intensity is 3.16×10^{-9} watts/cm^2.

25. Strategy: To find out the half life of a radioactive material use the exponential decay formula.

$A_0 = 10,\ A = 9,\ t = 5$

Solution: $A = A_0(0.5)^{t/k}$

$9 = 10(0.5)^{5/k}$

$0.9 = 0.5^{5/k}$

$\log 0.9 = \log 0.5^{5/k}$

$\log 0.9 = \dfrac{5}{k} \log 0.5$

$k = \dfrac{5 \log 0.5}{\log 0.9}$

$k \approx 33$

The half life is 33 h.

Cumulative Review Exercises

1. $4 - 2[x - 3(2 - 3x) - 4x] = 2x$

$4 - 2[x - 6 + 9x - 4x] = 2x$

$4 - 2[6x - 6] = 2x$

$4 - 12x + 12 = 2x$

$16 = 14x$

$x = \dfrac{16}{14} = \dfrac{8}{7}$

The solution is $\dfrac{8}{7}$.

2. $2x - y = 5$

$\quad -y = -2x + 5$

$\quad y = 2x - 5$

$m = 2$ and $(2, -2)$

$y - y_1 = m(x - x_1)$

$y - (-2) = 2(x - 2)$

$y + 2 = 2x - 4$

$y = 2x - 6$

3. $4x^{2n} + 7x^n + 3 = (4x^n + 3)(x^n + 1)$

4. $\dfrac{1 - \dfrac{5}{x} + \dfrac{6}{x^2}}{1 + \dfrac{1}{x} - \dfrac{6}{x^2}} = \dfrac{1 - \dfrac{5}{x} + \dfrac{6}{x^2}}{1 + \dfrac{1}{x} - \dfrac{6}{x^2}} \cdot \dfrac{x^2}{x^2}$

$\quad = \dfrac{x^2 - 5x + 6}{x^2 + x - 6} = \dfrac{(x-2)(x-3)}{(x-2)(x+3)}$

$\quad = \dfrac{x-3}{x+3}$

5. $\dfrac{\sqrt{xy}}{\sqrt{x} - \sqrt{y}} = \dfrac{\sqrt{xy}}{\sqrt{x} - \sqrt{y}} \cdot \dfrac{\sqrt{x} + \sqrt{y}}{\sqrt{x} + \sqrt{y}}$

$\quad = \dfrac{\sqrt{x^2 y} + \sqrt{xy^2}}{\sqrt{x^2} - \sqrt{y^2}}$

$\quad = \dfrac{x\sqrt{y} - y\sqrt{x}}{x - y}$

6. $x^2 - 4x - 6 = 0$

$x^2 - 4x + 4 = 6 + 4$

$(x - 2)^2 = 10$

$\sqrt{(x-2)^2} = \sqrt{10}$

$x - 2 = \pm\sqrt{10}$

$x = 2 \pm \sqrt{10}$

The solutions are $2 + \sqrt{10}$ and $2 - \sqrt{10}$.

7. $(x - r_1)(x - r_2) = 0$

$\left(x - \dfrac{1}{3}\right)(x - (-3)) = 0$

$\left(x - \dfrac{1}{3}\right)(x + 3) = 0$

$x^2 + \dfrac{8}{3}x - 1 = 0$

$3x^2 + 8x - 3 = 0$

8. $2x - y < 3 \qquad x + y < 1$

$\quad -y < -2x + 3 \qquad y < -x + 1$

$\quad y > 2x - 3$

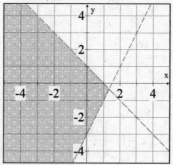

9. (1) $\quad 3x - y + z = 3$

(2) $\quad x + y + 4z = 7$

(3 $\quad 3x - 2y + 3z = 8$

Eliminate y. Add equations (1) and (2).

$3x - y + z = 3$

$\quad x + y + 4z = 7$

(4) $\quad 4x + 5z = 10$

Multiply equation (2) by 2 and add to equation (3).

$\quad 2(x + y + 4z) = 2(7)$

$\quad 3x - 2y + 3z = 8$

$2x + 2y + 8z = 14$

$3x - 2y + 3z = 8$

(5) $\quad 5x + 11z = 22$

Multiply equation (4) by 5 and equation (5) by -4, and add.

$5(4x + 5z) = 5(10)$

$-4(5x + 11z) = -4(22)$

$20x + 25z = 50$

$-20x - 44z = -88$

$\quad\quad\quad -19z = -38$

$\quad\quad\quad\quad z = 2$

Replace z with 2 in equation (4).

$4x + 5(2) = 10$
$\qquad 4x = 0$
$\qquad\quad x = 0$

Replace x with 0 and z with 2 in equation (1).

$3(0) - y + 2 = 3$
$\qquad -y + 2 = 3$
$\qquad\qquad -y = 1$
$\qquad\qquad\quad y = -1$

The solution is $(0, -1, 2)$.

10. $\dfrac{x-4}{2-x} - \dfrac{1-6x}{2x^2 - 7x + 6}$

$= \dfrac{x-4}{2-x} - \dfrac{1-6x}{(2x-3)(x-2)}$

$= \dfrac{x-4}{2-x} + \dfrac{1-6x}{(2x-3)(2-x)}$

$= \dfrac{(x-4)}{(2-x)} \cdot \dfrac{(2x-3)}{(2x-3)} + \dfrac{1-6x}{(2x-3)(2-x)}$

$= \dfrac{2x^2 - 11x + 12 + 1 - 6x}{(2-x)(2x-3)} = \dfrac{2x^2 - 17x + 13}{(2-x)(2x-3)}$

$= -\dfrac{2x^2 - 17x + 13}{(x-2)(2x-3)}$

11. $x^2 + 4x - 5 \leq 0$
$\quad (x+5)(x-1) \leq 0$
$\quad \{x \mid -5 \leq x \leq 1\}$

12. $|2x - 5| \leq 3$
$\qquad -3 \leq 2x - 5 \leq 3$
$\quad -3+5 \leq 2x - 5 + 5 \leq 3 + 5$
$\qquad\quad 2 \leq 2x \leq 8$
$\qquad\quad 1 \leq x \leq 4$
$\quad \{x \mid 1 \leq x \leq 4\}$

13. $f(x) = \left(\dfrac{1}{2}\right)^x + 1$

x	y
-1	3
0	2
2	$5/4$

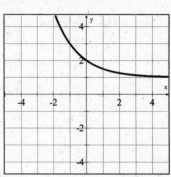

14. $f(x) = \log_2 x - 1$
$\quad y + 1 = \log_2 x$
$\quad x = 2^{y+1}$

x	y
1	-1
2	0
4	1

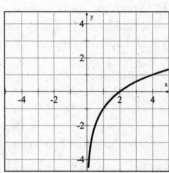

15. $f(-3) = 2^{-(-3)-1} = 2^2 = 4$

16. $\log_5 x = 3$
$\quad x = 5^3 = 125$

17. $3\log_b x - 5\log_b y = \log_b x^3 - \log_b y^5$

$\qquad = \log_b \dfrac{x^3}{y^5}$

18. $\log_3 7 = x$

$\quad x = \dfrac{\log 7}{\log 3} \approx 1.7712$

19. $4^{5x-2} = 4^{3x+2}$
$\quad 5x - 2 = 3x + 2$
$\quad 2x = 4$
$\quad x = 2$
The solution is 2.

20. $\log x + \log(2x + 3) = \log 2$

$\log(x(2x + 3)) = \log 2$

$x(2x + 3) = 2$

$2x^2 + 3x - 2 = 0$

$(2x - 1)(x + 2) = 0$

$2x - 1 = 0 \quad x + 2 = 0$

$2x = 1 \quad\quad x = -2$

$x = \dfrac{1}{2}$

-2 does not check as a solution.

The solution is $\dfrac{1}{2}$.

21. Strategy: Let c represent the number of checks. To find the number of checks, write and solve an inequality.

Solution: $5.00 + 0.02c > 2.00 + 0.08c$

$\quad\quad\quad\quad 5 > 2 + 0.06c$

$\quad\quad\quad\quad 3 > 0.06c$

$\quad\quad\quad\quad 50 > c$

The customer can write at most 49 checks.

22. Strategy: Let x represent the cost per pound of the mixture.

	Amount	Cost	Value
$4.00	16	4.00	16(4.00)
$2.50	24	2.50	24(2.50)
Mixture	40	x	$40x$

The sum of the values before mixing is equal to the value after mixing.

Solution:

$16(4.00) + 24(2.50) = 40x$

$\quad\quad\quad 64 + 60 = 40x$

$\quad\quad\quad\quad\; 124 = 40x$

$\quad\quad\quad\quad\quad\; x = 3.1$

The cost per pound of the mixture is $3.10.

23. Strategy: Let x represent the rate of the wind.

	Distance	Rate	Time
With wind	1000	$225 + x$	$\dfrac{1000}{225 + x}$
Against wind	800	$225 - x$	$\dfrac{800}{225 - x}$

The flying time with the wind is the same as the flying time against the wind.

Solution:

$\dfrac{1000}{225 + x} = \dfrac{800}{225 - x}$

$(225 + x)(225 - x)\dfrac{1000}{225 + x} = \dfrac{800}{225 - x}(225 + x)(225 - x)$

$(225 - x)1000 = 800(225 + x)$

$225000 - 1000x = 180{,}000 + 800x$

$45{,}000 = 1800x$

$25 = x$

The rate of the wind is 25 mph.

24. Strategy: Write the basic direct variation equation replacing the variable with the given values. Solve for k.
Write the direct variation equation replacing k with its value. Substitute 34 for f and solve for d.

Solution:

$d = kf \quad\quad\quad\quad d = 0.3f$

$6 = k(20) \quad\quad\quad\; = 0.3(34)$

$0.3 = k \quad\quad\quad\quad\; = 10.2$

The string will stretch 10.2 in.

25. Strategy: Let x represent the cost of redwood. The cost of fir is y.

First purchase:

	Amount	Cost	Value
Redwood	80	x	$80x$
Fir	140	y	$140y$

Second purchase:

	Amount	Cost	Value
Redwood	140	x	$140x$
Fir	100	y	$100y$

The total cost of the first purchase is $67.
The total cost of the second purchase is $81.

Solution:

$80x + 140y = 67$
$140x + 100y = 81$

$-5(80x + 140y) = -5(67)$
$7(140x + 100y) = 7(81)$

$-400x - 700y = -335$
$980x + 700y = 567$

$580x = 232$
$x = 0.40$

$80(0.40) + 140y = 67$
$32 + 140y = 67$
$140y = 35$
$y = 0.25$

The cost of the redwood is $.40 per foot.
The cost of the fir is $.25 per foot.

26. Strategy: To find how many years it will take for the investment to double in value, use the compound interest formula.

$A = 10,000, P = 5000, i = \dfrac{7\%}{2} = \dfrac{0.07}{2} = 0.035$

Solution: $A = P(1 + i)^n$

$10000 = 5000(1 + 0.035)^n$

$2 = (1.035)^n$

$\log 2 = \log(1.035)^n$

$\log 2 = n \log 1.035$

$\dfrac{\log 2}{\log 1.035} = n$

$n \approx 20$

$\dfrac{n}{2} = \dfrac{20}{2} = 10$

The investment will take 10 years to double in value.

Chapter 11: Conic Sections

Prep Test

1. $d = \sqrt{(x_1 - x_2)^2 + (y_1 - y_2)^2}$

$d = \sqrt{(-2 - 4)^2 + (3 - (-1))^2}$

$= \sqrt{(-6)^2 + (4)^2}$

$= \sqrt{36 + 16}$

$= \sqrt{52} = 7.21$

2. $x^2 - 8x + \left[\dfrac{1}{2}(-8)\right]^2 = x^2 - 8x + (-4)^2$

$= x^2 - 8x + 16$

$= (x - 4)^2$

3. $\dfrac{x^2}{16} + \dfrac{y^2}{9} = 1$

When $y = 3$:

$\dfrac{x^2}{16} + \dfrac{(3)^2}{9} = 1$

$\dfrac{x^2}{16} + 1 = 1$

$\dfrac{x^2}{16} = 0$

$x^2 = 0$

$x = 0$

When $y = 0$:

$\dfrac{x^2}{16} + \dfrac{(0)^2}{9} = 1$

$\dfrac{x^2}{16} + 0 = 1$

$\dfrac{x^2}{16} = 1$

$x^2 = 16$

$x = \pm 4$

4. $7x + 4y = 3$

$\quad y = x - 2$

$7x + 4(x - 2) = 3$

$7x + 4x - 8 = 3$

$11x = 11$

$x = 1$

$y = x - 2 = 1 - 2 = -1$

The solution is $(1, -1)$.

5. (1) $4x - y = 9$

(2) $2x + 3y = -13$

Eliminate y.

$3(4x - y) = 3(9)$

$2x + 3y = -13$

$12x - 3y = 27$

$2x + 3y = -13$

$14x = 14$

$x = 1$

Replace x with 1 in equation (1).

$4(1) - y = 9$

$4 - y = 9$

$-y = 5$

$y = -5$

The solution is $(1, -5)$.

6. $y = x^2 - 4x + 2$

$x = -\dfrac{b}{2a} = -\dfrac{-4}{2(1)} = -(-2) = 2$

$y = (2)^2 - 4(2) + 2 = 4 - 8 + 2 = -2$

The axis of symmetry is $x = 2$.

The vertex is $(2, -2)$.

7. $f(x) = -2x^2 + 4x$

x	y
-1	-6
0	0
1	2
2	0
3	-6

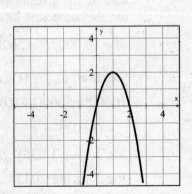

8. $x + 2y \le 4$

$2y \le -x + 4$

$y \le -\dfrac{1}{2}x + 2$

$x - y \le 2$

$-y \le -x + 2$

$y \ge x - 2$

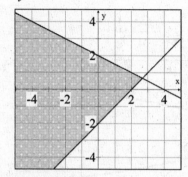

Section 11.1

Objective A Exercises

1. a) The axis of symmetry is a vertical line.
   b) Since $a > 0$, the parabola opens up.

3. a) The axis of symmetry is a horizontal line.
   b) Since $a > 0$, the parabola opens right.

5. a) The axis of symmetry is a horizontal line.
   b) Since $a < 0$, the parabola opens left.

7. $x = y^2 - 3y - 4$

$y = -\dfrac{b}{2a} = -\dfrac{-3}{2(1)} = -\left(-\dfrac{3}{2}\right) = \dfrac{3}{2}$

$x = \left(\dfrac{3}{2}\right)^2 - 3\left(\dfrac{3}{2}\right) - 4 = \dfrac{9}{4} - \dfrac{9}{2} - 4 = -\dfrac{25}{4}$

The axis of symmetry is $y = \dfrac{3}{2}$.

The vertex is $\left(-\dfrac{25}{4}, \dfrac{3}{2}\right)$.

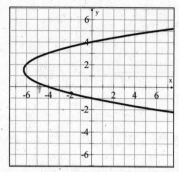

9. $y = x^2 + 2$

$x = -\dfrac{b}{2a} = -\dfrac{0}{2(1)} = 0$

$y = (0)^2 + 2 = 2$

The axis of symmetry is $x = 0$.
The vertex is $(0, 2)$.

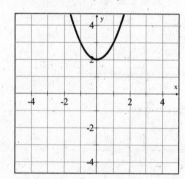

11. $x = -\dfrac{1}{4}y^2 - 1$

$y = -\dfrac{b}{2a} = \dfrac{0}{2\left(-\dfrac{1}{4}\right)} = 0$

$x = -\left(\dfrac{1}{4}\right)0^2 - 1 = -1$

The axis of symmetry is $y = 0$.
The vertex is $(-1, 0)$.

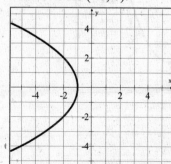

13. $x = -\dfrac{1}{2}y^2 + 2y - 3$

$$y = -\frac{b}{2a} = -\frac{2}{2\left(-\dfrac{1}{2}\right)} = -\frac{2}{-1} = 2$$

$$x = \left(-\frac{1}{2}\right)(2)^2 + 2(2) - 3 = -2 + 4 - 3 = -1$$

The axis of symmetry is $y = 2$.
The vertex is $(-1, 2)$.

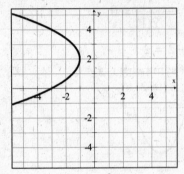

15. $y = \dfrac{1}{2}x^2 + x - 3$

$$x = -\frac{b}{2a} = -\frac{1}{2\left(\dfrac{1}{2}\right)} = -\frac{1}{1} = -1$$

$$y = \frac{1}{2}(-1)^2 + (-1) - 3 = \frac{1}{2} - 1 - 3 = -\frac{7}{2}$$

The axis of symmetry is $x = -1$.

The vertex is $\left(-1, -\dfrac{7}{2}\right)$.

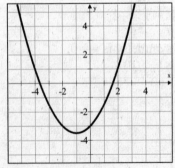

17. $x = y^2 - y - 6$

$$y = -\frac{b}{2a} = -\frac{-1}{2(1)} = -\left(-\frac{3}{2}\right) = \frac{1}{2}$$

$$x = \left(\frac{1}{2}\right)^2 - \left(\frac{1}{2}\right) - 6 = \frac{1}{4} - \frac{1}{2} - 6 = -\frac{25}{4}$$

The axis of symmetry is $y = \dfrac{1}{2}$.

The vertex is $\left(-\dfrac{25}{4}, \dfrac{1}{2}\right)$.

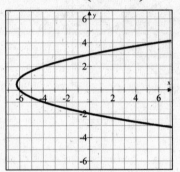

19. $y = 2x^2 + 4x - 5$

$$x = -\frac{b}{2a} = -\frac{4}{2(2)} = -1$$

$$y = 2(-1)^2 + 4(-1) - 5 = 2 - 4 - 5 = -7$$

The axis of symmetry is $x = -1$.
The vertex is $(-1, -7)$.

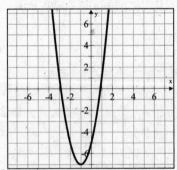

21. $y = 2x^2 - x - 3$

$$x = -\frac{b}{2a} = -\frac{-1}{2(2)} = \frac{1}{4}$$

$$y = 2\left(\frac{1}{4}\right)^2 - \frac{1}{4} - 3 = \frac{1}{8} - \frac{1}{4} - 3 = -\frac{25}{8}$$

The axis of symmetry is $x = \frac{1}{4}$.

The vertex is $\left(\frac{1}{4}, -\frac{25}{8}\right)$.

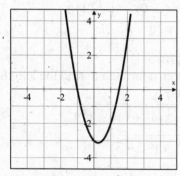

23. $y = x^2 + 5x + 6$

$$x = -\frac{b}{2a} = -\frac{5}{2(1)} = -\frac{5}{2}$$

$$y = \left(-\frac{5}{2}\right)^2 + 5\left(-\frac{5}{2}\right) + 6 = \frac{25}{4} - \frac{25}{2} + 6 = -\frac{1}{4}$$

The axis of symmetry is $x = -\frac{5}{2}$.

The vertex is $\left(-\frac{5}{2}, -\frac{1}{4}\right)$.

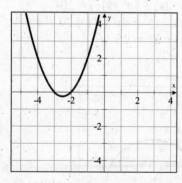

Applying the Concepts

25. $p = \frac{1}{4a}; \quad a = 2$

$$p = \frac{1}{4(2)} = \frac{1}{8}$$

The focus is $\left(0, \frac{1}{8}\right)$.

Section 11.2

Objective A Exercises

1. $(x - 2)^2 + (y + 2)^2 = 9$
Center: $(2, -2)$
Radius: $\sqrt{9} = 3$

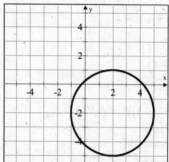

3. $(x + 3)^2 + (y - 1)^2 = 25$
Center: $(-3, 1)$
Radius: $\sqrt{25} = 5$

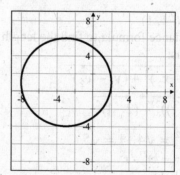

5. $(x + 2)^2 + (y + 2)^2 = 4$
Center: $(-2, -2)$
Radius: $\sqrt{4} = 2$

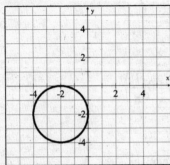

7. $(x - h)^2 + (y - k)^2 = r^2$
Center: $(2, -1)$
Radius: $r = 2$
$(x - 2)^2 + (y - (-1))^2 = 2^2$
$(x - 2)^2 + (y + 1)^2 = 4$

9. $(x - h)^2 + (y - k)^2 = r^2$
Center: $(0, -4)$
Radius: $r = \sqrt{7}$
$(x - 0)^2 + (y - (-4))^2 = (\sqrt{7})^2$
$x^2 + (y + 4)^2 = 7$

11. $(x - h)^2 + (y - k)^2 = r^2$
Center: $(-3, 0)$
Radius: $r = 9$
$(x - (-3))^2 + (y - 0)^2 = (9)^2$
$(x + 3)^2 + y^2 = 81$

13. $(1, 2)$ and $(-1, 1)$
Use the distance formula to find the length of the radius:
$$d = \sqrt{(x_1 - x_2)^2 + (y_1 - y_2)^2}$$
$$= \sqrt{(1 - (-1))^2 + (2 - 1)^2}$$
$$= \sqrt{2^2 + 1^2} = \sqrt{4 + 1} = \sqrt{5}$$
$(x - h)^2 + (y - k)^2 = r^2$
Center: $(-1, 1)$
Radius: $r = \sqrt{5}$
$(x - (-1))^2 + (y - 1)^2 = (\sqrt{5})^2$
$(x + 1)^2 + (y - 1)^2 = 5$

15. $(-3, -4)$ and $(-1, 0)$
Use the distance formula to find the length of the radius:
$$d = \sqrt{(x_1 - x_2)^2 + (y_1 - y_2)^2}$$
$$= \sqrt{(-3 - (-1))^2 + (-4 - 0)^2}$$
$$= \sqrt{(-2)^2 + (-4)^2} = \sqrt{4 + 16} = \sqrt{20}$$
$(x - h)^2 + (y - k)^2 = r^2$
Center: $(-1, 0)$
Radius: $r = \sqrt{20}$
$(x - (-1))^2 + (y - 0)^2 = (\sqrt{20})^2$
$(x + 1)^2 + y^2 = 20$

17. $(1, -5)$ and $(0, 3)$
Use the distance formula to find the length of the radius:
$$d = \sqrt{(x_1 - x_2)^2 + (y_1 - y_2)^2}$$
$$= \sqrt{(1 - 0)^2 + (-5 - 3)^2}$$
$$= \sqrt{1^2 + (-8)^2} = \sqrt{1 + 64} = \sqrt{65}$$
$(x - h)^2 + (y - k)^2 = r^2$
Center: $(0, 3)$
Radius: $r = \sqrt{65}$
$(x - 0)^2 + (y - 3)^2 = (\sqrt{65})^2$
$x^2 + (y - 3)^2 = 65$

19. $x^2 + y^2 = r^2$

Objective B Exercises

21. $x^2 + y^2 - 2x + 4y - 20 = 0$
$(x^2 - 2x) + (y^2 + 4y) = 20$
$(x^2 - 2x + 1) + (y^2 + 4y + 4) = 20 + 1 + 4$
$(x - 1)^2 + (y + 2)^2 = 25$
Center: $(1, -2)$
Radius: 5

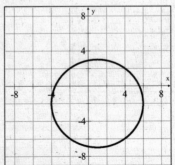

23.
$$x^2 + y^2 + 6x + 8y + 9 = 0$$
$$(x^2 + 6x) + (y^2 + 8y) = -9$$
$$(x^2 + 6x + 9) + (y^2 + 8y + 16) = -9 + 9 + 16$$
$$(x + 3)^2 + (y + 4)^2 = 16$$
Center: $(-3, -4)$
Radius: 4

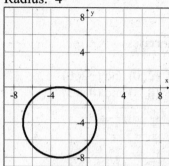

25.
$$x^2 + y^2 - 10x - 8y - 9 = 0$$
$$(x^2 - 10x) + (y^2 - 8y) = 9$$
$$(x^2 - 10x + 25) + (y^2 - 8y + 16) = 9 + 25 + 16$$
$$(x - 5)^2 + (y - 4)^2 = 50$$
Center: $(5, 4)$
Radius: $\sqrt{50}$

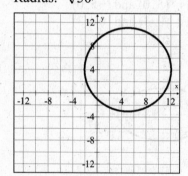

27. $x^2 + y^2 - 12x + 12 = 0$
$$(x^2 - 12x) + y^2 = -12$$
$$(x^2 - 12x + 36) + y^2 = -12 + 36$$
$$(x - 6)^2 + y^2 = 24$$
Center: $(6, 0)$
Radius: $\sqrt{24}$

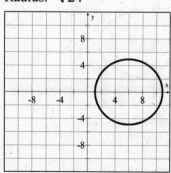

29. $x^2 + y^2 + 14y + 13 = 0$
$$x^2 + (y^2 + 14y) = -13$$
$$x^2 + (y^2 + 14y + 49) = -13 + 49$$
$$x^2 + (y + 7)^2 = 36$$
Center: $(0, -7)$
Radius: 6

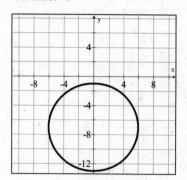

31.
$$x^2 + y^2 - 4x - 7y - 5 = 0$$
$$(x^2 - 4x) + (y^2 - 7y) = 5$$
$$(x^2 - 4x + 4) + (y^2 - 7y + \frac{49}{4}) = 5 + 4 + \frac{49}{4}$$
$$(x - 2)^2 + (y - \frac{7}{2})^2 = \frac{85}{4}$$

Center: $\left(2, \frac{7}{2}\right)$

Radius: $\sqrt{\frac{85}{4}} = \frac{\sqrt{85}}{2} \approx 4.6$

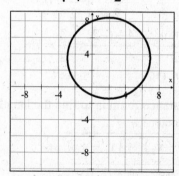

33.
$$x^2 + y^2 + 5x + 5y - 8 = 0$$
$$(x^2 + 5x) + (y^2 + 5y) = 8$$
$$\left(x^2 + 5x + \frac{25}{4}\right) + \left(y^2 + 5y + \frac{25}{4}\right) = 8 + \frac{25}{4} + \frac{25}{4}$$
$$\left(x + \frac{5}{2}\right)^2 + \left(y + \frac{5}{2}\right)^2 = \frac{41}{2}$$

Center: $\left(-\frac{5}{2}, -\frac{5}{2}\right)$

Radius: $\sqrt{\frac{41}{2}} \approx 4.5$

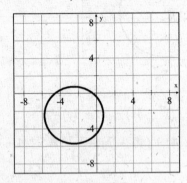

Applying the Concepts

35. $(-2, 4)$ and $(2, -2)$

Use the midpoint formula to find the coordinates for the center of the circle.

$$x_m = \frac{x_1 + x_2}{2} = \frac{-2 + 2}{2} = \frac{0}{2} = 0$$

$$y_m = \frac{y_1 + y_2}{2} = \frac{4 + (-2)}{2} = \frac{2}{2} = 1$$

Center: $(0, 1)$

Use the distance formula to find the length of the diameter:

$$d = \sqrt{(x_1 - x_2)^2 + (y_1 - y_2)^2}$$
$$= \sqrt{(-2 - 2)^2 + (4 - (-2))^2}$$
$$= \sqrt{(-4)^2 + (6)^2} = \sqrt{16 + 36} = \sqrt{52}$$

Radius: $\frac{\sqrt{52}}{2} = \frac{2\sqrt{13}}{2} = \sqrt{13}$

$$(x - h)^2 + (y - k)^2 = r^2$$
$$(x - 0)^2 + (y - 1)^2 = (\sqrt{13})^2$$
$$x^2 + (y - 1)^2 = 13$$

37. If the circle is tangent to the x-axis its radius must be 3.
$$(x - h)^2 + (y - k)^2 = r^2$$
$$(x - 3)^2 + (y - 3)^2 = 3^2$$
$$(x - 3)^2 + (y - 3)^2 = 9$$

Section 11.3

Objective A Exercises

1. x-intercepts: $(2, 0)$ and $(-2, 0)$
y-intercepts: $(0, 3)$ and $(0, -3)$

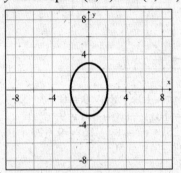

3. x-intercepts: $(5, 0)$ and $(-5, 0)$
y-intercepts: $(0, 3)$ and $(0, -3)$

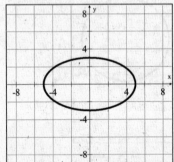

5. x-intercepts: $(6, 0)$ and $(-6, 0)$
y-intercepts: $(0, 4)$ and $(0, -4)$

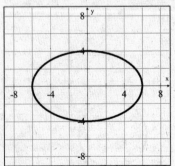

7. *x*-intercepts: (4, 0) and (−4, 0)
   y-intercepts: (0, 7) and (0, −7)

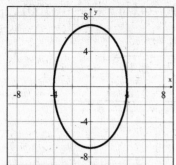

9. *x*-intercepts: (2, 0) and (−2, 0)
   y-intercepts: (0, 5) and (0, −5)

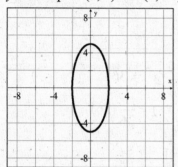

11. The graph of the ellipse gets rounder.

Objective B Exercises

13. Axis of symmetry: *x*-axis
   Vertices: (5, 0) and (−5, 0)

   Asymptotes: $y = \dfrac{2}{5}x$ and $y = -\dfrac{2}{5}x$

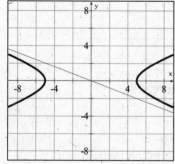

15. Axis of symmetry: *y*-axis
   Vertices: (0, 4) and (0, −4)

   Asymptotes: $y = \dfrac{4}{5}x$ and $y = -\dfrac{4}{5}x$

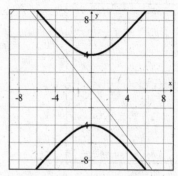

17. Axis of symmetry: *x*-axis
   Vertices: (3, 0) and (−3, 0)

   Asymptotes: $y = \dfrac{7}{3}x$ and $y = -\dfrac{7}{3}x$

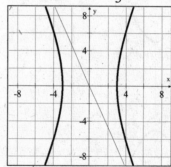

19. Axis of symmetry: *y*-axis
   Vertices: (0, 2) and (0, −2)

   Asymptotes: $y = \dfrac{1}{2}x$ and $y = -\dfrac{1}{2}x$

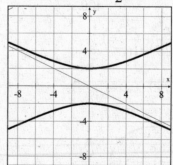

21. Axis of symmetry: x-axis
Vertices: $(6, 0)$ and $(-6, 0)$

Asymptotes: $y = \dfrac{1}{2}x$ and $y = -\dfrac{1}{2}x$

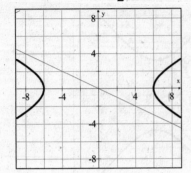

23. Axis of symmetry: y-axis
Vertices: $(0, 5)$ and $(0, -5)$

Asymptotes: $y = \dfrac{5}{2}x$ and $y = -\dfrac{5}{2}x$

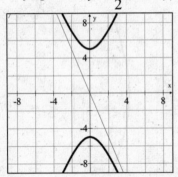

Applying the Concepts

25. $16x^2 + 25y^2 = 400$

$\dfrac{1}{400}(16x^2 + 25y^2) = \dfrac{1}{400}(400)$

$\dfrac{x^2}{25} + \dfrac{y^2}{16} = 1$

ellipse

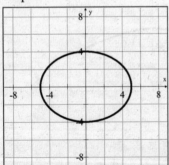

27. $25y^2 - 4x^2 = -100$

$-\dfrac{1}{100}(25y^2 - 4x^2) = -\dfrac{1}{100}(-100)$

$\dfrac{x^2}{25} - \dfrac{y^2}{4} = 1$

hyperbola

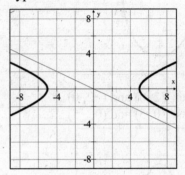

29. $4y^2 - x^2 = 36$

$\dfrac{1}{36}(4y^2 - x^2) = \dfrac{1}{36}(36)$

$\dfrac{y^2}{9} - \dfrac{x^2}{36} = 1$

hyperbola

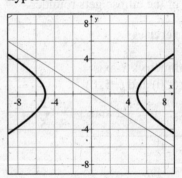

Section 11.4

Objective A Exercises

1. (1) $y = x^2 - x - 1$
(2) $y = 2x + 9$
Use the substitution method:
$y = x^2 - x - 1$
$2x + 9 = x^2 - x - 1$
$0 = x^2 - 3x - 10$
$0 = (x - 5)(x + 2)$
$x - 5 = 0 \qquad x + 2 = 0$
$\quad x = 5 \qquad\qquad x = -2$

Substitute the values of x into equation (2).

$y = 2(5) + 9$ $y = 2(-2) + 9$
$y = 10 + 9$ $y = -4 + 9$
$y = 19$ $y = 5$
The solutions are $(5, 19)$ and $(-2, 5)$.

3. (1) $y^2 = -x + 3$
(2) $x - y = 1$
Solve equation (2) for x.

$x - y = 1$
$x = y + 1$
Use the substitution method:

$y^2 = -x + 3$
$y^2 = -(y + 1) + 3$
$y^2 = -y - 1 + 3$
$y^2 = -y + 2$
$y^2 + y - 2 = 0$
$(y - 1)(y + 2) = 0$
$y - 1 = 0$ $y + 2 = 0$
$y = 1$ $y = -2$
Substitute the values of y into equation (2).

$x - y = 1$ $x - y = 1$
$x - 1 = 1$ $x - (-2) = 1$
$x = 2$ $x + 2 = 1$
                   $x = -1$
The solutions are $(-1, -2)$ and $(2, 1)$.

5. (1) $y^2 = 2x$
(2) $x + 2y = -2$
Solve equation (2) for x.

$x + 2y = -2$
$x = -2y - 2$
Use the substitution method:

$y^2 = 2x$
$y^2 = 2(-2y - 2)$
$y^2 = -4y - 4$
$y^2 + 4y + 4 = 0$
$(y + 2)(y + 2) = 0$
$y + 2 = 0$ $y + 2 = 0$
$y = -2$ $y = -2$
Substitute the value of y into equation (2).

$x + 2y = -2$
$x + 2(-2) = -2$
$x - 4 = -2$
$x = 2$
The solution is $(2, -2)$.

7. (1) $x^2 + 2y^2 = 12$
(2) $2x - y = 2$
Solve equation (2) for y.

$2x - y = 2$
$2x = y + 2$
$y = 2x - 2$
Use the substitution method:

$$x^2 + 2y^2 = 12$$
$$x^2 + 2(2x - 2)^2 = 12$$
$$x^2 + 2(4x^2 - 8x + 4) = 12$$
$$x^2 + 8x^2 - 16x + 8 = 12$$
$$9x^2 - 16x - 4 = 0$$
$$(x - 2)(9x + 2) = 0$$
$x - 2 = 0$ $9x + 2 = 0$
$x = 2$ $9x = -2$
               $x = -\dfrac{2}{9}$

Substitute the values of x into equation (2).

$2x - y = 2$ $2x - y = 2$

$2(2) - y = 2$ $2(-\dfrac{2}{9}) - y = 2$

$4 - y = 2$ $-\dfrac{4}{9} - y = 2$

$-y = -2$ $-y = \dfrac{22}{9}$

$y = 2$ $y = -\dfrac{22}{9}$

The solutions are $(2, 2)$ and $\left(-\dfrac{2}{9}, -\dfrac{22}{9}\right)$.

9. (1) $x^2 + y^2 = 13$
(2) $x + y = 5$
Solve equation (2) for y.

$x + y = 5$
$y = -x + 5$
Use the substitution method:

$$x^2 + y^2 = 13$$
$$x^2 + (-x + 5)^2 = 13$$
$$x^2 + x^2 - 10x + 25 = 13$$
$$2x^2 - 10x + 12 = 0$$
$$2(x^2 - 5x + 6) = 0$$
$$(x - 3)(x - 2) = 0$$
$x - 3 = 0$ $x - 2 = 0$
$x = 3$ $x = 2$
Substitute the values of x into equation (2).

$3 + y = 5$ $2 + y = 5$
$y = 2$ $y = 3$
The solutions are $(3, 2)$ and $(2, 3)$.

11. (1) $4x^2 + y^2 = 12$
(2) $y = 4x^2$
Use the substitution method:
$$4x^2 + y^2 = 12$$
$$4x^2 + (4x^2)^2 = 12$$
$$4x^2 + 16x^4 = 12$$
$$16x^4 + 4x^2 - 12 = 0$$
$$4(x^4 + x^2 - 3) = 0$$
$$4(4x^2 - 3)(x^2 + 1) = 0$$

$$4x^2 - 3 = 0 \qquad x^2 + 1 = 0$$
$$4x^2 = 3 \qquad x^2 = -1$$

$$x^2 = \frac{3}{4} \qquad x = \pm\sqrt{-1}$$

$$x = \pm\frac{\sqrt{3}}{2}$$

Substitute the real number values of x into equation (2).
$y = 4x^2$

$$y = 4\left(\frac{\sqrt{3}}{2}\right)^2 \qquad y = 4\left(-\frac{\sqrt{3}}{2}\right)^2$$

$$y = 4\left(\frac{3}{4}\right) \qquad y = 4\left(\frac{3}{4}\right)$$

$$y = 3 \qquad\qquad y = 3$$

The solutions are $\left(\dfrac{\sqrt{3}}{2}, 3\right)$ and $\left(-\dfrac{\sqrt{3}}{2}, 3\right)$.

13. (1) $y = x^2 - 2x - 3$
(2) $y = x - 6$
Use the substitution method:
$$y = x^2 - 2x - 3$$
$$x - 6 = x^2 - 2x - 3$$
$$0 = x^2 - 3x + 3$$

$$x = \frac{-b \pm \sqrt{b^2 - 4ac}}{2a}$$

$$= \frac{-(-3) \pm \sqrt{(-3)^2 - 4(1)(3)}}{2(1)}$$

$$= \frac{3 \pm \sqrt{9 - 12}}{2} = \frac{3 \pm \sqrt{-3}}{2}$$

The system of equations has no real number solution.

15. (1) $3x^2 - y^2 = -1$
(2) $x^2 + 4y^2 = 17$
Use the addition method. Multiply equation (1) by 4.
$$12x^2 - 4y^2 = -4$$
$$x^2 + 4y^2 = 17$$
$$13x^2 = 13$$
$$x^2 = 1$$
$$x = \pm\sqrt{1} = \pm 1$$

Substitute the values of x into equation (2).

$$x^2 + 4y^2 = 17 \qquad\qquad x^2 + 4y^2 = 17$$
$$1^2 + 4y^2 = 17 \qquad\qquad (-1)^2 + 4y^2 = 17$$
$$1 + 4y^2 = 16 \qquad\qquad 1 + 4y^2 = 17$$
$$4y^2 = 16 \qquad\qquad 4y^2 = 16$$
$$y^2 = 4 \qquad\qquad y^2 = 4$$
$$y = \pm\sqrt{4} = \pm 2 \qquad y = \pm\sqrt{4} = \pm 2$$

The solutions are $(1, 2)$, $(1, -2)$, $(-1, 2)$ and $(-1, -2)$.

17. (1) $2x^2 + 3y^2 = 30$
(2) $x^2 + y^2 = 13$
Use the addition method. Multiply equation (2) by -2.
$$2x^2 + 3y^2 = 30$$
$$-2x^2 - 2y^2 = -26$$
$$y^2 = 4$$
$$y = \pm\sqrt{4} = \pm 2$$

Substitute the values of y into equation (2).

$$x^2 + y^2 = 13 \qquad\qquad x^2 + y^2 = 13$$
$$x^2 + 2^2 = 13 \qquad\qquad x^2 + (-2)^2 = 13$$
$$x^2 + 4 = 13 \qquad\qquad x^2 + 4 = 13$$
$$x^2 = 9 \qquad\qquad x^2 = 9$$

$$x = \pm\sqrt{9} = \pm 3 \qquad x = \pm\sqrt{9} = \pm 3$$

The solutions are $(3, 2)$, $(-3, 2)$, $(3, -2)$ and $(-3, -2)$.

19. (1) $y = 2^{x2} - x + 1$
(2) $y = {}^{x2} - x + 5$
Use the substitution method:
$$y = 2^{x2} - x + 1$$
$${}^{x2} - x + 5 = 2^{x2} - x + 1$$
$$0 = {}^{x2} - 4$$
$$0 = (x - 2)(x + 2)$$
$$x - 2 = 0 \quad x + 2 = 0$$
$$x = 2 \quad\quad x = -2$$
Substitute the values of x into equation (2).
$$y = {}^{x2} - x + 5 \quad\quad y = {}^{x2} - x + 5$$
$$y = {}^{22} - 2 + 5 \quad\quad y = (-2)^{2} - (-2) + 5$$
$$y = 4 - 2 + 5 \quad\quad y = 4 + 2 + 5$$
$$y = 7 \quad\quad\quad\quad y = 11$$
The solutions are $(2, 7)$ and $(-2, 11)$.

21. (1) $2x^2 + 3y^2 = 24$
(2) $x^2 - y^2 = 7$
Use the addition method. Multiply equation
(2) by -2.
$$2x^2 + 3y^2 = 24$$
$$-2x^2 + 2y^2 = -14$$
$$5y^2 = 10$$
$$y^2 = 2$$
$$y = \pm\sqrt{2}$$
Substitute the values of y into equation (2).
$$x^2 - y^2 = 7 \quad\quad x^2 - y^2 = 7$$
$$x^2 - \left(\sqrt{2}\right)^2 = 7 \quad x^2 - \left(-\sqrt{2}\right)^2 = 7$$
$$x^2 - 2 = 7 \quad\quad x^2 - 2 = 7$$
$$x^2 = 9 \quad\quad\quad x^2 = 9$$
$$x = \pm\sqrt{9} = \pm 3 \quad\quad x = \pm\sqrt{9} = \pm 3$$
The solutions are $(3, \sqrt{2})$, $(-3, \sqrt{2})$,
$(3, -\sqrt{2})$ and $(-3, -\sqrt{2})$.

23. (1) $x^2 + y^2 = 36$
(2) $4x^2 + 9y^2 = 36$
Use the addition method. Multiply equation
(1) by -4.
$$-4x^2 - 4y^2 = -144$$
$$4x^2 + 9y^2 = 36$$
$$5y^2 = -108$$
$$y^2 = -\frac{108}{5}$$
$$y = \pm\sqrt{-\frac{108}{5}}$$
The system of equations has no real number
solution.

25. (1) $11x^2 - 2y^2 = 4$
(2) $3x^2 + y^2 = 15$
Use the addition method. Multiply equation
(2) by 2.
$$11x^2 - 2y^2 = 4$$
$$6x^2 + 2y^2 = 30$$
$$17x^2 = 34$$
$$x^2 = 2$$
$$x = \pm\sqrt{2}$$
Substitute the values of x into equation (2).
$$3x^2 + y^2 = 15 \quad\quad 3x^2 + y^2 = 15$$
$$3\left(\sqrt{2}\right)^2 + y^2 = 15 \quad 3\left(-\sqrt{2}\right)^2 + y^2 = 15$$
$$6 + y^2 = 15 \quad\quad 6 + y^2 = 15$$
$$y^2 = 9 \quad\quad\quad y^2 = 9$$
$$y = \pm\sqrt{9} = \pm 3 \quad\quad y = \pm\sqrt{9} = \pm 3$$
The solutions are $\left(3, \sqrt{2}\right)$, $\left(-3, \sqrt{2}\right)$,
$\left(3, -\sqrt{2}\right)$ and $\left(-3, -\sqrt{2}\right)$.

27. (1) $2x^2 - y^2 = 7$
(2) $2x - y = 5$
Solve equation (2) for y.
$$2x - y = 5$$
$$-y = -2x + 5$$
$$y = 2x - 5$$
Use the substitution method:
$$2x^2 - y^2 = 7$$
$$2x^2 - (2x - 5)^2 = 7$$
$$2x^2 - (4x^2 - 20x + 25) = 7$$
$$2x^2 - 4x^2 + 20x - 25 = 7$$
$$-2x^2 + 20x - 32 = 0$$
$$-2(x^2 - 10x + 16) = 0$$
$$-2(x - 8)(x - 2) = 0$$
$$x - 8 = 0 \quad x - 2 = 0$$
$$x = 8 \quad\quad x = 2$$
Substitute the values of x into equation (2).
$$2x - y = 5 \quad\quad 2x - y = 5$$
$$2(8) - y = 5 \quad\quad 2(2) - y = 5$$
$$-y = -11 \quad\quad -y = 1$$
$$y = 11 \quad\quad\quad y = -1$$
The solutions are $(8, 11)$ and $(2, -1)$.

29. (1) $y = 3^{x2} + x - 4$
 (2) $y = 3^{x2} - 8x + 5$
Use the substitution method:
$$y = 3^{x2} + x - 4$$
$$3^{x2} - 8x + 5 = 3^{x2} + x - 4$$
$$0 = 9x - 9$$
$$9 = 9x$$
$$x = 1$$
Substitute the value of x into equation (1).
$$y = 3^{x2} + x - 4$$
$$y = 3(1)^{2} + 1 - 4$$
$$y = 3 + 1 - 4$$
$$y = 0$$
The solution is $(1, 0)$.

31. (1) $x = y + 3$
 (2) $x^2 + y^2 = 5$
Use the substitution method:
$$(y + 3)^2 + y^2 = 5$$
$$y^2 + 6y + 9 + y^2 = 5$$
$$2y^2 + 6y + 4 = 0$$
$$2(y^2 + 3y + 2) = 0$$
$$2(y + 2)(y + 1) = 0$$
$$y + 2 = 0 \qquad y + 1 = 0$$
$$y = -2 \qquad\quad y = -1$$
Substitute the values of y into equation (1).
$$x = y + 3 \qquad\quad x = y + 3$$
$$x = -2 + 3 \qquad x = -1 + 3$$
$$x = 1 \qquad\qquad x = 2$$
The solutions are $(1, -2)$ and $(2, -1)$.

33. (1) $y = x^2 + 4x + 4$
 (2) $x + 2y = 4$
Use the substitution method:
$$x + 2y = 4$$
$$x + 2(x^2 + 4x + 4) = 4$$
$$x + 2x^2 + 8x + 8 = 4$$
$$2x^2 + 9x + 4 = 0$$
$$(2x + 1)(x + 4) = 0$$
$$2x + 1 = 0 \qquad x + 4 = 0$$
$$2x = -1 \qquad\qquad x = -4$$
$$x = -\frac{1}{2}$$

Substitute the values of x into equation (2).
$$x + 2y = 4 \qquad\qquad x + 2y = 4$$
$$-\frac{1}{2} + 2y = 4 \qquad -4 + 2y = 4$$
$$2y = \frac{9}{2} \qquad\qquad 2y = 8$$
$$y = \frac{9}{4} \qquad\qquad y = 4$$
The solutions are $\left(-\frac{1}{2}, \frac{9}{4}\right)$ and $(-4, 4)$.

Applying the Concepts

35. $y = 2^x$
 $x + y = 3$

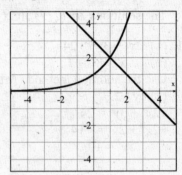

The solution is $(1, 2)$.

37. $y = \log_2 x$
$$\frac{x^2}{9} + \frac{y^2}{1} = 1$$

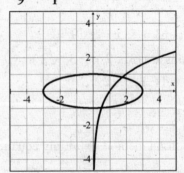

The solutions are $(1.755, 0.811)$
and $(0.505, -0.986)$.

39. $y = -\log_3 x$
    $x + y = 4$

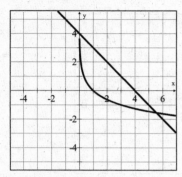

The solutions are $(5.562, -1.562)$
and $(0.013, 3.987)$.

Section 11.5

Objective A Exercises

1. A solid curve is used for the boundaries of inequalities that use $\leq$ or $\geq$, and a dashed curve is used for the boundaries of inequalities that use $<$ or $>$.

3. $y \leq x^2 - 4x + 3$

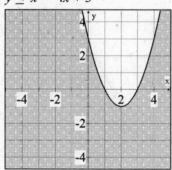

5. $(x-1)^2 + (y+2)^2 \leq 9$

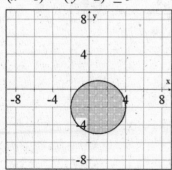

7. $(x+3)^2 + (y-2)^2 \geq 9$

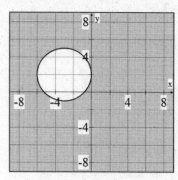

9. $\dfrac{x^2}{16} + \dfrac{y^2}{25} < 1$

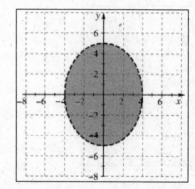

11. $\dfrac{x^2}{25} - \dfrac{y^2}{9} \leq 1$

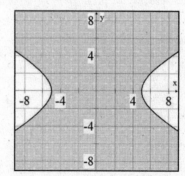

13. $\dfrac{x^2}{4} + \dfrac{y^2}{16} \geq 1$

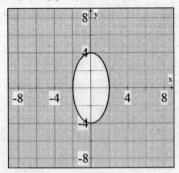

15. $y \leq x^2 - 2x + 3$

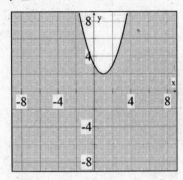

17. $\dfrac{y^2}{9} - \dfrac{x^2}{16} \leq 1$

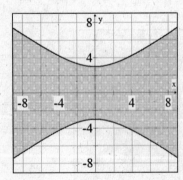

19. $\dfrac{x^2}{9} + \dfrac{y^2}{1} \leq 1$

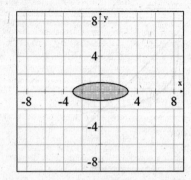

21. $(x-1)^2 + (y+3)^2 \leq 25$

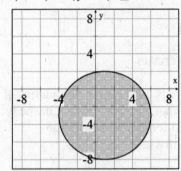

23. $\dfrac{y^2}{25} - \dfrac{x^2}{4} \leq 1$

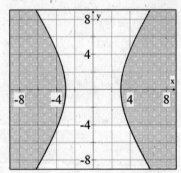

25. $\dfrac{x^2}{25} + \dfrac{y^2}{9} \leq 1$

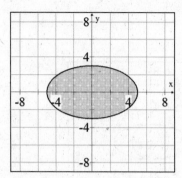

27. No, (0,0) cannot be used to determine whether to shade inside or outside of the graph of $y \leq x^2 - 4x$ because $(0, 0)$ is on the graph of $y = x^2 - 4x$.

Objective B Exercises

29. $y \leq (x - 2)^2$
$\quad\, y + x > 4$

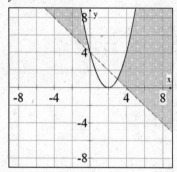

31. $x^2 + y^2 < 16$
$\quad\, y > x + 1$

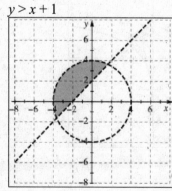

33. $\dfrac{x^2}{4} + \dfrac{y^2}{16} \leq 1$

$\quad y \leq -\dfrac{1}{2}x + 2$

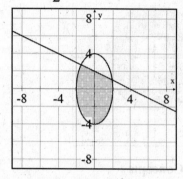

35. $x \geq y^2 - 3y + 2$
$\quad\, y \geq 2x - 2$

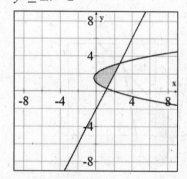

37. $x^2 + y^2 < 25$

$\quad \dfrac{x^2}{9} + \dfrac{y^2}{36} < 1$

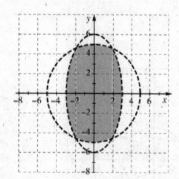

39. $x^2 + y^2 > 4$
$x^2 + y^2 < 25$

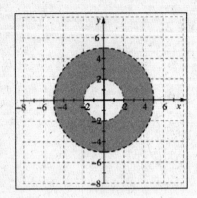

Applying the Concepts

41. $y > x^2 - 3$
$y < x + 3$
$x \leq 0$

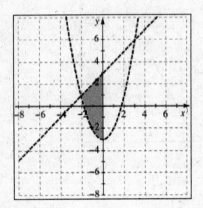

43. $x^2 + y^2 < 3$
$x > y^2 - 1$
$y \geq 0$

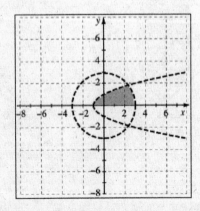

45. $\dfrac{x^2}{16} + \dfrac{y^2}{4} \leq 1$
$x^2 + y^2 \leq 4$
$x \geq 0$
$y \leq 0$

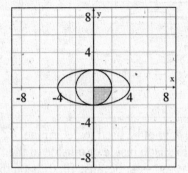

47. $y > 2^x$
$x + y < 4$

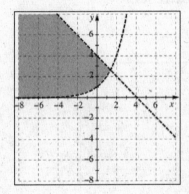

49. $y \geq \log_2 x$
$x^2 + y^2 < 9$

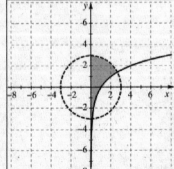

Chapter 11 Review Exercises

1. $y = -2x^2 + x - 2$

$x = -\dfrac{b}{2a} = -\dfrac{1}{2(-2)} = -\dfrac{1}{-4} = \dfrac{1}{4}$

$y = -2\left(\dfrac{1}{4}\right)^2 + \dfrac{1}{4} - 2 = -\dfrac{1}{8} + \dfrac{1}{4} - 2 = -\dfrac{15}{8}$

The axis of symmetry is $x = \dfrac{1}{4}$.

The vertex is $\left(\dfrac{1}{4}, -\dfrac{15}{8}\right)$.

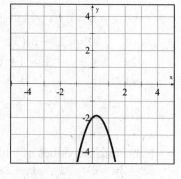

2. x-intercepts: $(1, 0)$ and $(-1, 0)$
y-intercepts: $(0, 3)$ and $(0, -3)$

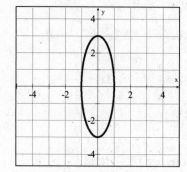

3. $\dfrac{x^2}{9} - \dfrac{y^2}{16} < 1$

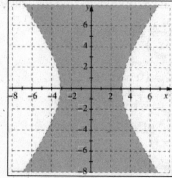

4. $(x + 3)^2 + (y + 1)^2 = 1$
Center: $(-3, -1)$
Radius: 1

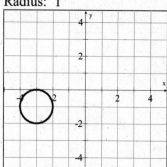

5. $y = x^2 - 4x + 8$

$x = -\dfrac{b}{2a} = -\dfrac{-4}{2(1)} = -\dfrac{-4}{2} = 2$

$y = (2)^2 - 4(2) + 8 = 4 - 8 + 8 = 4$

The axis of symmetry is $x = 2$.
The vertex is $(2, 4)$.

6. (1) $y^2 = 2x^2 - 3x + 6$
(2) $y^2 = 2x^2 + 5x - 2$
Use the substitution method.
$$y^2 = 2x^2 - 3x + 6$$
$$2x^2 + 5x - 2 = 2x^2 - 3x + 6$$
$$8x = 8$$
$$x = 1$$

Substitute the value of x into equation (1).
$y^2 = 2x^2 - 3x + 6$
$y^2 = 2(1)^2 - 3(1) + 6$
$\quad = 2 - 3 + 6 = 5$
$y = \pm\sqrt{5}$

The solutions are $\left(1, \sqrt{5}\right)$ and $\left(1, -\sqrt{5}\right)$.

7. $\dfrac{x^2}{25} + \dfrac{y^2}{16} \le 1$

$\dfrac{y^2}{4} - \dfrac{x^2}{4} \ge 1$

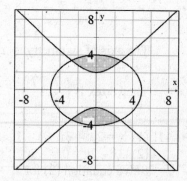

8. Axis of symmetry: x-axis
Vertices: $(5, 0)$ and $(-5, 0)$
Asymptotes: $y = \dfrac{1}{5}x$ and $y = -\dfrac{1}{5}x$

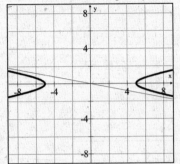

9. $(2, -1)$ and $(-1, 2)$
Use the distance formula to find the length of the radius:

$$d = \sqrt{(x_1 - x_2)^2 + (y_1 - y_2)^2}$$
$$= \sqrt{(2 - (-1))^2 + (-1 - 2)^2}$$
$$= \sqrt{3^2 + (-3)^2} = \sqrt{9 + 9} = \sqrt{18}$$

$(x - h)^2 + (y - k)^2 = r^2$
Center: $(-1, 2)$
Radius: $r = \sqrt{18}$
$(x - (-1))^2 + (y - 2)^2 = (\sqrt{18})^2$
$(x + 1)^2 + (y - 2)^2 = 18$

10. $y = -x^2 + 7x - 8$

$$x = -\frac{b}{2a} = -\frac{7}{2(-1)} = -\frac{7}{-2} = \frac{7}{2}$$

$$y = -\left(\frac{7}{2}\right)^2 + 7\left(\frac{7}{2}\right) - 8 = -\frac{49}{4} + \frac{49}{2} - 8 = \frac{17}{4}$$

The axis of symmetry is $x = \dfrac{7}{2}$.

The vertex is $\left(\dfrac{7}{2}, \dfrac{17}{4}\right)$.

11. (1) $x = 2y^2 - 3y + 1$
(2) $3x - 2y = 0$
Solve equation (2) for x.
$3x - 2y = 0$
$3x = 2y$

$x = \dfrac{2}{3}y$

Use the substitution method:

$$x = 2y^2 - 3y + 1$$
$$\frac{2}{3}y = 2y^2 - 3y + 1$$
$$2y = 6y^2 - 9y + 3$$
$$0 = 6y^2 - 11y + 3$$
$$0 = (2y - 3)(3y - 1)$$

$2y - 3 = 0 \qquad 3y - 1 = 0$
$2y = 3 \qquad\quad 3y = 1$

$y = \dfrac{3}{2} \qquad\quad y = \dfrac{1}{3}$

Substitute the values of y into equation (2).

$3x - 2y = 0 \qquad\qquad 3x - 2y = 0$

$3x - 2\left(\dfrac{3}{2}\right) = 0 \qquad 3x - 2\left(\dfrac{1}{3}\right) = 0$

$3x - 3 = 0 \qquad\qquad 3x - \left(\dfrac{2}{3}\right) = 0$

$3x = 3 \qquad\qquad\quad 3x = \dfrac{2}{3}$

$x = 1 \qquad\qquad\quad\; x = \dfrac{2}{9}$

The solutions are $\left(1, \dfrac{3}{2}\right)$ and $\left(\dfrac{2}{9}, \dfrac{1}{3}\right)$.

12. $(4, 6)$ and $(0, -3)$
Use the distance formula to find the length of the radius:

$$d = \sqrt{(x_1 - x_2)^2 + (y_1 - y_2)^2}$$
$$= \sqrt{(4 - 0)^2 + (6 - (-3))^2}$$
$$= \sqrt{4^2 + 9^2} = \sqrt{16 + 81} = \sqrt{97}$$

$(x - h)^2 + (y - k)^2 = r^2$
Center: $(0, -3)$
Radius: $r = \sqrt{97}$
$(x - 0)^2 + (y - (-3))^2 = (\sqrt{97})^2$
$x^2 + (y + 3)^2 = 97$

13. Center: $(-1, 5)$
Radius: 6
$(x - h)^2 + (y - k)^2 = r^2$
$(x - (-1))^2 + (y - 5)^2 = 6^2$
$(x + 1)^2 + (y - 5)^2 = 36$

14. (1) $2x^2 + y^2 = 19$
(2) $3x^2 - y^2 = 6$
Use the addition method.
$$2x^2 + y^2 = 19$$
$$3x^2 - y^2 = 6$$
$$5x^2 = 25$$
$$x^2 = 5$$
$$x = \pm\sqrt{5}$$
Substitute the values of x into equation (1).

$2x^2 + y^2 = 19$	$2x^2 + y^2 = 19$
$2(\sqrt{5})^2 + y^2 = 19$	$2(-\sqrt{5})^2 + y^2 = 19$
$10 + y^2 = 19$	$10 + y^2 = 19$
$y^2 = 9$	$y^2 = 9$
$y = \pm\sqrt{9} = \pm 3$	$y = \pm\sqrt{9} = \pm 3$

The solutions are $(\sqrt{5}, 3)$, $(-\sqrt{5}, 3)$, $(\sqrt{5}, -3)$ and $(-\sqrt{5}, -3)$.

15. $x^2 + y^2 + 4x - 2y = 4$
$(x^2 + 4x) + (y^2 - 2y) = 4$
$x^2 + 4x + 4 + y^2 - 2y + 1 = 4 + 4 + 1$
$(x + 2)^2 + (y - 1)^2 = 9$

16. (1) $y = x^2 + 5x - 6$
(2) $y = x - 10$
Use the substitution method:
$$y = x^2 + 5x - 6$$
$$x - 10 = x^2 + 5x - 6$$
$$0 = x^2 + 4x + 4$$
$$0 = (x + 2)^2$$

$x + 2 = 0$	$x + 2 = 0$
$x = -2$	$x = -2$

Substitute the value of x into equation (2).
$y = x - 10$
$y = -2 - 10$
$y = -12$
The solution is $(-2, -12)$.

17. $\dfrac{x^2}{16} + \dfrac{y^2}{4} > 1$

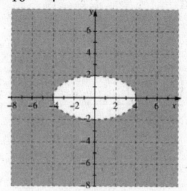

18. Axis of symmetry: y-axis
Vertices: $(0, 4)$ and $(0, -4)$
Asymptotes: $y = \dfrac{4}{3}x$ and $y = -\dfrac{4}{3}x$

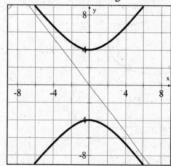

19. $(x - 2)^2 + (y + 1)^2 \leq 16$

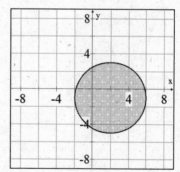

20. $x^2 + (y-2)^2 = 9$
Center: $(0, 2)$
Radius: 3

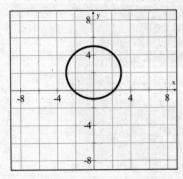

21. x-intercepts: $(5, 0)$ and $(-5, 0)$
y-intercepts: $(0, 3)$ and $(0, -3)$

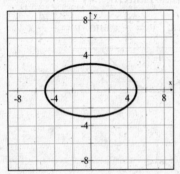

22. $y \geq -x^2 - 2x + 3$

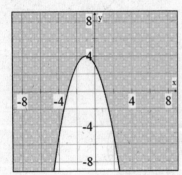

23. $\dfrac{x^2}{16} + \dfrac{y^2}{4} < 1$
$x^2 + y^2 > 9$

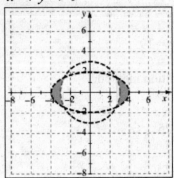

24. $y \geq x^2 - 4x + 2$
$y \leq \dfrac{1}{3}x - 1$

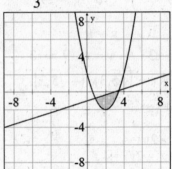

25. $x = 2y^2 - 6y + 5$

The axis of symmetry is $y = \dfrac{3}{2}$.

The vertex is $\left(\dfrac{1}{2}, \dfrac{3}{2} \right)$.

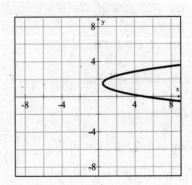

26. $\dfrac{x^2}{9} + \dfrac{y^2}{1} \geq 1$

$\dfrac{x^2}{4} - \dfrac{y^2}{1} \leq 1$

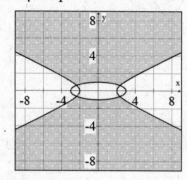

27. $y = x^2 - 4x - 1$

$x = -\dfrac{b}{2a} = -\dfrac{-4}{2(1)} = -\dfrac{-4}{2} = -(-2) = 2$

$y = (2)^2 - 4(2) - 1 = 4 - 8 - 1 = -5$

The axis of symmetry is $x = 2$.

The vertex is $(2, -5)$.

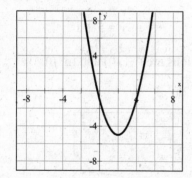

28. $x = y^2 - 1$

$y = -\dfrac{b}{2a} = -\dfrac{0}{2(1)} = 0$

$x = (0)^2 - 1 = -1$

The axis of symmetry is $y = 0$.

The vertex is $(-1, 0)$.

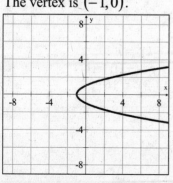

29. Center: $(3, -4)$

Radius: 5

$(x - h)^2 + (y - k)^2 = r^2$

$(x - 3)^2 + (y - (-4))^2 = 5^2$

$(x - 3)^2 + (y + 4)^2 = 25$

30. $x^2 + y^2 - 6x + 4y - 23 = 0$

$x^2 + y^2 - 6x + 4y = 23$

$(x^2 - 6x) + (y^2 + 4y) = 23$

$x^2 - 6x + 9 + y^2 + 4y + 4 = 23 + 9 + 4$

$(x - 3)^2 + (y + 2)^2 = 36$

31. $(x + 1)^2 + (y - 4)^2 = 36$

Center: $(-1, 4)$

Radius: 6

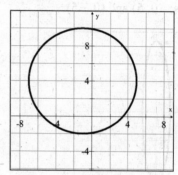

32. $x^2 + y^2 - 8x + 4y + 16 = 0$

$x^2 + y^2 - 8x + 4y = -16$

$(x^2 - 8x) + (y^2 + 4y) = -16$

$x^2 - 8x + 16 + y^2 + 4y + 4 = -16 + 16 + 4$

$(x - 4)^2 + (y + 2)^2 = 4$

Center: $(4, -2)$

Radius: 2

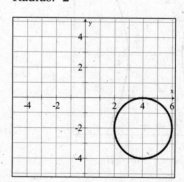

33. *x*-intercepts: (6, 0) and (−6, 0)
y-intercepts: (0, 4) and (0, −4)

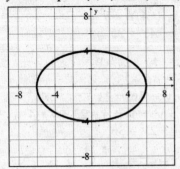

34. *x*-intercepts: (1, 0) and (−1, 0)
y-intercepts: (0, 5) and (0, −5)

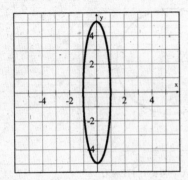

35. Axis of symmetry: *x*-axis
Vertices: (3, 0) and (−3, 0)
Asymptotes: $y = 2x$ and $y = -2x$

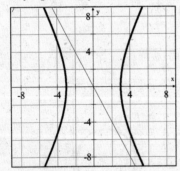

36. Axis of symmetry: *y*-axis
Vertices: (0, 6) and (0, −6)
Asymptotes: $y = 3x$ and $y = -3x$

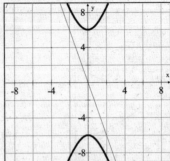

37. (1) $y = x^2 - 3x - 4$
(2) $y = x + 1$
Use the substitution method:
$$y = x^2 - 3x - 4$$
$$x + 1 = x^2 - 3x - 4$$
$$0 = x^2 - 4x - 5$$
$$0 = (x + 1)(x - 5)$$
$$x + 1 = 0 \qquad x - 5 = 0$$
$$x = -1 \qquad x = 5$$
Substitute the value of *x* into equation (2).
$$y = x + 1 \qquad\qquad y = x + 1$$
$$y = -1 + 1 \qquad\qquad y = 5 + 1$$
$$y = 0 \qquad\qquad y = 6$$
The solutions are (−1, 0) and (5, 6).

38. $x \leq y^2 + 2y - 3$

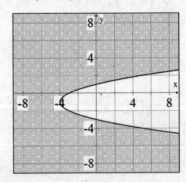

Chapter 11 Test

1. Center: (−3, −3)
Radius: 4
$$(x - h)^2 + (y - k)^2 = r^2$$
$$(x - (-3))^2 + (y - (-3))^2 = 4^2$$
$$(x + 3)^2 + (y + 3)^2 = 16$$

2. $(2, 5)$ and $(-2, 1)$

Use the distance formula to find the length of the radius:

$$d = \sqrt{(x_1 - x_2)^2 + (y_1 - y_2)^2}$$

$$= \sqrt{(2 - (-2))^2 + (5 - 1)^2}$$

$$= \sqrt{4^2 + 4^2} = \sqrt{16 + 16} = \sqrt{32}$$

$(x - h)^2 + (y - k)^2 = r^2$

Center: $(-2, 1)$

Radius: $r = \sqrt{32}$

$(x - (-2))^2 + (y - 1)^2 = (\sqrt{32})^2$

$(x + 2)^2 + (y - 1)^2 = 32$

3. Axis of symmetry: y-axis

Vertices: $(0, 5)$ and $(0, -5)$

Asymptotes: $y = \dfrac{5}{4}x$ and $y = -\dfrac{5}{4}x$

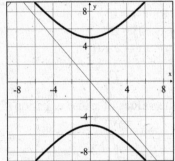

4. $x^2 + y^2 < 36$

$x + y > 4$

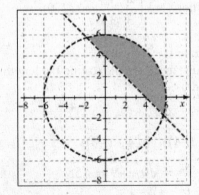

5. $y = -x^2 + 6x - 5$

$$x = -\frac{b}{2a} = -\frac{6}{2(-1)} = -\frac{6}{-2} = -(-3) = 3$$

The axis of symmetry is $x = 3$.

6. (1) $x^2 - y^2 = 24$

(2) $2x^2 + 5y^2 = 55$

Use the addition method. Multiply equation (1) by -2.

$-2x^2 + 2y^2 = -48$

$\underline{2x^2 + 5y^2 = 55}$

$7y^2 = 7$

$y^2 = 1$

$y = \pm 1$

Substitute the values of y into equation (1).

$\begin{array}{ll} x^2 - y^2 = 24 & x^2 - y^2 = 24 \\ x^2 - (1)^2 = 24 & x^2 - (-1)^2 = 24 \\ x^2 - 1 = 24 & x^2 - 1 = 24 \\ x^2 = 25 & x^2 = 25 \\ x = \pm\sqrt{25} = \pm 5 & x = \pm\sqrt{25} = \pm 5 \end{array}$

The solutions are $(5, 1)$, $(-5, 1)$, $(5, -1)$ and $(-5, -1)$.

7. $y = -x^2 + 3x - 2$

$$x = -\frac{b}{2a} = -\frac{3}{2(-1)} = -\frac{3}{-2} = \frac{3}{2}$$

$$y = -\left(\frac{3}{2}\right)^2 + 3\left(\frac{3}{2}\right) - 2 = -\frac{9}{4} + \frac{9}{2} - 2 = \frac{1}{4}$$

The vertex is $\left(\dfrac{3}{2}, \dfrac{1}{4}\right)$.

8. $x = y^2 - y - 2$

$$y = -\frac{b}{2a} = -\frac{-1}{2(1)} = \frac{1}{2}$$

$$x = \left(\frac{1}{2}\right)^2 - \frac{1}{2} - 2 = -\frac{9}{4}$$

The axis of symmetry is $y = \dfrac{1}{2}$.

The vertex is $\left(-\dfrac{9}{4}, \dfrac{1}{2}\right)$.

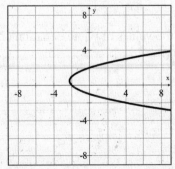

9. *x*-intercepts: $(4, 0)$ and $(-4, 0)$
y-intercepts: $(0, 2)$ and $(0, -2)$

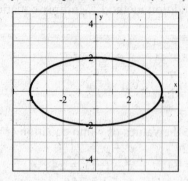

10. $\dfrac{x^2}{25} + \dfrac{y^2}{4} \le 1$

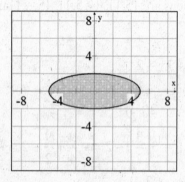

11. Center: $(-2, 4)$
Radius: 3
$(x - h)^2 + (y - k)^2 = r^2$
$(x - (-2))^2 + (y - 4)^2 = 3^2$
$(x + 2)^2 + (y - 4)^2 = 9$

12. (1) $x = 3y^2 + 2y - 4$
  (2) $x = y^2 - 5y$
Use the substitution method:
$$x = 3y^2 + 2y - 4$$
$$y^2 - 5y = 3y^2 + 2y - 4$$
$$0 = 2y^2 + 7y - 4$$
$$0 = (y + 4)(2y - 1)$$
$y + 4 = 0 \qquad 2y - 1 = 0$

$y = -4 \qquad y = \dfrac{1}{2}$

Substitute the values of *y* into equation (2).
$x = y^2 - 5y \qquad\qquad x = y^2 - 5y$

$x = (-4)^2 - 5(-4) \qquad x = \left(\dfrac{1}{2}\right)^2 - 5\left(\dfrac{1}{2}\right)$

$x = 36 \qquad\qquad\qquad x = -\dfrac{9}{4}$

The solutions are $\left(-\dfrac{9}{4}, \dfrac{1}{2}\right)$ and $(36, -4)$.

13. (1) $x^2 + 2y^2 = 4$
  (2) $\quad x + y = 2$
Solve equation (2) for *y*.
$x + y = 2$
$y = -x + 2$
Use the substitution method:
$$x^2 + 2y^2 = 4$$
$$x^2 + 2(-x + 2)^2 = 4$$
$$x^2 + 2(x^2 - 4x + 4) = 4$$
$$x^2 + 2x^2 - 8x + 8 = 4$$
$$3x^2 - 8x + 4 = 0$$
$$(3x - 2)(x - 2) = 0$$
$3x - 2 = 0 \qquad x - 2 = 0$

$x = \dfrac{2}{3} \qquad\qquad x = 2$

Substitute the values of *x* into equation (2).
$x + y = 2 \qquad\qquad x + y = 2$

$\dfrac{2}{3} + y = 2 \qquad\qquad 2 + y = 2$

$y = \dfrac{4}{3} \qquad\qquad\quad y = 0$

The solutions are $\left(\dfrac{2}{3}, \dfrac{4}{3}\right)$ and $(2, 0)$.

14. $(2, 4)$ and $(-1, -3)$
Use the distance formula to find the length of the radius:
$$d = \sqrt{(x_1 - x_2)^2 + (y_1 - y_2)^2}$$
$$= \sqrt{(2 - (-1))^2 + (4 - (-3))^2}$$
$$= \sqrt{3^2 + 7^2} = \sqrt{9 + 49} = \sqrt{58}$$
$(x - h)^2 + (y - k)^2 = r^2$
Center: $(-1, -3)$
Radius: $r = \sqrt{58}$
$(x - (-1))^2 + (y - (-3))^2 = (\sqrt{58})^2$
$(x + 1)^2 + (y + 3)^2 = 58$

15. $\dfrac{x^2}{25} - \dfrac{y^2}{16} > 1$

$x^2 + y^2 < 3$

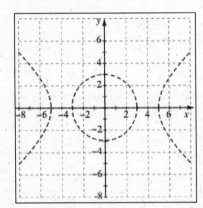

The solution sets of these inequalities do not intersect, so the system has no real number solution.

16. $x^2 + y^2 - 4x + 2y + 1 = 0$
$x^2 + y^2 - 4x + 2y = -1$
$(x^2 - 4x) + (y^2 + 2y) = -1$
$x^2 - 4x + 4 + y^2 + 2y + 1 = -1 + 4 + 1$
$(x - 2)^2 + (y + 1)^2 = 4$
Center: $(2, -1)$
Radius: 2

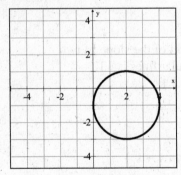

17. $y = \dfrac{1}{2}x^2 + x - 4$

$x = -\dfrac{b}{2a} = -\dfrac{1}{2\left(\dfrac{1}{2}\right)} = -\dfrac{1}{1} = -1$

$y = \left(\dfrac{1}{2}\right)(-1)^2 + (-1) - 4 = \dfrac{1}{2} - 1 - 4 = -\dfrac{9}{2}$

The axis of symmetry is $x = -1$.

The vertex is $\left(-1, -\dfrac{9}{2}\right)$.

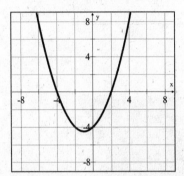

18. $(x - 2)^2 + (y + 1)^2 = 9$
Center: $(2, -1)$
Radius: 3

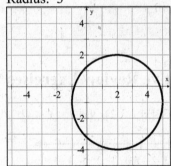

19. Axis of symmetry: x-axis

Vertices: $(3, 0)$ and $(-3, 0)$

Asymptotes: $y = \dfrac{2}{3}x$ and $y = -\dfrac{2}{3}x$

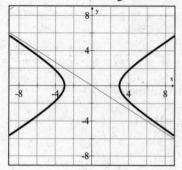

20. $\dfrac{x^2}{16} - \dfrac{y^2}{25} < 1$

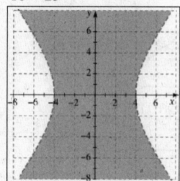

Cumulative Review Exercises

1. $\{x \mid x < 4\} \cap \{x \mid x > 2\} = \{x \mid 2 < x < 4\}$

$$\xleftarrow{\qquad}\underset{-5\ -4\ -3\ -2\ -1\ \ 0\ \ 1\ \ 2\ \ 3\ \ 4\ \ 5}{+\!+\!+\!+\!+\!+\!+\!(\!+\!)\!+}\xrightarrow{\qquad}$$

2. $\dfrac{5x-2}{3} - \dfrac{1-x}{5} = \dfrac{x+4}{10}$

$30\left(\dfrac{5x-2}{3} - \dfrac{1-x}{5}\right) = 30\left(\dfrac{x+4}{10}\right)$

$10(5x-2) - 6(1-x) = 3(x+4)$

$50x - 20 - 6 + 6x = 3x + 12$

$56x - 26 = 3x + 12$

$53x = 38$

$x = \dfrac{38}{53}$

The solution is $\dfrac{38}{53}$.

3. $4 + |3x + 2| < 6$

$\quad |3x + 2| < 2$

$\quad -2 < 3x + 2 < 2$

$\quad -4 < 3x < 0$

$\quad -\dfrac{4}{3} < x < 0$

$\quad \left\{ x \mid -\dfrac{4}{3} < x < 0 \right\}$

4. $m = -\dfrac{3}{2}$ and $(2, -3)$

$\quad y - y_1 = m(x - x_1)$

$\quad y - (-3) = -\dfrac{3}{2}(x - 2)$

$\quad y + 3 = -\dfrac{3}{2}x + 3$

$\quad y = -\dfrac{3}{2}x$

5. The product of the slope of two perpendicular lines is -1.

$\quad m_1 \cdot m_2 = -1$

$\quad m_1 \cdot (-1) = -1$

$\quad m_1 = 1$

$\quad y - y_1 = m(x - x_1)$

$\quad y - (-2) = 1(x - 4)$

$\quad y + 2 = x - 4$

$\quad y = x - 6$

6. $3a^2 b(4a - 3b^2) = 12a^3 b - 9a^2 b^3$

7. $(x-1)^3 - y^3$
$= [(x-1) - y][(x-1)^2 + (x-1)y + y^2]$
$= (x-1-y)(x^2 - 2x + 1 + xy - y + y^2)$
$= (x-y-1)(x^2 - 2x + 1 + xy - y + y^2)$

8. $\dfrac{3x-2}{x+4} \le 1$

$\dfrac{3x-2}{x+4} - 1 \le 0$

$\dfrac{3x-2}{x+4} - \dfrac{x+4}{x+4} \le 0$

$\dfrac{2x-6}{x+4} \le 0$

$\{x \mid -4 < x \le 3\}$

9. $\dfrac{ax-bx}{ax+ay-bx-by} = \dfrac{x(a-b)}{a(x+y)-b(x+y)}$

$= \dfrac{x(a-b)}{(x+y)(a-b)} = \dfrac{x}{x+y}$

10. $\dfrac{x-4}{3x-2} - \dfrac{1+x}{3x^2+x-2} = \dfrac{x-4}{3x-2} - \dfrac{1+x}{(3x-2)(x+1)}$

$= \dfrac{x-4}{3x-2} \cdot \dfrac{x+1}{x+1} - \dfrac{1+x}{(3x-2)(x+1)}$

$= \dfrac{x^2-3x-4}{(3x-2)(x+1)} - \dfrac{1+x}{(3x-2)(x+1)}$

$= \dfrac{x^2-4x-5}{(3x-2)(x+1)} = \dfrac{(x-5)(x+1)}{(3x-2)(x+1)}$

$= \dfrac{x-5}{3x-2}$

11. $\dfrac{6x}{2x-3} - \dfrac{1}{2x-3} = 7$

$(2x-3)\left(\dfrac{6x}{2x-3} - \dfrac{1}{2x-3}\right) = 7(2x-3)$

$6x - 1 = 14x - 21$

$20 = 8x$

$\dfrac{5}{2} = x$

The solution is $\dfrac{5}{2}$.

12. $5x + 2y > 10$
$2y > -5x + 10$
$y > -\dfrac{5}{2}x + 5$

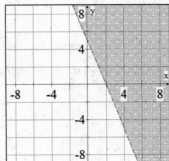

13. $\left(\dfrac{12a^2b^2}{a^{-3}b^{-4}}\right)^{-1} \left(\dfrac{ab}{4^{-1}a^{-2}b^4}\right)^2 = (12a^5b^6)^{-1}\left(\dfrac{4a^3}{b^3}\right)^2$

$= (12^{-1}a^{-5}b^{-6})\left(\dfrac{4^2a^6}{b^6}\right)$

$= \dfrac{1}{12a^5b^6} \cdot \dfrac{16a^6}{b^6} = \dfrac{16a^6}{12a^5b^{12}}$

$= \dfrac{4a}{3b^{12}}$

14. $2x^{3/4}$

15. $(3 + 5i)(4 - 6i) = 12 - 18i + 20i - 30i^2$
$= 12 + 2i - 30(-1)$
$= 12 + 2i + 30$
$= 42 + 2i$

16. $2x^2 + 2x - 3 = 0$

$x = \dfrac{-b \pm \sqrt{b^2 - 4ac}}{2a}$

$= \dfrac{-2 \pm \sqrt{2^2 - 4(2)(-3)}}{2(2)}$

$= \dfrac{-2 \pm \sqrt{4 + 24}}{4} = \dfrac{-2 \pm \sqrt{28}}{4} = \dfrac{-2 \pm 2\sqrt{7}}{4}$

$= \dfrac{-1 \pm \sqrt{7}}{2}$

The solutions are $\dfrac{-1 + \sqrt{7}}{2}$ and $\dfrac{-1 - \sqrt{7}}{2}$.

17. (1) $x^2 + y^2 = 20$
(2) $x^2 - y^2 = 12$
Use the addition method.
$x^2 + y^2 = 20$
$x^2 - y^2 = 12$
$\quad 2x^2 = 32$
$\quad\; x^2 = 16$
$\quad\;\; x = \pm 4$
Substitute the values of x into equation (1).

$x^2 + y^2 = 20 \qquad x^2 + y^2 = 20$
$(4)^2 + y^2 = 20 \qquad (-4)^2 + y^2 = 20$
$\quad\;\; y^2 = 4 \qquad\qquad\;\; y^2 = 4$
$\quad\;\;\; y = \pm 2 \qquad\qquad\; y = \pm 2$

The solutions are $(4, 2)$, $(-4, 2)$,
$(4, -2)$ and $(-4, -2)$.

18. $x - \sqrt{2x - 3} = 3$
$-\sqrt{2x - 3} = -x + 3$
$\sqrt{2x - 3} = x - 3$
$\left(\sqrt{2x - 3}\right)^2 = (x - 3)^2$
$2x - 3 = x^2 - 6x + 9$
$0 = x^2 - 8x + 12$
$0 = (x - 6)(x - 2)$
$x - 6 = 0 \qquad x - 2 = 0$
$\quad\; x = 6 \qquad\quad x = 2$
The solution is 6.

19. $f(-3) = -(-3)^2 + 3(-3) - 2$
$\qquad\quad = -9 - 9 - 2$
$\qquad\quad = -20$

20. $f(x) = 4x + 8$
$y = 4x + 8$
$x = 4y + 8$
$x - 8 = 4y$
$y = \dfrac{1}{4}x - 2$
$f^{-1}(x) = \dfrac{1}{4}x - 2$

21. $f(x) = -2x^2 + 4x - 2$
$x = -\dfrac{b}{2a} = -\dfrac{4}{2(-2)} = -\dfrac{4}{-4} = -(-1) = 1$
$f(1) = -2(1)^2 + 4(1) - 2 = -2 + 4 - 2 = 0$
The maximum value of the function is 0.

22. Strategy: Let x represent one number. The
other number is $40 - x$. To find the first
number, find the x-coordinate of the vertex
of the given function.
Solution: $f(x) = -x^2 + 40x$
$x = -\dfrac{b}{2a} = -\dfrac{40}{2(-1)} = -\dfrac{40}{-2} = -(-20) = 20$
$40 - x = 40 - 20 = 20$
The maximum product of the two numbers
whose sum is 40 is 400.

23. $(2, 4)$ and $(-1, 0)$
$d = \sqrt{(x_1 - x_2)^2 + (y_1 - y_2)^2}$
$\;\; = \sqrt{(2 - (-1))^2 + (4 - 0)^2}$
$\;\; = \sqrt{3^2 + 4^2} = \sqrt{9 + 16} = \sqrt{25}$
$\;\; = 5$

24. $(3, 1)$ and $(-1, 2)$
Use the distance formula to find the length
of the radius:
$d = \sqrt{(x_1 - x_2)^2 + (y_1 - y_2)^2}$
$\;\; = \sqrt{(3 - (-1))^2 + (1 - 2)^2}$
$\;\; = \sqrt{4^2 + (-1)^2} = \sqrt{16 + 1} = \sqrt{17}$
$(x - h)^2 + (y - k)^2 = r^2$
Center: $(-1, 2)$
Radius: $r = \sqrt{17}$
$(x - (-1))^2 + (y - 2)^2 = (\sqrt{17})^2$
$(x + 1)^2 + (y - 2)^2 = 17$

25. $x = y^2 - 2y + 3$

$$y = -\frac{b}{2a} = -\frac{-2}{2(1)} = -(-1) = 1$$

$$x = (1)^2 - 2(1) + 3 = 1 - 2 + 3 = 2$$

The axis of symmetry is $y = 1$.
The vertex is (2, 1).

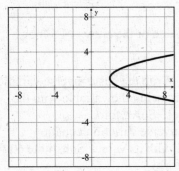

26. x-intercepts: (5, 0) and (−5, 0)
y-intercepts: (0, 2) and (0, −2)

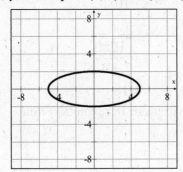

27. Axis of symmetry: y-axis
Vertices: (0, 2) and (0, −2)

Asymptotes: $y = \dfrac{2}{5}x$ and $y = -\dfrac{2}{5}x$

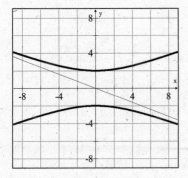

28. $(x - 1)^2 + y^2 \le 25$
$\quad\quad y^2 < x$

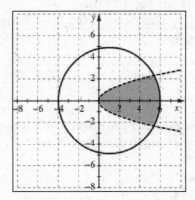

29. Strategy: Let x represent the number of adult tickets.
The number of children's tickets is $192 - x$.

	Amount	Cost	Value
Adult	x	12.00	$12x$
Children	192 - x	4.50	$4.5(192 - x)$

The sum of the values of each type of ticket sold equals the total value of all of the tickets sold ($1479).
Solution:

$$12x + 4.5(192 - x) = 1479$$
$$12x + 864 - 4.5x = 1479$$
$$7.5x + 864 = 1479$$
$$7.5x = 615$$
$$x = 82$$

There were 82 adult tickets sold.

30. Strategy: Let x represent the rate of the motorcycle.
The rate of the car is $x - 12$.

	Distance	Rate	Time
Motorcycle	180	x	$\dfrac{180}{x}$
Car	144	$x - 12$	$\dfrac{144}{x-12}$

The time traveled by the motorcycle is the same as the time traveled by the car.

Solution:
$$\frac{180}{x} = \frac{144}{x-12}$$
$$(x)(x-12)\frac{180}{x} = \frac{144}{x-12}(x)(x-12)$$
$$(x-12)180 = 144(x)$$
$$180x - 2160 = 144x$$
$$36x = 2160$$
$$x = 60$$
The rate of the motorcycle is 60 mph.

31. Strategy: Let x represent the rowing rate of the crew in calm water.

	Distance	Rate	Time
With current	12	$x + 1.5$	$\dfrac{12}{x+1.5}$
Against current	12	$x - 1.5$	$\dfrac{12}{x-1.5}$

The sum of the time traveling upriver and the time traveling downriver equals the total time traveled (6 h).

Solution:
$$\frac{12}{x+1.5} + \frac{12}{x-1.5} = 6$$
$$(x+1.5)(x-1.5)\left(\frac{12}{x+1.5} + \frac{12}{x-1.5}\right) = 6(x+1.5)(x-1.5)$$
$$12(x-1.5) + 12(x+1.5) = 6(x+1.5)(x-1.5)$$
$$12x - 18 + 12x + 18 = 6x^2 - 13.5$$
$$24x = 6x^2 - 13.5$$
$$0 = 6x^2 - 24x - 13.5$$

$$0 = 12x^2 - 48x - 27$$
$$0 = 3(4x^2 - 16x - 9)$$
$$0 = 3(2x + 1)(2x - 9)$$
$$2x + 1 = 0 \qquad 2x - 9 = 0$$
$$x = -\frac{1}{2} \qquad\qquad x = \frac{9}{2}$$

The rowing rate of the crew is 4.5 mph.

32. Strategy: Write the basic inverse variation equation replacing the variable with the given values. Solve for k.
Write the inverse variation equation replacing k with its value. Substitute 60 for t and solve for v.

Solution:
$$v = \frac{k}{t} \qquad\qquad v = \frac{1080}{t}$$
$$30 = \frac{k}{36} \qquad\qquad = \frac{1080}{60}$$
$$1080 = k \qquad\qquad = 18$$
The gear will make 18 revolutions/min.

Chapter 12: Sequences and Series

Prep Test

1. $[3(1) - 2] + [3(2) - 2] + [3(3) - 2]$
$= [3 - 2] + [6 - 2] + [9 - 2]$
$= 1 + 4 + 7$
$= 12$

2. $f(n) = \dfrac{n}{n+2}$

$f(6) = \dfrac{6}{6+2} = \dfrac{6}{8} = \dfrac{3}{4}$

3. $a_1 + (n-1)d$
$2 + (5-1)4 = 2 + (4)(4)$
$= 2 + 16 = 18$

4. $a_1 r^{n-1}$
$(-3)(-2)^{6-1} = (-3)(-2)^5$
$= (-3)(-32) = 96$

5. $\dfrac{a_1(1 - r^n)}{1 - r}$

$\dfrac{(-2)(1 - (-4)^5)}{1 - (-4)} = \dfrac{(-2)(1 - (-1024))}{1 + 4}$

$= \dfrac{(-2)(1025)}{5} = \dfrac{-2050}{5}$

$= -410$

6. $\dfrac{\dfrac{4}{10}}{1 - \dfrac{1}{10}} = \dfrac{\dfrac{4}{10}}{\dfrac{9}{10}} = \dfrac{4}{10} \div \dfrac{9}{10}$

$= \dfrac{4}{10} \cdot \dfrac{10}{9} = \dfrac{4}{9}$

7. $(x + y)(x + y) = x^2 + 2xy + y^2$

8. $(x + y)^3 = (x + y)^2(x + y)$
$= (x^2 + 2xy + y^2)(x + y)$
$= x(x^2 + 2xy + y^2) + y(x^2 + 2xy + y^2)$
$= x^3 + 2x^2y + xy^2 + x^2y + 2xy^2 + y^3$
$= x^3 + 3x^2y + 3xy^2 + y^3$

Section 12.1

Objective A Exercises

1. A sequence is an ordered list of numbers.

3. 8

5. $a_n = n + 1$
$a_1 = 1 + 1 = 2$ The first term is 2.
$a_2 = 2 + 1 = 3$ The second term is 3.
$a_3 = 3 + 1 = 4$ The third term is 4.
$a_4 = 4 + 1 = 5$ The fourth term is 5.

7. $a_n = 2n + 1$
$a_1 = 2(1) + 1 = 3$ The first term is 3.
$a_2 = 2(2) + 1 = 5$ The second term is 5.
$a_3 = 2(3) + 1 = 7$ The third term is 7.
$a_4 = 2(4) + 1 = 9$ The fourth term is 9.

9. $a_n = 2 - 2n$
$a_1 = 2 - 2(1) = 0$ The first term is 0.
$a_2 = 2 - 2(2) = -2$ The second term is -2.
$a_3 = 2 - 2(3) = -4$ The third term is -4.
$a_4 = 2 - 2(4) = -6$ The fourth term is -6.

11. $a_n = 2^n$
$a_1 = 2^1 = 2$ The first term is 2.
$a_2 = 2^2 = 4$ The second term is 4.
$a_3 = 2^3 = 8$ The third term is 8.
$a_4 = 2^4 = 16$ The fourth term is 16.

13. $a_n = n^2 + 1$
$a_1 = 1^2 + 1 = 2$ The first term is 2.
$a_2 = 2^2 + 1 = 5$ The second term is 5.
$a_3 = 3^2 + 1 = 10$ The third term is 10.
$a_4 = 4^2 + 1 = 17$ The fourth term is 17.

15. $a_n = n^2 - \dfrac{1}{n}$

$a_1 = 1^2 - \dfrac{1}{1} = 0$ The first term is 0.

$a_2 = 2^2 - \dfrac{1}{2} = \dfrac{7}{2}$ The second term is $\dfrac{7}{2}$.

$a_3 = 3^2 - \dfrac{1}{3} = \dfrac{26}{3}$ The third term is $\dfrac{26}{3}$.

$a_4 = 4^2 - \dfrac{1}{4} = \dfrac{63}{4}$ The fourth term is $\dfrac{63}{4}$.

17. $a_n = 3n + 4$
$a_{12} = 3(12) + 4 = 40$
The twelfth term is 40.

19. $a_n = n(n-1)$
$a_{11} = 11(11-1) = 110$
The eleventh term is 110.

21. $a_n = (-1)^{n-1}n^2$
$a_{15} = (-1)^{14}(15)^2 = 225$
The fifteenth term is 225.

23. $a_n = \left(\dfrac{1}{2}\right)^n$

$a_8 = \left(\dfrac{1}{2}\right)^8 = \dfrac{1}{256}$

The eighth term is $\dfrac{1}{256}$.

25. $a_n = (n+2)(n+3)$
$a_{17} = (17+2)(17+3) = (19)(20) = 380$
The seventeenth term is 380.

27. $a_n = \dfrac{(-1)^{2n-1}}{n^2}$

$a_6 = \dfrac{(-1)^{11}}{6^2} = -\dfrac{1}{36}$

The sixth term is $-\dfrac{1}{36}$.

29. $a_n = n^2$

Objective B Exercises

31. $\displaystyle\sum_{n=1}^{5}(2n+3) = (2\cdot1+3)+(2\cdot2+3)+(2\cdot3+3)+(2\cdot4+3)+(2\cdot5+3)$
$= 5+7+9+11+13 = 45$

33. $\displaystyle\sum_{i=1}^{4}(2i) = (2\cdot1)+(2\cdot2)+(2\cdot3)+(2\cdot4) = 2+4+6+8 = 20$

35. $\displaystyle\sum_{i=1}^{6}i^2 = 1^2+2^2+3^2+4^2+5^2+6^2 = 1+4+9+16+25+36 = 91$

37. $\displaystyle\sum_{n=1}^{6}(-1)^n = (-1)^1+(-1)^2+(-1)^3+(-1)^4+(-1)^5+(-1)^6$
$= -1+1-1+1-1+1 = 0$

39. $\displaystyle\sum_{i=3}^{6}i^3 = 3^3+4^3+5^3+6^3 = 27+64+125+216 = 432$

41. $\displaystyle\sum_{n=3}^{5}\frac{(-1)^{n-1}}{n-2}=\frac{(-1)^{2}}{3-2}+\frac{(-1)^{3}}{4-2}+\frac{(-1)^{4}}{5-2}=\frac{1}{1}+\frac{-1}{2}+\frac{1}{3}$

   $=\dfrac{6-3+2}{6}=\dfrac{5}{6}$

43. $\displaystyle\sum_{n=1}^{5}2x^{n}=2x+2x^{2}+2x^{3}+2x^{4}+2x^{5}$

45. $\displaystyle\sum_{i=1}^{5}\frac{x^{i}}{i}=\frac{x}{1}+\frac{x^{2}}{2}+\frac{x^{3}}{3}+\frac{x^{4}}{4}+\frac{x^{5}}{5}=x+\frac{x^{2}}{2}+\frac{x^{3}}{3}+\frac{x^{4}}{4}+\frac{x^{5}}{5}$

47. $\displaystyle\sum_{n=1}^{5}x^{2n}=x^{2}+x^{4}+x^{6}+x^{8}+x^{10}$

49. $\displaystyle\sum_{i=1}^{4}\frac{x^{i}}{i^{2}}=\frac{x}{1^{2}}+\frac{x^{2}}{2^{2}}+\frac{x^{3}}{3^{2}}+\frac{x^{4}}{4^{2}}=x+\frac{x^{2}}{4}+\frac{x^{3}}{9}+\frac{x^{4}}{16}$

51. $\displaystyle\sum_{i=1}^{5}x^{-i}=x^{-1}+x^{-2}+x^{-3}+x^{-4}+x^{-5}=\frac{1}{x}+\frac{1}{x^{2}}+\frac{1}{x^{3}}+\frac{1}{x^{4}}+\frac{1}{x^{5}}$

53. $\displaystyle\left(\sum_{i=1}^{n}i\right)^{2}$

Applying the Concepts

55. The first six terms of the Fibonacci Sequence are 1, 1, 2, 3, 5 and 8. Each remaining term is the sum of the previous two terms. $a_{n}=a_{n-2}+a_{n-1}$.
    The Fibonacci sequence was the outcome of a mathematical problem about rabbit breeding that was discussed in Leonard Fibonacci's book, *Liber Abaci*. The sequence is the solution to the following problem: Beginning with a single pair of rabbits (one male and one female), how many pairs of rabbits will be born in a year, assuming that every month each male and female rabbit gives birth to a new pair of rabbits, and the new pair of rabbits itself starts giving birth to additional pairs of rabbits after the first month of their birth?

Section 12.2

Objective A Exercises

1. $d = a_2 - a_1 = 11 - 1 = 10$
$a_n = a_1 + (n-1)d$
$a_{15} = 1 + (15-1)(10) = 1 + 14(10) = 1 + 140$
$a_{15} = 141$

3. $d = a_2 - a_1 = -2 - (-6) = 4$
$a_n = a_1 + (n-1)d$
$a_{15} = -6 + (15-1)(4) = -6 + 14(4) = -6 + 56$
$a_{15} = 50$

5. $d = a_2 - a_1 = \dfrac{5}{2} - 2 = \dfrac{1}{2}$
$a_n = a_1 + (n-1)d$
$a_{31} = 2 + (31-1)\left(\dfrac{1}{2}\right) = 2 + 30\left(\dfrac{1}{2}\right) = 2 + 15$
$a_{31} = 17$

7. $d = a_2 - a_1 = -\dfrac{5}{2} - (-4) = \dfrac{3}{2}$
$a_n = a_1 + (n-1)d$
$a_{12} = -4 + (12-1)\left(\dfrac{3}{2}\right) = -4 + 11\left(\dfrac{3}{2}\right) = -4 + \dfrac{33}{2}$
$a_{12} = \dfrac{25}{2}$

9. $d = a_2 - a_1 = 5 - 8 = -3$
$a_n = a_1 + (n-1)d$
$a_{40} = 8 + (40-1)(-3) = 8 + 39(-3) = 8 - 117$
$a_{40} = -109$

11. $d = a_2 - a_1 = 4 - 1 = 3$
$a_n = a_1 + (n-1)d$
$a_n = 1 + (n-1)(3) = 1 + 3n - 3$
$a_n = 3n - 2$

13. $d = a_2 - a_1 = 0 - 3 = -3$
$a_n = a_1 + (n-1)d$
$a_n = 3 + (n-1)(-3) = 3 - 3n + 3$
$a_n = -3n + 6$

15. $d = a_2 - a_1 = 4.5 - 7 = -2.5$
$a_n = a_1 + (n-1)d$
$a_n = 7 + (n-1)(-2.5) = 7 - 2.5n + 2.5$
$a_n = -2.5n + 9.5$

17. $d = a_2 - a_1 = 8 - 3 = 5$
$a_n = a_1 + (n-1)d$
$98 = 3 + (n-1)(5)$
$98 = 3 + 5n - 5$
$98 = 5n - 2$
$100 = 5n$
$20 = n$
There are 20 terms in the sequence.

19. $d = a_2 - a_1 = -3 - 1 = -4$
$a_n = a_1 + (n-1)d$
$-75 = 1 + (n-1)(-4)$
$-75 = 1 - 4n + 4$
$-75 = -4n + 5$
$-80 = -4n$
$20 = n$
There are 20 terms in the sequence.

21. $d = a_2 - a_1 = \dfrac{13}{3} - \dfrac{7}{3} = \dfrac{6}{3} = 2$

$a_n = a_1 + (n-1)d$

$\dfrac{79}{3} = \dfrac{7}{3} + (n-1)2$

$\dfrac{79}{3} = \dfrac{7}{3} + 2n - 2$

$\dfrac{79}{3} = 2n + \dfrac{1}{3}$

$\dfrac{78}{3} = 2n$

$26 = 2n$

$13 = n$

There are 13 terms in the sequence.

23. $d = a_2 - a_1 = 2 - 3.5 = -1.5$

$a_n = a_1 + (n-1)d$

$-25 = 3.5 + (n-1)(-1.5)$

$-25 = 3.5 - 1.5n + 1.5$

$-25 = -1.5n + 5$

$-30 = -1.5n$

$20 = n$

There are 20 terms in the sequence.

25. No

27. Yes

Objective B Exercises

29. $d = a_2 - a_1 = 3 - 1 = 2$

$a_n = a_1 + (n-1)d$

$a_{50} = 1 + (50-1)2 = 1 + 49(2) = 99$

$S_n = \dfrac{n}{2}(a_1 + a_n)$

$S_{50} = \dfrac{50}{2}(1 + 99) = 25(100) = 2500$

31. $d = a_2 - a_1 = 18 - 20 = -2$

$a_n = a_1 + (n-1)d$

$a_{40} = 20 + (40-1)(-2) = 20 + 39(-2) = -58$

$S_n = \dfrac{n}{2}(a_1 + a_n)$

$S_{40} = \dfrac{40}{2}(20 - 58) = 20(-38) = -760$

33. $d = a_2 - a_1 = 1 - \dfrac{1}{2} = \dfrac{1}{2}$

$a_n = a_1 + (n-1)d$

$a_{27} = \dfrac{1}{2} + (27-1)\dfrac{1}{2} = \dfrac{1}{2} + (26)\dfrac{1}{2}$

$a_{27} = \dfrac{1}{2} + 13 = \dfrac{27}{2}$

$S_n = \dfrac{n}{2}(a_1 + a_n)$

$S_{27} = \dfrac{27}{2}\left(\dfrac{1}{2} + \dfrac{27}{2}\right) = \dfrac{27}{2}\left(\dfrac{28}{2}\right) = 189$

35. $a_i = 3i - 1$

$a_1 = 3(1) - 1 = 2$

$a_{15} = 3(15) - 1 = 44$

$S_i = \dfrac{i}{2}(a_1 + a_i)$

$S_{15} = \dfrac{15}{2}(2 + 44) = \left(\dfrac{15}{2}\right)(46) = 345$

37. $a_n = \dfrac{1}{2}n + 1$

$a_1 = \dfrac{1}{2}(1) + 1 = \dfrac{3}{2}$

$a_{17} = \dfrac{1}{2}(17) + 1 = \dfrac{19}{2}$

$S_n = \dfrac{n}{2}(a_1 + a_n)$

$S_{17} = \dfrac{17}{2}\left(\dfrac{3}{2} + \dfrac{19}{2}\right) = \left(\dfrac{17}{2}\right)(11) = \dfrac{187}{2}$

39. $a_i = 4 - 2i$

$a_1 = 4 - 2(1) = 2$

$a_{15} = 4 - 2(15) = -26$

$S_i = \dfrac{i}{2}(a_1 + a_i)$

$S_{15} = \dfrac{15}{2}\left(2 + (-26)\right) = \left(\dfrac{15}{2}\right)(-24) = -180$

41. $d = a_2 - a_1 = 3 - 1 = 2$

$a_n = a_1 + (n-1)d$

$a_n = 1 + (n-1)2 = 1 + 2n - 2 = 2n - 1$

$S_n = \dfrac{n}{2}(a_1 + a_n)$

$S_n = \dfrac{n}{2}(1 + 2n - 1) = \dfrac{n}{2}(2n) = n^2$

Objective C Exercises

43. Strategy: Write the arithmetic sequence.
Find the common difference, d.
Use the formula for the nth Term of an Arithmetic Sequence to find the number of terms in the sequence.

Solution: 10, 15, 20, …, 60

$d = a_2 - a_1 = 15 - 10 = 5$

$a_n = a_1 + (n-1)d$

$60 = 10 + (n-1)(5)$

$60 = 10 + 5n - 5$

$60 = 5n + 5$

$55 = 5n$

$11 = n$

In 11 weeks, the person will walk 60 min per day.

45. Strategy: Write the arithmetic sequence.
Find the common difference, d.
Use the formula for the nth Term of an Arithmetic Sequence to find the number of rows.
Use the Formula for the Sum of n Terms of an Arithmetic Sequence.

Solution: 73, 79, 85 …

$d = a_2 - a_1 = 79 - 73 = 6$

$a_n = a_1 + (n-1)d$

$a_{27} = 73 + (27-1)(6) = 73 + (26)(6)$

$a_{27} = 229$

$S_n = \dfrac{n}{2}(a_1 + a_n)$

$S_{27} = \dfrac{27}{2}(73 + 229) = \dfrac{27}{2}(302) = 4077$

There are 4077 seats in the loge seating section.

47. $\dfrac{1225 \text{ calories}}{1 \text{ hour}} \cdot \dfrac{1 \text{ lb}}{3500 \text{ calories}} = 0.35 \text{ lb/h}$

$a_n = 200 - 0.35n$

$186 = 200 - 0.35n$

$-14 = -0.35n$

$40 = n$

It will take 40 h.

Applying the Concepts

49. $\displaystyle\sum_{i=1}^{2} \log 2i = \log 2(1) + \log 2(2) = \log 2 + \log 4$

$= \log(2 \cdot 4) = \log 8$

Section 12.3

Objective A Exercises

1. An arithmetic sequence is one in which the difference between any two consecutive terms is constant.
A geometric sequence is one in which each successive term in the sequence is the same nonzero constant multiple of the preceding term.

3. $r = \dfrac{a_2}{a_1} = \dfrac{8}{2} = 4$

$a_n = a_1 r^{n-1}$

$a_9 = 2(4)^8 = 2(65,536)$

$a_9 = 131,072$

5. $r = \dfrac{a_2}{a_1} = \dfrac{-4}{6} = -\dfrac{2}{3}$

$a_n = a_1 r^{n-1}$

$a_7 = 6\left(-\dfrac{2}{3}\right)^6 = 6\left(\dfrac{64}{729}\right)$

$a_7 = \dfrac{128}{243}$

7. $r = \dfrac{a_2}{a_1} = \dfrac{\frac{1}{8}}{-\frac{1}{16}} = -2$

$a_n = a_1 r^{n-1}$

$a_{10} = -\dfrac{1}{16}(-2)^9 = -\dfrac{1}{16}(-512)$

$a_{10} = 32$

9. $a_n = a_1 r^{n-1}$

$a_4 = 9r^3$

$\dfrac{8}{3} = 9r^3$

$\dfrac{8}{27} = r^3$

$\dfrac{2}{3} = r$

$a_n = a_1 r^{n-1}$

$a_2 = 9\left(\dfrac{2}{3}\right)^1 = 6$

$a_3 = 9\left(\dfrac{2}{3}\right)^2 = 9\left(\dfrac{4}{9}\right) = 4$

11. $a_n = a_1 r^{n-1}$

$a_4 = 3r^3$

$-\dfrac{8}{9} = 3r^3$

$-\dfrac{8}{27} = r^3$

$-\dfrac{2}{3} = r$

$a_n = a_1 r^{n-1}$

$a_2 = 3\left(-\dfrac{2}{3}\right)^1 = -2$

$a_3 = 3\left(-\dfrac{2}{3}\right)^2 = 3\left(\dfrac{4}{9}\right) = \dfrac{4}{3}$

13. $a_n = a_1 r^{n-1}$

$a_4 = -3r^3$

$192 = -3r^3$

$-64 = r^3$

$-4 = r$

$a_n = a_1 r^{n-1}$

$a_2 = -3(-4)^1 = 12$

$a_3 = -3(-4)^2 = -3(16) = -48$

15. No

17. Yes

Objective B Exercises

19. $r = \dfrac{a_2}{a_1} = \dfrac{6}{2} = 3$

$S_n = \dfrac{a_1(1 - r^n)}{1 - r}$

$S_7 = \dfrac{2(1 - 3^7)}{1 - 3} = \dfrac{2(1 - 2187)}{-2} = \dfrac{2(-2186)}{-2}$

$S_7 = 2186$

21. $r = \dfrac{a_2}{a_1} = \dfrac{-2}{3} = -\dfrac{2}{3}$

$S_n = \dfrac{a_1(1-r^n)}{1-r}$

$S_5 = \dfrac{3\left(1-\left(-\dfrac{2}{3}\right)^5\right)}{1-\left(-\dfrac{2}{3}\right)} = \dfrac{3\left(1+\left(\dfrac{32}{243}\right)\right)}{\dfrac{5}{3}} = \dfrac{3\left(\dfrac{275}{243}\right)}{\dfrac{5}{3}}$

$S_5 = \dfrac{\dfrac{275}{81}}{\dfrac{5}{3}} = \dfrac{275}{81} \cdot \dfrac{3}{5} = \dfrac{55}{27}$

23. $r = \dfrac{a_2}{a_1} = \dfrac{9}{12} = \dfrac{3}{4}$

$S_n = \dfrac{a_1(1-r^n)}{1-r}$

$S_5 = \dfrac{12\left(1-\left(\dfrac{3}{4}\right)^5\right)}{1-\left(\dfrac{3}{4}\right)} = \dfrac{12\left(1-\left(\dfrac{243}{1024}\right)\right)}{\dfrac{1}{4}} = \dfrac{12\left(\dfrac{781}{1024}\right)}{\dfrac{1}{4}}$

$S_5 = \dfrac{\dfrac{2343}{256}}{\dfrac{1}{4}} = \dfrac{2343}{256} \cdot \dfrac{4}{1} = \dfrac{2343}{64}$

25. $a_n = 2^i$

$a_1 = 2^1 = 2$

$a_2 = 2^2 = 4$

$r = \dfrac{a_2}{a_1} = \dfrac{4}{2} = 2$

$S_n = \dfrac{a_1(1-r^n)}{1-r}$

$S_5 = \dfrac{2(1-2^5)}{1-2} = \dfrac{2(1-32)}{-1} = \dfrac{2(-31)}{-1}$

$S_5 = 62$

27. $a_i = \left(\dfrac{1}{3}\right)^i$

$a_1 = \left(\dfrac{1}{3}\right)^1 = \dfrac{1}{3}$

$a_2 = \left(\dfrac{1}{3}\right)^2 = \dfrac{1}{9}$

$r = \dfrac{a_2}{a_1} = \dfrac{\dfrac{1}{9}}{\dfrac{1}{3}} = \dfrac{1}{3}$

$S_n = \dfrac{a_1(1-r^n)}{1-r}$

$S_5 = \dfrac{\dfrac{1}{3}\left(1-\left(\dfrac{1}{3}\right)^5\right)}{1-\dfrac{1}{3}} = \dfrac{\dfrac{1}{3}\left(1-\dfrac{1}{243}\right)}{\dfrac{2}{3}} = \dfrac{\dfrac{1}{3}\left(\dfrac{242}{243}\right)}{\dfrac{2}{3}}$

$S_6 = \dfrac{1}{3} \cdot \dfrac{242}{243} \cdot \dfrac{3}{2} = \dfrac{121}{243}$

29. Positive

Objective C Exercises

31. $r = \dfrac{a_2}{a_1} = \dfrac{2}{3}$

$S = \dfrac{a_1}{1-r} = \dfrac{3}{1-\dfrac{2}{3}} = \dfrac{3}{\dfrac{1}{3}}$

$S = 9$

33. $r = \dfrac{a_2}{a_1} = \dfrac{\dfrac{7}{100}}{\dfrac{7}{10}} = \dfrac{1}{10}$

$S = \dfrac{a_1}{1-r} = \dfrac{\dfrac{7}{10}}{1-\dfrac{1}{10}} = \dfrac{\dfrac{7}{10}}{\dfrac{9}{10}} = \dfrac{7}{9}$

35. $0.8\overline{88} = 0.8 + 0.08 + 0.008 + \dots$

$0.8\overline{88} = \dfrac{8}{10} + \dfrac{8}{100} + \dfrac{8}{1000} + \dots$

$S = \dfrac{a_1}{1-r} = \dfrac{\dfrac{8}{10}}{1 - \dfrac{1}{10}} = \dfrac{\dfrac{8}{10}}{\dfrac{9}{10}} = \dfrac{8}{9}$

The equivalent fraction is $\dfrac{8}{9}$.

37. $0.2\overline{22} = 0.2 + 0.02 + 0.002 + \dots$

$0.2\overline{22} = \dfrac{2}{10} + \dfrac{2}{100} + \dfrac{2}{1000} + \dots$

$S = \dfrac{a_1}{1-r} = \dfrac{\dfrac{2}{10}}{1 - \dfrac{1}{10}} = \dfrac{\dfrac{2}{10}}{\dfrac{9}{10}} = \dfrac{2}{9}$

The equivalent fraction is $\dfrac{2}{9}$.

39. $0.45\overline{45} = 0.45 + 0.0045 + 0.000045 + \dots$

$0.45\overline{45} = \dfrac{45}{100} + \dfrac{45}{10,000} + \dfrac{45}{1,000,000} + \dots$

$S = \dfrac{a_1}{1-r} = \dfrac{\dfrac{45}{100}}{1 - \dfrac{1}{100}} = \dfrac{\dfrac{45}{100}}{\dfrac{99}{100}} = \dfrac{45}{99} = \dfrac{9}{11}$

The equivalent fraction is $\dfrac{9}{11}$.

41. $0.1\overline{66} = 0.1 + 0.06 + 0.006 + \dots$

$0.1\overline{66} = \dfrac{1}{10} + \dfrac{6}{100} + \dfrac{6}{1000} + \dots$

$S = \dfrac{a_1}{1-r} = \dfrac{\dfrac{6}{100}}{1 - \dfrac{1}{10}} = \dfrac{\dfrac{6}{100}}{\dfrac{9}{10}} = \dfrac{6}{90} = \dfrac{1}{15}$

$0.1\overline{66} = \dfrac{1}{10} + \dfrac{1}{15} = \dfrac{5}{30} = \dfrac{1}{6}$

The equivalent fraction is $\dfrac{1}{6}$.

43. Yes

45. No

Objective D Exercises

47. Strategy: To find the height of the ball on the sixth bounce, use the formula for the nth Term of a Geometric Sequence.

Solution: $n = 6$, $a_1 = 75\%$ of $10 = 7.5$

$r = 75\% = \dfrac{3}{4}$

$a_n = a_1 r^{n-1}$

$a_6 = 7.5\left(\dfrac{3}{4}\right)^5$

$a_6 = 7.5\left(\dfrac{243}{1024}\right) = 1.8$

The ball bounces to a height of 1.8 ft on the sixth bounce.

49. Strategy: Find the overall percent increase in waste generated. Then divide by 5 years to find the average annual percent increase.

Solution: $\dfrac{3.7 - 2.85}{3.7} = \dfrac{0.85}{3.7} = 0.22$ or 22% from 2001 to 2006.

$\dfrac{0.22}{5} = 0.044$ or 4.4% average annual increase in waste generated.

51. Strategy: To find the total amount of waste generated, use the formula for the Sum of a Finite Geometric Sequence.

Solution: $n = 5$, $a_1 = 2.85$, $r = 1.044$

$S_n = \dfrac{a_1(1 - r^n)}{1 - r}$

$S_6 = \dfrac{2.85(1 - 1.044^6)}{1 - 1.044} = 19.1$

19.1 million tons of waste were generated from 2001 to 2006.

Applying the Concepts

53. a) False
b) False
c) False
d) False

Section 12.4

Objective A Exercises

1. $3! = 3 \cdot 2 \cdot 1 = 6$

3. $8! = 8 \cdot 7 \cdot 6 \cdot 5 \cdot 4 \cdot 3 \cdot 2 \cdot 1 = 40,320$

5. $0! = 1$

7. $\dfrac{5!}{2!3!} = \dfrac{5 \cdot 4 \cdot 3 \cdot 2 \cdot 1}{(2 \cdot 1)(3 \cdot 2 \cdot 1)} = 10$

9. $\dfrac{6!}{6!0!} = \dfrac{6 \cdot 5 \cdot 4 \cdot 3 \cdot 2 \cdot 1}{(6 \cdot 5 \cdot 4 \cdot 3 \cdot 2 \cdot 1)(1)} = 1$

11. $\dfrac{9!}{6!3!} = \dfrac{9 \cdot 8 \cdot 7 \cdot 6 \cdot 5 \cdot 4 \cdot 3 \cdot 2 \cdot 1}{(6 \cdot 5 \cdot 4 \cdot 3 \cdot 2 \cdot 1)(3 \cdot 2 \cdot 1)} = 84$

13. $\dbinom{7}{2} = \dfrac{7!}{(7-2)!2!} = \dfrac{7!}{5!2!} = \dfrac{7 \cdot 6 \cdot 5 \cdot 4 \cdot 3 \cdot 2 \cdot 1}{(5 \cdot 4 \cdot 3 \cdot 2 \cdot 1)(2 \cdot 1)} = 21$

15. $\dbinom{10}{2} = \dfrac{10!}{(10-2)!2!} = \dfrac{10!}{8!2!} = \dfrac{10 \cdot 9 \cdot 8 \cdot 7 \cdot 6 \cdot 5 \cdot 4 \cdot 3 \cdot 2 \cdot 1}{(8 \cdot 7 \cdot 6 \cdot 5 \cdot 4 \cdot 3 \cdot 2 \cdot 1)(2 \cdot 1)} = 45$

17. $\dbinom{9}{0} = \dfrac{9!}{(9-0)!0!} = \dfrac{9!}{9!0!} = \dfrac{9 \cdot 8 \cdot 7 \cdot 6 \cdot 5 \cdot 4 \cdot 3 \cdot 2 \cdot 1}{(9 \cdot 8 \cdot 7 \cdot 6 \cdot 5 \cdot 4 \cdot 3 \cdot 2 \cdot 1)(1)} = 1$

19. $\dbinom{6}{3} = \dfrac{6!}{(6-3)!3!} = \dfrac{6!}{3!3!} = \dfrac{6 \cdot 5 \cdot 4 \cdot 3 \cdot 2 \cdot 1}{(3 \cdot 2 \cdot 1)(3 \cdot 2 \cdot 1)} = 20$

21. $\dbinom{11}{1} = \dfrac{11!}{(11-1)!1!} = \dfrac{11!}{10!1!} = \dfrac{11 \cdot 10 \cdot 9 \cdot 8 \cdot 7 \cdot 6 \cdot 5 \cdot 4 \cdot 3 \cdot 2 \cdot 1}{(10 \cdot 9 \cdot 8 \cdot 7 \cdot 6 \cdot 5 \cdot 4 \cdot 3 \cdot 2 \cdot 1)(1)} = 11$

23. $\dbinom{4}{4} = \dfrac{4!}{(4-2)!2!} = \dfrac{4!}{2!2!} = \dfrac{4 \cdot 3 \cdot 2 \cdot 1}{(2 \cdot 1)(2 \cdot 1)} = 6$

25. $(x+y)^4 = \binom{4}{0}x^4 + \binom{4}{1}x^3 y + \binom{4}{2}x^2 y^2 + \binom{4}{3}xy^3 + \binom{4}{4}y^4 = x^4 + 4x^3 y + 6x^2 y^2 + 4xy^3 + y^4$

27. $(x-y)^5 = \binom{5}{0}x^5 + \binom{5}{1}x^4(-y) + \binom{5}{2}x^3(-y)^2 + \binom{5}{3}x^2(-y)^3 + \binom{5}{4}x(-y)^4 + \binom{5}{5}(-y)^5$

$= x^5 - 5x^4 y + 10x^3 y^2 - 10x^2 y^3 + 5xy^4 - y^5$

29. $(2m+1)^4 = \binom{4}{0}(2m)^4 + \binom{4}{1}(2m)^3(1) + \binom{4}{2}(2m)^2(1)^2 + \binom{4}{3}(2m)(1)^3 + \binom{4}{4}(1)^4$

$= 16m^4 + 4(8m^3) + 6(4m^2) + 4(2m) + 1$

$= 16m^4 + 32m^3 + 24m^2 + 8m + 1$

31. $(2r-3)^5 = \binom{5}{0}(2r)^5 + \binom{5}{1}(2r)^4(-3) + \binom{5}{2}(2r)^3(-3)^2 + \binom{5}{3}(2r)^2(-3)^3 + \binom{5}{4}(2r)(-3)^4 + \binom{5}{5}(-3)^5$

$= 32r^5 + 5(16r^4)(-3) + 10(8r^3)(9) + 10(4r^2)(-27) + 5(2r)(81) - 243$

$= 32r^5 - 240r^4 + 720r^3 - 1080r^2 + 810r - 243$

33. $(a+b)^{10} = \binom{10}{0}a^{10} + \binom{10}{1}a^9 b + \binom{10}{2}a^8 b^2 + \ldots$

$= a^{10} + 10a^9 b + 45a^8 b^2 + \ldots$

35. $(a-b)^{11} = \binom{11}{0}a^{11} + \binom{11}{1}a^{10}(-b) + \binom{11}{2}a^9(-b)^2 + \ldots$

$= a^{11} + 11a^{10}(-b) + 55a^9(-b)^2 + \ldots$

$= a^{11} - 11a^{10}b + 55a^9 b^2 + \ldots$

37. $(2x+y)^8 = \binom{8}{0}(2x)^8 + \binom{8}{1}(2x)^7(y) + \binom{8}{2}(2x)^6(y)^2 + \ldots$

$= 256x^8 + 8(128x^7)(y) + 28(64x^6)y^2 + \ldots$

$= 256x^8 + 1024x^7 y + 1792x^6 y^2 + \ldots$

39. $(4x-3y)^8 = \binom{8}{0}(4x)^8 + \binom{8}{1}(4x)^7(-3y) + \binom{8}{2}(4x)^6(-3y)^2 + \ldots$

$= 65{,}536x^8 + 8(16{,}384x^7)(-3y) + 28(4096x^6)(9y^2) + \ldots$

$= 65{,}536x^8 - 393{,}216x^7 y + 1{,}032{,}192x^6 y^2 + \ldots$

41. $\left(x+\dfrac{1}{x}\right)^7 = \dbinom{7}{0}x^7 + \dbinom{7}{1}x^6\left(\dfrac{1}{x}\right) + \dbinom{7}{2}x^5\left(\dfrac{1}{x}\right)^2 + \ldots$

$= x^7 + 7x^6\left(\dfrac{1}{x}\right) + 21x^5\left(\dfrac{1}{x}\right)^2 + \ldots$

$= x^7 + 7x^5 + 21x^3 + \ldots$

43. $n = 7, a = 2x, b = -1, r = 4$

$\dbinom{7}{4-1}(2x)^{7-4+1}(-1)^{4-1} = \dbinom{7}{3}(2x)^4(-1)^3 = 35(16x^4)(-1) = -560x^4$

45. $n = 6, a = x^2, b = -y^2, r = 2$

$\dbinom{6}{2-1}(x^2)^{6-2+1}(-y^2)^{2-1} = \dbinom{6}{1}(x^2)^5(-y^2) = -6x^{10}y^2$

47. $n = 9, a = y, b = -1, r = 5$

$\dbinom{9}{5-1}y^{9-5+1}(-1)^{5-1} = \dbinom{9}{4}y^5(-1)^4 = 126y^5(1) = 126y^5$

49. $n = 5, a = n, b = \dfrac{1}{n}, r = 2$

$\dbinom{5}{2-1}n^{5-2+1}\left(\dfrac{1}{n}\right)^{2-1} = \dbinom{5}{1}n^4\left(\dfrac{1}{n}\right) = 5n^4\left(\dfrac{1}{n}\right) = 5n^3$

51. False

Applying the Concepts

53. a) False
b) False
c) False
d) False
e) False
f) True

Chapter 12 Review Exercises

1. $\displaystyle\sum_{i=1}^{4} 2x^{i-1} = 2x^0 + 2x^1 + 2x^2 + 2x^3$

$= 2 + 2x + 2x^2 + 2x^3$

2. $a_n = 3n - 2$
$a_{10} = 3(10) - 2 = 30 - 2 = 28$
The tenth term is 28.

3. $0.6\overline{33} = 0.6 + 0.03 + 0.003 + \ldots$

$0.6\overline{33} = \dfrac{6}{10} + \dfrac{3}{100} + \dfrac{3}{1000} + \ldots$

$S = \dfrac{a_1}{1-r} = \dfrac{\dfrac{3}{100}}{1 - \dfrac{1}{10}} = \dfrac{\dfrac{3}{100}}{\dfrac{9}{10}} = \dfrac{1}{30}$

$0.6\overline{33} = \dfrac{6}{10} + \dfrac{1}{30} = \dfrac{19}{30}$

The equivalent fraction is $\dfrac{19}{30}$.

4. $r = \dfrac{a_2}{a_1} = \dfrac{\frac{3}{100}}{\frac{3}{10}} = \dfrac{1}{10}$

$S = \dfrac{a_1}{1-r} = \dfrac{\frac{3}{10}}{1-\frac{1}{10}} = \dfrac{\frac{3}{10}}{\frac{9}{10}} = \dfrac{1}{3}$

5. $\displaystyle\sum_{i=1}^{30} 4i - 1 = 3, 7, 11, 15 \ldots 119$

$S_n = \dfrac{30}{2}(3+119) = 15(122) = 1830$

6. $\displaystyle\sum_{n=1}^{5}(3n-2) = (3\cdot 1 - 2) + (3\cdot 2 - 2)$

$\qquad\qquad + (3\cdot 3 - 2) + (3\cdot 4 - 2) + (3\cdot 5 - 2)$

$= 1 + 4 + 7 + 10 + 13 = 35$

7. $0.23\overline{23} = 0.23 + 0.0023 + 0.000023 + \ldots$

$0.23\overline{23} = \dfrac{23}{100} + \dfrac{23}{10,000} + \dfrac{23}{1,000,000} + \ldots$

$S = \dfrac{a_1}{1-r} = \dfrac{\frac{23}{100}}{1-\frac{1}{100}} = \dfrac{\frac{23}{100}}{\frac{99}{100}} = \dfrac{23}{99}$

The equivalent fraction is $\dfrac{23}{99}$.

8. $\dbinom{9}{3} = \dfrac{9!}{(9-3)!3!} = \dfrac{9!}{6!3!}$

$= \dfrac{9\cdot 8\cdot 7\cdot 6\cdot 5\cdot 4\cdot 3\cdot 2\cdot 1}{(6\cdot 5\cdot 4\cdot 3\cdot 2\cdot 1)(3\cdot 2\cdot 1)} = 84$

9. $\displaystyle\sum_{n=1}^{5} 2(3)^n = 6, 18, 54 \ldots$

$r = \dfrac{a_2}{a_1} = \dfrac{18}{6} = 3$

$S_n = \dfrac{a_1(1-r^n)}{1-r}$

$S_5 = \dfrac{6(1-3^5)}{1-3} = \dfrac{6(1-243)}{-2} = \dfrac{6(-242)}{-2}$

$S_5 = 726$

10. $d = a_2 - a_1 = 2 - 8 = -6$

$a_n = a_1 + (n-1)d$

$-118 = 8 + (n-1)(-6)$

$-118 = 8 - 6n + 6$

$-118 = -6n + 14$

$-132 = -6n$

$22 = n$

There are 22 terms in the sequence.

11. $r = \dfrac{a_2}{a_1} = \dfrac{\frac{1}{4}}{\frac{1}{2}} = \dfrac{1}{2}$

$S_n = \dfrac{a_1(1-r^n)}{1-r}$

$S_8 = \dfrac{\frac{1}{2}\left(1-\left(\frac{1}{2}\right)^8\right)}{1-\frac{1}{2}} = \dfrac{\frac{1}{2}\left(1-\frac{1}{256}\right)}{\frac{1}{2}} = \dfrac{255}{256}$

$S_8 = 0.996$

12. $n = 8,\ a = 3x,\ b = -y,\ r = 5$

$\dbinom{8}{5-1}(3x)^{8-5+1}(-y)^{5-1} = \dbinom{8}{4}(3x)^4(-y)^4$

$= 70(81x^4)y^4 = 5670x^4 y^4$

13. $a_n = \dfrac{(-1)^{2n-1}n}{n^2+2}$

$a_5 = \dfrac{(-1)^9(5)}{5^2+2} = -\dfrac{5}{27}$

The sixth term is $-\dfrac{5}{27}$.

14. $r = \dfrac{a_2}{a_1} = \dfrac{5\sqrt{5}}{5} = \sqrt{5}$

$S_n = \dfrac{a_1(1 - r^n)}{1 - r}$

$S_7 = \dfrac{5(1 - (\sqrt{5})^7)}{1 - \sqrt{5}} = \dfrac{5(1 - 125\sqrt{5})}{1 - \sqrt{5}}$

$= \dfrac{5 - 625\sqrt{5}}{1 - \sqrt{5}} = \dfrac{5 - 625\sqrt{5}}{1 - \sqrt{5}} \cdot \dfrac{1 + \sqrt{5}}{1 + \sqrt{5}}$

$= \dfrac{5 + 5\sqrt{5} - 625\sqrt{5} - 3125}{1 - 5} = \dfrac{-620\sqrt{5} - 3120}{-4}$

$S_7 = 1127$

15. $d = a_2 - a_1 = -2 - (-7) = -2 + 7 = 5$

$a_n = a_1 + (n - 1)d$

$a_n = -7 + (n - 1)(5)$

$\quad = -7 + 5n - 5$

$\quad = 5n - 12$

16. $n = 11, a = x, b = -2y, r = 8$

$\dbinom{11}{8-1}(x)^{11-8+1}(-2y)^{8-1} = \dbinom{11}{7}x^4(-2y)^7$

$= 330x^4(-128y^7) = -42{,}240x^4y^7$

17. $r = \dfrac{a_2}{a_1} = \dfrac{\sqrt{3}}{1} = \sqrt{3}$

$a_n = a_1 r^{n-1}$

$a_{12} = 1(\sqrt{3})^{11} = 243\sqrt{3}$

18. $d = a_2 - a_1 = 13 - 11 = 2$

$a_n = a_1 + (n - 1)d$

$a_{40} = 11 + (40 - 1)(2) = 11 + (39)(2)$

$a_{40} = 89$

$S_n = \dfrac{n}{2}(a_1 + a_n)$

$S_{40} = \dfrac{40}{2}(11 + 89) = 20(100) = 2000$

19. $\dbinom{12}{9} = \dfrac{12!}{(12-9)!9!} = \dfrac{12!}{3!9!}$

$= \dfrac{12 \cdot 11 \cdot 10 \cdot 9 \cdot 8 \cdot 7 \cdot 6 \cdot 5 \cdot 4 \cdot 3 \cdot 2 \cdot 1}{(3 \cdot 2 \cdot 1)(9 \cdot 8 \cdot 7 \cdot 6 \cdot 5 \cdot 4 \cdot 3 \cdot 2 \cdot 1)}$

$= 220$

20. $\displaystyle\sum_{n=1}^{4} \frac{(-1)^{n-1} n}{n+1} = \frac{(-1)^0 (1)}{1+1}$

$\qquad + \dfrac{(-1)^1 (2)}{2+1} + \dfrac{(-1)^2 (3)}{3+1} + \dfrac{(-1)^3 (4)}{4+1}$

$\qquad = \dfrac{1}{2} + \left(-\dfrac{2}{3}\right) + \dfrac{3}{4} + \left(-\dfrac{4}{5}\right) = -\dfrac{13}{60}$

21. $\displaystyle\sum_{i=1}^{5} \frac{(2x)^i}{i} = \frac{(2x)^1}{1}$

$\qquad + \dfrac{(2x)^2}{2} + \dfrac{(2x)^3}{3} + \dfrac{(2x)^4}{4} + \dfrac{(2x)^5}{5}$

$\qquad = 2x + \dfrac{4x^2}{2} + \dfrac{8x^3}{3} + \dfrac{16x^4}{4} + \dfrac{32x^5}{5}$

$\qquad = 2x + 2x^3 + \dfrac{8x^3}{3} + 4x^4 + \dfrac{32x^5}{5}$

22. $r = \dfrac{a_2}{a_1} = \dfrac{1}{3}$

$\quad a_n = a_1 r^{n-1}$

$\quad a_8 = 3\left(\dfrac{1}{3}\right)^7 = \dfrac{1}{729}$

23. $(x - 3y^2)^5 = \dbinom{5}{0} x^5 + \dbinom{5}{1} x^4 (-3y^2)$

$\qquad + \dbinom{5}{2} x^3 (-3y^2)^2 + \dbinom{5}{3} x^2 (-3y^2)^3$

$\qquad + \dbinom{5}{4} x(-3y^2)^4 + \dbinom{5}{5} (-3y^2)^5$

$\quad = x^5 + (5x^4)(-3y^2) + 10x^3 (9y^4)$

$\quad + 10x^2 (-27y^6) + 5x(81y^8) - 243y^{10}$

$\quad = x^5 - 15x^4 y^2 + 90x^3 y^4 - 270x^2 y^6$

$\quad + 405xy^8 - 243y^{10}$

24. $r = \dfrac{a_2}{a_1} = -\dfrac{1}{4}$

$\quad S = \dfrac{a_1}{1-r} = \dfrac{4}{1 - \left(-\dfrac{1}{4}\right)} = \dfrac{4}{\dfrac{5}{4}} = \dfrac{16}{5}$

25. $r = \dfrac{a_2}{a_1} = \dfrac{\dfrac{4}{3}}{2} = \dfrac{2}{3}$

$\quad S = \dfrac{a_1}{1-r} = \dfrac{2}{1 - \left(\dfrac{2}{3}\right)} = \dfrac{2}{\dfrac{1}{3}} = 6$

26. $\dfrac{12!}{5! \, 8!} = \dfrac{12 \cdot 11 \cdot 10 \cdot 9 \cdot 8 \cdot 7 \cdot 6 \cdot 5 \cdot 4 \cdot 3 \cdot 2 \cdot 1}{(5 \cdot 4 \cdot 3 \cdot 2 \cdot 1)(8 \cdot 7 \cdot 6 \cdot 5 \cdot 4 \cdot 3 \cdot 2 \cdot 1)}$

$\quad = 99$

27. $d = a_2 - a_1 = 5 - 1 = 4$

$\quad a_n = a_1 + (n-1)d$

$\quad a_{20} = 1 + (20 - 1)(4) = 1 + (19)(4)$

$\quad a_{20} = 77$

28. $\displaystyle\sum_{i=1}^{6} 4x^i = 4x + 4x^2 + 4x^3 + 4x^4 + 4x^5 + 4x^6$

29. $d = a_2 - a_1 = -7 - (-3) = -4$

$\quad a_n = a_1 + (n-1)d$

$\quad -59 = -3 + (n-1)(-4)$

$\quad -59 = -3 - 4n + 4$

$\quad -59 = -4n + 1$

$\quad -60 = -4n$

$\quad 15 = n$

There are 15 terms in the sequence.

30. $r = \dfrac{a_2}{a_1} = \dfrac{5\sqrt{3}}{5} = \sqrt{3}$

$\quad a_n = a_1 r^{n-1}$

$\quad a_7 = 5\left(\sqrt{3}\right)^6 = 135$

31. $r = \dfrac{a_2}{a_1} = \dfrac{2}{3}$

$\quad S = \dfrac{a_1}{1-r} = \dfrac{3}{1 - \left(\dfrac{2}{3}\right)} = \dfrac{3}{\dfrac{1}{3}} = 9$

32. $a_n = \dfrac{10}{n+3}$

$\quad a_{17} = \dfrac{10}{17+3} = \dfrac{10}{20} = \dfrac{1}{2}$

33. $d = a_2 - a_1 = -4 - (-10) = 6$

$\quad a_n = a_1 + (n-1)d$

$\quad a_{10} = -10 + (10 - 1)(6) = -10 + (9)(6)$

$\quad a_{10} = 44$

34. $d = a_2 - a_1 = -12 - (-19) = 7$

$a_n = a_1 + (n-1)d$

$a_{18} = -19 + (18-1)(7) = -19 + (17)(7)$

$a_{18} = 100$

$S_n = \dfrac{n}{2}(a_1 + a_n)$

$S_{18} = \dfrac{18}{2}(-19 + 100) = 9(81) = 729$

35. $r = \dfrac{a_2}{a_1} = \dfrac{-16}{8} = -2$

$S_n = \dfrac{a_1(1-r^n)}{1-r}$

$S_6 = \dfrac{8(1-(-2)^6)}{1-(-2)} = \dfrac{8(1-64)}{3}$

$= \dfrac{8(-63)}{3} = 8(-21)$

$S_6 = -168$

36. $\dfrac{7!}{5!\,2!} = \dfrac{7 \cdot 6 \cdot 5 \cdot 4 \cdot 3 \cdot 2 \cdot 1}{(5 \cdot 4 \cdot 3 \cdot 2 \cdot 1)(2 \cdot 1)} = 21$

37. $n = 9, a = 3x, b = y, r = 7$

$\dbinom{9}{7-1}(3x)^{9-7+1}(y)^{7-1} = \dbinom{9}{6}(3x)^3 y^6$

$= 84(27x^3)y^6 = 2268x^3 y^6$

38. $\displaystyle\sum_{n=1}^{5}(4n-3) = 1 + 5 + 9 + 13 + 17 = 45$

39. $a_n = \dfrac{n+2}{2n}$

$a_8 = \dfrac{8+2}{2(8)} = \dfrac{10}{16} = \dfrac{5}{8}$

40. $d = a_2 - a_1 = 0 - (-4) = 4$

$a_n = a_1 + (n-1)d$

$a_n = -4 + (n-1)(4) = -4 + 4n - 4$

$a_n = 4n - 8$

41. $r = \dfrac{a_2}{a_1} = \dfrac{-2}{10} = -\dfrac{1}{5}$

$a_n = a_1 r^{n-1}$

$a_5 = 10\left(-\dfrac{1}{5}\right)^4 = 10\left(\dfrac{1}{625}\right) = \dfrac{10}{625} = \dfrac{2}{125}$

42. $0.3\overline{666} = 0.3 + 0.06 + 0.006 + \dots$

$0.3\overline{666} = \dfrac{3}{10} + \dfrac{6}{100} + \dfrac{6}{1000} + \dots$

$S = \dfrac{a_1}{1-r} = \dfrac{\dfrac{6}{100}}{1 - \dfrac{1}{10}} = \dfrac{\dfrac{6}{100}}{\dfrac{9}{10}} = \dfrac{6}{90} = \dfrac{2}{30}$

$0.3\overline{666} = \dfrac{3}{10} + \dfrac{2}{30} = \dfrac{11}{30}$

The equivalent fraction is $\dfrac{11}{30}$.

43. $d = a_2 - a_1 = -32 - (-37) = 5$

$a_n = a_1 + (n-1)d$

$a_{37} = -37 + (37-1)(5) = -37 + (36)(5)$

$a_{37} = -37 + 180$

$a_{37} = 143$

44. $r = \dfrac{a_2}{a_1} = \dfrac{4}{1} = 4$

$S_n = \dfrac{a_1(1-r^n)}{1-r}$

$S_5 = \dfrac{1(1-4^5)}{1-4} = \dfrac{1(1-1024)}{-3}$

$S_5 = \dfrac{-1023}{-3} = 341$

45. Strategy: Write the arithmetic sequence. Find the common difference, d.
Use the formula for the nth Term of an Arithmetic Sequence to find the number of terms in the sequence.
Solution: $d = 3, a_1 = 15, a_n = 60$

$a_n = a_1 + (n-1)d$

$60 = 15 + (n-1)(3)$

$60 = 15 + 3n - 3$

$60 = 3n + 12$

$48 = 3n$

$16 = n$

In 16 weeks, the person will walk 60 min per day.

46. Strategy: Use the formula for the nth Term of a Geometric Sequence to find the temperature after 8 h.

Solution:

$n = 8, a_1 = 0.95(102) = 96.9, r = 0.95$

$a_n = a_1 r^{n-1}$

$a_8 = 96.9(0.95)^7 = 67.7$

The temperature is $67.7°$ F after 8 h.

47. Strategy: Write the arithmetic sequence. Find the common difference, d. Use the formula for the nth Term of an Arithmetic Sequence to find the salary after 9 months. Use the Formula for the Sum of n Terms of an Arithmetic Sequence.

Solution: $2400, 2480, 2560, \ldots$

$d = a_2 - a_1 = 2480 - 2400 = 80$

$a_n = a_1 + (n-1)d$

$a_9 = 2400 + (9-1)(80) = 2400 + (8)(80)$

$a_9 = 3040$

$S_n = \dfrac{n}{2}(a_1 + a_n)$

$S_9 = \dfrac{9}{2}(2400 + 3040) = 24{,}480$

The total salary for the nine-month period is \$24,480.

48. Strategy: To find the amount of radioactive material at the beginning of the seventh hour use the Formula for the nth term of a Geometric Sequence.

Solution: $n = 7, a_1 = 200, r = \dfrac{1}{2}$

$a_n = a_1 r^{n-1}$

$a_7 = 200\left(\dfrac{1}{2}\right)^6 = \dfrac{200}{64} = 3.125$

There will be 3.125 mg of radioactive material in the sample at the beginning of the seventh hour.

Chapter 12 Test

1. $\displaystyle\sum_{n=1}^{4}(3n+1) = 4 + 7 + 10 + 13 = 34$

2. $\dbinom{9}{6} = \dfrac{9!}{(9-6)!\,6!} = \dfrac{9!}{3!\,6!}$

$= \dfrac{9 \cdot 8 \cdot 7 \cdot 6 \cdot 5 \cdot 4 \cdot 3 \cdot 2 \cdot 1}{(3 \cdot 2 \cdot 1)(6 \cdot 5 \cdot 4 \cdot 3 \cdot 2 \cdot 1)} = 84$

3. $r = \dfrac{a_2}{a_1} = \dfrac{4\sqrt{2}}{4} = \sqrt{2}$

$a_n = a_1 r^{n-1}$

$a_7 = 4\left(\sqrt{2}\right)^6 = 32$

4. $a_n = \dfrac{8}{n+2}$

$a_{14} = \dfrac{8}{14+2} = \dfrac{8}{16} = \dfrac{1}{2}$

5. $\displaystyle\sum_{i=1}^{4} 3x^i = 3x + 3x^2 + 3x^3 + 3x^4$

6. $d = a_2 - a_1 = -19 - (-25) = 6$

$a_n = a_1 + (n-1)d$

$a_{18} = -25 + (18-1)(6) = -25 + (17)(6)$

$a_{18} = 77$

$S_n = \dfrac{n}{2}(a_1 + a_n)$

$S_{18} = \dfrac{18}{2}(-25 + 77) = 9(52) = 468$

7. $r = \dfrac{a_2}{a_1} = \dfrac{3}{4}$

$S_n = \dfrac{a_1}{1-r}$

$S = \dfrac{4}{1-\dfrac{3}{4}} = \dfrac{4}{\dfrac{1}{4}} = 16$

8. $d = a_2 - a_1 = -8 - (-5) = -3$

$a_n = a_1 + (n-1)d$

$-50 = -5 + (n-1)(-3)$

$-50 = -5 - 3n + 3$

$-50 = -3n - 2$

$-48 = -3n$

$16 = n$

9. $0.23\overline{3} = 0.2 + 0.03 + 0.003 + \ldots$

$0.23\overline{3} = \dfrac{2}{10} + \dfrac{3}{100} + \dfrac{3}{1000} + \ldots$

$S = \dfrac{a_1}{1-r} = \dfrac{\dfrac{3}{100}}{1-\dfrac{1}{10}} = \dfrac{\dfrac{3}{100}}{\dfrac{9}{10}} = \dfrac{3}{90} = \dfrac{1}{30}$

$0.23\overline{3} = \dfrac{2}{10} + \dfrac{1}{30} = \dfrac{7}{30}$

The equivalent fraction is $\dfrac{7}{30}$.

10. $\dfrac{8!}{4!\,4!} = \dfrac{8 \cdot 7 \cdot 6 \cdot 5 \cdot 4 \cdot 3 \cdot 2 \cdot 1}{(4 \cdot 3 \cdot 2 \cdot 1)(4 \cdot 3 \cdot 2 \cdot 1)} = 70$

11. $r = \dfrac{a_2}{a_1} = \dfrac{2}{6} = \dfrac{1}{3}$

$a_n = a_1 r^{n-1}$

$a_5 = 6\left(\dfrac{1}{3}\right)^4 = \dfrac{6}{81} = \dfrac{2}{27}$

12. $n = 7,\ a = x,\ b = -2y,\ r = 4$

$\begin{pmatrix} 7 \\ 4-1 \end{pmatrix}(x)^{7-4+1}(-2y)^{4-1} = \begin{pmatrix} 7 \\ 3 \end{pmatrix}x^4(-2y)^3$

$= 35x^4(-8y^3) = -280x^4 y^3$

13. $d = a_2 - a_1 = -16 - (-13) = -3$

$a_n = a_1 + (n-1)d$

$a_{35} = -13 + (35-1)(-3) = -13 + (34)(-3)$

$a_{35} = -115$

14. $d = a_2 - a_1 = 9 - 12 = -3$

$a_n = a_1 + (n-1)d$

$a_n = 12 + (n-1)(-3) = 12 - 3n + 3$

$a_n = -3n + 15$

15. $r = \dfrac{a_2}{a_1} = \dfrac{12}{-6} = -2$

$S_n = \dfrac{a_1(1-r^n)}{1-r}$

$S_5 = \dfrac{-6(1-(-2)^5)}{1-(-2)} = \dfrac{-6(1+32)}{3}$

$= \dfrac{-6(33)}{3} = -2(-33)$

$S_5 = -66$

16. $a_n = \dfrac{n+1}{n}$

$a_6 = \dfrac{6+1}{6} = \dfrac{7}{6}$

$a_7 = \dfrac{7+1}{7} = \dfrac{8}{7}$

17. $r = \dfrac{a_2}{a_1} = \dfrac{\frac{3}{2}}{1} = \dfrac{3}{2}$

$S_n = \dfrac{a_1(1-r^n)}{1-r}$

$S_6 = \dfrac{1\left(1-\left(\frac{3}{2}\right)^6\right)}{1-\dfrac{3}{2}} = \dfrac{1-\dfrac{729}{64}}{-\dfrac{1}{2}} = \dfrac{665}{32}$

18. $d = a_2 - a_1 = 12 - 5 = 7$

$a_n = a_1 + (n-1)d$

$a_{21} = 5 + (21-1)(7) = 5 + (20)(7)$

$a_{21} = 145$

$S_n = \dfrac{n}{2}(a_1 + a_n)$

$S_{21} = \dfrac{21}{2}(5+145) = 1575$

19. Strategy: Write the arithmetic sequence. Find the common difference, d. Use the formula for the nth Term of an Arithmetic Sequence to find the number skeins in stock in October.

Solution: 7500, 6950, 6400, ...

$d = a_2 - a_1 = 6950 - 7500 = -550$

$a_n = a_1 + (n-1)d$

$a_{10} = 7500 + (10-1)(-550) = 75000 + (9)(-550)$

$a_{10} = 2550$

The inventory after the October 1st shipment was 2550 skeins.

20. Strategy: To find the amount of radioactive material at the beginning of the fifth day, use the Formula for the nth term of a Geometric Sequence.

Solution: $n = 5, a_1 = 320, r = \dfrac{1}{2}$

$a_n = a_1 r^{n-1}$

$a_5 = 320\left(\dfrac{1}{2}\right)^4 = \dfrac{320}{16} = 20$

There will be 20 mg of radioactive material in the sample at the beginning of the fifth day.

Cumulative Review Exercises

1. $3x - 2y = -4$

$\qquad -2y = -3x - 4$

$\qquad y = \dfrac{3}{2}x + 2$

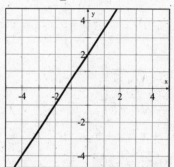

2. $2x^6 + 16 = 2(x^6 + 8)$

$\qquad = 2((x^2)^3 + 2^3)$

$\qquad = 2(x^2 + 2)(x^4 - 2x^2 + 4)$

3. $\dfrac{4x^2}{x^2 + x - 2} - \dfrac{3x-2}{x+2}$

$= \dfrac{4x^2}{(x+2)(x-1)} - \dfrac{3x-2}{x+2} \cdot \dfrac{x-1}{x-1}$

$= \dfrac{4x^2 - (3x^2 - 5x + 2)}{(x+2)(x-1)}$

$= \dfrac{x^2 + 5x - 2}{(x+2)(x-1)}$

4. $f(-2) = 2(-2)^2 - 3(-2) = 2(4) + 6 = 14$

5. $\sqrt{2y}(\sqrt{8xy} - \sqrt{y}) = \sqrt{16xy^2} - \sqrt{2y^2}$

$= \sqrt{16y^2x} - \sqrt{y^2(2)}$

$= 4y\sqrt{x} - y\sqrt{2}$

6. $2x^2 - x + 7 = 0$

$a = 2 \quad b = -1 \quad c = 7$

$$x = \frac{-b \pm \sqrt{b^2 - 4ac}}{2a}$$

$$x = \frac{-(-1) \pm \sqrt{(-1)^2 - 4(2)(7)}}{2(2)}$$

$$x = \frac{1 \pm \sqrt{1 - 56}}{4} = \frac{1 \pm \sqrt{-55}}{4}$$

$$x = \frac{1}{4} \pm \frac{\sqrt{55}}{4} i$$

The solutions are $\frac{1}{4} + \frac{\sqrt{55}}{4} i$ and

$\frac{1}{4} - \frac{\sqrt{55}}{4} i$.

7. $\quad 5 - \sqrt{x} = \sqrt{x + 5}$

$\left(5 - \sqrt{x}\right)^2 = \left(\sqrt{x + 5}\right)^2$

$25 - 10\sqrt{x} + x = x + 5$

$-10\sqrt{x} = -20$

$\sqrt{x} = 2$

$\left(\sqrt{x}\right)^2 = 2^2$

$x = 4$

The solution is 4.

8. $(4, 2)$ and $(-1, -1)$

$d = \sqrt{(x_1 - x_2)^2 + (y_1 - y_2)^2}$

$= \sqrt{(4 - (-1))^2 + (2 - (-1))^2}$

$= \sqrt{(5)^2 + 3^2} = \sqrt{25 + 9} = \sqrt{34}$

$(x - h)^2 + (y - k)^2 = r^2$

Center: $(-1, -1)$

Radius: $r = \sqrt{34}$

9. $(x - (-1))^2 + (y - (-1))^2 = (\sqrt{34})^2$

$(x + 1)^2 + (y + 1)^2 = 34$

10. (1) $3x - 3y = 2$

(2) $6x - 4y = 5$

Eliminate x. Multiply equation (1) by -2.

$-6x + 6y = -4$

$6x - 4y = 5$

$2y = 1$

$y = \frac{1}{2}$

Substitute $\frac{1}{2}$ for y in equation (2).

$6x - (4)\frac{1}{2} = 5$

$6x - 2 = 5$

$6x = 7$

$x = \frac{7}{6}$

The solution is $\left(\frac{7}{6}, \frac{1}{2}\right)$.

11. $\begin{vmatrix} -3 & 1 \\ 4 & 2 \end{vmatrix} = -3(2) - 4(1) = -6 - 4 = -10$

12. $2x - 1 > 3 \quad$ or $\quad 1 - 3x > 7$

$2x > 4 \qquad\qquad -3x > 6$

$x > 2 \qquad\qquad x < -2$

$\{x \mid x > 2\}$ or $\{x \mid x < -2\}$

$\{x \mid x < -2 \text{ or } x > 2\}$

13. $2x - 3y < 9$

$-3y < -2x + 9$

$y > \frac{2}{3}x - 3$

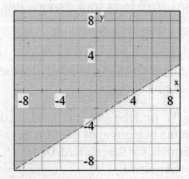

14. $\log_5 \sqrt{\dfrac{x}{y}} = \dfrac{1}{2}[\log_5 x - \log_5 y]$

$= \dfrac{1}{2}\log_5 x - \dfrac{1}{2}\log_5 y$

15. $4^x = 8^{x-1}$

$(2^2)^x = (2^3)^{x-1}$

$2x = 3(x-1)$

$2x = 3x - 3$

$3 = x$

The solution is 3.

16. $a_n = n(n-1)$

$a_5 = 5(5-1) = 5(4) = 20$

$a_6 = 6(6-1) = 6(5) = 30$

17. $\displaystyle\sum_{n=1}^{7}(-1)^{n-1}(n+2) = (-1)^0(1+2) + (-1)^1(2+2) + (-1)^2(3+2) + (-1)^3(4+2) + (-1)^4(5+2) + (-1)^5(6+2) + (-1)^6(7+2)$

$= 1(3) + (-1)(4) + 1(5) + (-1)(6) + 1(7) + (-1)(8) + 1(9)$

$= 3 - 4 + 5 - 6 + 7 - 8 + 9 = 6$

18. $d = a_2 - a_1 = -10 - (-7) = -3$

$a_n = a_1 + (n-1)d$

$a_{33} = -7 + (33-1)(-3) = -7 + (32)(-3)$

$a_{35} = -103$

21. $n = 6,\ a = 2x,\ b = y,\ r = 6$

$\dbinom{6}{6-1}(2x)^{6-6+1}(y)^{6-1} = \dbinom{6}{5}(2x)^1(y)^5$

$= 6(2x)y^5 = 12xy^5$

19. $r = \dfrac{a_2}{a_1} = -\dfrac{2}{3}$

$S_n = \dfrac{a_1}{1-r}$

$S = \dfrac{3}{1-\left(-\dfrac{2}{3}\right)} = \dfrac{3}{\dfrac{5}{3}} = \dfrac{9}{5}$

22. Strategy: Let x represent the amount of water.

	Amount	Percent	Value
Water	x	0	$0x$
8%	200	0.08	0.08(200)
5%	$200 + x$	0.05	$0.05(200 + x)$

The sum of the values before mixing is equal to the value after mixing.

Solution:

$0x + 0.08(200) = 0.05(200 + x)$

$16 = 10 + 0.05x$

$6 = 0.05x$

$x = 120$

120 oz of water must be added.

20. $0.4\overline{6} = 0.4 + 0.06 + 0.006 + \ldots$

$0.4\overline{6} = \dfrac{4}{10} + \dfrac{6}{100} + \dfrac{6}{1000} + \ldots$

$S = \dfrac{a_1}{1-r} = \dfrac{\dfrac{6}{100}}{1-\dfrac{1}{10}} = \dfrac{\dfrac{6}{100}}{\dfrac{9}{10}} = \dfrac{1}{15}$

$0.4\overline{6} = \dfrac{4}{10} + \dfrac{1}{15} = \dfrac{14}{30} = \dfrac{7}{15}$

The equivalent fraction is $\dfrac{7}{15}$.

23. Strategy: Let x represent the time required for the older computer.

The time required for the new computer is $x - 16$.

Rate   Time   Part

Older computer	$\dfrac{1}{x}$	15	$\dfrac{15}{x}$
New computer	$\dfrac{1}{x-16}$	15	$\dfrac{15}{x-16}$

The sum of the parts of the task completed must equal 1.

Solution:

$$\frac{15}{x}+\frac{15}{x-16}=1$$

$$x(x-16)\left(\frac{15}{x}+\frac{15}{x-16}\right)=1(x)(x-16)$$

$$15(x-16)+15x=x^2-16x$$

$$15x-240+15x=x^2-16x$$

$$0=x^2-46x+240$$

$$0=(x-40)(x-6)$$

$$x-40=0 \quad x-6=0$$

$$x=40 \qquad x=6$$

$x=6$ does not check as a solution since $x-16=6-16=-10$; $x=40$, $40-16=24$
The new computer takes 24 minutes to complete the payroll while the older computer takes 40 minutes.

24. **Strategy:** Let x represent the rate of the boat in calm water.
    The rate of the current is y.

	Rate	Time	Distance
With current	$x+y$	2	$2(x+y)$
Against current	$x-y$	3	$3(x-y)$

The distance traveled with the current is 15 mi. The distance traveled against the current is 15 mi.

Solution:

$$2(x+y)=15$$

$$3(x-y)=15$$

$$\frac{1}{2}\cdot 2(x+y)=15\cdot\frac{1}{2}$$

$$\frac{1}{3}\cdot 3(x-y)=15\cdot\frac{1}{3}$$

$$x+y=7.5$$

$$x-y=5$$

$$2x=12.5$$

$$x=6.25$$

$$x+y=7.5$$

$$6.25+y=7.5$$

$$y=1.25$$

The boat was traveling at 6.25 mph and the current was moving at 1.25 mph.

25. **Strategy:** To find the half-life use the exponential decay formula.
    $A_0=80$, $A=55$, $t=30$

Solution: $A=A_0(0.5)^{t/k}$

$$55=80(0.5)^{30/k}$$

$$0.6875=0.5^{30/k}$$

$$\log 0.6875=\log 0.5^{30/k}$$

$$\log 0.6875=\frac{30}{k}\log 0.5$$

$$k\log 0.6875=30\log 0.5$$

$$k=\frac{30\log 0.5}{\log 0.6875}=55.49$$

The half-life is 55 days.

26. **Strategy:** Write the arithmetic sequence. Find the common difference, d.
    Use the formula for the nth Term of an Arithmetic Sequence to find the number of rows.
    Use the Formula for the Sum of n Terms of an Arithmetic Sequence.

Solution: 62, 74, 86 ...
$$d=a_2-a_1=74-62=12$$

$$a_n = a_1 + (n-1)d$$

$$a_{12} = 62 + (12-1)(12) = 62 + (11)(12)$$

$$a_{12} = 194$$

$$S_n = \frac{n}{2}(a_1 + a_n)$$

$$S_{12} = \frac{12}{2}(62 + 194) = 6(256) = 1536$$

There are 1536 seats in the theater.

27. **Strategy:** To find the height of the ball on the fifth bounce, use the formula for the nth Term of a Geometric Sequence.

Solution: $n = 5$, $a_1 = 80\%$ of $8 = 6.4$
$r = 80\% = 0.8$

$$a_n = a_1 r^{n-1}$$

$$a_5 = 6.4(0.8)^4$$

$$a_5 = 6.4(0.4096) = 2.62144$$

The ball bounces to a height of 2.6 ft on the fifth bounce.

Final Exam

1. $12 - 8[3 - (-2)]^2 \div 5 - 3$

$= 12 - 8(3 + 2)^2 \div 5 - 3$

$= 12 - 8(5)^2 \div 5 - 3$

$= 12 - 8(25) \div 5 - 3$

$= 12 - 200 \div 5 - 3$

$= 12 - 40 - 3$

$= -31$

2. $\dfrac{a^2 - b^2}{a - b}$

$\dfrac{3^2 - (-4)^2}{3 - (-4)} = \dfrac{9 - 16}{3 + 4} = \dfrac{-7}{7}$

$= -1$

3. $f(x) = 3x - 7 \quad g(x) = x^2 - 4x$

$(f \circ g)(3) = f(g(3))$

$g(3) = 3^3 - 4(3) = 9 - 12 = -3$

$f(-3) = 3(-3) - 7 = -9 - 7 = -16$

$(f \circ g)(3) = -16$

4. $\dfrac{3}{4}x - 2 = 4$

$\dfrac{3}{4}x = 6$

$\dfrac{4}{3} \cdot \dfrac{3}{4}x = 6 \cdot \dfrac{4}{3}$

$x = 8$

The solution is 8.

5. $\dfrac{2 - 4x}{3} - \dfrac{x - 6}{12} = \dfrac{5x - 2}{6}$

$12\left(\dfrac{2 - 4x}{3} - \dfrac{x - 6}{12}\right) = 12\left(\dfrac{5x - 2}{6}\right)$

$4(2 - 4x) - (x - 6) = 2(5x - 2)$

$8 - 16x - x + 6 = 10x - 4$

$-17x + 14 = 10x - 4$

$-27x = -18$

$x = \dfrac{18}{27} = \dfrac{2}{3}$

The solution is $\dfrac{2}{3}$.

6. $8 - |5 - 3x| = 1$

$-|5 - 3x| = -7$

$|5 - 3x| = 7$

$5 - 3x = 7 \qquad 5 - 3x = -7$

$-3x = 2 \qquad\quad -3x = -12$

$x = -\dfrac{2}{3} \qquad\qquad x = 4$

The solutions are $-\dfrac{2}{3}$ and 4.

7. $2x - 3y = 9$

$2x - 3(0) = 9$

$2x = 9$

$x = \dfrac{9}{2}$

The x-intercept is $\left(\dfrac{9}{2}, 0\right)$.

$2(0) - 3y = 9$

$-3y = 9$

$y = -3$

The y-intercept is $(0, -3)$.

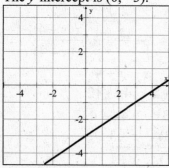

8. $(3, -2)$ and $(1, 4)$

$m = \dfrac{y_2 - y_1}{x_2 - x_1} = \dfrac{4 - (-2)}{1 - 3} = \dfrac{4 + 2}{-2} = \dfrac{6}{-2}$

$m = -3$

$y - y_1 = m(x - x_1)$

$y - 4 = -3(x - 1)$

$y - 4 = -3x + 3$

$y = -3x + 7$

9. $3x - 2y = 6$

$-2y = -3x + 6$

$y = \dfrac{3}{2}x - 3$

$m_1 = \dfrac{3}{2}$

$m_1 \cdot m_2 = -1$

$\dfrac{3}{2}m_2 = -1$

$m_2 = -\dfrac{2}{3} \quad (-2, 1)$

$y - y_1 = m(x - x_1)$

$y - 1 = -\dfrac{2}{3}(x - (-2))$

$y - 1 = -\dfrac{2}{3}(x + 2)$

$y - 1 = -\dfrac{2}{3}x - \dfrac{4}{3}$

$y = -\dfrac{2}{3}x - \dfrac{1}{3}$

10. $2a[5 - a(2 - 3a) - 2a] + 3a^2$

$= 2a[5 - 2a + 3a^2 - 2a] + 3a^2$

$= 2a[5 - 4a + 3a^2] + 3a^2$

$= 10a - 8a^2 + 6a^3 + 3a^2$

$= 6a^3 - 5a^2 + 10a$

11. $8 - x^3y^3 = 2^3 - (xy)^3$

$= (2 - xy)(4 + 2xy + x^2y^2)$

12. $x - y - x^3 + x^2y = x - y - x^2(x - y)$

$= 1(x - y) - x^2(x - y)$

$= (x - y)(1 - x^2)$

$= (x - y)(1 - x)(1 + x)$

13.

$$
\begin{array}{r}
x^2 - 2x - 3 \\
2x-3\overline{\smash{\big)}\,2x^3 - 7x^2 + 0x + 4} \\
\underline{2x^3 - 3x^2} \\
-4x^2 + 0x \\
\underline{-4x^2 + 6x} \\
-6x + 4 \\
\underline{-6x + 9} \\
-5
\end{array}
$$

$$(2x^3 - 7x^2 + 4) \div (2x - 3) = x^2 - 2x - 3 + \dfrac{-5}{2x-3}$$

14.

$$\dfrac{x^2 - 3x}{2x^2 - 3x - 5} \div \dfrac{4x - 12}{4x^2 - 4}$$

$$= \dfrac{x^2 - 3x}{2x^2 - 3x - 5} \cdot \dfrac{4x^2 - 4}{4x - 12}$$

$$= \dfrac{x(x-3)}{(2x-5)(x+1)} \cdot \dfrac{4(x+1)(x-1)}{4(x-3)}$$

$$= \dfrac{x(x-3) \cdot 4(x+1)(x-1)}{(2x-5)(x+1) \cdot 4(x-3)}$$

$$= \dfrac{x(x+1)}{2x-5}$$

15. The LCM is $(x-3)(x+2)$.

$$\dfrac{x-2}{x+2} - \dfrac{x+3}{x-3} = \dfrac{x-2}{x+2} \cdot \dfrac{x-3}{x-3} - \dfrac{x+3}{x-3} \cdot \dfrac{x+2}{x+2}$$

$$= \dfrac{(x-2)(x-3) - (x+3)(x+2)}{(x-3)(x+2)}$$

$$= \dfrac{x^2 - 5x + 6 - x^2 - 5x - 6}{(x-3)(x+2)}$$

$$= \dfrac{-10x}{(x-3)(x+2)}$$

16. The LCM is $x(x+4)$.

$$\dfrac{\dfrac{3}{x} + \dfrac{1}{x+4}}{\dfrac{1}{x} + \dfrac{3}{x+4}} = \dfrac{\dfrac{3}{x} + \dfrac{1}{x+4}}{\dfrac{1}{x} + \dfrac{3}{x+4}} \cdot \dfrac{x(x+4)}{x(x+4)}$$

$$= \dfrac{3x + 12 + x}{x + 4 + 3x} = \dfrac{4x + 12}{4x + 4}$$

$$= -\dfrac{4(x+3)}{4(x+1)} = \dfrac{x+3}{x+1}$$

17.

$$\dfrac{5}{x-2} - \dfrac{5}{x^2 - 4} = \dfrac{1}{x+2}$$

$$\dfrac{5}{x-2} - \dfrac{5}{(x+2)(x-2)} = \dfrac{1}{x+2}$$

$$(x-2)(x+2)\left(\dfrac{5}{x-2} - \dfrac{5}{(x+2)(x-2)}\right) = (x-2)(x+2)\left(\dfrac{1}{x+2}\right)$$

$$5(x+2) - 5 = x - 2$$

$$5x + 10 - 5 = x - 2$$

$$4x + 5 = -2$$

$$4x = -7$$

$$x = -\dfrac{7}{4}$$

The solution is $-\dfrac{7}{4}$.

18. $a_n = a_1 + (n-1)d$

$a_n - a_1 = (n-1)d$

$d = \dfrac{a_n - a_1}{n-1}$

19. $\left(\dfrac{4x^2 y^{-1}}{3x^{-1} y}\right)^{-2} \left(\dfrac{2x^{-1} y^2}{9x^{-2} y^2}\right)^3$

$\dfrac{4^{-2} x^{-4} y^2}{3^{-2} x^2 y^{-2}} \cdot \dfrac{2^3 x^{-3} y^6}{9^3 x^{-6} y^6}$

$= \dfrac{2^3 \cdot 3^2 \, y^4 x^3}{4^2 \cdot 9^3 x^6}$

$= \dfrac{y^4}{162 x^3}$

20. $\left(\dfrac{3x^{2/3} y^{1/2}}{6x^2 y^{4/3}}\right)^6 = \dfrac{3^6 x^4 y^3}{6^6 x^{12} y^8} = \dfrac{1}{64 x^8 y^5}$

21. $x\sqrt{18x^2 y^3} - y\sqrt{50x^4 y}$

$= x\sqrt{3^2 x^2 y^2 (2y)} - y\sqrt{5^2 x^4 (2y)}$

$= 3x^2 y\sqrt{2y} - 5x^2 y\sqrt{2y}$

$= -2x^2 y\sqrt{2y}$

22. $\dfrac{\sqrt{16x^5 y^4}}{\sqrt{32xy^7}} = \sqrt{\dfrac{16x^5 y^4}{32xy^7}} = \sqrt{\dfrac{x^4}{2y^3}}$

$= \sqrt{\dfrac{x^4}{y^2(2y)}} = \dfrac{x^2}{y}\sqrt{\dfrac{1}{2y}} \cdot \sqrt{\dfrac{2y}{2y}}$

$= \dfrac{x^2}{y}\sqrt{\dfrac{2y}{(2y)^2}}$

$= \dfrac{x^2\sqrt{2y}}{2y^2}$

23. $\dfrac{5-2i}{3+4i} \cdot \dfrac{3-4i}{3-4i} = \dfrac{15-26i+8i^2}{9-16i^2} = \dfrac{15-26i-8}{9+16}$

$= \dfrac{7-26i}{25} = \dfrac{7}{25} - \dfrac{26}{25}i$

24. $(x-r_1)(x-r_2) = 0$

$\left(x - \left(-\dfrac{1}{2}\right)\right)(x-2) = 0$

$\left(x + \dfrac{1}{2}\right)(x-2) = 0$

$x^2 - \dfrac{3}{2}x - 1 = 0$

$2\left(x^2 - \dfrac{3}{2}x - 1\right) = 0(2)$

$2x^2 - 3x - 2 = 0$

25. $2x^2 - 3x - 1 = 0$

$a = 2, \, b = -3, \, c = -1$

$x = \dfrac{-b \pm \sqrt{b^2 - 4ac}}{2a}$

$x = \dfrac{-(-3) \pm \sqrt{(-3)^2 - 4(2)(-1)}}{2(2)}$

$x = \dfrac{3 \pm \sqrt{9+8}}{4}$

$x = \dfrac{3 \pm \sqrt{17}}{4}$

The solutions are $\dfrac{3+\sqrt{17}}{4}$ and $\dfrac{3-\sqrt{17}}{4}$.

26. $x^{2/3} - x^{1/3} - 6 = 0$

$\left(x^{1/3}\right)^2 - x^{1/3} - 6 = 0$

$u^2 - u - 6 = 0$

$(u-3)(u+2) = 0$

$u - 3 = 0 \quad u + 2 = 0$

$u = 3 \qquad u = -2$

Replace u with $x^{1/3}$.

$x^{1/3} = 3 \qquad\qquad x^{1/3} = -2$

$\left(x^{1/3}\right)^3 = (3)^3 \quad \left(x^{1/3}\right)^3 = (-2)^3$

$x = 27 \qquad\qquad x = -8$

The solutions are -8 and 27.

27. $\dfrac{2}{x} - \dfrac{2}{2x+3} = 1$

$x(2x+3)\left(\dfrac{2}{x} - \dfrac{2}{2x+3}\right) = 1x(2x+3)$

$2(2x+3) - 2x = 2x^2 + 3x$

$4x + 6 - 2x = 2x^2 + 3x$

$0 = 2x^2 + x - 6$

$(2x-3)(x+2) = 0$

$2x - 3 = 0 \quad x + 2 = 0$

$2x = 3 \qquad x = -2$

$x = \dfrac{3}{2}$

The solutions are -2 and $\dfrac{3}{2}$.

28. $x = y^2 - 6y + 6$

$y = -\dfrac{b}{2a} = -\dfrac{-6}{2(1)} = 3$

$x = (3)^2 - 6(3) + 6 = 9 - 18 + 6 = -3$

The axis of symmetry is $y = 3$.

The vertex is $(-3, 3)$.

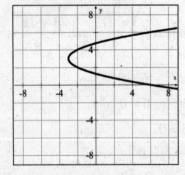

29. x-intercepts: $(4, 0)$ and $(-4, 0)$

y-intercepts: $(0, 2)$ and $(0, -2)$

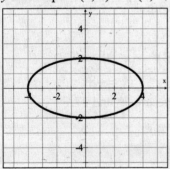

30. $f(x) = \dfrac{2}{3}x - 4$

$y = \dfrac{2}{3}x - 4$

$x = \dfrac{2}{3}y - 4$

$x + 4 = \dfrac{2}{3}y$

$\dfrac{3}{2}(x+4) = \dfrac{3}{2}\left(\dfrac{2}{3}y\right)$

$y = \dfrac{3}{2}x + 6$

$f^{-1}(x) = \dfrac{3}{2}x + 6$

31. (1) $3x - 2y = 1$

(2) $5x - 3y = 3$

Eliminate y. Multiply equation (1) by -3 and equation (2) by 2. Add the two equations.

$-3(3x - 2y) = -3(1)$

$2(5x - 3y) = 2(3)$

$-9x + 6y = -3$

$10x - 6y = 6$

$x = 3$

Substitute 3 for x in equation (1).

$3(3) - 2y = 1$

$9 - 2y = 1$

$-2y = -8$

$y = 4$

The solution is $(3, 4)$.

32. $\begin{vmatrix} 3 & 4 \\ -1 & 2 \end{vmatrix} = 3(2) - (-1)(4) = 6 + 4 = 10$

33. (1) $x^2 - y^2 = 4$

(2) $x + y = 1$

Solve equation (2) for y and substitute into equation (1) by 2.

$x + y = 1$

$y = -x + 1$

$x^2 - (-x + 1)^2 = 4$

$x^2 - (x^2 - 2x + 1) = 4$

$x^2 - x^2 + 2x - 1 = 4$

$2x - 1 = 4$

$2x = 5$

$x = \dfrac{5}{2}$

Substitute $\dfrac{5}{2}$ for x in equation (2).

$\dfrac{5}{2} + y = 1$

$y = -\dfrac{3}{2}$

The solutions is $\left(\dfrac{5}{2}, -\dfrac{3}{2}\right)$.

34. $2 - 3x < 6 \qquad 2x + 1 > 4$

$\quad -3x < 4 \qquad\quad 2x > 3$

$\quad x > -\dfrac{4}{3} \qquad\quad x > \dfrac{3}{2}$

The solution is $\left(\dfrac{3}{2}, \infty\right)$.

35. $|2x + 5| < 3$

$-3 < 2x + 5 < 3$

$-3 - 5 < 2x + 5 - 5 < 3 - 5$

$-8 < 2x < -2$

$-4 < x < -1$

$\{x \mid -4 < x < -1\}$

36. $3x + 2y > 6$

$2y > -3x + 6$

$y > -\dfrac{3}{2}x + 3$

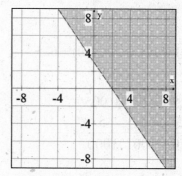

37. $f(x) = \log_2(x + 1)$

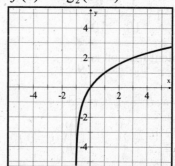

38. $2(\log_2 a - \log_2 b) = 2\log_2 \dfrac{a}{b} = \log_2 \dfrac{a^2}{b^2}$

39. $\log_3(x) - \log_3(x - 3) = 2$

$\log_3 \dfrac{x}{x - 3} = 2$

$\dfrac{x}{x - 3} = 3^2$

$(x - 3)\dfrac{x}{x - 3} = 9(x - 3)$

$x = 9x - 27$

$-8x = -27$

$x = \dfrac{-27}{-8} = \dfrac{27}{8}$

The solution is $\dfrac{27}{8}$.

40. $a_i = 2i - 1$

$a_1 = 2(1) - 1 = 1$

$a_{25} = 2(25) - 1 = 49$

$S_i = \dfrac{i}{2}(a_1 + a_i)$

$S_{25} = \dfrac{25}{2}(1 + 49) = \left(\dfrac{25}{2}\right)(50) = 625$

41. $0.55\overline{5} = 0.5 + 0.05 + 0.005 + \ldots$

$0.55\overline{5} = \dfrac{5}{10} + \dfrac{5}{100} + \dfrac{5}{1000} + \ldots$

$S = \dfrac{a_1}{1 - r} = \dfrac{\dfrac{5}{10}}{1 - \dfrac{1}{10}} = \dfrac{\dfrac{5}{10}}{\dfrac{9}{10}} = \dfrac{5}{9}$

The equivalent fraction is $\dfrac{5}{9}$.

42. $n = 9,\, a = x,\, b = -2y,\, r = 3$

$\dbinom{9}{3-1} x^{9-3+1}(-2y)^{3-1} = \dbinom{9}{2} x^7 (-2y)^2$

$= 36x^7 (4y^2)$

$= 144x^7 y^2$

43. Strategy: Let x represent the score on the last test.
To find the range of scores, solve the inequality.

Solution:

$70 \leq \dfrac{64 + 58 + 82 + 77 + x}{5} \leq 79$

$70 \leq \dfrac{281 + x}{5} \leq 79$

$350 \leq 281 + x \leq 395$

$69 \leq x \leq 114$

Since 100 is the maximum score, the range of scores is $69 \leq x \leq 100$.

44. Strategy: Let x represent the average speed of the jogger.
The average speed of the cyclist is $2.5x$.

	Rate	Time	Distance
Jogger	x	2	$2(x)$
Cyclist	$2.5x$	2	$2(2.5x)$

The distance traveled by the cyclist is 24 mi more than the distance traveled by the jogger.

Solution: $2x + 24 = 2(2.5x)$

$\ 2x + 24 = 5x$

$\ 24 = 3x$

$\ x = 8$

$2(2.5x) = 5(8) = 40$

The cyclist traveled 40 mi.

45. Strategy: Let x represent the amount invested at 8.5%.
The amount invested at 6.4% is $12000 - x$.

	Principal	Rate	Interest
8.5%	x	0.085	$0.085x$
6.4%	$12000 - x$	0.064	$0.064(12000 - x)$

The sum of the interest earned from the two investments is $936.

Solution:

$0.085x + 0.064(12000 - x) = 936$

$0.085x + 768 - 0.064x = 936$

$0.021x + 768 = 936$

$0.021x = 168$

$x = 8000$

$12000 - x = 12000 - 8000 = 4000$

$8,000 is invested at 8.5% and $4,000 is invested at 6.4%.

46. Strategy: Let x represent the width of the rectangle.
The length of the rectangle is $3x - 1$.
The area of the rectangle is 140 ft^2.

Solution: $A = LW$
$140 = x(3x - 1)$
$140 = 3x^2 - x$
$0 = 3x^2 - x - 140$
$0 = (3x + 20)(x - 7)$
$3x + 20 = 0 \quad x - 7 = 0$
$3x = -20 \quad x = 7$
$x = -\dfrac{20}{3}$

The width cannot be a negative number.
$3x - 1 = 3(7) - 1 = 20$
The width is 7 ft.
The length is 20 ft.

47. Strategy: Let x represent the number of additional shares. Write and solve a proportion.

Solution: $\dfrac{300}{486} = \dfrac{300 + x}{810}$
$(810 \cdot 486)\dfrac{300}{486} = \dfrac{300 + x}{810}(810 \cdot 486)$
$300(810) = 486(300 + x)$
$243{,}000 = 145{,}800 + 486x$
$97200 = 486x$
$x = 200$

200 additional shares would need to be purchased.

48. Strategy: Let x represent the rate of the car.
The rate of the plane is $7x$.

	Distance	Rate	Time
Car	45	x	$\dfrac{45}{x}$
Plane	1050	$7x$	$\dfrac{1050}{7x}$

The total time traveled is $3\dfrac{1}{4}$ h.

Solution: $\dfrac{45}{x} + \dfrac{1050}{7x} = 3\dfrac{1}{4}$
$\dfrac{45}{x} + \dfrac{150}{x} = \dfrac{13}{4}$
$4x\left(\dfrac{45}{x} + \dfrac{150}{x}\right) = \left(\dfrac{13}{4}\right)4x$
$180 + 600 = 13x$
$780 = 13x$
$x = 60$
$7x = 7(60) = 420$
The rate of the plane is 420 mph.

49. Strategy: To find the distance the object has fallen, substitute 75 ft/s for v and solve for d.

Solution: $v = \sqrt{64d}$
$75 = \sqrt{64d}$
$75^2 = \left(\sqrt{64d}\right)^2$
$5625 = 64d$
$d = 87.89$
The distance traveled is 88 ft.

50. Strategy: Let x represent the rate traveled during the first 360 mi.
The rate traveled during the next 300 mi is $x + 30$.

	Distance	Rate	Time
First part of the trip	360	x	$\dfrac{360}{x}$
Second part of the trip	300	$x + 30$	$\dfrac{300}{x+30}$

The total time traveled was 5 h.

Solution: $\dfrac{360}{x} + \dfrac{300}{x+30} = 5$

$x(x+30)\left(\dfrac{360}{x} + \dfrac{300}{x+30}\right) = 5(x)(x+30)$

$360(x+30) + 300x = 5x^2 + 150x$

$360x + 10800 + 300x = 5x^2 + 150x$

$0 = 5x^2 - 510x - 10800$

$0 = 5(x^2 + 102x - 2160)$

$0 = 5(x+18)(x-120)$

$x + 18 = 0 \quad x - 120 = 0$

$x = -18 \qquad x = 120$

The rate cannot be a negative number.
The rate of the plane for the first 360 mi is
120 mph.

51. Strategy: Write the basic inverse variation
equation, replacing the variable with the
given values. Solve for k.
Write the inverse variation equation,
replacing k with its value. Substitute 4 for d
and solve for I.

Solution:

$I = \dfrac{k}{d^2}$

$8 = \dfrac{k}{20^2}$

$8 = \dfrac{k}{400}$

$3200 = k$

$I = \dfrac{3200}{d^2} = \dfrac{3200}{4^2} = \dfrac{3200}{16} = 200$

The intensity is 200 foot-candles.

52. Strategy: Let x represent the rate of the
boat in calm water.
The rate of the current is y.

	Rate	Time	Distance
With current	$x+y$	2	$2(x+y)$
Against current	$x-y$	3	$3(x-y)$

The distance traveled with the current is 30
mi. The distance traveled against the current
is 30 mi.
$2(x+y) = 30$
$3(x-y) = 30$

Solution:
$2(x+y) = 30$

$3(x-y) = 30$

$\dfrac{1}{2} \cdot 2(x+y) = \dfrac{1}{2} \cdot 30$

$\dfrac{1}{3} \cdot 3(x-y) = \dfrac{1}{3} \cdot 30$

$x + y = 15$

$x - y = 10$

$2x = 25$

$x = 12.5$

$x + y = 15$

$12.5 + y = 15$

$y = 2.5$

The rate of the boat in calm water is 12.5
mph. The rate of the current is 2.5 mph.

53. Strategy: To find the value of the
investment after two years, use the
compound interest formula.

$A = 4000, n = 24, i = \dfrac{9\%}{12} = \dfrac{0.09}{12} = 0.0075$

Solution: $P = A(1+i)^n$

$P = 4000(1 + 0.0075)^{24}$

$P = 4000(1.0075)^{24}$

$P \approx 4785.65$

The value of the investment after 2 years is
$4785.65.

54. Strategy: To find the height of the ball on the fifth bounce, use the formula for the nth Term of a Geometric Sequence.

Solution: $n = 5$, $a_1 = 5$

$r = 80\% = \dfrac{4}{5}$

$a_n = a_1 r^{n-1}$

$a_5 = 5\left(\dfrac{4}{5}\right)^4$

$a_5 = 5\left(\dfrac{256}{625}\right) \approx 2.05$

The ball bounces to a height of 2.05 ft on the fifth bounce.